Theory of Charges

This is a volume in
PURE AND APPLIED MATHEMATICS

A Series of Monographs and Textbooks

Editors: SAMUEL EILENBERG AND HYMAN BASS

A list of recent titles in this series appears at the end of this volume.

Theory of Charges

A Study of Finitely Additive Measures

K. P. S. Bhaskara Rao
Indian Statistical Institute, Calcutta, India

M. Bhaskara Rao
University of Sheffield, UK

1983

Academic Press
A Subsidiary of Harcourt Brace Jovanovich, Publishers
London   New York
Paris   San Diego   San Francisco
São Paulo   Sydney   Tokyo   Toronto

ACADEMIC PRESS INC. (LONDON) LTD
24/28 Oval Road,
London NW1 7DX

United States Edition published by
ACADEMIC PRESS INC.
111 Fifth Avenue,
New York, New York 10003

Copyright © 1983 by ACADEMIC PRESS INC. (LONDON) LTD

All Rights Reserved

No part of this book may be reproduced in any form
by photostat, microfilm, or any other means
without written permission from the publishers

British Library Cataloguing in Publication Data
Bhaskara Rao, K.P.S.
Theory of Charges—(Pure and Applied Mathematics)
1. Algebraic Topology
I. Title II. Bhaskara Rao, M. III. Series
514′.2 QA612
ISBN 0-12-095780-9

Typeset and printed by J. W. Arrowsmith Ltd

Foreword

Many years ago, S. Bochner remarked to me that, contrary to popular mathematical opinion, finitely additive measures were more interesting, more difficult to handle, and perhaps more important than countably additive ones. At that time, I held the popular point of view, but since then I have come around to Bochner's opinion. Apparently, many other mathematicians have also done so, as is indicated by the large number of papers listed in the bibliography of this book. I, for one, had not realized how much research had been done on finitely additive measures, at least partly because the material is scattered in isolated papers. The authors have done the mathematical community a real service by providing easy access to this research (to which they themselves have made significant contributions). This service is all the greater in that not only is the material that they cover interesting in itself, but the presentation is very clear and is enlivened with many illustrative examples. Two especially valuable features of their work are an *annotated bibliography* and a section of notes and comments. But perhaps the most valuable feature of the work to the working measure theorist or functional analyst lies in exhibiting clearly where countable additivity of a measure is used, and what can and what cannot be done without it. Roughly speaking, without countable additivity most of the measure theoretic examples are "counter", but a great deal of functional analysis can be done—with more work!

No one book in an area as large as this can do justice to all the material that deserves coverage, and I certainly do not blame the authors for omitting or treating too briefly some topics which I think are important. I only regret the necessity.

I understand that the authors expect to write a book on finitely additive probability also, and perhaps they will include some of the topics so omitted. In any case, I look forward to seeing a continuation of the excellent work they have done in this book.

December 1982   Dorothy Maharam Stone
                                                     University of Rochester
                                                     Rochester, New York

Preface

According to S. Bochner, finitely additive measures are more interesting, and perhaps more important, than countably additive ones (see Maharam (1976)). Finitely additive measures arise quite naturally in many areas of analysis. Over the years, there has been a sustained growth of activity in finitely additive measures propelled by mathematicians and statisticians. The case for finitely additive probability is put forward strongly by Dubins and Savage (1965) in their book "How to Gamble If You Must". They refer to de Finetti, who, in a large number of papers published as early as 1930, "has always insisted that countable additivity is not an integral part of the probability concept but is rather in the nature of a regularity hypothesis." In fact, Dubins and Savage "view countably additive measures much as one views analytic functions—as a particularly important special case." But not much attention is paid to finitely additive measures in text-books on Measure Theory. (Books on Functional Analysis do a bit better.) One reason could be that countably additive measures are more tractable than finitely additive ones.

A need was felt to have a book on finitely additive measures which could serve as a reference book as well as a text-book. Cultivation of our interest in finitely additive measures started ten years ago. Our sustained interest in this area over the years led us to write this book.

In this book we have made an attempt to present a systematic and detailed study of finitely additive measures as we understand them, filling in any gaps that we discerned. This study of finitely additive measures as a mathematical object, in many of its manifestations, is like a study that a botanist would carry on a particular plant, or that a zoologist would launch on a particular species of mammals, or that a sociologist would initiate about a certain tribe, delving deep into various facets of the subject of interest. We look at the finitely additive measure (i) as a single entity (extension, nonatomicity and purity); (ii) in relation to another of its own kind (absolute continuity and singularity); (iii) in an introspective mood (decomposition theorems); (iv) as a member of a community (Nikodym theorem and Vitali–Hahn–Saks theorem); (v) as a member of a community in motion (weak convergence); (vi) in interaction with objects of different kind (integration); (vii) in association with related external communities (V_p-spaces); (viii) and its behaviour in external environment (range); (ix) in its internal environment (lifting).

Measure Theory (The Study of Countably Additive Measures) is an integral part of this wider study and the contrast between finite additivity and countable additivity is brought into sharp focus at various junctures in this work.

This book contains a good number of examples illustrating various aspects of finitely additive measures. A special feature of this book is the Selected Annotated Bibliography provided at the end of the book listing research papers we have come across in our pursuit of finitely additive measures.

We hope that this book serves practising analysts well and stimulates further research.

K.P.S. Bhaskara Rao gave a series of lectures on some of the topics covered in this book at the University of Lecce (Italy) in 1980 and at the University of Naples in 1981. He acknowledges gratefully the help given by these universities in making the visits possible. We also thank the Indian Statistical Institute for rendering help in making reciprocal visits of the authors possible in connection with this work.

Finally, a word of appreciation and gratitude to Surekha for her monumental patience in putting up with one of the most taxing and demanding spouses while this work was in progress. We also thank B. R. Marepalli for typing the entire manuscript.

December 1982   K.P.S. Bhaskara Rao
Calcutta
M. Bhaskara Rao
Sheffield

Contents

	Foreword	v
	Preface	vii
CHAPTER 1	**PRELIMINARIES**	1
1.1	Classes of sets	1
1.2	Set theoretical concepts	13
1.3	Topological concepts	15
1.4	Boolean algebras	18
1.5	Functional analytic concepts	23
CHAPTER 2	**CHARGES**	35
2.1	Basic concepts	35
2.2	The space of all bounded charges, $ba(\Omega, \mathscr{F})$	43
2.3	Measures	47
2.4	The space of all bounded measures, $ca(\Omega, \mathscr{F})$	50
2.5	Jordan Decomposition theorem	52
2.6	Hahn Decomposition theorem	56
CHAPTER 3	**EXTENSIONS OF CHARGES**	58
3.1	Real valued set functions and induced functionals	58
3.2	Real partial charges and their extensions	64
3.3	Extension procedure of Łos and Marczewski	70
3.4	Extension of partial charges in the general case	76
3.5	Miscellaneous extensions	78
3.6	Common extensions	82
CHAPTER 4	**INTEGRATION**	85
4.1	Total variation and outer charges	85
4.2	Null sets and null functions	87
4.3	Hazy convergence	92
4.4	D-integral	96
4.5	S-integral	115
4.6	L_p-spaces	121
4.7	$ba(\Omega, \mathscr{F})$ as a dual space	133
CHAPTER 5	**NONATOMIC CHARGES**	141
5.1	Basic concepts	141
5.2	Sobczyk-Hammer Decomposition theorem	144
5.3	Existence of nonatomic charges	150
5.4	Denseness	156

CHAPTER 6	**ABSOLUTE CONTINUITY**	159
6.1	Absolute continuity and singularity	159
6.2	Lebesgue Decomposition theorem	166
6.3	Radon-Nikodym theorem	169

CHAPTER 7	**V_p-SPACES**	178
7.1	L_p-spaces—An overview	178
7.2	V_p-spaces	185
7.3	Duals of V_p-spaces	193
7.4	Strong Convergence	197
7.5	Weak Convergence	200

CHAPTER 8	**NIKODYM THEOREM, WEAK CONVERGENCE AND VITALI–HAHN–SAKS THEOREM**	203
8.1	Nikodym and Vitali–Hahn–Saks theorems in the classical case	204
8.2	Examples	205
8.3	Phillips' lemma	206
8.4	Nikodym theorem	209
8.5	Norm bounded sets in the presence of uniform absolute continuity	213
8.6	A decomposition theorem	216
8.7	Weak convergence	218
8.8	Vitali–Hahn–Saks theorem	226

CHAPTER 9	**THE DUAL OF $ba(\Omega, \mathscr{F})$ AND THE REFINEMENT INTEGRAL**	231
9.1	Refinement integral	231
9.2	The dual of $ba(\Omega, \mathscr{F})$	234

CHAPTER 10	**PURE CHARGES**	240
10.1	Definitions and properties	240
10.2	A decomposition theorem	241
10.3	Pure charges on σ-fields	243
10.4	Examples	244
10.5	Pure charges on Boolean algebras	246

CHAPTER 11	**RANGES OF CHARGES**	249
11.1	Ranges of bounded charges on fields	249
11.2	Ranges of charges on σ-fields	252
11.3	Cardinalities of ranges of charges	256
11.4	Charges with closed range	257
11.5	Charges whose ranges are neither Lebesgue measurable nor have the property of Baire	264

CHAPTER 12	**ON LIFTING**	268
Appendix 1: Notes and Comments		272
Appendix 2: Selected Annotated Bibliography		282
Appendix 3: Some Set Theoretic Nomenclature		305
Index of Symbols and Function Spaces		306
Subject Index		309

CHAPTER 1

Preliminaries

The only prerequisite that is needed for understanding a substantial part of this book is a knowledge of Real Analysis, Set Theory and General Topology at a rudimentary level. The purpose of this chapter is to collect, in succinct form, various basic notions and results that are needed in this book. Section 1.1 presents various classes of sets and their properties. Section 1.2 briefly touches on some notions from Set Theory. Section 1.3 makes a sojourn with General Topology. Section 1.4 briefly dwells on Boolean Algebras. Finally, Section 1.5 presents vector lattices in some detail adequate for our needs.

A word of advice; before entering the terrain of finitely additive measures, the reader is urged to ensure a good degree of familiarity with the concepts presented in this chapter.

1.1 CLASSES OF SETS

Various types of classes of sets are presented in this section. The most important concept is the field of subsets of a set. This collection is, usually, the domain of definition for finitely additive measures.

Throughout this book, Ω is always understood to be a non-empty set. The set theoretic operations we use are standard. For the reader's convenience, a list is appended at the end of this book.

1.1.1 Definitions. Let Ω be a set and $\mathscr{F}$ a collection of subsets of Ω.
(1). $\mathscr{F}$ is said to be a *lattice* on Ω if the following conditions are satisfied.
   (i). $A, B \in \mathscr{F} \Rightarrow A \cup B \in \mathscr{F}$.
   (ii). $A, B \in \mathscr{F} \Rightarrow A \cap B \in \mathscr{F}$.
(2). $\mathscr{F}$ is said to be a *semi-ring* on Ω if the following conditions are satisfied.
   (i). $\emptyset \in \mathscr{F}$.
   (ii). $A, B \in \mathscr{F} \Rightarrow A \cap B \in \mathscr{F}$.
   (iii). If $A, B \in \mathscr{F}$ and $A \subset B$, then there exists a finite number $A_0, A_1, \ldots, A_n$ of sets in $\mathscr{F}$ such that $A = A_0 \subset A_1 \subset A_2 \subset \cdots \subset A_n = B$ and $A_i - A_{i-1} \in \mathscr{F}$ for $i = 1, 2, \ldots, n$.

(3). $\mathscr{F}$ is said to be a *semi-field* on Ω if $\mathscr{F}$ is a semi-ring and $\Omega \in \mathscr{F}$.
(4). $\mathscr{F}$ is said to be a *ring* on Ω if the following conditions are satisfied.
   (i). $\varnothing \in \mathscr{F}$.
   (ii). $A, B \in \mathscr{F} \Rightarrow A \cup B \in \mathscr{F}$.
   (iii). $A, B \in \mathscr{F} \Rightarrow A - B \in \mathscr{F}$.
(5). $\mathscr{F}$ is said to be a *field* on Ω if $\mathscr{F}$ is a ring and $\Omega \in \mathscr{F}$.
(6). $\mathscr{F}$ is said to be an *additive-class* on Ω if the following conditions are satisfied.
   (i). $\varnothing \in \mathscr{F}$.
   (ii). $A, B \in \mathscr{F}$ and $A \cap B = \varnothing \Rightarrow A \cup B \in \mathscr{F}$.
   (iii). $A \in \mathscr{F} \Rightarrow A^c \in \mathscr{F}$.
(7). $\mathscr{F}$ is said to be a σ-*ring* on Ω if the following conditions are satisfied.
   (i). $\varnothing \in \mathscr{F}$.
   (ii). $\{A_n; n \geq 1\} \subset \mathscr{F} \Rightarrow \bigcup_{n \geq 1} A_n \in \mathscr{F}$.
   (iii). $A, B \in \mathscr{F} \Rightarrow A - B \in \mathscr{F}$.
(8). $\mathscr{F}$ is said to be a σ-*field* on Ω if $\mathscr{F}$ is a σ-ring on Ω and $\Omega \in \mathscr{F}$.
(9). $\mathscr{F}$ is said to be a σ-*class* on Ω if the following conditions are satisfied.
   (i). $\varnothing \in \mathscr{F}$.
   (ii). If $A_n, n \geq 1$ is a sequence of pairwise disjoint sets in $\mathscr{F}$, then $\bigcup_{n \geq 1} A_n \in \mathscr{F}$.
   (iii). $A \in \mathscr{F} \Rightarrow A^c \in \mathscr{F}$.

One could form a comprehensive picture of the interrelations between various types of classes introduced above. We shall not go into details. We present some important ones in the following.

1.1.2 Remarks.
(1). A lattice of sets need not be a semi-ring.
(2). A semi-ring need not be a lattice.
(3). Every ring is a semi-ring.
(4). Every ring is a lattice.
(5). Every field is a ring.
(6). Every field is a semi-field.
(7). Every field is an additive-class.
(8). Every σ-field is a field.
(9). Every σ-field is a σ-ring.
(10). Every σ-field is a σ-class.

The above statements can be easily verified. The converse of any of the implications in (3) to (10) does not hold. Examples can easily be constructed. A sample of examples to illustrate some of the definitions is presented here. Many more are to come later.

1.1.3 Examples.
(1). Let Ω be any infinite set. A set $A \subset \Omega$ is said to be cofinite if A^c is a finite subset of Ω. Let $\mathscr{F}$ be the collection of all finite and cofinite subsets of Ω. Then $\mathscr{F}$ is a field on Ω, but not a σ-field on Ω.
(2). Let $\Omega = [0, 1)$ and $\mathscr{C} = \{[a, b); 0 \leq a \leq b \leq 1\}$. Then $\mathscr{C}$ is a semi-field on Ω.
(3). Let $\Omega = [0, 1)$ and $\mathscr{F} = \{\bigcup_{i=1}^n [a_i, b_i); [a_i, b_i) \cap [a_j, b_j) = \varnothing$ for all $i \neq j$, $0 \leq a_i \leq b_i \leq 1$ for all i, $n \geq 1\}$. Then $\mathscr{F}$ is a field on Ω. $\mathscr{F}$ is precisely the collection of all those subsets of Ω each of which is a finite disjoint union of sets from $\mathscr{C}$ of (2) above. It will follow from Theorem 1.1.9(2) that $\mathscr{F}$ is precisely the smallest field on Ω containing $\mathscr{C}$.
(4). Let $\mathscr{P}(\Omega)$ denote the class of all subsets of Ω. ($\mathscr{P}(\Omega)$ is called the power set of Ω.) Then $\mathscr{P}(\Omega)$ is an example of each type of class presented in Definition 1.1.1. $\mathscr{P}(\Omega)$ is also called discrete field or discrete σ-field on Ω.

In the following, we give salient features of some important types of classes presented in Definition 1.1.1. These are not hard to discern.

1.1.4 Properties.
(1). If $\mathscr{F}$ is a ring on Ω, then $\bigcup_{i=1}^n A_i$ and $\bigcap_{i=1}^n A_i \in \mathscr{F}$ for any finite number $A_1, A_2, \ldots, A_n$ of sets in $\mathscr{F}$, i.e. $\mathscr{F}$ is closed under finite unions and finite intersections.
(2). If $\mathscr{F}$ is a ring on Ω, then $\mathscr{F}$ is closed under symmetric differences, i.e. $A \Delta B = (A - B) \cup (B - A) \in \mathscr{F}$ whenever $A, B \in \mathscr{F}$.
(3). If $\mathscr{F}$ is a semi-ring on Ω and is closed under finite disjoint unions, then $\mathscr{F}$ is a ring on Ω.
(4). If $\mathscr{F}$ is an additive-class on Ω, then $\mathscr{F}$ is closed under proper differences, i.e. $A - B \in \mathscr{F}$ whenever $A, B \in \mathscr{F}$ and $B \subset A$.
(5). If $\mathscr{F}$ is an additive-class on Ω and is closed under finite intersections or differences, then $\mathscr{F}$ is a field on Ω.
(6). If $\mathscr{F}$ is a σ-ring on Ω, then $\mathscr{F}$ is closed under countable intersections, i.e. $\bigcap_{n \geq 1} A_n \in \mathscr{F}$ whenever A_n, $n \geq 1$ is a sequence of sets in $\mathscr{F}$.
(7). If $\mathscr{F}$ is a σ-class on Ω and is closed under finite intersections or differences, then $\mathscr{F}$ is a σ-field on Ω.

A given collection $\mathscr{C}$ of subsets of a set Ω may not be of a particular type P listed in Definition 1.1.1. It is natural to enquire about the existence of a smallest collection $\mathscr{F}$ of subsets of Ω of the type P containing $\mathscr{C}$. The following results are designed to answer this query.

1.1.5 Lemma. *Let Ω be any set and P be any of the types listed in Definition 1.1.1 with the exception of P being a semi-ring or a semi-field. Let $\mathscr{F}_\alpha$, $\alpha \in \Gamma$ be a family of collections of subsets of Ω such that each $\mathscr{F}_\alpha$ is of type P. Then $\bigcap_{\alpha \in \Gamma} \mathscr{F}_\alpha$ is of type P.*

Proof. The proof easily follows from the definitions of each type. □

1.1.6 Remark. The following example explains why we have made exceptions of certain types of classes of sets in the above lemma. Let

$$\Omega = \{1, 2, 3, 4\},$$

$$\mathscr{F}_1 = \{\varnothing, \{1\}, \{2\}, \{1, 2\}, \{3, 4\}, \Omega\}$$

and

$$\mathscr{F}_2 = \{\varnothing, \{1\}, \{2, 3, 4\}, \Omega\}.$$

$\mathscr{F}_1$ and $\mathscr{F}_2$ are semi-fields on Ω but
$\mathscr{C} = \mathscr{F}_1 \cap \mathscr{F}_2 = \{\varnothing, \{1\}, \Omega\}$ is not a semi-field on Ω.

1.1.7 Theorem. *Let Ω be any set and P be any of the types listed in Definition 1.1.1 with the exception of P being a semi-ring or a semi-field. Let $\mathscr{C}$ be a collection of subsets of Ω. Then there exists a smallest collection $\mathscr{F}$ of subsets of Ω of type P containing $\mathscr{C}$.*

Proof. Let $\mathscr{F}_\alpha$, $\alpha \in \Gamma$ be the family of all collections of subsets of Ω each of which is of type P and contains $\mathscr{C}$. $\Gamma \neq \varnothing$ since $\mathscr{P}(\Omega)$ is of type P and contains $\mathscr{C}$. Then, by Lemma 1.1.5, $\mathscr{F} = \bigcap_{\alpha \in \Gamma} \mathscr{F}_\alpha$ is of type P and contains $\mathscr{C}$. It is not hard to see that $\mathscr{F}$ is the desired collection of sets. □

In the above, $\mathscr{C}$ is called a generator of $\mathscr{F}$ with respect to the type P or, simply a generator of $\mathscr{F}$ if P is understood.

1.1.8 Remark. Let $\Omega = \{1, 2, 3, 4\}$ and $\mathscr{C} = \{\varnothing, \{1\}, \Omega\}$. Then there is no smallest semi-field $\mathscr{F}$ on Ω containing $\mathscr{C}$. See Remark 1.1.6.

In some special cases, we can explicitly construct the smallest collection $\mathscr{F}$ of subsets of Ω of a given type containing a given collection $\mathscr{C}$ of subsets of Ω. The following theorem gives some examples of such cases.

1.1.9 Theorem. *Let Ω be any set.*
(1). Let $\mathscr{C}$ be a lattice on Ω and $\varnothing \in \mathscr{C}$. Let $\mathscr{F} = \{F - E;\ E, F \in \mathscr{C}$ and $E \subset F\}$. Then $\mathscr{F}$ is the smallest semi-ring on Ω containing $\mathscr{C}$.
(2). Let $\mathscr{C}$ be a semi-ring on Ω and $\mathscr{F} = \{\bigcup_{i=1}^{n} C_i;\ C_1, C_2, \ldots, C_n$ are pairwise disjoint sets in $\mathscr{C}$ and $n \geq 1\}$. Then $\mathscr{F}$ is the smallest ring on Ω containing $\mathscr{C}$.
(3). Let $\mathscr{C}$ be a semi-field on Ω and $\mathscr{F} = \{\bigcup_{i=1}^{n} C_i;\ C_1, C_2, \ldots, C_n$ are pairwise disjoint sets in $\mathscr{C}$ and $n \geq 1\}$. Then $\mathscr{F}$ is the smallest field on Ω containing $\mathscr{C}$.
(4). Let $\mathscr{C}$ be a ring on Ω. Let $\mathscr{C}_1 = \{A \subset \Omega;\ A^c \in \mathscr{C}\}$. Then $\mathscr{F} = \mathscr{C} \cup \mathscr{C}_1$ is the smallest field on Ω containing $\mathscr{C}$.
(5). Let $\mathscr{C}$ be a σ-ring on Ω. Let $\mathscr{C}_1 = \{A \subset \Omega;\ A^c \in \mathscr{C}\}$. Then $\mathscr{F} = \mathscr{C} \cup \mathscr{C}_1$ is the smallest σ-field on Ω containing $\mathscr{C}$.

Proof. (1). Let $F_1 - E_1$ and $F_2 - E_2 \in \mathscr{F}$, where $E_1, F_1, E_2, F_2 \in \mathscr{C}$, $E_1 \subset F_1$ and $E_2 \subset F_2$. Then $(F_1 - E_1) \cap (F_2 - E_2) = (F_1 \cap F_2) - (E_1 \cup E_2) \cap (F_1 \cap F_2) \in \mathscr{F}$. So, $\mathscr{F}$ is closed under finite intersections. Let $F_1 - E_1, F_2 - E_2 \in \mathscr{F}$ and $F_1 - E_1 \subset F_2 - E_2$, where $E_1, F_1, E_2, F_2 \in \mathscr{C}$, $E_1 \subset F_1$ and $E_2 \subset F_2$. Let $C = (F_1 \cap F_2) - (F_1 \cap E_2)$. Then $C \in \mathscr{F}, F_1 - E_1 \subset C \subset F_2 - E_2, C - (F_1 - E_1) \in \mathscr{F}$ and $(F_2 - E_2) - C \in \mathscr{F}$. Hence $\mathscr{F}$ is a semi-ring on Ω. Since $\varnothing \in \mathscr{C}, \mathscr{C} \subset \mathscr{F}$. It is not hard to see that $\mathscr{F}$ is indeed the smallest semi-ring on Ω containing $\mathscr{C}$.

(2). From the definition of $\mathscr{F}$, it is clear that $\mathscr{F}$ is closed under finite disjoint unions. We show that $\mathscr{F}$ is closed under finite intersections. let $\bigcup_{i=1}^{m} C_i \in \mathscr{F}$ and $\bigcup_{j=1}^{n} D_j \in \mathscr{F}$, where $C_1, C_2, \ldots, C_m$ are pairwise disjoint sets in $\mathscr{C}$ and $D_1, D_2, \ldots, D_n$ are pairwise disjoint sets in $\mathscr{C}$. Then

$$\left(\bigcup_{i=1}^{m} C_i\right) \cap \left(\bigcup_{j=1}^{n} D_j\right) = \bigcup_{i=1}^{m} \bigcup_{j=1}^{n} (C_i \cap D_j) \in \mathscr{F},$$

since $C_i \cap D_j, i = 1, 2, \ldots, m$ and $j = 1, 2, \ldots, n$ are pairwise disjoint sets in $\mathscr{C}$. Now, we show that $\mathscr{F}$ is closed under differences. Suppose $E, F \in \mathscr{C}$ and $E \subset F$. Then there exist $E_0, E_1, \ldots, E_n$ in $\mathscr{C}$ such that $E = E_0 \subset E_1 \subset \cdots \subset E_n = F$ and $E_i - E_{i-1} \in \mathscr{C}$ for every $i = 1, 2, \ldots, n$. So, $F - E = \bigcup_{i=1}^{n} (E_i - E_{i-1}) \in \mathscr{F}$, since $E_i - E_{i-1}$'s are pairwise disjoint. Let $\bigcup_{i=1}^{m} C_i$ and $\bigcup_{j=1}^{n} D_j$ be any two sets in $\mathscr{F}$, where $C_1, C_2, \ldots, C_m$ are pairwise disjoint sets from $\mathscr{C}$ and $D_1, D_2, \ldots, D_n$ are pairwise disjoint sets from $\mathscr{C}$. Note that

$$\bigcup_{j=1}^{n} D_j - \bigcup_{i=1}^{m} C_i = \bigcup_{j=1}^{n} \left[D_j - \left(\bigcup_{i=1}^{m} C_i\right)\right]$$
$$= \bigcup_{j=1}^{n} \bigcap_{i=1}^{m} (D_j - (C_i \cap D_j)).$$

From what we have proved above, $D_j - (C_i \cap D_j) \in \mathscr{F}$ for every i and j. Since $\mathscr{F}$ is closed under finite intersections, $\bigcap_{i=1}^{m} (D_j - (C_i \cap D_j)) \in \mathscr{F}$ for every j. Since $\mathscr{F}$ is closed under finite disjoint unions,

$$\bigcup_{j=1}^{n} \bigcap_{i=1}^{m} (D_j - (C_i \cap D_j)) \in \mathscr{F}.$$

If we show that $\mathscr{F}$ is closed under finite unions, it would imply that $\mathscr{F}$ is a ring on Ω. Let $A, B \in \mathscr{F}$. Then $A \cup B = A \cup (B - A)$. Since $B - A \in \mathscr{F}$ and A and $B - A$ are disjoint, it follows that $A \cup B \in \mathscr{F}$.

It is obvious that $\mathscr{C} \subset \mathscr{F}$. It is not hard to see that $\mathscr{F}$ is indeed the smallest ring on Ω containing $\mathscr{C}$.

(3). This is similar to (2).

(4). We show that $\mathscr{F}$ is a field on Ω. It is obvious that $\mathscr{F}$ is closed under complementation. Let $E, F \in \mathscr{F}$. *Case* (i). $E, F \in \mathscr{C}$. Then $E \cup F \in \mathscr{C} \subset \mathscr{F}$.

Case (ii). $E \in \mathscr{C}$ and $F \in \mathscr{C}_1$. Then $E \cup F \in \mathscr{C}_1$. For, $(E \cup F)^c = E^c \cap F^c = F^c - E \in \mathscr{C}$. *Case (iii)*. $E \in \mathscr{C}_1$ and $F \in \mathscr{C}$. This case is similar to Case (ii). *Case (iv)*. $E, F \in \mathscr{C}_1$. Then $E \cup F \in \mathscr{C}_1$. For, $(E \cup F)^c = E^c \cap F^c \in \mathscr{C}$. In any case, we have $E \cup F \in \mathscr{F}$. Thus $\mathscr{F}$ is a field on Ω containing $\mathscr{C}$. It is obvious that $\mathscr{F}$ is the smallest field on Ω containing $\mathscr{C}$.
(5). This can be proved as in (4). □

Next, we describe a constructive procedure for obtaining the smallest field $\mathscr{F}$ on Ω containing a given class $\mathscr{C}$ of subsets of Ω in a finite number of steps. First, we introduce a special notation.

1.1.10 Notation. For any subset A of Ω, let $A^1 = A$ and $A^0 = A^c$.

1.1.11 Theorem. *Let $\mathscr{C}$ be any class of subsets of a set Ω. Form the following classes of sets successively.*

$\mathscr{C}_1 = \{\varnothing, \Omega\} \cup \mathscr{C} \cup \{A \subset \Omega; A^c \in \mathscr{C}\}$.

$\mathscr{C}_2 = $ *The collection of all subsets of Ω each of which is a finite intersection of sets from $\mathscr{C}_1$.*

$= \left\{ \bigcap_{i=1}^{n} A_i; A_i \in \mathscr{C}_1 \text{ for } i = 1, 2, \ldots, n \text{ and } n \geq 1 \right\}$.

$\mathscr{C}_3 = $ *The collection of all subsets of Ω each of which is a finite disjoint union of sets from $\mathscr{C}_2$.*

$= \left\{ \bigcup_{j=1}^{m} B_j; B_j \in \mathscr{C}_2 \text{ for all } j, B_i \cap B_j = \varnothing \text{ for all } i \neq j \text{ and } m \geq 1 \right\}$.

Then $\mathscr{C}_3$ is the smallest field on Ω containing $\mathscr{C}$.

Proof. We note the following obvious facts. (i). $\mathscr{C} \subset \mathscr{C}_1 \subset \mathscr{C}_2 \subset \mathscr{C}_3$. (ii). $\varnothing \in \mathscr{C}_3$. (iii). $\mathscr{C}_1$ is closed under complementation, i.e. if $A \in \mathscr{C}_1$, then $A^c \in \mathscr{C}_1$. (iv). $\mathscr{C}_2$ is closed under finite intersections. (v). $\mathscr{C}_3$ is closed under finite disjoint unions. (vi). If $\mathscr{C}_3$ is a field on Ω, then it is indeed the smallest field on Ω containing $\mathscr{C}$.

We show that $\mathscr{C}_3$ is a field on Ω. This is carried out in the following steps.
Step 1. Note that if a collection $\mathscr{F}$ of subsets of a set Ω is closed under complementation and finite intersections, then $\mathscr{F}$ is a field on Ω. (If A, B $\in \mathscr{F}$, then $A \cup B = (A^c \cap B^c)^c \in \mathscr{F}$.)
Step 2. It is clear that $\mathscr{C}_3$ is closed under finite intersections in view of the fact that $\mathscr{C}_2$ is closed under finite intersections.
Step 3. Let $A \in \mathscr{C}_2$. We claim that $A^c \in \mathscr{C}_3$. We can write $A = \bigcap_{i=1}^{n} A_i$ for some $A_1, A_2, \ldots, A_n$ in $\mathscr{C}_1$. Note that $A^c = \bigcup A_1^{\delta_1} \cap A_2^{\delta_2} \cap \cdots \cap A_n^{\delta_n}$, where the union is taken over all $\delta_1, \delta_2, \ldots, \delta_n$ in $\{0, 1\}$ with the exception that

$(\delta_1, \delta_2, \ldots, \delta_n) = (1, 1, \ldots, 1)$. Each $A_1^{\delta_1} \cap A_2^{\delta_2} \cap \cdots \cap A_n^{\delta_n}$ belongs to $\mathscr{C}_2$ and these sets are pairwise disjoint. So, $A^c \in \mathscr{C}_3$.

Step 4. As a final step, let $B \in \mathscr{C}_3$. Then $B = \bigcup_{i=1}^{m} B_i$ for some pairwise disjoint sets $B_1, B_2, \ldots, B_m$ in $\mathscr{C}_2$. Then $B^c = \bigcap_{i=1}^{m} B_i^c$. Then each $B_i^c \in \mathscr{C}_3$, by Step 3. By Step 2, $B^c \in \mathscr{C}_3$.

This completes the proof. □

We obtain some important consequences of this result.

1.1.12 Corollary. *Let $\mathscr{C}$ be any class of subsets of a set Ω and $\mathscr{F}$ the smallest field on Ω containing $\mathscr{C}$. Then the following statements are true.*

(1). $A \in \mathscr{F}$ *if and only if there exist sets* A_{ij}, $j = 1, 2, \ldots, n_i$ *and* $i = 1, 2, \ldots, m$ *such that each* A_{ij} *or* $A_{ij}^c \in \mathscr{C}$ *and*

$$A = \bigcup_{i=1}^{m} \bigcap_{j=1}^{n_i} A_{ij}.$$

(2). $A \in \mathscr{F}$ *if and only if there exist sets* $B_{ij}, j = 1, 2, \ldots, k_i$ *and* $i = 1, 2, \ldots, n$, *such that each* B_{ij} *or* $B_{ij}^c \in \mathscr{C}$ *and*

$$A = \bigcap_{i=1}^{n} \bigcup_{j=1}^{k_i} B_{ij}.$$

Proof. (1). This follows from Theorem 1.1.11. (2). This follows from (1) and the distributive laws of the operations $\cup$ and $\cap$. □

1.1.13 Corollary. *Let $\mathscr{F}$ be a field of subsets of a set Ω and $A \subset \Omega$. Then the smallest field $\mathscr{F}_1$ on Ω containing $\mathscr{F}$ and A is given by*

$$\mathscr{F}_1 = \{(B_1 \cap A) \cup (B_2 \cap A^c); B_1, B_2 \in \mathscr{F}\}.$$

Proof. This follows from Corollary 1.1.12(1). (One can also show that $\mathscr{F}_1$ is a field on Ω directly.) □

1.1.14 Corollary. *Let $\mathscr{C}$ be any countable collection of subsets of a set Ω. Then the smallest field $\mathscr{F}$ on Ω containing $\mathscr{C}$ is also countable.*

Proof. This follows from Theorem 1.1.11. (Note that each $\mathscr{C}_i$ constructed in the proof of Theorem 1.1.11 is countable.) □

1.1.15 Remark. Given a class $\mathscr{C}$ of subsets of a set Ω, there is no simple way of constructing the smallest σ-field on Ω containing $\mathscr{C}$. The collection of all countable unions of sets each of which either belongs to $\mathscr{C}$ or its complement belongs to $\mathscr{C}$ need not be a σ-field on Ω.

Now, we introduce the notion of an atom of a field.

1.1.16 Definitions. Let $\mathscr{F}$ be a field of subsets of a set Ω.

(1). A set A in $\mathscr{F}$ is said to be an *atom* of $\mathscr{F}$ if the following conditions are satisfied.
  (i). $A \neq \emptyset$.
  (ii). $B \in \mathscr{F}$, $B \subset A \Rightarrow B = \emptyset$ or $B = A$.
(2). $\mathscr{F}$ is said to be *atomic* if every non-empty set in $\mathscr{F}$ contains an atom of $\mathscr{F}$.
(3). $\mathscr{F}$ is said to be *nonatomic* if $\mathscr{F}$ has no atoms.

An atom A of $\mathscr{F}$ is, intuitively, a minimal non-empty element of $\mathscr{F}$. If A and B are atoms of $\mathscr{F}$, then either $A = B$ or $A \cap B = \emptyset$. The following remarks give some more information about these notions.

1.1.17 Remarks.
(1). If Ω is the union of all atoms of $\mathscr{F}$, then $\mathscr{F}$ is atomic. But the converse is not true. As an example, let $\Omega = \{1, 2, 3, \ldots, \infty\}$ and $\mathscr{F}$ be the collection of all finite subsets of $\{1, 2, 3, \ldots\}$ and their complements. Then $\mathscr{F}$ is atomic. But the union of all atoms of $\mathscr{F} = \{1, 2, 3, \ldots\} \neq \Omega$.
(2). If $\mathscr{F}$ is a finite field on Ω, then $\mathscr{F}$ is atomic. For each ω in Ω, let $A_\omega = \bigcap_{\omega \in A \in \mathscr{F}} A$. Then A_ω is an atom of $\mathscr{F}$ containing ω.
(3). For any set Ω, $\mathscr{P}(\Omega)$ is atomic.
(4). The field given in Example 1.1.3(3) is nonatomic.

If $\mathscr{C}$ is a finite collection of subsets of a set Ω, then the smallest field $\mathscr{F}$ on Ω containing $\mathscr{C}$ can be described in a simple way. The following proposition amplifies this point.

1.1.18 Proposition.
(1). *If $\mathscr{C} = \{A_1, A_2, \ldots, A_n\}$ is a finite partition of Ω, i.e. $A_i \cap A_j = \emptyset$ for all $i \neq j$, $\bigcup_{i=1}^n A_i = \Omega$ and $A_i \neq \emptyset$ for every i, then the smallest field $\mathscr{F}$ on Ω containing $\mathscr{C}$ is the collection of all possible unions of sets from $\mathscr{C}$. The atoms of $\mathscr{F}$ are precisely $A_1, A_2, \ldots, A_n$.*
(2). *If $\mathscr{C} = \{A_1, A_2, \ldots, A_n\}$ is any finite collection of subsets of a set Ω, then the smallest field $\mathscr{F}$ on Ω containing $\mathscr{C}$ is the collection of all possible unions of sets from $\mathscr{C}_1 = \{A_1^{\delta_1} \cap A_2^{\delta_2} \cap \cdots \cap A_n^{\delta_n}; \delta_1, \delta_2, \ldots, \delta_n \in \{0, 1\}\}$. The atoms of $\mathscr{F}$ are precisely the non-empty sets in $\mathscr{C}_1$.*
(3). *If $\mathscr{F} = \{B_1, B_2, \ldots, B_m\}$ is a finite field on a set Ω, then the atoms of $\mathscr{F}$ are the non-empty sets from*

$$\{B_1^{\delta_1} \cap B_2^{\delta_2} \cap \cdots \cap B_m^{\delta_m}; \delta_1, \delta_2, \ldots, \delta_m \in \{0, 1\}\}.$$

(4). *If $\mathscr{F}$ is a finite field on Ω, then the number of sets in $\mathscr{F}$ is 2^k for some $k \geq 1$.*

The above statements are easy to check and the details are left to the reader.

Now, we collate the two notions, an additive-class and a field, in relation to generators. The following results are in this direction.

1.1.19 Theorem. *Let $\mathscr{C}$ be a class of subsets of a set Ω, $\mathscr{F}_0$ the smallest additive-class on Ω containing $\mathscr{C}$ and $\mathscr{F}_1$ the smallest field on Ω containing $\mathscr{C}$. Suppose $\mathscr{C}$ has one of the following properties.*
  (i). $A, B \in \mathscr{C} \Rightarrow A \cap B \in \mathscr{F}_0$.
  (ii). $A, B \in \mathscr{C} \Rightarrow A - B \in \mathscr{F}_0$.
Then $\mathscr{F}_0 = \mathscr{F}_1$.

Proof. Since $\mathscr{F}_0$ is the smallest additive-class containing $\mathscr{C}$ and $\mathscr{F}_1$ is an additive-class containing $\mathscr{C}$, it follows that $\mathscr{F}_0 \subset \mathscr{F}_1$. If we show that $\mathscr{F}_0$ is a field, it would then follow that $\mathscr{F}_1 \subset \mathscr{F}_0$ and hence $\mathscr{F}_0 = \mathscr{F}_1$. For this, it suffices to show that $\mathscr{F}_0$ is closed under finite intersections or differences. See Property 1.1.4(5). Suppose (i) holds. For each A in $\mathscr{F}_0$, define $\mathscr{F}_A = \{B \in \mathscr{F}_0 ; A \cap B \in \mathscr{F}_0\}$. We show that $\mathscr{F}_A$ is an additive-class. Clearly, $\varnothing \in \mathscr{F}_A$. If $B \in \mathscr{F}_A$, then $A \cap B^c = A - B = A - (A \cap B) \in \mathscr{F}_0$ as $\mathscr{F}_0$ is closed under proper differences. See Property 1.1.4(4). So, $B^c \in \mathscr{F}_A$. It is easy to check that $\mathscr{F}_A$ is closed under finite disjoint unions. Thus $\mathscr{F}_A$ is an additive-class contained in $\mathscr{F}_0$. If $A \in \mathscr{C}$, then $\mathscr{C} \subset \mathscr{F}_A$, by (i). Therefore, $\mathscr{F}_A = \mathscr{F}_0$. Now, if $A \in \mathscr{F}_0$, even then $\mathscr{C} \subset \mathscr{F}_A$. For, if $C \in \mathscr{C}$, then $\mathscr{F}_C = \mathscr{F}_0$ and so, $A \cap C \in \mathscr{F}_0$ or $C \in \mathscr{F}_A$. Consequently, $\mathscr{F}_A = \mathscr{F}_0$ for any A in $\mathscr{F}_0$. This implies that $\mathscr{F}_0$ is closed under finite intersections. One can show that the properties (i) and (ii) are equivalent using Property 1.1.4(4). This completes the proof. □

An easy consequence of the above result is the following observation. If $\mathscr{C}$ is a class of subsets of a set Ω closed under finite intersections or differences, then the smallest additive-class on Ω containing $\mathscr{C}$ and the smallest field on Ω containing $\mathscr{C}$ are identical.

The following results are in the spirit of Theorem 1.1.19 but in the setting of σ-classes and σ-fields.

1.1.20 Theorem. *Let $\mathscr{C}$ be a class of subsets of a set Ω and $\mathscr{F}_0$ the smallest σ-class on Ω containing $\mathscr{C}$. Suppose $\mathscr{C}$ has one of the following properties (which are equivalent anyway).*
  (i). $A, B \in \mathscr{C} \Rightarrow A \cap B \in \mathscr{F}_0$.
  (ii). $A, B \in \mathscr{C} \Rightarrow A - B \in \mathscr{F}_0$.
Let $\mathscr{F}_1$ be the smallest σ-field on Ω containing $\mathscr{C}$. Then $\mathscr{F}_0 = \mathscr{F}_1$. □

1.1.21 Corollary. *Let $\mathscr{C}$ be a collection of subsets of a set Ω closed under finite intersections or differences. Let $\mathscr{F}_0$ be the smallest σ-class on Ω containing $\mathscr{C}$ and $\mathscr{F}_1$ the smallest σ-field on Ω containing $\mathscr{C}$. Then $\mathscr{F}_0 = \mathscr{F}_1$.* □

A proof of Theorem 1.1.20 can be given by a slight modification of the proof of Theorem 1.1.19. Corollary 1.1.21 follows easily from Theorem 1.1.20.

Next, we introduce two very useful notions, namely "Ideals and Filters" which are complementary to each other.

1.1.22 Definitions. Let $\mathscr{F}$ be a field of subsets of a set Ω.
(1). $\mathscr{I} \subset \mathscr{F}$ is said to be an *ideal* in $\mathscr{F}$ if the following conditions are satisfied.
  (i). $\Omega \notin \mathscr{I}$.
  (ii). $A, B \in \mathscr{I} \Rightarrow A \cup B \in \mathscr{I}$.
  (iii). $A \in \mathscr{I}, B \in \mathscr{F}, B \subset A \Rightarrow B \in \mathscr{I}$.

An ideal $\mathscr{I}$ in $\mathscr{F}$ is said to be a *maximal ideal* in $\mathscr{F}$ if there is no ideal in $\mathscr{F}$ properly containing $\mathscr{I}$.

(2). $\mathscr{J} \subset \mathscr{F}$ is said to be a *filter* in $\mathscr{F}$ if the following conditions are satisfied.
  (i). $\varnothing \notin \mathscr{J}$.
  (ii). $A, B \in \mathscr{J} \Rightarrow A \cap B \in \mathscr{J}$.
  (iii). $A \in \mathscr{J}, B \in \mathscr{F}, A \subset B \Rightarrow B \in \mathscr{J}$.

A filter $\mathscr{J}$ in $\mathscr{F}$ is said to be a *maximal filter* in $\mathscr{F}$ if there is no filter in $\mathscr{F}$ properly containing $\mathscr{J}$.

1.1.23 Remarks. The following statements follow from the above definitions.

(1). If $\mathscr{I}$ is an ideal in $\mathscr{F}$, then $\mathscr{J} = \{A \in \mathscr{F}; A^c \in \mathscr{I}\}$ is a filter in $\mathscr{F}$. If $\mathscr{I}$ is a maximal ideal in $\mathscr{F}$, then the filter $\mathscr{J}$ defined above is a maximal filter in $\mathscr{F}$.

(2). If $\Omega = \{1, 2, 3, \ldots\}$ and $\mathscr{I} = \{A \subset \Omega; A \text{ is finite}\}$, then $\mathscr{I}$ is an ideal in $\mathscr{F} = \mathscr{P}(\Omega)$. $\mathscr{I}$ is not a maximal ideal in $\mathscr{F}$.

(3). If $\mathscr{C} \subset \mathscr{F}$ has the property that $\bigcup_{i=1}^{n} C_i \neq \Omega$ for any finite number $C_1, C_2, \ldots, C_n$ of sets in $\mathscr{C}$, then there exists a smallest ideal $\mathscr{I}$ in $\mathscr{F}$ containing $\mathscr{C}$. In fact, $\mathscr{I}$ is given by

$$\mathscr{I} = \left\{ A \in \mathscr{F}; A \subset \bigcup_{i=1}^{n} C_i \text{ for some } C_1, C_2, \ldots, C_n \text{ in } \mathscr{C} \right\}.$$

(4). If $\mathscr{C} \subset \mathscr{F}$ has the property that $\bigcap_{i=1}^{n} C_i \neq \varnothing$ for any finite number $C_1, C_2, \ldots, C_n$ of sets in $\mathscr{C}$, then there exists a smallest filter $\mathscr{J}$ in $\mathscr{F}$ containing $\mathscr{C}$. In fact, $\mathscr{J}$ is given by

$$\mathscr{J} = \left\{ A \in \mathscr{F}; \bigcap_{i=1}^{n} C_i \subset A \text{ for some } C_1, C_2, \ldots, C_n \text{ in } \mathscr{C} \right\}.$$

(5). Let $\mathscr{I}$ be an ideal in $\mathscr{F}$ and $A \in \mathscr{F}$. Suppose $A \cup B \neq \Omega$ for any B in $\mathscr{I}$. Then

$$\mathscr{I}_1 = \{C \in \mathscr{F}; C \subset A \cup B \text{ for some } B \text{ in } \mathscr{I}\}$$

1. PRELIMINARIES   11

is the smallest ideal in $\mathscr{F}$ containing $\mathscr{I}$ and A. We call $\mathscr{I}_1$ the ideal generated by $\mathscr{I}$ and A.

(6). Let $\mathscr{G}$ be a filter in $\mathscr{F}$ and $D \in \mathscr{F}$. Suppose $D \cap B \neq \varnothing$ for every B in $\mathscr{G}$. Then

$$\mathscr{G}_1 = \{C \in \mathscr{F}; D \cap B \subset C \text{ for some B in } \mathscr{G}\}$$

is the smallest filter containing $\mathscr{G}$ and D. We call $\mathscr{G}_1$ the filter generated by $\mathscr{G}$ and D.

(7). An ideal $\mathscr{I}$ in $\mathscr{F}$ is a maximal ideal in $\mathscr{F}$ if and only if for every A in $\mathscr{F}$ either A or $A^c \in \mathscr{I}$. (This follows from (5).)

(8). A filter $\mathscr{G}$ in $\mathscr{F}$ is a maximal filter in $\mathscr{F}$ if and only if for every A in $\mathscr{F}$ either A or $A^c \in \mathscr{G}$. (This follows from (6).)

(9). If $\mathscr{G}_1$ and $\mathscr{G}_2$ are two distinct maximal filters in $\mathscr{F}$, then there exists A in $\mathscr{F}$ such that $A \in \mathscr{G}_1$ and $A^c \in \mathscr{G}_2$.

We define limit supremum and limit infimum of a sequence of sets and give some properties of these notions.

1.1.24 Definitions. Let A_n, $n \geq 1$ be a sequence of subsets of a set Ω. Define

$$\limsup_{n \to \infty} A_n = \bigcap_{n \geq 1} \bigcup_{k \geq n} A_k$$

and

$$\liminf_{n \to \infty} A_n = \bigcup_{n \geq 1} \bigcap_{k \geq n} A_k.$$

1.1.25 Properties.
(1). $\limsup_{n \to \infty} A_n = \{\omega; \omega \in A_n \text{ for infinitely many } n\text{'s}\}$.
(2). $\liminf_{n \to \infty} A_n = \{\omega; \omega \in A_n \text{ for all but a finite number of } n\text{'s}\}$.
(3). $\liminf_{n \to \infty} A_n \subset \limsup_{n \to \infty} A_n$.
(4). $(\limsup_{n \to \infty} A_n)^c = \liminf_{n \to \infty} A_n^c$.
(5). $(\liminf_{n \to \infty} A_n)^c = \limsup_{n \to \infty} A_n^c$.

Now, we introduce some operations in the setting of real valued functions defined on a set. Let f and g be two real valued functions defined on a set Ω. $f \vee g, f \wedge g, f^-$ and $|f|$ are functions on Ω defined by

$$(f \vee g)(\omega) = \max\{f(\omega), g(\omega)\}, \quad \omega \in \Omega,$$

$$(f \wedge g)(\omega) = \min\{f(\omega), g(\omega)\}, \quad \omega \in \Omega,$$

$$f^+ = f \vee 0,$$

$$f^- = (-f) \vee 0$$

and
$$|f|(\omega) = |f(\omega)|, \quad \omega \in \Omega.$$

We give some identities involving these operations which actually stem from the corresponding properties of real numbers.

1.1.26 Identities.
(1). $f = f^+ - f^-$.
(2). $|f| = f^+ + f^-$.
(3). $f \vee g = \frac{1}{2}(f + g + |f - g|)$.
(4). $f \wedge g = \frac{1}{2}(f + g - |f - g|)$.
(5). $f + g = (f \vee g) + (f \wedge g)$.
(6). $|f - g| = (f \vee g) - (f \wedge g)$.
(7). $fg = \frac{1}{4}[(f + g)^2 - (f - g)^2]$.
(8). $|f| = f^+ \vee f^- = f \vee (-f)$.
(9). $||f| - |g|| \leq |f + g| \leq |f| + |g|$.
(10). $|(f_1 \vee g_1) - (f_2 \vee g_2)| \leq |f_1 - f_2| + |g_1 - g_2|$ for any real valued functions f_1, f_2, g_1, g_2 on Ω.
(11). $|(f_1 \wedge g_1) - (f_2 \wedge g_2)| \leq |f_1 - f_2| + |g_1 - g_2|$ for any real valued functions f_1, f_2, g_1, g_2 on Ω.
(12). $-|f| - |g| \leq f \wedge g \leq f \vee g \leq |f| + |g|$.

Finally, we end this section with some notes on measurable functions.

1.1.27 Definitions.
(1). The *Borel σ-field* on the real line R is defined to be the smallest σ-field on R containing all intervals.
(2). Let Ω be a set and $\mathfrak{A}$ a σ-field on Ω. A real valued function f on Ω is said to be *measurable* with respect to $\mathfrak{A}$ if $f^{-1}(B) = \{\omega \in \Omega; f(\omega) \in B\} \in \mathfrak{A}$ for every Borel subset B of R.
(3). For any subset A of Ω, the *indicator function* I_A of A is a map from Ω to R given by
$$I_A(\omega) = 1, \quad \text{if } \omega \in A,$$
$$= 0, \quad \text{if } \omega \in A^c.$$

1.1.28 Properties. Let $\mathfrak{A}$ be a σ-field of subsets of a set Ω. Then the following statements are true.
(1). For $A \subset \Omega$, I_A is measurable with respect to $\mathfrak{A}$ if and only if $A \in \mathfrak{A}$.
(2). A real valued function f on Ω is measurable with respect to $\mathfrak{A}$ if and only if $f^{-1}\{[k, \infty)\} \in \mathfrak{A}$ for every $-\infty < k < \infty$.
(3). If f, g are measurable functions with respect to $\mathfrak{A}$ and c, d are real numbers, then $cf + dg$, $|f|$, fg, f^2, $f \vee g$, $f \wedge g$, f^+ and f^- are all measurable functions with respect to $\mathfrak{A}$.
(4). A real valued function f on Ω is a measurable function with respect to $\mathfrak{A}$ if and only if there exists a sequence f_n, $n \geq 1$ of functions on Ω such

that $f(\omega) = \lim_{n \to \infty} f_n(\omega)$ for every ω in Ω, where each f_n is of the form

$$f_n = \sum_{i=1}^{k_n} c_{ni} I_{A_{ni}}$$

for some real numbers $c_{n1}, c_{n2}, \ldots, c_{nk_n}$ and for some sets $A_{n1}, A_{n2}, \ldots, A_{nk_n}$ from $\mathfrak{A}$.

1.2 SET THEORETICAL CONCEPTS

In this section, we present, in a concise form, some of the set theoretical notions we have used in this book.

One of the basic concepts in set theory is the notion of a relation on a set X. A *relation on X* is any subset Z of $X \times X$, where $X \times X$ is the cartesian product of X with itself, i.e. $X \times X = \{(x, y); x, y \in X\}$. A relation $Z \subset X \times X$ is said to be *reflexive* if $(x, x) \in Z$ for every x in X, *symmetric* if $(x, y) \in Z$ whenever $(y, x) \in Z$, *antisymmetric* if $(x, y), (y, x) \in Z$ only when $x = y$, *transitive* if $(x, z) \in Z$ whenever (x, y) and $(y, z) \in Z$ for some y in X.

A *partial order* on a set X is any relation Z which is reflexive, antisymmetric and transitive. We usually denote a partial order $Z \subset X \times X$ by $\leq$, i.e. for x, y in X, $x \leq y$ if and only if $(x, y) \in Z$. In terms of the notation $\leq$, the axioms of a partial order can be restated as follows. (i). $x \leq x$ for every x in X. (ii). $x \leq y$ and $y \leq x \Rightarrow x = y$ (antisymmetry). (iii). $x \leq y$ and $y \leq z \Rightarrow x \leq z$ (transitivity). The pair $(X, \leq)$ is called a partially ordered set.

A partial order $\leq$ on a set X is said to be a *linear order* on X if $x \leq y$ or $y \leq x$ for any x, y in X. In this case, we call $(X, \leq)$ a linearly ordered set.

Let $(X, \leq)$ be a partially ordered set. A subset C of X is called a *chain* in X if the partial order $\leq$ restricted to C is a linear order on C.

A partial order $\leq$ on a set X is said to be a *well-ordering* on X, if for every non-empty subset A of X, there exists an element a in A such that $a \leq x$ for all x in A, i.e. a is the smallest element in A. In this case, the pair $(X, \leq)$ is called a well-ordered set.

A relation $Z \subset X \times X$ is said to direct X if the following are satisfied. (i). $(x, x) \in Z$ for every x in X (reflexivity). (ii). $(x, z) \in Z$ whenever (x, y) and $(y, z) \in Z$ for some y in X (transitivity). (iii). For every x, y in X, there exists z in X such that $(z, x) \in Z$ and $(z, y) \in Z$. The relation $Z \subset X \times X$ satisfying (i), (ii) and (iii) above is usually denoted by $\geq$, i.e. $x \geq y$ if and only if $(x, y) \in Z$. In terms of this notation $\geq$, the above three conditions can be restated as follows. The relation $\geq$ on X is said to direct X if the following are satisfied. (i). $x \geq x$ for every x in X. (ii). $x \geq z$ whenever $x \geq y$ and $y \geq z$ for some y in X. (iii). For every x, y in X, there exists z in X such that $z \geq x$ and $z \geq y$. The pair $(X, \geq)$ is called a *directed set*.

A relation $Z \subset X \times X$ is said to be an *equivalence relation* if the following are satisfied. (i). $(x, x) \in Z$ for every x in X (reflexivity). (ii). $(x, y) \in Z$ whenever $(y, x) \in Z$ (symmetry). (iii). $(x, z) \in Z$ whenever (x, y) and $(y, z) \in Z$ for some y in X (transitivity). An equivalence relation on X is usually denoted by $\sim$. In terms of the notation $\sim$, the above conditions can be rephrased as follows. (i). $x \sim x$ for every x in X. (ii). $x \sim y$ if $y \sim x$. (iii). $x \sim z$ whenever $x \sim y$ and $y \sim z$. If $\sim$ is an equivalence relation on X, let $[x]$ denote the equivalence class in X containing x for any x in X, i.e. $[x] = \{y \subset X; y \sim x\}$. For x, y in X, $[x] = [y]$ or $[x] \cap [y] = \varnothing$. Under the equivalence relation $\sim$ on X, X thus can be written as the union of all its equivalence classes.

On some occasions, we use cardinals and ordinals. The reader is advised to refer to Kamke (1950) or any other book on Set Theory.

Transfinite induction is often used in inductive definitions, inductive constructions and in proofs. Let $(X, \leq)$ be a well-ordered set. Let a_0 be the smallest element in X. For each x in X, let $A(x)$ be a proposition.

1.2.1 Principle of Transfinite Induction. *If (i) $A(a_0)$ is valid, and (ii) for any a in X, $A(a)$ is valid whenever $A(x)$ is valid for every $x < a$ (i.e. $x \leq a$ and $x \neq a$) hold, then $A(x)$ is valid for every x in X.*

The above principle is also adopted for inductive definitions and inductive constructions.

We will be using Zorn's lemma in this book to exhibit objects with specified properties. This lemma which is equivalent to the *Axiom of Choice* is stated below.

1.2.2 Zorn's Lemma. *Let $(X, \leq)$ be a partially ordered set in which every chain C has an upper bound, i.e. there exists a in X such that $x \leq a$ for every x in C. Then X has a maximal element a_0, i.e. if $a_1 \in X$ and $a_0 \leq a_1$, then $a_1 = a_0$.*

1.2.3 *An Application of Zorn's Lemma.* Let $\mathscr{F}$ be a field of subsets of a set Ω and $\mathscr{I}$ an ideal in $\mathscr{F}$. Then there exists a maximal ideal $\mathscr{I}^*$ in $\mathscr{F}$ containing $\mathscr{I}$. Existence of $\mathscr{I}^*$ can be established by Zorn's lemma as follows. Let X be the collection of all ideals in $\mathscr{F}$ containing $\mathscr{I}$. For $\mathscr{I}_1, \mathscr{I}_2$ in X, say $\mathscr{I}_1 \leq \mathscr{I}_2$ if $\mathscr{I}_1 \subset \mathscr{I}_2$. $\leq$ is a partial order on X. Let $\{\mathscr{I}_\alpha; \alpha \in \Gamma\}$ be a chain in X. Then $\bigcup_{\alpha \in \Gamma} \mathscr{I}_\alpha \in X$ and is an upper bound of the chain $\{\mathscr{I}_\alpha; \alpha \in \Gamma\}$. Consequently, there exists a maximal element $\mathscr{I}^*$ in X. Similarly, if $\mathscr{J}$ is a filter in $\mathscr{F}$, one can show that there exists a maximal filter $\mathscr{J}^*$ in $\mathscr{F}$ containing $\mathscr{J}$, using Zorn's lemma.

Finally, we end this section with an important example of a directed set. Let $\mathscr{F}$ be a field of subsets of a set Ω. *A finite partition of Ω in $\mathscr{F}$* is a finite family $P = \{E_1, E_2, \ldots, E_m\}$ of pairwise disjoint sets in $\mathscr{F}$ whose union is Ω. Let $\mathscr{P}$ denote the collection of all finite partitions of Ω in $\mathscr{F}$. For P_1, P_2 in $\mathscr{P}$, say $P_1 \geq P_2$ if every set in P_1 is contained in some set of P_2. Another expression that is commonly used in this case is that P_1 is a refinement of P_2. We claim that the relation $\geq$ on $\mathscr{P}$ directs $\mathscr{P}$. It is obvious that $P \geq P$ for every P in $\mathscr{P}$. If $P_1 \geq P_2$ and $P_2 \geq P_3$, then it is clear that $P_1 \geq P_3$. If $P_1 = \{E_1, E_2, \ldots, E_m\} \in \mathscr{P}$ and $P_2 = \{F_1, F_2, \ldots, F_n\} \in \mathscr{P}$, then $P = \{E_i \cap F_j; 1 \leq i \leq m, 1 \leq j \leq n\} \in \mathscr{P}$ and $P \geq P_1$, $P \geq P_2$. Hence the relation $\geq$ directs $\mathscr{P}$. In other words, $(\mathscr{P}, \geq)$ is a directed set. If $F \in \mathscr{F}$, let $\mathscr{P}_F$ denote the collection of all finite partitions of F in $\mathscr{F}$. Under the same relation $\geq$ as above, $(\mathscr{P}_F, \geq)$ is also a directed set.

1.3 TOPOLOGICAL CONCEPTS

In this section, we briefly review some of the topological concepts needed in the subsequent chapters. For any other unexplained terminology in the text which slipped inadvertently from our compilation here, the reader is advised to refer to Kelley (1955).

A *topological space* is a pair $(X, \mathscr{T})$, where X is a set and $\mathscr{T}$ is a collection of subsets of X closed under unions and finite intersections and contains $\emptyset$ and X. Members of $\mathscr{T}$ are called *open sets* and the complements of open sets are called *closed sets*. For any set $A \subset X$, the *closure of* A is defined to be the set $\bigcap \{C; C \text{ closed}, A \subset C\}$ and is denoted by $\bar{A}$. $\bar{A}$ is the smallest closed set containing A. *Interior of* A is defined to be the set $\bigcup \{V; V \text{ open}, V \subset A\}$ and is denoted by A°. A° is the largest open set contained in A. A subset A of X is said to be a G_δ-set if A is a countable intersection of open sets. A subset B of X is said to be an F_σ-set if B is a countable union of closed sets. A point x in X is an *accumulation point* of a subset A of X if $(A - \{x\}) \cap U \neq \emptyset$ for every open set U containing x. A point x in $A \subset X$ is an *isolated point* of A if there exists an open set U containing x such that $A \cap U = \{x\}$.

A *net* in a topological space $(X, \mathscr{T})$ is any function from a directed set $(D, \geq)$ into X. Nets are usually denoted by x_α, $\alpha \in D$. A net x_α, $\alpha \in D$ is said to converge to an element x in X if given any open set V containing x, there exists α_0 in D such that $x_\alpha \in V$ whenever $\alpha \geq \alpha_0$. A subset A of X is closed if and only if $x \in A$ whenever there is a net in A converging to x. Let $(D, \geq)$ be a directed set. Suppose for every α in D, there is a β_α in D such that $\beta_\alpha \geq \alpha$. If x_α, $\alpha \in D$ is a net in X, then x_{β_α}, $\alpha \in D$ is called a *subnet* of x_α, $\alpha \in D$. If a net x_α, $\alpha \in D$ is convergent in X, then every subnet of x_α, $\alpha \in D$ is also convergent and converges to the same limit.

A topological space $(X, \mathcal{T})$ is said to be a *Hausdorff* topological space if for every distinct x, y in X, there exist open sets V_1 and V_2 such that $x \in V_1$, $y \in V_2$ and $V_1 \cap V_2 = \varnothing$. An *open cover* of X is any family of open sets whose union is X. A topological space $(X, \mathcal{T})$ is said to be *compact* if every open cover of X admits a finite subcover of X. A topological space $(X, \mathcal{T})$ is compact if and only if every net in X admits a convergent subnet in X. Equivalently, $(X, \mathcal{T})$ is compact if and only if for every collection of closed sets $\{A_\alpha ; \alpha \in \Gamma\}$ with finite intersection property, i.e. $\bigcap_{\alpha \in \Gamma_1} A_\alpha \neq \varnothing$ for every finite subset Γ_1 of Γ, $\bigcap_{\alpha \in \Gamma} A_\alpha \neq \varnothing$. The notion of compactness of a subset A of X can also be introduced in the same way as above for X. If A is a closed subset of X and X is compact, then A is compact. If $(X, \mathcal{T})$ is any Hausdorff topological space, then any compact subset of X is closed. A subset A of X is said to be *clopen* if A is open as well as closed. A topological space $(X, \mathcal{T})$ is said to be *totally disconnected* if the family of all clopen subsets of X forms a *base* for the topology of X, i.e. every open set is a union of clopen sets.

Let X be any set. Let $\mathscr{E}$ be a family of subsets of X closed under finite intersections and containing X. Then there is a smallest topology on X for which $\mathscr{E}$ is a base. This smallest topology is precisely the collection of all unions of sets from $\mathscr{E}$.

Let $(X, \mathcal{T})$ be a topological space. A real valued function f on X is said to be *continuous* if $f^{-1}(U) \in \mathcal{T}$ for every open subset U of the real line R, i.e. U is a union of open intervals. f is continuous if and only if $f(x_\alpha)$, $\alpha \in D$ converges to $f(x)$ whenever x_α, $\alpha \in D$ converges to x in X. Every real valued continuous function on a compact space X is bounded. If A is a compact subset of X and f is a continuous function on X, then the image of A under f, i.e. $f(A)$, is a compact subset of the real line R.

Let $(X, \mathcal{T})$ be a topological space and A a subset of X. Then $A \cap \mathcal{T} = \{A \cap V; V \in \mathcal{T}\}$ is called the *relative topology* on A. $(A, A \cap \mathcal{T})$ is a topological space. A is a compact subset of X if and only if $(A, A \cap \mathcal{T})$ is a compact topological space.

Let $(X, \mathcal{T})$ be a topological space. A subset A of X is said to be *dense-in-itself* if there are no isolated points in A. A subset A of X is said to be *perfect* if it is closed and there are no isolated points in A. X is said to be *scattered* if no non-empty closed subset of X is perfect, i.e. every non-empty closed subset A of X contains an isolated point of A.

Let X be any set. A *pseudo-metric* on X is a map p from the cartesian product space $X \times X$ to $[0, \infty)$ satisfying the following. (i). $p(x, x) = 0$ for every x in X. (ii). $p(x, y) = p(y, x)$ for every x, y in X. (iii). $p(x, z) \leq p(x, y) + p(y, z)$ for all x, y, z in X. A pseudo-metric p on X is said to be a *metric* on X if $x = y$ whenever $p(x, y) = 0$. A pseudo-metric space (X, p) is said to be *complete* if every Cauchy sequence in X is convergent, i.e.

1. PRELIMINARIES 17

whenever x_n, $n \geq 1$ is a sequence in X and $\lim_{m,n\to\infty} p(x_m, x_n) = 0$, there exists x in X such that $\lim_{n\to\infty} p(x_n, x) = 0$. If (X, p) is a pseudo-metric space, there exists a complete pseudo-metric space $(\tilde{X}, \tilde{p})$ with the following properties. (i). $X \subset \tilde{X}$. (ii). $\tilde{p}(x, y) = p(x, y)$ for all x, y in X. (iii). X is a *dense* subset of $\tilde{X}$, i.e. for every $\tilde{x}$ in $\tilde{X}$, there exists a sequence x_n, $n \geq 1$ in X converging to x, i.e. $\lim_{n\to\infty} \tilde{p}(x_n, \tilde{x}) = 0$. $(\tilde{X}, \tilde{p})$ is called a *completion* of the pseudo-metric space (X, p). Every pseudo-metric space (X, p) induces a natural topology $\mathcal{T}$ on X. $\mathcal{T}$ is the smallest topology on X containing $\mathcal{E} = \{B(x, r); x \in X, r > 0\}$, where $B(x, r) = \{y \in X; p(x, y) < r\}$.

Let $(X, \mathcal{T})$ be a topological space. A subset A of X is said to be *nowhere dense* if $(\bar{A})^\circ = \varnothing$. A subset B of X is said to be of *first category* if B is a countable union of nowhere dense subsets of X. *Baire Category Theorem*: If (X, p) is a complete metric space, then X is not of first category. A subset A of X is said to have the *property of Baire* if there exists an open set V such that $A \Delta V$ is of first category. The collection of all subsets of X each of which has the property of Baire is a σ-field on X containing all open subsets of X.

The product space $C = \{0, 1\}^{\aleph_0} = \{(x_1, x_2, \ldots); x_i = 0 \text{ or } 1 \text{ for every } i \geq 1\}$ is called *Cantor Set*. There is a natural metric p on C defined by

$$p((x_1, x_2, \ldots), (y_1, y_2, \ldots)) = \sum_{n \geq 1} \frac{1}{2^n} |x_n - y_n|$$

for $(x_1, x_2, \ldots)$ and $(y_1, y_2, \ldots)$ in C. (C, p) is a compact Hausdorff totally disconnected perfect metric space. Every clopen subset of C is a finite union of sets of the form $\{x_1\} \times \{x_2\} \times \cdots \times \{x_n\} \times \{0, 1\} \times \{0, 1\} \times \cdots$, for $x_1, x_2, \ldots, x_n$ in $\{0, 1\}$ and $n \geq 1$.

Let $(X, \mathcal{T})$ be a compact Hausdorff space. The *Baire σ-field* $\mathcal{B}_0$ on X is the smallest σ-field on X with respect to which every real valued continuous function on X is measurable. The *Borel σ-field* $\mathcal{B}$ on X is the smallest σ-field on X containing all open subsets of X. Obviously, $\mathcal{B}_0 \subset \mathcal{B}$. We now show that $\mathcal{B}_0$ is the smallest σ-field on X containing all compact G_δ subsets of X. Let $\mathcal{B}_{00}$ be the smallest σ-field on X containing all compact G_δ subsets of X. Let A be a compact G_δ subset of X. By *Urysohn's lemma*, there exists a real valued continuous function f on X such that $A = f^{-1}(\{0\})$. Therefore, $A \in \mathcal{B}_0$. Consequently, $\mathcal{B}_{00} \subset \mathcal{B}_0$. If f is a real valued continuous function on X, then $f^{-1}\{[k, \infty)\}$ is a compact G_δ subset of X and so $f^{-1}\{[k, \infty)\} \in \mathcal{B}_{00}$. Therefore, f is measurable with respect to $\mathcal{B}_{00}$. Hence $\mathcal{B}_0 \subset \mathcal{B}_{00}$. Thus we have shown that $\mathcal{B}_0 = \mathcal{B}_{00}$.

Let $(X, \mathcal{T})$ be a compact Hausdorff totally disconnected space, $\mathcal{C}$ the field of all clopen subsets of X and $\mathcal{B}_0$ the Baire σ-field on X. We show that the smallest σ-field $\mathcal{B}_0'$ on X containing $\mathcal{C}$ is $\mathcal{B}_0$. Since every C in $\mathcal{C}$ is a compact G_δ subset of X, it follows that $\mathcal{B}_0' \subset \mathcal{B}_0$. We now show that every real valued continuous function f on X is measurable with respect to $\mathscr{B}_0'$. For $n \geq 1$ and x in X, let $U_x = \{y \in X; |f(x)-f(y)| < 1/n\}$. U_x contains a clopen set V_x containing x. $\{V_x; x \in X\}$ is an open cover for X. Let $\{V_{x_1}, V_{x_2}, \ldots, V_{x_k}\}$ be a sub-cover of X. Let $D_1 = V_{x_1}$, $D_2 = V_{x_2} - V_{x_1}$, $D_3 = V_{x_3} - (V_{x_1} \cup V_{x_2}), \ldots, D_k = V_{x_k} - (V_{x_1} \cup V_{x_2} \cup \cdots \cup V_{x_{k-1}})$. $D_1, D_2, \ldots, D_k$ are pairwise disjoint clopen subsets of X whose union is X. Choose and fix y_i in D_i for each $i = 1, 2, \ldots, k$. Let

$$f_n = \sum_{i=1}^{k} f(y_i) I_{D_i}.$$

Then $|f_n(x) - f(x)| \leq 1/n$ for every x in X. Consequently, $\lim_{n \to \infty} f_n(x) = f(x)$ for every x in X. Each f_n is measurable with respect to $\mathscr{B}_0'$. Therefore, f is measurable with respect to $\mathscr{B}_0'$. Hence $\mathscr{B}_0' = \mathscr{B}_0$.

1.4 BOOLEAN ALGEBRAS

The aim of this section is to present some basic ideas on Boolean algebras and prove Stone Representation Theorem for Boolean algebras.

1.4.1 Definition. A *Boolean algebra* is a non-empty set $\mathbb{B}$ in which two binary operations $\vee, \wedge$ (join and meet) and one unary operation ^{c} (complementation) are defined satisfying the following identities.
(i). $a \vee b = b \vee a$, $a \wedge b = b \wedge a$ for all a, b in $\mathbb{B}$.
(ii). $a \vee (b \vee c) = (a \vee b) \vee c$, $a \wedge (b \wedge c) = (a \wedge b) \wedge c$ for all a, b, c in $\mathbb{B}$.
(iii). $(a \wedge b) \vee b = b$, $(a \vee b) \wedge b = b$ for all a, b in $\mathbb{B}$.
(iv). $a \wedge (b \vee c) = (a \wedge b) \vee (a \wedge c)$, $a \vee (b \wedge c) = (a \vee b) \wedge (a \vee c)$ for all a, b, c in $\mathbb{B}$.
(v). $(a \wedge a^c) \vee b = b$, $(a \vee a^c) \wedge b = b$ for all a, b in $\mathbb{B}$.

One can show that $a \wedge a^c$ does not depend on a in $\mathbb{B}$ and $b \vee b^c$ does not depend on b in $\mathbb{B}$. Denote $a \wedge a^c$ by 0 and $a \vee a^c$ by 1.

If $\mathscr{F}$ is a field of subsets of a set Ω, then $\mathscr{F}$ is a Boolean algebra in the above sense if we identify the operation of join by union of sets, meet by intersection of sets and complementation by complementation of sets with respect to Ω. 0 identifies with the empty set $\emptyset$ and 1 identifies with the whole space Ω.

We introduce a partial order $\leq$ on any given Boolean algebra $\mathbb{B}$ as follows. For a, b in $\mathbb{B}$, say $a \leq b$ if $a \wedge b = a$, or equivalently, $a \vee b = b$. The relation $\leq$ is a partial order on $\mathbb{B}$, i.e. it is reflexive ($a \leq a$ for every a in $\mathbb{B}$), antisymmetric (if $a \leq b$ and $b \leq a$, then $a = b$), and transitive (if $a \leq b$ and $b \leq c$, then $a \leq c$). With respect to this partial order, $a \vee b$ is the smallest

1. PRELIMINARIES

element in $\mathbb{B} \geq a$ and b, $a \wedge b$ is the largest element in $\mathbb{B} \leq$ both a and b. In other words, $a \vee b = \text{Sup}\{a, b\}$ and $a \wedge b = \text{Inf}\{a, b\}$. For any subset $\{a_\alpha; \alpha \in \Gamma\}$ of $\mathbb{B}$, $\bigvee_{\alpha \in \Gamma} a_\alpha$ denotes the smallest element in $\mathbb{B}$, if it exists, $\geq a_\alpha$ for every α in Γ, and $\bigwedge_{\alpha \in \Gamma} a_\alpha$ the largest element in $\mathbb{B}$, if it exists, $\leq a_\alpha$ for every α in Γ.

1.4.2 Definition. A Boolean algebra $\mathbb{B}$ is said to be a *Boolean σ-algebra* if for any sequence b_n, $n \geq 1$ in $\mathbb{B}$, $\bigvee_{n \geq 1} b_n$ exists in $\mathbb{B}$.

In a Boolean σ-algebra $\mathbb{B}$, one can show that $\bigwedge_{n \geq 1} b_n$ exists in $\mathbb{B}$ for any sequence b_n, $n \geq 1$ in $\mathbb{B}$. Further, $(\bigvee_{n \geq 1} b_n)^c = \bigwedge_{n \geq 1} b_n^c$ and $(\bigwedge_{n \geq 1} b_n)^c = \bigvee_{n \geq 1} b_n^c$.

1.4.3 Definition. A Boolean algebra $\mathbb{B}$ is said to be a *complete Boolean algebra* if for any subset $\{b_\alpha; \alpha \in \Gamma\}$ of $\mathbb{B}$, $\bigvee_{\alpha \in \Gamma} b_\alpha$ exists in $\mathbb{B}$.

In a complete Boolean algebra $\mathbb{B}$, one can show that $\bigwedge_{\alpha \in \Gamma} b_\alpha$ exists in $\mathbb{B}$. Further, $(\bigvee_{\alpha \in \Gamma} b_\alpha)^c = \bigwedge_{\alpha \in \Gamma} b_\alpha^c$ and $(\bigwedge_{\alpha \in \Gamma} b_\alpha)^c = \bigvee_{\alpha \in \Gamma} b_\alpha^c$.

1.4.4 Definitions. Let $\mathbb{A}$ and $\mathbb{B}$ be two Boolean algebras and h a map from $\mathbb{A}$ to $\mathbb{B}$.
(1). h is said to be a *homomorphism* from $\mathbb{A}$ to $\mathbb{B}$ if $h(a \vee b) = h(a) \vee h(b)$, $h(a \wedge b) = h(a) \wedge h(b)$ and $h(a^c) = (h(a))^c$ for all a, b in $\mathbb{A}$.
(2). h is said to be an *isomorphism* between $\mathbb{A}$ and $\mathbb{B}$ if it is a homomorphism from $\mathbb{A}$ to $\mathbb{B}$, one-to-one and onto.
(3). Two Boolean algebras $\mathbb{A}$ and $\mathbb{B}$ are said to be *isomorphic* if there is an isomorphism from $\mathbb{A}$ to $\mathbb{B}$.

The notions of an ideal and a filter introduced in the following definition are analogous to those introduced in Section 1.1.

1.4.5 Definitions. (1) A subset I of a Boolean algebra $\mathbb{B}$ is said to be an *ideal* in $\mathbb{B}$ if the following conditions are satisfied.
  (i). $1 \notin \text{I}$.
  (ii). $a \vee b \in \text{I}$ whenever $a, b \in \text{I}$.
  (iii). If $b \in \text{I}$, $a \in \mathbb{B}$ and $a \leq b$, then $a \in \text{I}$.
(2). A subset J of a Boolean algebra $\mathbb{B}$ is said to be a *filter* in $\mathbb{B}$ if the following conditions are satisfied.
  (i). $0 \notin \text{J}$.
  (ii). $a \wedge b \in \text{J}$ whenever $a, b \in \text{J}$.
  (iii). If $b \in \text{J}$, $a \in \mathbb{B}$ and $b \leq a$, then $a \in \text{J}$.
(3). An ideal I in a Boolean algebra $\mathbb{B}$ is said to be a *maximal ideal* in $\mathbb{B}$ if there is no ideal in $\mathbb{B}$ properly containing I.
(4). A filter J in a Boolean algebra $\mathbb{B}$ is said to be a *maximal filter* in $\mathbb{B}$ if there is no filter in $\mathbb{B}$ properly containing J.

If I is an ideal in a Boolean algebra $\mathbb{B}$, then $J = \{b \in \mathbb{B}; b^c \in I\}$ is a filter in $\mathbb{B}$. If I is a maximal ideal in $\mathbb{B}$, then the J defined above is a maximal filter in $\mathbb{B}$. Using Zorn's lemma, one can show that there exists a maximal ideal in $\mathbb{B}$ containing any given ideal in $\mathbb{B}$ and a maximal filter in $\mathbb{B}$ containing any given filter in $\mathbb{B}$. If J_1 is a subset of $\mathbb{B}$ satisfying $\bigwedge_{i=1}^n a_i \neq 0$ for any finite number $a_1, a_2, \ldots, a_n$ of elements in J_1, then there exists a filter in $\mathbb{B}$ containing J_1. If I_1 is a subset of $\mathbb{B}$ satisfying $\bigvee_{j=1}^m b_j \neq 1$ for any finite number $b_1, b_2, \ldots, b_m$ of elements in I_1, then there exists an ideal in $\mathbb{B}$ containing I_1. As in Section 1.1, one can show that an ideal I in $\mathbb{B}$ is a maximal ideal in $\mathbb{B}$ if and only if either a or $a^c \in I$ for every a in $\mathbb{B}$. Similarly, a filter J in $\mathbb{B}$ is a maximal filter in $\mathbb{B}$ if and only if either b or $b^c \in J$ for every b in $\mathbb{B}$.

Now, we describe quotient Boolean algebras. Let I be an ideal in a Boolean algebra $\mathbb{B}$. Introduce an equivalence relation $\sim$ on $\mathbb{B}$ as follows. For a, b in $\mathbb{B}$, say $a \sim b$ if $a \Delta b = (a \wedge b^c) \vee (a^c \wedge b) \in I$. Let $\mathbb{B}/I$ denote the collection of all equivalence classes in $\mathbb{B}$. For a in $\mathbb{B}$, let $[a]$ denote the equivalence class in $\mathbb{B}$ containing a, i.e. $[a] = \{b \in \mathbb{B}; b \sim a\}$. Note that $[0] = I$. $\mathbb{B}/I$ can be made into a Boolean algebra. The operations $\vee$, $\wedge$ and ^{c} on $\mathbb{B}/I$ are introduced as follows. For a, b in $\mathbb{B}$,

$$[a] \vee [b] = [a \vee b],$$
$$[a] \wedge [b] = [a \wedge b],$$
$$[a]^c = [a^c].$$

One can check that with these operations, $\mathbb{B}/I$ becomes a Boolean algebra. $\mathbb{B}/I$ is called the *quotient Boolean algebra* with respect to the ideal I. There is a natural map h from $\mathbb{B}$ to $\mathbb{B}/I$ defined by, for a in $\mathbb{B}$, $h(a) = [a]$. This map is a homomorphism from the Boolean algebra $\mathbb{B}$ *onto* the Boolean algebra $\mathbb{B}/I$.

Now, we prove Stone Representation Theorem for Boolean algebras.

1.4.6 Stone Representation Theorem. *Let $\mathbb{B}$ be a Boolean algebra. Then there exists a compact Hausdorff totally disconnected space X such that $\mathbb{B}$ and the field $\mathscr{F}$ of all clopen subsets of X are isomorphic.*

Proof. Let X be the collection of all maximal filters in $\mathbb{B}$. For each a in $\mathbb{B}$, let $h(a)$ be the collection of all maximal filters in $\mathbb{B}$ each of which contains a. h has the following properties.

(i). $h(0) = \emptyset$.
(ii). $h(a_1 \vee a_2) = h(a_1) \cup h(a_2)$, $a_1, a_2 \in \mathbb{B}$.
(iii). $h(a_1 \wedge a_2) = h(a_1) \cap h(a_2)$, $a_1, a_2 \in \mathbb{B}$.
(iv). $h(a_1^c) = (h(a_1))^c$, $a_1 \in \mathbb{B}$.
(v). $h(a_1) \neq h(a_2)$, if $a_1 \neq a_2$.

Look at the collection $\mathscr{F} = \{h(a); a \in \mathbb{B}\}$. By (iii), $\mathscr{F}$ is closed under finite intersections. Consequently, there exists a smallest topology $\mathscr{T}$ on X for which $\mathscr{F}$ is a base. $\mathscr{T}$ is precisely unions of sets from $\mathscr{F}$. We claim that $(X, \mathscr{T})$ is a compact Hausdorff totally disconnected space and $\mathscr{F}$ is precisely the collection of all clopen subsets of X. We first show that $(X, \mathscr{T})$ is a Hausdorff space. Let J_1 and J_2 be two distinct maximal filters in X. Then there exists a in $\mathbb{B}$ such that $a \in J_1$ and $a^c \in J_2$. $J_1 \in h(a)$ and $J_2 \in h(a^c)$. Further, $h(a) \cap h(a^c) = \emptyset$. $h(a)$ and $h(a^c)$ are obviously open sets. Next, we show that X is compact. Since every closed set in X is an intersection of sets of the type $h(a)$, $a \in \mathbb{B}$, it suffices to show that for any family $\{h(a_\alpha); \alpha \in \Gamma\}$ with finite intersection property, i.e. $\bigcap_{\alpha \in \Gamma_1} h(a_\alpha) \neq \emptyset$ for any finite subset Γ_1 of Γ, $\bigcap_{\alpha \in \Gamma} h(a_\alpha) \neq \emptyset$. For any such family $\{h(a_\alpha); \alpha \in \Gamma\}$, it follows that $\bigwedge_{\alpha \in \Gamma_1} a_\alpha \neq 0$ for any finite subset Γ_1 of Γ. Consequently, there exists a maximal filter $J \supset \{a_\alpha; \alpha \in \Gamma\}$. Obviously, $J \in \bigcap_{\alpha \in \Gamma} h(a_\alpha)$. Hence X is compact. Since $\mathscr{F}$ is closed under complementation, every set in $\mathscr{F}$ is clopen. On the other hand, let V be a clopen subset of X. Since V is open, it is a union of sets from $\mathscr{F}$. Since V is closed (hence compact), this union admits a finite subcover for V. Since $\mathscr{F}$ is closed under finite unions, $V \in \mathscr{F}$. This shows that $\mathscr{F}$ is precisely the collection of all clopen subsets of X. It is obvious that $(X, \mathscr{T})$ is totally disconnected because $\mathscr{F}$ is a base for $\mathscr{T}$.

The map h defined from $\mathbb{B}$ to $\mathscr{F}$ is an isomorphism. Hence $\mathbb{B}$ and $\mathscr{F}$ are isomorphic.

This completes the proof. $\square$

We call X given above the Stone space of $\mathscr{F}$.

Let $\mathbb{B}$ be a Boolean algebra. a and b in $\mathbb{B}$ are said to be disjoint if $a \wedge b = 0$. A subset $\{a_\alpha; \alpha \in \Gamma\}$ of $\mathbb{B}$ is said to be pairwise disjoint if $a_\alpha \wedge a_\beta = 0$ for every $\alpha \neq \beta$ in Γ. We give a sufficient condition for a Boolean σ-algebra to be complete. First, we need a definition.

1.4.7 Definition. A Boolean algebra $\mathbb{B}$ is said to satisfy the *countable chain condition* if every collection $\{a_\alpha; \alpha \in \Gamma\}$ of pairwise disjoint elements in $\mathbb{B}$ is at most countable.

1.4.8 Theorem. *Let $\mathbb{B}$ be a Boolean σ-algebra satisfying the countable chain condition. Then $\mathbb{B}$ is a complete Boolean algebra.*

Proof. Let $\{b_\alpha; \alpha \in \Gamma\} \subset \mathbb{B}$. We use Zorn's lemma to show that $\bigvee_{\alpha \in \Gamma} b_\alpha$ exists in $\mathbb{B}$. Let $\mathscr{C} = \{C \subset \mathbb{B};$ the elements of C are pairwise disjoint, each element of C is non-zero and $\leq b_\alpha$ for some α in $\Gamma\}$. Clearly, $\mathscr{C}$ is non-empty. On $\mathscr{C}$, we introduce a partial order as follows. For C_1, C_2 in $\mathscr{C}$, say $C_1 \leq C_2$ if $C_1 \subset C_2$. Let $\{C_\delta; \delta \in T\}$ be a chain in $\mathscr{C}$. Then $\bigcup_{\delta \in T} C_\delta \in \mathscr{C}$ and is an upper bound of the chain $\{C_\delta; \delta \in T\}$. Let C_0 be a maximal element in $\mathscr{C}$. Since

$\mathbb{B}$ satisfies the countable chain condition, C_0 is countable. Let $C_0 = \{c_1, c_2, \ldots\}$. Since $\mathbb{B}$ is a Boolean σ-algebra, $z = \bigvee_{n \geq 1} c_n$ exists in $\mathbb{B}$. We show that z is indeed the supremum of all b_α, $\alpha \in \Gamma$. For this, we first show that $z \geq b_\alpha$ for every α in Γ. Suppose this is not true. Then there exists α_0 in Γ such that $d = b_{\alpha_0} - z = b_{\alpha_0} - \bigvee_{n \geq 1} c_n \neq 0$. This implies that $d \wedge c_n = 0$ for every $n \geq 1$. So, $C_0 \cup \{d\} \in \mathscr{C}$ contradicting the maximality of C_0. Thus z is an upper bound of all b_α, $\alpha \in \Gamma$. Let $z' \geq b_\alpha$ for every α in Γ for some z' in $\mathbb{B}$. We show that $z' \geq z$. Since $C_0 \in \mathscr{C}$, for every $n \geq 1$, there exists α_n in Γ such that $c_n \leq b_{\alpha_n}$ for every $n \geq 1$. So,

$$z = \bigvee_{n \geq 1} c_n \leq \bigvee_{n \geq 1} b_{\alpha_n} \leq z'.$$

Hence $z = \bigvee_{\alpha \in \Gamma} b_\alpha$. □

1.4.9 Definitions. Let $\mathbb{B}$ be a Boolean algebra.
(1). An element a in $\mathbb{B}$ is said to be an *atom* of $\mathbb{B}$ if the following conditions are satisfied.
 (i). $a \neq 0$.
 (ii). $b \in \mathbb{B}$, $b \leq a \Rightarrow b = 0$ or $b = a$.
(2). $\mathbb{B}$ is said to be *atomic* if for every non-zero element b in $\mathbb{B}$, there exists an atom a of $\mathbb{B}$ such that $a \leq b$.
(3). $\mathbb{B}$ is said to be *nonatomic* if there are no atoms in $\mathbb{B}$.

1.4.10 Theorem. *Let $\mathbb{B}$ be a Boolean algebra. Then $\mathbb{B}$ is nonatomic if and only if its Stone space X is perfect.*

Proof. Let $\mathscr{F}$ be the field of all clopen subsets of X and h the isomorphism from $\mathbb{B}$ to $\mathscr{F}$. Note that $a \in \mathbb{B}$ is an atom of $\mathbb{B}$ if and only if the clopen set $h(a)$ does not contain properly any non-empty clopen subset of X. $h(a)$ has such a property if and only if $h(a)$ is a singleton set, i.e. $h(a)$ is an isolated point of X. If x in X is an isolated point of X, then $\{x\}$ is a clopen subset of X and hence there exists a in $\mathbb{B}$ such that $h(a) = \{x\}$. Since x is isolated, a is an atom of $\mathbb{B}$. Hence $\mathbb{B}$ is nonatomic if and only if X is a perfect set. □

We end this section with some observations on homomorphisms between Boolean algebras and induced maps on their Stone spaces. Let $\mathbb{A}$ and $\mathbb{B}$ be two Boolean algebras and h a homomorphism from $\mathbb{A}$ to $\mathbb{B}$. Let X and Y be the Stone spaces of $\mathbb{A}$ and $\mathbb{B}$ respectively. We define a map T from Y to X as follows. Let $J \in Y$. J is a maximal filter in $\mathbb{B}$. Let $T(J) = \{d \in \mathbb{A}; h(d) \in J\} = h^{-1}(J)$. It can be checked quite easily that $T(J)$ is a maximal filter in $\mathbb{A}$. (See Remark 1.1.23 (8).) Thus $T(J) \in X$. Thus we have defined a map from Y to X. We now show that T is a continuous map from Y to X. Let C be any clopen subset of X. Then there exists a in $\mathbb{A}$ such that $C = \{I \in X; a \in I\}$. See Theorem 1.4.6. Then

1. PRELIMINARIES 23

$$T^{-1}(C) = \{J \in Y; T(J) \in C\} = \{J \in Y; h^{-1}(J) \in C\}$$
$$= \{J \in Y; a \in h^{-1}(J)\} = \{J \in Y; h(a) \in J\}.$$

From the description of the clopen subsets of Y, it follows that $T^{-1}(C)$ is a clopen subset of Y. Hence T is continuous. The following theorem describes the duality between h and T.

1.4.11 Theorem. *The following statements are true.*
(1). *h is onto if and only if T is a one-to-one map.*
(2). *h is one-to-one if and only if T is an onto map.*

Proof. (1). Let J_1 and J_2 be two distinct elements in Y. We show that $T(J_1)$ and $T(J_2)$ are distinct. By Remark 1.1.23(9), there exists b in $\mathbb{B}$ such that $b \in J_1$ and $b^c \in J_2$. Since h is onto, there exist a_1, a_2 in $\mathbb{A}$ such that $h(a_1) = b$ $h(a_2) = b^c$. Therefore, $a_1 \in T(J_1)$ and $a_2 \in T(J_2)$. If $T(J_1) = T(J_2)$, then a_1, $a_2 \in T(J_1)$ and so, $a_1 \wedge a_2 \in T(J_1)$. This means that $h(a_1 \wedge a_2) = h(a_1) \wedge h(a_2) = b \wedge b^c = 0 \in J_1$. Then J_1 cannot be a maximal filter in $\mathbb{B}$. This contradiction shows that $T(J_1)$ and $T(J_2)$ are distinct. The converse can be proved analogously.

(2). Suppose h is one-to-one. Let $I \in X$. We will exhibit J in Y such that $I = T(J)$. Let $J_1 = \{h(a); a \in I\}$. J_1 has the following property. $h(a_1) \wedge h(a_2) \wedge \cdots \wedge h(a_n) \neq 0$ for any finite set $\{a_1, a_2, \ldots, a_n\} \subset I$. For, $\bigwedge_{i=1}^{n} h(a_i) = h(\bigwedge_{i=1}^{n} a_i) \neq 0$ as h is one-to-one and $\bigwedge_{i=1}^{n} a_i \neq 0$. By Remark 1.1.23(4) and Zorn's lemma, there exists a maximal filter J in $\mathbb{B}$ containing J_1. It is now obvious that $T(J) = I$. Hence T is onto. The converse can be proved analogously.

This completes the proof. □

1.5 FUNCTIONAL ANALYTIC CONCEPTS

In this section, our main goal is to present some ideas on vector lattices and prove Riesz Decomposition Theorem. Riesz Decomposition Theorem is one of the main tools we use in the study of finitely additive measures. The other relevant functional analytic concepts used in this book are also compiled in this section mostly without proofs. We begin with vector lattices.

1.5.1 Definition. Let L be a vector space over the real line R which is endowed with an order structure defined by a reflexive, transitive and anti-symmetric binary relation denoted by $\leq$. $(L, \leq)$ is called an *ordered vector space* if the following hold.
(i). $x, y \in L, x \leq y \Rightarrow x + z \leq y + z$ for all z in L.
(ii). $x, y \in L, x \leq y \Rightarrow cx \leq cy$ for every c in $[0, \infty)$.

Let (L, $\leq$) be an ordered vector space. For x, y in L, an element z in L with the properties (i) $x \leq z$, (ii) $y \leq z$, and (iii) $z \leq z'$ whenever $z' \in L$, $x \leq z'$ and $y \leq z'$ (i.e. z is the least element in L dominating both x and y) is called the supremum of x and y and is denoted by Sup $\{x, y\}$ or by $x \vee y$. Clearly, Sup $\{x, y\}$, when it exists, is well defined. For any subset $\{x_\alpha; \alpha \in \Gamma\}$ of L, supremum of $\{x_\alpha; \alpha \in \Gamma\}$ is defined analogously and is denoted by $\bigvee_{\alpha \in \Gamma} x_\alpha$ if the supremum exists. Similarly, for x, y in L, an element u in L with the properties (i) $u \leq x$, (ii) $u \leq y$ and (iii) $u' \leq u$ whenever $u' \in L$, $u' \leq x$ and $u' \leq y$, is called the infimum of x and y. It is denoted by Inf $\{x, y\}$ or by $x \wedge y$. Clearly, Inf $\{x, y\}$ is well defined when it exists. For any subset $\{x_\alpha; \alpha \in \Gamma\}$ of L, infimum of $\{x_\alpha; \alpha \in \Gamma\}$ is analogously defined and if the infimum exists, it is denoted by $\bigwedge_{\alpha \in \Gamma} x_\alpha$.

1.5.2 Definition. An ordered vector space (L, $\leq$) is said to be a *vector lattice* if Sup $\{x, y\}$ and Inf $\{x, y\}$ exist for all x and y in L.

1.5.3 Definition. Let (L, $\leq$) be a vector lattice. For x in L, define $x^+ = x \vee 0$, $x^- = (-x) \vee 0$ and $|x| = x^+ + x^-$. x^+, x^- and $|x|$ are called the *positive part*, *negative part* and the *modulus* of x respectively. For x, y in L, x and y are said to be *orthogonal* if $|x| \wedge |y| = 0$. In such a case, we use the notation $x \perp y$.

The following theorem gives some elementary identities that hold in any vector lattice.

1.5.4 Theorem. *Let* (L, $\leq$) *be a vector lattice and* $\{x_\alpha; \alpha \in \Gamma\}$ *a non-empty subset of* L. *Then the following are true.*
(1). *If, for x in* L, *one of* $\bigvee_{\alpha \in \Gamma} x_\alpha$ *and* $\bigvee_{\alpha \in \Gamma} (x + x_\alpha)$ *exists, then the other exists and*
$$x + \bigvee_{\alpha \in \Gamma} x_\alpha = \bigvee_{\alpha \in \Gamma} (x + x_\alpha).$$

(2). *If, for x in* L, *one of* $\bigwedge_{\alpha \in \Gamma} x_\alpha$ *and* $\bigwedge_{\alpha \in \Gamma} (x + x_\alpha)$ *exists, then the other exists and*
$$x + \bigwedge_{\alpha \in \Gamma} x_\alpha = \bigwedge_{\alpha \in \Gamma} (x + x_\alpha).$$

(3). *If one of* $\bigvee_{\alpha \in \Gamma} x_\alpha$ *and* $\bigwedge_{\alpha \in \Gamma} (-x_\alpha)$ *exists, then the other exists and*
$$\bigvee_{\alpha \in \Gamma} x_\alpha = - \bigwedge_{\alpha \in \Gamma} (-x_\alpha).$$

(4). *For $\beta > 0$, if one of* $\bigvee_{\alpha \in \Gamma} x_\alpha$ *and* $\bigvee_{\alpha \in \Gamma} \beta x_\alpha$ *exists, then the other exists and*
$$\beta \left(\bigvee_{\alpha \in \Gamma} x_\alpha \right) = \bigvee_{\alpha \in \Gamma} \beta x_\alpha.$$

(5). *For $\beta > 0$, if one of* $\bigwedge_{\alpha \in \Gamma} x_\alpha$ *and* $\bigwedge_{\alpha \in \Gamma} \beta x_\alpha$ *exists, then the other exists and*
$$\beta \left(\bigwedge_{\alpha \in \Gamma} x_\alpha \right) = \bigwedge_{\alpha \in \Gamma} \beta x_\alpha.$$

(6). $(x \vee y) + (x \wedge y) = x + y$ for every x, y in L.
(7). $x = x^+ - x^-$ for every x in L.
(8). $x^+ \wedge x^- = 0$ for every x in L.
(9). $|x| = x^+ \vee x^- = x \vee (-x) = x^+ + x^-$ for every x in L.
(10). For x, y, z in L, if $x = y - z$, $y \geq 0$, $z \geq 0$, then $y \geq x^+$ and $z \geq x^-$.
(11). For x, y, z in L, if $x = y - z$ and $y \wedge z = 0$, then $y = x^+$ and $z = x^-$.
(12). For x in L, $|x| = 0$ if and only if $x = 0$.
    For λ in R and x in L, $|\lambda x| = |\lambda||x|$.
    For x, y in L, $|x + y| \leq |x| + |y|$.
(13). For x in L, $x^+ = \frac{1}{2}(|x| + x)$ and $x^- = \frac{1}{2}(|x| - x)$.
(14). For x, y in L, $-|x| - |y| \leq x \wedge y \leq x \vee y \leq |x| + |y|$.
(15). For x, y in L, $|x - y| = (x \vee y) - (x \wedge y)$.
(16). For x, y in L, $x \vee y = \frac{1}{2}(x + y + |x - y|)$.
(17). For x, y in L, $x \wedge y = \frac{1}{2}(x + y - |x - y|)$.
(18). For x, y in L, $|x| \vee |y| = \frac{1}{2}(|x + y| + |x - y|)$.
(19). For x, y in L, $|x| \wedge |y| = \frac{1}{2}(||x + y| - |x - y||)$.
(20). For x, y in L, $x \leq y$ if and only if $x^+ \leq y^+$ and $y^- \leq x^-$.
(21). For x, y in L, $x \perp y$ if and only if $|x + y| = |x - y|$.
(22). For x, y in L, $x \perp y$ if and only if $|x| \vee |y| = |x| + |y|$.
(23). For x, y in L, if $x \perp y$, then $(x + y)^+ = x^+ + y^+$ and $|x + y| = |x| + |y|$.
(24). *Distributive laws hold.* If $\bigvee_{\alpha \in \Gamma} x_\alpha$ exists, then for any x in L, $\bigvee_{\alpha \in \Gamma}(x \wedge x_\alpha)$ exists and

$$x \wedge \left(\bigvee_{\alpha \in \Gamma} x_\alpha\right) = \bigvee_{\alpha \in \Gamma}(x \wedge x_\alpha).$$

If $\bigwedge_{\alpha \in \Gamma} x_\alpha$ exists, then for any x in L, $\bigwedge_{\alpha \in \Gamma}(x \vee x_\alpha)$ exists and

$$x \vee \left(\bigwedge_{\alpha \in \Gamma} x_\alpha\right) = \bigwedge_{\alpha \in \Gamma}(x \vee x_\alpha).$$

(25). For x, y in L, $|x + y| \leq 2|x| \vee 2|y|$.
(26). For x, y in L, if $x \perp y$ and $\lambda \in R$, then $\lambda x \perp y$.
(27). For x, y, z in L, $|(x \vee z) - (y \vee z)| + |(x \wedge z) - (y \wedge z)| = |x - y|$.
(28). For x, y, x_1, y_1 in L, $|(x \vee y) - (x_1 \vee y_1)| \leq |x - x_1| + |y - y_1|$.
(29). For x, y, x_1, y_1 in L, $|(x \wedge y) - (x_1 \wedge y_1)| \leq |x - x_1| + |y - y_1|$.

Proof. (1), (2), (3), (4) and (5) follow directly from the definitions involved.

(6). $x + y - (x \wedge y) = x + y + [(-x) \vee (-y)]$ (by (3))

$$= [(x + y) - x] \vee [(x + y) - y]$$ (by (1))

$$= y \vee x = x \vee y.$$

(7). $x^+ - x^- = (x \vee 0) - [(-x) \vee 0]$

$$= (x \vee 0) + (x \wedge 0)$$ (by (3))

$$= x + 0 = x$$ (by (4))

(8). $(x^+ \vee x^-) + (x^+ \wedge x^-) = x^+ + x^-$ (by (6))

It suffices to show that $x^+ \vee x^- = x^+ + x^-$. Observe that

$$x^+ + x^- = x + x^- + x^- \quad \text{(by (7))}$$
$$= x + 2[(-x) \vee 0]$$
$$= x + [(-2x) \vee 0] \quad \text{(by (4))}$$
$$= (x - 2x) \vee x \quad \text{(by (1))}$$
$$= x \vee (-x)$$
$$= x \vee (-x) \vee 0 = (x \vee 0) \vee [(-x) \vee 0]$$
$$= x^+ \vee x^-.$$

(Note that $x \vee (-x) \geq 0$. For, $x \leq x \vee (-x)$ and $-x \leq x \vee (-x)$, and so $0 \leq 2[x \vee (-x)]$.)

(9). The proof of this is included in the proof of (8).

(10). $x = y - z$ implies that $y - x = z \geq 0$. So, $y \geq x$. Since $y \geq 0$, $y \geq x \vee 0 = x^+$. Similarly, one can show that $z \geq x^-$.

(11). $y \wedge z = 0$ implies that $y \geq 0$ and $z \geq 0$. By (10), $0 \leq y - x^+ \leq y$ and $0 \leq z - x^- \leq z$. Since $y \wedge z = 0$, $(y - x^+) \wedge (z - x^-) = 0$. Since $x = y - z = x^+ - x^-$, $y - x^+ = z - x^-$. Hence $y - x^+ = 0 = z - x^-$.

(12). This follows from what we have proved above.

(13). This follows from (7) and the definition of $|x|$.

(14). It is enough to observe that $x \vee y \leq |x| + |y|$ which is obvious.

(15). By (1), (2) and (3),

$$x + (y - x)^+ = x + [(y - x) \vee 0]$$
$$= [x + (y - x)] \vee (x + 0)$$
$$= x \vee y.$$

$$x - (x - y)^+ = x - [(x - y) \vee 0] = x + [(y - x) \wedge 0]$$
$$= [x + (y - x)] \wedge (x + 0)$$
$$= x \wedge y.$$

Consequently,

$$(x \vee y) - (x \wedge y) = (y - x)^+ + (x - y)^+$$
$$= (x - y)^+ + (x - y)^-$$
$$= |x - y|.$$

(16). This is a consequence of (6) and (15).

(17). This is a consequence of (6) and (15).

(18). $|x| \vee |y| = x \vee (-x) \vee y \vee (-y)$

$\qquad = [x \vee (-y)] \vee [y \vee (-x)]$ (by (9))

$\qquad = \frac{1}{2}\{[x - y + |x + y|] \vee [(y - x) + |x + y|]\}$ (by (16) & (17))

$\qquad = \frac{1}{2}(|x + y| + |x - y|)$ (by (16))

(19). $(|x| \vee |y|) + (|x| \wedge |y|) = |x| + |y|$ (by (6))

Therefore,

$2(|x| \wedge |y|) = 2[|x| + |y| - (|x| \vee |y|)]$

$\qquad = 2|x| + 2|y| - |x + y| - |x - y|$ (by (18))

$\qquad = 2|p + q| + 2|p - q| - 2|p| - 2|q|$

(by letting $x + y = 2p$ and $x - y = 2q$ and using (12))

$\qquad = 4(|p| \vee |q|) - 2|p| - 2|q|$ (by (18))

$\qquad = 2(|p| + |q| + ||p| - |q||) - 2|p| - 2|q|$ (by (16))

$\qquad = 2||p| - |q||$ (by (12))

$\qquad = ||2p| - |2q||$

$\qquad = ||x + y| - |x - y||.$

(20). $x \leq y$ implies that $(x \vee 0) \leq (y \vee 0)$. This means that $x^+ \leq y^+$. Similarly, one can show that $y^- \leq x^-$. The converse is trivial.

(21). This follows from (19).

(22). $(|x| \vee |y|) + (|x| \wedge |y|) = |x| + |y|$, (by (6))

From this identity, (22) follows obviously.

(23). If $x \perp y$, then

$0 = |x| \wedge |y| = |x| + |y| - (|x| \vee |y|)$ (by (6))

$\qquad = |x| + |y| - \frac{1}{2}(|x + y| + |x - y|)$ (by (18))

$\qquad = |x| + |y| - |x + y|$ (by (21))

Hence, we have the equality $|x| + |y| = |x + y|$. Further,

$(x + y)^+ = (x + y) \vee 0 = \frac{1}{2}(x + y + |x + y|)$ (by (16))

$\qquad = \frac{1}{2}(x + y + |x| + |y|)$

(by what we have proved above)

$\qquad = x^+ + y^+.$

(24). Let $y = \bigvee_{\alpha \in \Gamma} x_\alpha$. Then $x \wedge x_\alpha \leq x \wedge y$ for every α in Γ. We show that $\bigvee_{\alpha \in \Gamma}(x \wedge x_\alpha) = x \wedge y$. Let $w \in L$ and $x \wedge x_\alpha \leq w$ for every α in Γ. By (6), $x + x_\alpha - (x \vee x_\alpha) \leq w$ for every α in Γ. By (1), $w + (x \vee y) \geq x + y$. So,

$w \geq x + y - (x \vee y) = x \wedge y$. Hence $\bigvee_{\alpha \in \Gamma}(x \wedge x_\alpha) = x \wedge y = x \wedge (\bigvee_{\alpha \in \Gamma} x_\alpha)$ indeed. The other law can be established analogously.

(25). This follows from (18).

(26). *Case* (i). $|\lambda| \geq 1$. Let $z = |\lambda x| \wedge |y| = |\lambda||x| \wedge |y|$. Then $z \geq 0$ and $z \leq |\lambda||x|$ and $z \leq |y|$. From this, it follows that $(1/|\lambda|)z \leq |x|$ and $(1/|\lambda|)z \leq |y|$. Since $|x| \wedge |y| = 0$, $z \leq 0$. Hence $z = 0$. *Case* (ii). $|\lambda| < 1$. If $\lambda = 0$, there is nothing to prove. Let $|\lambda| > 0$. Find $c > 1$ such that $c|\lambda| = 1$. Then $x \perp cy$, by case (i). This means that $|x| \wedge c|y| = 0 = c|\lambda||x| \wedge c|y| = c(|\lambda||x| \wedge |y|)$ which implies that $|\lambda||x| \wedge |y| = 0$.

(27). Using (15) and the distributive laws, we obtain

$|(x \vee z) - (y \vee z)| + |(x \wedge z) - (y \wedge z)|$

$= (x \vee z) \vee (y \vee z) - (x \vee z) \wedge (y \vee z) + (x \wedge z) \vee (y \wedge z) - (x \wedge z) \wedge (y \wedge z)$

$= (x \vee y \vee z) - (x \wedge y) \vee z + (x \vee y) \wedge z - (x \wedge y \wedge z)$

$= [(x \vee y) \vee z + (x \vee y) \wedge z] - [(x \wedge y) \vee z + (x \wedge y) \wedge z]$

$= [(x \vee y) + z] - [(x \wedge y) + z]$ (by (6))

$= (x \vee y) - (x \wedge y) = |x - y|$ (by (15)).

(28). $|(x \vee y) - (x_1 \vee y_1)| = |(x \vee y) - (x_1 \vee y) + (x_1 \vee y) - (x_1 \vee y_1)|$

$\leq |(x \vee y) - (x_1 \vee y)| + |(x_1 \vee y) - (x_1 \vee y_1)|$ (by 12))

$\leq |x - x_1| + |y - y_1|$ (by (27)).

(29). This is similar to the above.

This completes the proof of all the identities. □

An important example of a vector lattice is the space of all bounded finitely additive measures to be presented in the next chapter. The above theorem brings into focus many facets of the main object of study in this book.

Now, we introduce some important concepts such as normal sublattices and boundedly complete lattices.

1.5.5 Definition. Let $(L, \leq)$ be a vector lattice and $W \subset L$. W is called a *sublattice* of L if W is a vector subspace of L and $x \vee y \in W$, $x \wedge y \in W$ for every x, y in W.

1.5.6 Definition. A vector sublattice W of a vector lattice $(L, \leq)$ is said to be *normal* if the following conditions are satisfied.

(i). $x \in W$, $0 \leq |y| \leq |x| \Rightarrow y \in W$.

(ii). If $\bigvee_{\alpha \in \Gamma} x_\alpha$ exists for a given non-empty $\{x_\alpha; \alpha \in \Gamma\} \subset W$, then $\bigvee_{\alpha \in \Gamma} x_\alpha \in W$.

1.5.7 Definition. Let $(L, \leq)$ be a vector lattice and $S \subset L$. The orthogonal complement of S is denoted by $S^\perp$ and is defined by

$$S^\perp = \{y \in L; y \perp x \text{ for every } x \text{ in } S\}.$$

It is easy to check that for any $S \subset L$, $S \cap S^\perp = \varnothing$ or $= \{0\}$. If $S_1 \subset S_2 \subset L$, then $S_2^\perp \subset S_1^\perp$. The following theorem shows that normal sublattices arise in a natural way.

1.5.8 Theorem. *Let $(L, \leq)$ be a vector lattice and $S \subset L$. Then $S^\perp$ is a normal sublattice of L.*

Proof. First, we show that $S^\perp$ is a vector subspace of L. Let $y, z \in S^\perp$ and c, d real numbers. Then for any x in S,

$$0 \leq |(cy + dz)| \wedge |x| \leq (2|c||y| \vee 2|d||z|) \wedge |x|$$
$$= 2[(|c||y| \wedge |x|) \vee (|d||z| \wedge |x|)].$$

The second inequality above follows from Theorem 1.5.4(25) and the equality from Theorem 1.5.4(24). Since $y \perp x$ and $z \perp x$, $cy \perp x$ and $dz \perp x$. See Theorem 1.5.4(26). Hence $(cy + dz) \perp x$. So, $cy + dz \in S^\perp$. Next, we show that for $y, z \in S^\perp$, $y \vee z \in S^\perp$ and $y \wedge z \in S^\perp$. But these follow from Distributive laws. Thus we have proved that $S^\perp$ is a vector sublattice of L. The normality of $S^\perp$ again follows from distributive laws. □

1.5.9 Definition. A vector lattice $(L, \leq)$ is said to be *boundedly complete* if for every non-empty subset $\{x_\alpha; \alpha \in \Gamma\}$ of L bounded above, i.e. there is an element x in L such that $x_\alpha \leq x$ for every α in Γ, $\bigvee_{\alpha \in \Gamma} x_\alpha$ exists.

If $(L, \leq)$ is a boundedly complete vector lattice and if $\{x_\alpha; \alpha \in \Gamma\}$ is a non-empty subset of L bounded below, then $\bigwedge_{\alpha \in \Gamma} x_\alpha$ exists. This follows from Theorem 1.5.4(3).

The following is the main theorem of this section.

1.5.10 Riesz Decomposition Theorem. *If S is a normal vector sublattice of a boundedly complete vector lattice $(L, \leq)$, then any element x in L can be written as a sum $x' + x''$ for some x' in S and x'' in $S^\perp$, and this decomposition is unique. Further, if $x \geq 0$, then $x' = \bigvee_{y \in S} (x \wedge |y|)$. For x in L, in general, $x' = (x^+)' - (x^-)'$.*

Proof. First, assume that $x \geq 0$. Then the set $\{x \wedge |y|; y \in S\}$ is bounded above, and as L is boundedly complete, $\bigvee_{y \in S} (x \wedge |y|)$ exists. Define $x' = \bigvee_{y \in S} (x \wedge |y|)$. As S is normal, $x \wedge |y| \in S$ for every y in S and consequently, $x' \in S$. Now, we show that $x - x' \in S^\perp$. Let $y \in S$ and $u = (x - x') \wedge |y|$. The problem reduces to showing that $u = 0$. Since $0 \leq u \leq |y|$ and $y \in S$, $u \in S$. Since S is a vector space, $u + x' \in S$. Note that $u + x' \leq (x - x') + x' = x$.

Consequently, $u + x' = (u + x') \wedge x = |u + x'| \wedge x = x \wedge |u + x'| \leq x'$ from the definition of x'. This implies that $u \leq 0$. Hence $u = 0$. Thus $x - x' \in S^\perp$. Denote $x - x'$ by x''. Thus, we have achieved in writing $x = x' + x''$ with $x' \in S$ and $x'' \in S^\perp$. Now, we obtain a decomposition for any x in L. Write

$$x = x^+ - x^- = (x^+)' + (x^+)'' - (x^-)' - (x^-)''$$
$$= [(x^+)' - (x^-)'] + [(x^+)'' - (x^-)''] = x' + x'', \quad \text{say}.$$

It is obvious that $x' \in S$ and $x'' \in S^\perp$. Thus $x = x' + x''$ is a decomposition of x with the desired properties. To prove uniqueness, let $x = x_1 + x_2 = y_1 + y_2$ with x_1, y_1 in S and x_2, y_2 in $S^\perp$. Then $x_1 - y_1 \in S$ and $x_2 - y_2 \in S^\perp$. Since $x_1 - y_1 = -(x_2 - y_2)$, $(x_1 - y_1) \in S^\perp$ as well. This implies that $|x_1 - y_1| = 0$ or $x_1 = y_1$. See Theorem 1.5.4(12). We also obtain $x_2 = y_2$. Hence the decomposition obtained above is unique with the stated properties. □

1.5.11 Corollary. *If S is a normal vector sublattice of a boundedly complete vector lattice* $(L, \leq)$, *then* $(S^\perp)^\perp = S$.

Proof. The result follows from the following relations.
  (i). $L = S \oplus S^\perp$.
  (ii). $L = S^\perp \oplus (S^\perp)^\perp$.
  (iii). $S \subset (S^\perp)^\perp$.
(Here, $\oplus$ denotes the direct sum of the sets involved.) □

Now, let $y \in L$ be fixed. We characterize the smallest normal vector sublattice of $(L, \leq)$ containing y.

1.5.12 Theorem. *Let* $(L, \leq)$ *be a boundedly complete vector lattice and* $y \in L$. *Then the smallest normal vector sublattice of L containing y is* $(S^\perp)^\perp$, *where* $S = \{y\}$. *Further,* $x \in (S^\perp)^\perp$ *if and only if*

$$x = \bigvee_{n \geq 1} (x^+ \wedge n|y|) - \bigvee_{n \geq 1} (x^- \wedge n|y|).$$

Proof. Since $S \subset (S^\perp)^\perp$, $(S^\perp)^\perp$ is a normal vector sublattice of L containing y. If S_1 is a normal vector sublattice of L containing y, then $S \subset S_1$. This implies that $S_1^\perp \subset S^\perp$. From this, it follows that $(S^\perp)^\perp \subset (S_1^\perp)^\perp = S_1$. Hence $(S^\perp)^\perp$ is the smallest normal vector sublattice of L containing y. Thus the first part of the theorem is proved.

Since $(S^\perp)^\perp$ is a normal vector sublattice of L, it follows that

$$x = \bigvee_{n \geq 1} (x^+ \wedge n|y|) - \bigvee_{n \geq 1} (x^- \wedge n|y|) \in (S^\perp)^\perp.$$

Conversely, let $x \in (S^\perp)^\perp$ and assume that $x \geq 0$. Let $u = \bigvee_{n \geq 1} (x \wedge n|y|)$. We show that $u = x$. If we show that $(x - u) \wedge |y| = 0$, it would then follow that $x - u \in S^\perp$ from the definition of $S^\perp$. Since $(S^\perp)^\perp$ is normal, $u \in (S^\perp)^\perp$.

Since $x \in (S^{\perp})^{\perp}$, $x - u \in (S^{\perp})^{\perp}$. Thus we find that $x - u$ is available in both $S^{\perp}$ and $(S^{\perp})^{\perp}$. This implies that $x - u = 0$ or $x = u$. In view of this argument, let us show that $(x - u) \wedge |y| = 0$. To this end, we note

$$(x - u) \wedge |y| \leq ((x - u) \wedge x) \wedge |y| = (x - u) \wedge (x \wedge |y|)$$
$$\leq (x - u) \wedge u.$$

By Theorem 1.5.4(2), we also note that $[(x - u) \wedge u] + u = x \wedge 2u$. So,

$$(x - u) \wedge u = -u + (x \wedge 2u)$$

$$= -u + \left[x \wedge \left(\bigvee_{n \geq 1} (2x \wedge 2n|y|) \right) \right], \qquad \text{by Theorem 1.5.4(4)}$$

$$= -u + \left(\bigvee_{n \geq 1} (x \wedge 2x \wedge 2n|y|) \right)$$

$$= -u + \left(\bigvee_{n \geq 1} (x \wedge 2xn|y|) \right)$$

$$= -u + u = 0.$$

Hence $0 \leq (x - u) \wedge |y| \leq (x - u) \wedge u = 0$. Thus $(x - u) \wedge |y| = 0$.

So, if $x \in (S^{\perp})'$ and $x \geq 0$, the above representation is valid. If $x \in (S^{\perp})^{\perp}$ is any general element, write $x = x^+ - x^-$. Note that $x^+, x^- \in (S^{\perp})^{\perp}$. By what we have proved above, $x^+ = \bigvee_{n \geq 1} (x^+ \wedge n|y|)$ and $x^- = \bigvee_{n \geq 1} (x^- \wedge n|y|)$. This completes the proof of the theorem. □

Now, we provide a brief review of other relevant functional analytic concepts used in the book. First, we introduce vector spaces over the field of rational numbers which is specially required in Chapter 3.

1.5.13 Definition. Let Q be the field of all rational numbers. *A vector space or a linear space over* Q is an additive group L (with the additive operation denoted by +) together with a map m from the product space $Q \times L$ into L written as $m(r, x) = rx$ for r in Q and x in L, which satisfies the following four conditions.
  (i). $r(x + y) = rx + ry$, $r \in Q$, $x, y \in L$.
  (ii). $(r + s)x = rx + sx$, $r, s \in Q$ and $x \in L$.
  (iii). $r(sx) = (rs)x$, $r, s \in Q$ and $x \in L$.
  (iv). $1(x) = x$, $x \in L$.

The usual concept of a vector space over the real line R has the same description as the above with Q replaced by R. If L is a vector space over Q, a linear functional on L is any map T from L to R satisfying $T(rx + sy) = rT(x) + sT(y)$ for all r, s in Q and x, y in L.

Let L be a vector space over Q. A susbset A of L is said to be a *Hamel Basis* for L if every non-zero x in L admits a unique representation

$$x = r_1 x_1 + r_2 x_2 + \cdots + r_n x_n$$

for some non-zero $r_1, r_2, \ldots, r_n$ in Q, $x_1, x_2, \ldots, x_n$ in L and $n \geq 1$. A subset B of L is said to be *linearly independent* if $r_1 = r_2 = \cdots = r_n = 0$ whenever $r_1 x_1 + r_2 x_2 + \cdots + r_n x_n = 0$ for some $r_1, r_2, \ldots, r_n$ in Q, $x_1, x_2, \ldots, x_n$ in B and $n \geq 1$. Given any linearly independent subset B of L, one can find a maximal linearly independent set A containing B. A subset A of L is a maximal linearly independent set in L if and only if A is a Hamel Basis for L. Consequently, given any linearly independent subset B of L, one can find a Hamel Basis for L containing B.

The following is a version of *Hahn–Banach Theorem*.

1.5.14 Theorem. *Let L be a vector space over Q. Let L_1 be a subspace of L, i.e. $rx + sy \in L_1$ whenever $x, y \in L_1$ and $r, s \in Q$. Let x be an element of L which is not a member of L_1. Let c be any real number. Let T_1 be any linear functional on L_1. Then there exists a linear functional T on L with the following properties.*
  (i). $T(y) = T_1(y)$ *for all y in L_1.*
  (ii). $T(x) = c$.

Proof. Let B_1 be a Hamel Basis for L_1 and $B = B_1 \cup \{x\}$. Note that B is a linearly independent set. Let A be a Hamel Basis for L containing B. Define T on A as follows.

$$\begin{aligned} T(y) &= T_1(y), &&\text{if } y \in B_1, \\ &= c, &&\text{if } y = x, \\ &= 0, &&\text{otherwise.} \end{aligned}$$

T can be extended to L in the obvious fashion and this extension, denoted again by T, is the desired linear functional. □

The following is generally known as *Hahn–Banach Theorem*.

1.5.15 Theorem. *Let L be a vector space over the field of all real numbers R. Let p be a real valued function on L satisfying $p(x+y) \leq p(x) + p(y)$ for all x, y in L and $p(cx) = cp(x)$ for all $c \geq 0$ and x in L. Let L_1 be a subspace of L and T_1 a linear functional on L_1 satisfying*

$$T_1(x) \leq p(x) \quad \text{for every } x \text{ in } L_1.$$

Then there exists a linear functional T on L having the following properties.
  (i). $T(x) \leq p(x)$ *for every x in L.*
  (ii). $T(x) = T_1(x)$ *for every x in L_1.*

Now, we introduce normed linear spaces. Let L be a vector space over R. A *pseudo-norm* on L is a map p from L to $[0, \infty)$ with the following properties. (i). $p(x) = 0$ if $x = 0$. (ii). $p(cx) = |c|p(x)$ for every c in R and x in L. (iii). $p(x+y) \leq p(x) + p(y)$ for all x, y in L. A pseudo-norm p on L induces a pseudo-metric ρ on L as follows. For x, y in L, let $\rho(x, y) = p(x-y)$. A vector space L equipped with a pseudo-norm p is said to be *complete* if the pseudo-metric space (L, ρ) is complete, i.e. every Cauchy sequence in L is convergent under the pseudo-metric ρ.

In what follows, we assume that all vector spaces are over R. A pseudo-norm p on a vector space L is said to be a *norm* if $x = 0$ whenever $p(x) = 0$. If p is a norm on L, it is customary to denote $p(x)$ by $\|x\|$ for x in L. The pair $(L, \|\cdot\|)$ is called a *normed linear space*. If L is complete under the norm $\|\cdot\|$, then $(L, \|\cdot\|)$ is called a *Banach space*.

Let $(L, \|\cdot\|)$ be a Banach space. Then the topology induced by the norm $\|\cdot\|$ on L is called *strong topology* on L. If x, x_n, $n \geq 1$ is a sequence in L such that x_n, $n \geq 1$ converges to x in the norm $\|\cdot\|$, i.e. $\lim_{n \to \infty} \|x_n - x\| = 0$, we say that x_n, $n \geq 1$ converges to x strongly or x_n, $n \geq 1$ converges to x in L.

If $(L, \|\cdot\|)$ is a Banach space, then the space of all continuous linear functionals on L is called the *dual space of L* and is denoted by L^*. For any linear functional T on L, one defines $\|T\| = \text{Sup}\{|T(x)|; x \in L, \|x\| \leq 1\}$. $T \in L^*$ if and only if $\|T\| < \infty$. In fact, T is continuous if and only if there exists a positive constant k such that $|T(x)| \leq k\|x\|$ for all x in L. If $T \in L^*$, then $|T(x)| \leq \|T\|\|x\|$ for all x in L. The function $\|\cdot\|$ on L^* is indeed a norm on L^*. (We use the same symbol $\|\cdot\|$ for both L and L^* and it should be clear from the context which norm we are dealing with.) Moreover, $(L^*, \|\cdot\|)$ is a Banach space.

If $(L, \|\cdot\|)$ is a Banach space, the *weak topology* on L is the smallest topology on L with respect to which every T in L^* is continuous. A net x_α, $\alpha \in D$ in L is said to be *weakly convergent* to a x in L if $\lim_{\alpha \in D} T(x_\alpha) = T(x)$ for every T in L^*. A sequence x_n, $n \geq 1$ in L is a *weak Cauchy sequence* if $\lim_{m,n \to \infty} T(x_n - x_m) = 0$ for every T in L^*. The following is an important result which we have an occasion to use.

1.5.16 Theorem. *Let $(L, \|\cdot\|)$ be a Banach space. Then every weak Cauchy sequence x_n, $n \geq 1$ in L is norm bounded*, i.e. $\text{Sup}_{n \geq 1} \|x_n\| < \infty$.

Let $(L, \|\cdot\|)$ be a Banach space. A subset A of L is said to be *weakly closed* if A is a closed set in the weak topology on L. Equivalently, if x_α, $\alpha \in D$ is a net in A converging weakly to x in L, then $x \in A$. In this context, we quote a result.

1.5.17 Theorem. *Let $(L, \|\cdot\|)$ be a Banach space and L_1 a closed (in the strong topology on L) subspace of L. Then L_1 is weakly closed.*

A Banach space $(L, \|\cdot\|)$ is said to be *weakly complete* if every weak Cauchy sequence in L is weakly convergent.

Finally, we close this section with Banach lattices.

1.5.18 Definitions. A vector lattice $(L, \leq)$ is said to be a *normed vector lattice* if there is a norm $\|\cdot\|$ defined on L such that $\|x\| \leq \|y\|$ whenever $x, y \in L$ and $|x| \leq |y|$. A *Banach lattice* is a normed vector lattice $(L, \leq)$ with a norm $\|\cdot\|$ such that $(L, \|\cdot\|)$ is a Banach space.

1.5.19 Theorem. *In any normed vector lattice* $(L, \leq, \|\cdot\|)$, *the maps* $(x, y) \to x \vee y$ *and* $(x, y) \to x \wedge y$ *from* $L \times L$ *to* L *are uniformly continuous. Hence for any subset* S *of* L, $S^{\perp}$ *is a closed subspace of* L. *In particular, any normal vector sublattice of* L *is closed.*

Proof. For any x in L, observe that $\|x\| = \|\,|x|\,\|$. Now, the uniform continuity of the maps $(x, y) \to x \vee y$ and $(x, y) \to x \wedge y$ from $L \times L$ to L follows from Theorem 1.5.4 (28) and (29). The rest of the assertions also follow from the same observations. □

CHAPTER 2

Charges

In this chapter, we introduce the main concern of this book, namely charges. They are usually known as finitely additive measures in the literature. We chronicle most of the rudimentary facts about charges in this chapter. In Section 2.1, basic concepts about charges are presented. The space of all bounded charges on a field of sets is shown to be a boundedly complete vector lattice in Section 2.2. Sections 2.3 and 2.4 deal with measures. Jordan Decomposition theorem and Hahn Decomposition theorem for charges not necessarily bounded are covered in Sections 2.5 and 2.6, respectively.

2.1 BASIC CONCEPTS

This section is mainly devoted to the study of various properties of charges.

2.1.1 Definitions. Let $\mathcal{F}$ be a field of subsets of a set Ω.
(1). A map $\mu : \mathcal{F} \to [-\infty, \infty]$ is said to be a *charge* on $\mathcal{F}$ if the following conditions are satisfied.
  (i). $\mu(\varnothing) = 0$.
  (ii). If A, B $\in \mathcal{F}$ and $A \cap B = \varnothing$, then $\mu(A \cup B) = \mu(A) + \mu(B)$.
(2). A charge μ on $\mathcal{F}$ is said to be a *real charge* if $-\infty < \mu(F) < \infty$ for every F in $\mathcal{F}$.
(3). A charge μ on $\mathcal{F}$ is said to be *bounded* if $\text{Sup}\{|\mu(F)|; F \in \mathcal{F}\} < \infty$.
(4). A charge μ on $\mathcal{F}$ is said to be *positive* if $\mu(F) \geq 0$ for every F in $\mathcal{F}$.
(5). A charge μ on $\mathcal{F}$ is said to be *positive bounded* if μ is positive and bounded.
(6). A charge μ on $\mathcal{F}$ is said to be 0-a *valued* ($a \neq 0$) if $\mu(F) = 0$ or a for every F in $\mathcal{F}$ and there is a set F_0 in $\mathcal{F}$ such that $\mu(F_0) = a$.
(7). A charge μ on $\mathcal{F}$ is said to be a *probability charge* if μ is positive and $\mu(\Omega) = 1$.

One could introduce the notion of a charge in exactly the same way as above on general domains such as semi-rings of sets or rings of sets or additive-classes of sets or Boolean algebras. In fact, there are some situations in the subsequent chapters where we deal with charges on these general domains.

If μ is a charge on a field $\mathscr{F}$ of subsets of a set Ω, μ cannot take both the values $+\infty$ and $-\infty$. For, if A, B in $\mathscr{F}$ are such that $\mu(A) = \infty$ and $\mu(B) = -\infty$, then $\mu(\Omega) = \mu(A) + \mu(A^c) = \infty = \mu(B) + \mu(B^c) = -\infty$, a contradiction.

The following proposition gives some properties of charges. In this proposition, μ stands for a charge on a given field $\mathscr{F}$ of subsets of a set Ω.

2.1.2 Proposition
(i). If $F_1, F_2, \ldots, F_n$ are any finite number of pairwise disjoint sets in $\mathscr{F}$, then
$$\mu\left(\bigcup_{i=1}^{n} F_i\right) = \sum_{i=1}^{n} \mu(F_i).$$

(ii). If $E, F \in \mathscr{F}$, $E \subset F$ and $-\infty < \mu(E) < \infty$, then $\mu(F - E) = \mu(F) - \mu(E)$.
(iii). If $E, F \in \mathscr{F}$, $E \subset F$ and μ positive, then $\mu(E) \leq \mu(F)$.
(iv). If $F_0, F_1, F_2, \ldots, F_n$ are any finite number of sets in $\mathscr{F}$ such that $F_0 \subset \bigcup_{i=1}^{n} F_i$ and μ is positive, then
$$\mu(F_0) \leq \sum_{i=1}^{n} \mu(F_i).$$

In particular,
$$\mu\left(\bigcup_{i=1}^{n} F_i\right) \leq \sum_{i=1}^{n} \mu(F_i).$$

(v). If $F \in \mathscr{F}$ and $\mu(F) < \infty$, then $\mu(E) < \infty$ for any E in $\mathscr{F}$ with $E \subset F$.
(vi). If $F \in \mathscr{F}$ and $\mu(F) > -\infty$, then $\mu(E) > -\infty$ for any E in $\mathscr{F}$ with $E \subset F$.
(vii). If $F \in \mathscr{F}$ and $-\infty < \mu(F) < \infty$, then $-\infty < \mu(E) < \infty$ for any E in $\mathscr{F}$ with $E \subset F$.
(viii). If E_n, $n \geq 1$ is a sequence of pairwise disjoint sets in $\mathscr{F}$, $E \in \mathscr{F}$, $\bigcup_{n \geq 1} E_n \subset E$ and μ is positive, then
$$\sum_{n \geq 1} \mu(E_n) \leq \mu(E).$$

(ix). μ is modular, i.e. $\mu(E) + \mu(F) = \mu(E \cup F) + \mu(E \cap F)$ for any E and F in $\mathscr{F}$.
(x). If $F_1, F_2, \ldots, F_n$ are any finite number of sets in $\mathscr{F}$, then
$$\sum_{i=1}^{n} \mu(F_i) = \mu\left(\bigcup_{i=1}^{n} F_i\right) + \mu\left(\bigcup_{\substack{i=1 \\ i<j}}^{n} \bigcup_{j=1}^{n} F_i \cap F_j\right)$$
$$+ \mu\left(\bigcup_{\substack{i=1 \\ i<j<k}}^{n} \bigcup_{j=1}^{n} \bigcup_{k=1}^{n} F_i \cap F_j \cap F_k\right) + \cdots + \mu\left(\bigcap_{i=1}^{n} F_i\right).$$

Proof. (i) follows by induction. (ii) and (iii) are easy to prove.
(iv). Let $E_1 = F_1$, $E_2 = F_2 - F_1$, $E_3 = F_3 - (F_1 \cup F_2), \ldots, E_n = F_n - (\bigcup_{i=1}^{n-1} F_i)$.
Then $E_1, E_2, \ldots, E_n$ are pairwise disjoint, $E_i \subset F_i$ for every $i = 1, 2, \ldots, n$ and $\bigcup_{i=1}^n F_i = \bigcup_{i=1}^n E_i$. Consequently,

$$\mu(F_0) \leq \mu\left(\bigcup_{i=1}^n F_i\right) = \mu\left(\bigcup_{i=1}^n E_i\right) = \sum_{i=1}^n \mu(E_i)$$

$$\leq \sum_{i=1}^n \mu(F_i).$$

(v), (vi) and (vii) are easy exercises.
(viii). For every $m \geq 1$, $\bigcup_{i=1}^m E_i \subset E$, and so $\sum_{i=1}^m \mu(E_i) = \mu(\bigcup_{i=1}^m E_i) \leq \mu(E)$. Hence $\sum_{i \geq 1} \mu(E_i) \leq \mu(E)$.
(ix). Note that $\mu(E) + \mu(F) = [\mu(E \cap F) + \mu(E - F)] + [\mu(E \cap F) + \mu(F - E)]$.
On the other hand, $\mu(E \cup F) + \mu(E \cap F) = \mu(E - F) + \mu(F - E) + \mu(E \cap F) + \mu(E \cap F)$. Hence (ix) follows.
(x). This is a generalization of (ix) and can be proved by induction. The result is true for $n = 2$ by (ix). Suppose the result is true for $n = m$. We prove the result for $n = m + 1$. Let $F_1, F_2, \ldots, F_{m+1}$ be any $m + 1$ sets from $\mathscr{F}$. Then

$$\sum_{i=1}^{m+1} \mu(F_i) = \sum_{i=1}^m \mu(F_i) + \mu(F_{m+1})$$

$$= \left[\mu\left(\bigcup_{i=1}^m F_i\right) + \mu\left(\bigcup_{\substack{i=1 \\ i<j}}^m \bigcup_{j=1}^m F_i \cap F_j\right) + \cdots\right.$$

$$\left. + \mu\left(\bigcap_{i=1}^m F_i\right)\right] + \mu(F_{m+1}) \quad \text{(by induction hypothesis)}$$

$$= \mu\left(\bigcup_{i=1}^{m+1} F_i\right) + \mu\left(\left(\bigcup_{i=1}^m F_i\right) \cap F_{m+1}\right)$$

$$+ \mu\left(\bigcup_{\substack{i=1 \\ i<j}}^m \bigcup_{j=1}^m F_i \cap F_j\right) + \cdots + \mu\left(\bigcap_{i=1}^m F_i\right)$$

(by using the induction hypothesis for the sum $\mu(\bigcup_{i=1}^m F_i) + \mu(F_{m+1})$ with $n = 2$).
Continuing this way, we get the desired equality. □

We give some examples of charges.

2.1.3 Examples.
(1). Let Ω be the set of all rational numbers in $[0, 1)$ and $\mathscr{F} = \{\bigcup_{i=1}^n [a_i, b_i) \cap \Omega; [a_i, b_i) \cap [a_j, b_j) = \emptyset$ for $i \neq j$, $0 \leq a_i \leq b_i \leq 1$ for every i,

a_i, b_i rational for every i and $n \geq 1$}. $\mathscr{F}$ is a field on Ω. For any set described above, let
$$\mu\left(\bigcup_{i=1}^{n} [a_i, b_i) \cap \Omega\right) = \sum_{i=1}^{n} (b_i - a_i).$$
Then μ is a positive bounded charge on $\mathscr{F}$.

(2). Let $\Omega = \{1, 2, 3, \ldots\}$, $\mathscr{F} = \{A \subset \Omega;\ A$ or A^c is finite$\}$, and μ on $\mathscr{F}$ be defined by
$$\mu(A) = n, \quad \text{if A is finite and has } n \text{ elements},$$
$$= -\mu(A^c), \text{ if } A^c \text{ is finite}.$$
Then μ is a real charge on $\mathscr{F}$ but not bounded.

(3). Let $\mathscr{F}$ be a field of subsets of a set Ω and $\mathscr{I}$ a maximal ideal in $\mathscr{F}$. Define μ on $\mathscr{F}$ by
$$\mu(F) = 0, \quad \text{if } F \in \mathscr{I},$$
$$= 1, \quad \text{if } F \notin \mathscr{I}, F \in \mathscr{F}.$$
Then μ is a charge on $\mathscr{F}$.

Conversely, if μ is a 0-1 valued charge on $\mathscr{F}$, then $\mathscr{I} = \{F \in \mathscr{F}; \mu(F) = 0\}$ is a maximal ideal in $\mathscr{F}$ and $\mathscr{J} = \{E \in \mathscr{F}; \mu(E) = 1\}$ is a maximal filter in $\mathscr{F}$. Consequently, the Stone space of $\mathscr{F}$ can be identified as the collection of all 0-1 valued charges on $\mathscr{F}$. See Theorem 1.4.6.

(4). Let $\Omega = \{1, 2, 3, \ldots\}$. There is a 0-1 valued charge μ on $\mathscr{P}(\Omega)$ such that $\mu(A) = 0$ if A is a finite subset of Ω. This can be shown as follows. Let $\mathscr{I}_1 = \{A \subset \Omega;\ A$ is finite$\}$. $\mathscr{I}_1$ is an ideal in $\mathscr{P}(\Omega)$. There is a maximal ideal $\mathscr{I}$ in $\mathscr{P}(\Omega)$ containing $\mathscr{I}_1$. (See Section 1.2.) Define μ on $\mathscr{P}(\Omega)$ by
$$\mu(A) = 0, \quad \text{if } A \in \mathscr{I},$$
$$= 1, \quad \text{if } A \notin \mathscr{I}.$$
μ is the desired 0-1 valued charge on $\mathscr{P}(\Omega)$.

(5). Let $\Omega = [0, 1)$ and $\mathscr{F}$ the collection of all sets each of which is a finite disjoint union of intervals of the type $[a, b)$ with $0 \leq a \leq b \leq 1$. Let φ be any real valued function defined on $[0, 1]$. Define μ_φ on $\mathscr{F}$ as follows. For $0 \leq a \leq b \leq 1$, $\mu_\varphi([a, b)) = \varphi(b) - \varphi(a)$. For any F in $\mathscr{F}$, $\mu_\varphi(F)$ is defined additively. Then μ_φ is a real charge on $\mathscr{F}$.

(6). A special case of the above is given by the function φ defined as follows.
$$\varphi(x) = 0, \quad \text{if } x \text{ is irrational}, x \in [0, 1],$$
$$= n, \quad \text{if } x \text{ is rational}, x = m/n, m$$
$$\text{and } n \text{ are mutually prime integers}, x \in (0, 1],$$
$$= 0, \quad \text{if } x = 0.$$

In this example, μ_φ is unbounded on every non-empty set in $\mathscr{F}$, i.e. Sup $\{|\mu(E)|;\ E \subset F,\ E \in \mathscr{F}\} = \infty$ for every non-empty set F in $\mathscr{F}$.

(7). Another special case of (5) is the following.

$$\varphi(0) = 0$$

and

$$\varphi(x) = 1/x, \quad \text{if } 0 < x \le 1.$$

In this case, μ_φ is unbounded on every set in $\mathscr{F}$ containing 0.

(8). *Banach Limits and Shift-invariant Charges.* Let $\Omega = \{1, 2, 3, \ldots\}$, $\mathscr{F} = \mathscr{P}(\Omega)$ and S the shift transformation from Ω to Ω defined by $S(n) = n+1$ for $n \in \Omega$. We show that there exists an S-invariant probability charge μ on $\mathscr{F}$, i.e. μ is a probability charge on $\mathscr{F}$ and $\mu(A) = \mu(S^{-1}(A))$ for every $A \subset \Omega$. We use Banach Limits, which we shall now describe, to show the existence of charges of the type described above. Let ℓ_∞ be the space of all bounded sequences of real numbers. For $x = (x_1, x_2, \ldots) \in \ell_\infty$, let $\|x\|_\infty = \text{Sup}_{n \ge 1} |x_n|$. $(\ell_\infty, \|\cdot\|_\infty)$ is a Banach space. Let

$$L = \{y = (y_1, y_2, \ldots);\ (y_1, y_1+y_2, y_1+y_2+y_3, \ldots) \in \ell_\infty\}.$$

It is clear that if $y = (y_1, y_2, \ldots) \in L$, then y_n, $n \ge 1$ is a bounded sequence of real numbers and that L is a subspace of ℓ_∞. For $x \in \ell_\infty$, let $p(x) = \text{Sup}_{n \ge 1} x_n$. We show that $p(\cdot)$ has the following properties.

(i). $p(cx) = cp(x)$ for $x \in \ell_\infty$ and $c \ge 0$.
(ii). $p(x+y) \le p(x) + p(y)$ for all $x, y \in \ell_\infty$.
(iii). $p(y) \ge 0$ for $y \in L$.

(i) and (ii) are obvious. We prove (iii). Suppose $p(y) = \text{Sup}_{n \ge 1} y_n = k < 0$ for some $y = (y_1, y_2, \ldots) \in L$. Then $y_1 + y_2 + \cdots + y_n \le nk$ for every $n \ge 1$. So, $y_1 + y_2 + \cdots + y_n$, $n \ge 1$ cannot be a bounded sequence of real numbers. This contradiction shows that $p(y) \ge 0$ for every y in L.

For y in L, let $T_1(y) = 0$. Then T_1 is a linear functional on L satisfying $T_1(y) \le p(y)$ for every y in L. By Hahn–Banach Theorem 1.5.15, there exists a linear functional T on ℓ_∞ such that

$$T(y) = T_1(y), \quad \text{if } y \in L,$$

and

$$T(x) \le p(x) \quad \text{for every } x \text{ in } \ell_\infty.$$

This functional T has the following properties.
1°. $T(cx + dy) = cT(x) + dT(y)$ for all x, y in ℓ_∞ and c and d real numbers.
2°. $T(x) \ge 0$ if $x = (x_1, x_2, \ldots) \ge 0$, i.e. T is a non-negative linear functional on ℓ_∞.

3°. $T(\underline{1}) = 1$, where $\underline{1} = (1, 1, 1, \ldots)$.
4°. $T((x_1, x_2, x_3, \ldots)) = T((x_2, x_3 \ldots))$ for every $x = (x_1, x_2, \ldots)$ in ℓ_∞.
5°. $|T(x)| \leq \|x\|_\infty$ for every x in ℓ_∞, i.e. T is a continuous linear functional on ℓ_∞.
6°. $\liminf_{n \to \infty} x_n \leq T((x_1, x_2, \ldots)) \leq \limsup_{n \to \infty} x_n$ for every $x = (x_1, x_2, \ldots)$ in ℓ_∞.
7°. If x_n, $n \geq 1$ is a convergent sequence of real numbers, then $T((x_1, x_2, \ldots)) = \lim_{n \to \infty} x_n$.

1° is clear. We prove 2°. Note that for any x in ℓ_∞, $-T(x) = T(-x) \leq p(-x) = \operatorname{Sup}_{n \geq 1} -x_n \leq 0$ if $x = (x_1, x_2, \ldots) \geq 0$. Consequently, $T(x) \geq 0$ if $x \geq 0$. For 3°, observe that $T(\underline{1}) \leq p(\underline{1}) = 1$ and $-T(\underline{1}) = T(-\underline{1}) \leq p(-\underline{1}) = -1$, so that $T(\underline{1}) \geq 1$. Hence $T(\underline{1}) = 1$. To prove 4°, we proceed as follows. For any $x = (x_1, x_2, \ldots)$ in ℓ_∞, it can be checked that $(x_2, x_3, \ldots) - (x_1, x_2, \ldots) = (x_2 - x_1, x_3 - x_2, \ldots)$ is in L. Consequently,

$$T((x_2, x_3, \ldots)) - T((x_1, x_2, \ldots)) = T((x_2, x_3, \ldots) - (x_1, x_2, \ldots))$$
$$= T((x_2 - x_1, x_3 - x_2, \ldots))$$
$$= T_1((x_2 - x_1, x_3 - x_2, \ldots))$$
$$= 0.$$

From this, 4° follows. For 5°, we observe the following inequalities.

$$T(x) \leq p(x) = \operatorname*{Sup}_{n \geq 1} x_n \leq \operatorname*{Sup}_{n \geq 1} |x_n| = \|x\|_\infty,$$

$$-T(x) = T(-x) \leq p(-x) = \operatorname*{Sup}_{n \geq 1} -x_n = -\operatorname*{Inf}_{n \geq 1} x_n,$$

and

$$-\|x\|_\infty = -\operatorname*{Sup}_{n \geq 1} |x_n| \leq \operatorname*{Inf}_{n \geq 1} x_n \leq T(x)$$

for any $x = (x_1, x_2, \ldots)$ in ℓ_∞. Hence $|T(x)| \leq \|x\|_\infty$. 6° is a consequence of 4° and the fact that $T(x) \leq p(x)$ for every x in ℓ_∞. 7° follows from 6°.

The functional obtained above is called a Banach Limit. One can define a Banach Limit, in abstract terms, as follows. A functional T on ℓ_∞ is said to be a Banach Limit if T has the properties 1°, 2°, 3° and 4° listed above. One can show that if T has properties 1°, 2°, 3° and 4°, then it has properties 5°, 6° and 7°.

Now, we come to the problem of existence of S-invariant charges on $\mathcal{P}(\Omega)$. For $A \subset \Omega$, let $x_A \in \ell_\infty$ be as defined below.

$$(x_A)_n = 1, \quad \text{if } n \in A,$$
$$= 0, \quad \text{if } n \notin A, n \in \Omega.$$

Let T be a Banach Limit on ℓ_∞. Define μ on $\mathcal{F}$ by $\mu(A) = T(x_A)$, $A \subset \Omega$. It is clear that μ is a probability charge on $\mathcal{F}$. That, μ is S-invariant follows from the observation that for any $A \subset \Omega$,

$$x_{S^{-1}(A)} = ((x_A)_2, (x_A)_3, \ldots)$$

and that T has property 4°, where x_A is as defined above.

(9). *Banach Limits and General Invariant Charges.* Let $\mathcal{F}$ be a field of subsets of a set Ω and S a map from Ω to Ω such that $S^{-1}(A) \in \mathcal{F}$ whenever $A \in \mathcal{F}$. We show that there exists an S-invariant probability charge μ on $\mathcal{F}$, i.e. μ is a probability charge on $\mathcal{F}$ and $\mu(A) = \mu(S^{-1}(A))$ for every A in $\mathcal{F}$. Let ν be any probability charge on $\mathcal{F}$ and T a Banach Limit on ℓ_∞. Define μ on $\mathcal{F}$ by $\mu(A) = T((\nu(A), \nu(S^{-1}(A)), \nu(S^{-2}(A)), \ldots))$ for A in $\mathcal{F}$. It is clear that μ is a probability charge on $\mathcal{F}$. From 4° of (8), it follows that μ is S-invariant.

If $\Omega = \{1, 2, 3, \ldots\}$, $\mathcal{F} = \mathcal{P}(\Omega)$, S the shift transformation on Ω and ν the charge on $\mathcal{F}$ defined by

$$\nu(A) = 0, \quad \text{if } 1 \notin A,$$
$$= 1, \quad \text{if } 1 \in A, A \subset \Omega,$$

then the charge μ defined above is precisely the S-invariant charge constructed in (8).

(10). *Banach Limits and Density Charges.* Let $\Omega = \{1, 2, 3, \ldots\}$ and $\mathcal{F} = \mathcal{P}(\Omega)$. A charge μ on $\mathcal{F}$ is said to be a *density charge* on $\mathcal{F}$ if

$$\mu(A) = \lim_{n \to \infty} \frac{\#(A \cap \{1, 2, 3, \ldots, n\})}{n}$$

whenever this limit exists for any set $A \subset \Omega$, where for any set B, #B is the number of elements in B. It is obvious that μ is positive and $\mu(\Omega) = 1$. We show the existence of a density charge. For each $n \geq 1$, let μ_n on $\mathcal{F}$ be defined by

$$\mu_n(A) = \frac{\#(A \cap \{1, 2, 3, \ldots, n\})}{n}, \quad A \subset \Omega.$$

Then each μ_n is a probability charge on $\mathcal{F}$. Let T be a Banach Limit on ℓ_∞. Define μ on $\mathcal{F}$ by

$$\mu(A) = T((\mu_1(A), \mu_2(A), \ldots)), \quad A \subset \Omega.$$

It is clear that μ is a probability charge on $\mathcal{F}$. From 7° of (8), it follows that μ is a density charge on $\mathcal{P}(\Omega)$.

Now, we introduce the concept of s-boundedness of charges.

2.1.4 Definition. Let μ be a charge on a field $\mathscr{F}$ of subsets of a set Ω. μ is said to be *s-bounded* if for every sequence A_n, $n \geq 1$ of pairwise disjoint sets in $\mathscr{F}$, $\lim_{n \to \infty} \mu(A_n) = 0$.

If μ is a real charge, then μ is s-bounded if and only if μ is bounded. The following results pave the way for the validity of this assertion.

2.1.5 Lemma. *Let μ be a real charge on a field $\mathscr{F}$ of subsets of a set Ω. If μ is unbounded, there exist two sets A_1 and A_2 in $\mathscr{F}$ satisfying the following conditions.*

(i). $A_1 \cap A_2 = \emptyset$.

(ii). $|\mu(A_1)| \geq 1$ *and* $|\mu(A_2)| \geq 1$.

(iii). μ *is unbounded either on* A_1 *or on* A_2, *i.e.* $\sup\{|\mu(B)|;\ B \subset A_1, B \in \mathscr{F}\} = \infty$ *or* $\sup\{|\mu(C)|;\ C \subset A_2, C \in \mathscr{F}\} = \infty$.

Proof. Since μ is unbounded, there exists B_1 in $\mathscr{F}$ such that $|\mu(B_1)| \geq |\mu(\Omega)| + 1$. Then $|\mu(B_1^c)| = |\mu(\Omega) - \mu(B_1)| \geq ||\mu(\Omega)| - |\mu(B_1)|| \geq |\mu(B_1)| - |\mu(\Omega)| \geq 1$. It is obvious that μ is unbounded either on B_1 or on B_1^c. Take $A_1 = B_1$ and $A_2 = B_1^c$, $\square$

2.1.6 Theorem. *Let μ be a real but unbounded charge on a field $\mathscr{F}$ of subsets of a set Ω. Then there exists a sequence A_n, $n \geq 1$ of pairwise disjoint sets in $\mathscr{F}$ such that $|\mu(A_n)| \geq 1$ for every $n \geq 1$.*

Proof. By Lemma 2.1.5, there are two disjoint sets A_1 and B_1 in $\mathscr{F}$ such that $|\mu(A_1)| \geq 1$, $|\mu(B_1)| \geq 1$ and μ is unbounded on B_1. Applying Lemma 2.1.5 again to B_1, there are two disjoint sets A_2 and B_2 contained in B_1 such that $|\mu(A_2)| \geq 1$, $|\mu(B_2)| \geq 1$ and μ is unbounded on B_2. Continuing this way, we obtain a sequence A_n, $n \geq 1$ of pairwise disjoint sets in $\mathscr{F}$ such that $|\mu(A_n)| \geq 1$ for every $n \geq 1$. This completes the proof. $\square$

An important consequence of this theorem is the following corollary.

2.1.7 Corollary. *Let μ be a real charge on a field $\mathscr{F}$ of subsets of a set Ω. Then μ is s-bounded if and only if μ is bounded.*

Proof. Suppose μ is bounded. Let $k = \sup\{|\mu(B)|;\ B \in \mathscr{F}\}$. Then k is finite. We show that $\sum_{n \geq 1} |\mu(A_n)| \leq 2k$ for any sequence A_n, $n \geq 1$ of pairwise disjoint sets in $\mathscr{F}$. Let $m \geq 1$ be fixed. Let $I_1 = \{1 \leq i \leq m;\ \mu(A_i) \geq 0\}$ and $I_2 = \{1 \leq i \leq m;\ \mu(A_i) < 0\}$. Then

$$\sum_{i=1}^{m} |\mu(A_i)| = \sum_{i \in I_1} \mu(A_i) - \sum_{j \in I_2} \mu(A_j)$$

$$= \mu\left(\bigcup_{i \in I_1} A_i\right) - \mu\left(\bigcup_{j \in I_2} A_j\right) \leq 2k.$$

Hence $\sum_{i\geq 1} |\mu(A_i)| \leq 2k$. From this, it follows that $\lim_{n\to\infty} \mu(A_n) = 0$. So, μ is s-bounded.

Conversely, let μ be s-bounded but not bounded. By Theorem 2.1.6, there exists a sequence A_n, $n \geq 1$ of pairwise disjoint sets in $\mathscr{F}$ such that $|\mu(A_n)| \geq 1$ for every $n \geq 1$. Then $\lim_{n\to\infty} \mu(A_n)$ cannot be zero. This contradiction shows that μ is bounded. This completes the proof. □

2.1.8 Remark. The assumption that μ be real valued in the statement of Corollary 2.1.7 cannot be dropped. As an example, let $\Omega = \{1, 2, 3, \ldots\}$ and $\mathscr{F}$ the finite–cofinite field on Ω. Let μ on $\mathscr{F}$ be defined by

$$\mu(A) = 0, \quad \text{if A is finite,}$$
$$= \infty, \quad \text{if A is cofinite.}$$

Then μ is an s-bounded charge on $\mathscr{F}$ but not bounded.

2.2 THE SPACE OF ALL BOUNDED CHARGES, ba($\Omega, \mathscr{F}$)

The object of study in this section is the space of all bounded charges on a field $\mathscr{F}$ of subsets of a set Ω. We denote this space by ba($\Omega, \mathscr{F}$). There is a natural partial order $\leq$ on ba($\Omega, \mathscr{F}$). For $\mu, \nu \in$ ba($\Omega, \mathscr{F}$), say $\mu \leq \nu$ if $\mu(F) \leq \nu(F)$ for every F in $\mathscr{F}$. The relation $\leq$ is reflexive, antisymmetric and transitive. So, $\leq$ is a partial order on ba($\Omega, \mathscr{F}$). The main result of this section is that ba($\Omega, \mathscr{F}$) is a boundedly complete vector lattice. We also introduce a norm $\|\cdot\|$ on ba($\Omega, \mathscr{F}$) so that (ba($\Omega, \mathscr{F}$), $\leq, \|\cdot\|$) becomes a Banach lattice.

2.2.1 Theorem. *Let $\mathscr{F}$ be a field of subsets of a set Ω. Then the following statements are true.*
(1). *If $\mu, \nu \in$ ba($\Omega, \mathscr{F}$) and c, d are real numbers, then $c\mu + d\nu \in$ ba($\Omega, \mathscr{F}$).*
(2). *ba($\Omega, \mathscr{F}$) is a vector space.*
(3). *(ba($\Omega, \mathscr{F}$), $\leq$) is an ordered vector space, where $\leq$ is the partial order on ba($\Omega, \mathscr{F}$) introduced above.*
(4). *Let $\mu, \nu \in$ ba($\Omega, \mathscr{F}$). Define λ on $\mathscr{F}$ by $\lambda(F) = \text{Sup}\{\mu(E) + \nu(F-E); E \subset F, E \in \mathscr{F}\}$, $F \in \mathscr{F}$. Then $\lambda \in$ ba($\Omega, \mathscr{F}$).*
(5). *Let $\mu, \nu \in$ ba($\Omega, \mathscr{F}$). Then $\mu \vee \nu$ exists in the partial order $\leq$ on ba($\Omega, \mathscr{F}$) and is equal to λ defined in (4).*
(6). *Let $\mu, \nu \in$ ba($\Omega, \mathscr{F}$). Define τ on $\mathscr{F}$ by $\tau(F) = \text{Inf}\{\mu(E) + \nu(F-E); E \subset F, E \in \mathscr{F}\}$, $F \in \mathscr{F}$. Then $\tau \in$ ba($\Omega, \mathscr{F}$).*
(7). *Let $\mu, \nu \in$ ba($\Omega, \mathscr{F}$). Then $\mu \wedge \nu$ exists in the partial order $\leq$ on ba($\Omega, \mathscr{F}$) and is equal to τ defined in (6).*
(8). *(ba($\Omega, \mathscr{F}$), $\leq$) is a vector lattice.*

(9). $(\mathrm{ba}(\Omega, \mathcal{F}), \leq)$ *is a boundedly complete vector lattice.*
(10). *For* $\mu \in \mathrm{ba}(\Omega, \mathcal{F})$, *let* $\|\mu\| = |\mu|(\Omega)$. *(Recall* $|\mu| = \mu^+ + \mu^-$ *from Section 1.5.) Then* $\|\cdot\|$ *is a norm on* $\mathrm{ba}(\Omega, \mathcal{F})$.
(11). $(\mathrm{ba}(\Omega, \mathcal{F}), \leq, \|\cdot\|)$ *is a Banach lattice.*

Proof. (1), (2) and (3) are obvious.
(4). It is obvious that $\lambda(\varnothing) = 0$. Since μ and ν are bounded, λ is a bounded function on $\mathcal{F}$. Let A_1 and A_2 be two disjoint sets in $\mathcal{F}$. We show that $\lambda(A_1 \cup A_2) = \lambda(A_1) + \lambda(A_2)$. Let $B \in \mathcal{F}$ and $B \subset A_1 \cup A_2$. Let $B_1 = B \cap A_1$ and $B_2 = B \cap A_2$. Then

$$\mu(B) + \nu[(A_1 \cup A_2) - B] = \mu(B_1) + \mu(B_2) + \nu(A_1 - B_1) + \nu(A_2 - B_2)$$
$$= [\mu(B_1) + \nu(A_1 - B_1)] + [\mu(B_2) + \nu(A_2 - B_2)]$$
$$\leq \lambda(A_1) + \lambda(A_2),$$

as $B_1 \subset A_1$, $B_2 \subset A_2$ and $B_1, B_2 \in \mathcal{F}$. Taking supremum over all $B \subset A_1 \cup A_2$ with B in $\mathcal{F}$, we obtain $\lambda(A_1 \cup A_2) \leq \lambda(A_1) + \lambda(A_2)$. On the other hand, let $B_1, B_2 \in \mathcal{F}$, $B_1 \subset A_1$ and $B_2 \subset A_2$. Then

$$[\mu(B_1) + \nu(A_1 - B_1)] + [\mu(B_2) + \nu(A_2 - B_2)]$$
$$= \mu(B_1 \cup B_2) + \nu[(A_1 \cup A_2) - (B_1 \cup B_2)] \leq \lambda(A_1 \cup A_2).$$

Taking the supremum over all $B_1 \subset A_1$ with $B_1 \in \mathcal{F}$ and then supremum over all $B_2 \subset A_2$ with $B_2 \in \mathcal{F}$, we obtain $\lambda(A_1) + \lambda(A_2) \leq \lambda(A_1 \cup A_2)$. Thus $\lambda(A_1 \cup A_2) = \lambda(A_1) + \lambda(A_2)$. Hence $\lambda \in \mathrm{ba}(\Omega, \mathcal{F})$.
(5). It is obvious that $\mu \leq \lambda$ and $\nu \leq \lambda$. Let $\psi \in \mathrm{ba}(\Omega, \mathcal{F})$ be such that $\mu \leq \psi$ and $\nu \leq \psi$. Then for any F in $\mathcal{F}$,

$$\lambda(F) = \mathrm{Sup}\{\mu(E) + \nu(F - E); E \subset F, E \in \mathcal{F}\}$$
$$\leq \mathrm{Sup}\{\psi(E) + \psi(F - E); E \subset F, E \in \mathcal{F}\}$$
$$= \psi(F).$$

So, $\lambda \leq \psi$. Consequently, $\mu \vee \nu$ exists and is equal to λ.
(6) and (7) can be proved along the same lines as those of (4) and (5). (8) is, by now, obvious.
(9). Let $\{\mu_\alpha; \alpha \in \Gamma\} \subset \mathrm{ba}(\Omega, \mathcal{F})$ be bounded above, i.e. there exists $\lambda \in \mathrm{ba}(\Omega, \mathcal{F})$ such that $\mu_\alpha \leq \lambda$ for every α in Γ. We show that $\bigvee_{\alpha \in \Gamma} \mu_\alpha$ exists. Let $\mathcal{A}$ be the collection of all finite subsets of Γ. On $\mathcal{A}$, we introduce a partial order $\geq^*$ as follows. For $C, D \in \mathcal{A}$, say $C \geq^* D$ if $C \supset D$. $(\mathcal{A}, \geq^*)$ is a directed set. For each C in $\mathcal{A}$, let $\mu_C = \bigvee_{\alpha \in C} \mu_\alpha$. μ_C, obviously, exists in $\mathrm{ba}(\Omega, \mathcal{F})$ as C is only a finite set. Then the net $\{\mu_C; C \in \mathcal{A}\}$ is an increasing net, i.e. if $C, D \in \mathcal{A}$ and $C \geq^* D$, then $\mu_C \geq \mu_D$ in $\mathrm{ba}(\Omega, \mathcal{F})$. Further, $\mu_C \leq \lambda$

for every C in $\mathcal{A}$. Let τ be the pointwise limit of μ_C, $C \in \mathcal{A}$, i.e. $\tau(F) = \lim_{C \in \mathcal{A}} \mu_C(F)$, $F \in \mathcal{F}$. This limit obviously exists for every F in $\mathcal{F}$. τ is also a charge on $\mathcal{F}$. Since $\mu_\alpha \leq \tau \leq \lambda$ for any fixed α in Γ, τ is also bounded. It is obvious that $\tau = \bigvee_{\alpha \in \Gamma} \mu_\alpha$. Hence $(\text{ba}(\Omega, \mathcal{F}), \leq)$ is a boundedly complete vector lattice.

(10). If $\mu = 0$, then $|\mu| = 0$. So, $\|\mu\| = 0$. Conversely, if $\|\mu\| = 0$, $|\mu|(\Omega) = 0$. Since $|\mu|$ is a positive charge on $\mathcal{F}$, $|\mu| = 0$. Hence $\mu = 0$. See Theorem 1.5.4(12). If α is any real number and $\mu \in \text{ba}(\Omega, \mathcal{F})$, then $\|\alpha\mu\| = |\alpha\mu|(\Omega) = |\alpha||\mu|(\Omega) = |\alpha|\|\mu\|$. See Theorem 1.5.4(12). Finally, let $\mu_1, \mu_2 \in \text{ba}(\Omega, \mathcal{F})$. Then $\|\mu_1 + \mu_2\| = |\mu_1 + \mu_2|(\Omega) \leq |\mu_1|(\Omega) + |\mu_2|(\Omega) = \|\mu_1\| + \|\mu_2\|$. See Theorem 1.5.4(12). Hence $\|\cdot\|$ is a norm on $\text{ba}(\Omega, \mathcal{F})$.

(11). If $\mu, \nu \in \text{ba}(\Omega, \mathcal{F})$ and $|\mu| \leq |\nu|$, then $|\mu|(\Omega) \leq |\nu|(\Omega)$. Consequently, $\|\mu\| \leq \|\nu\|$. Hence $\text{ba}(\Omega, \mathcal{F})$ is a normed vector lattice. Now, we show that $\text{ba}(\Omega, \mathcal{F})$ is complete under the norm $\|\cdot\|$. Let μ_n, $n \geq 1$ be a Cauchy sequence in $\text{ba}(\Omega, \mathcal{F})$, i.e. $\lim_{m,n \to \infty} \|\mu_m - \mu_n\| = 0$. Since $\mu_m - \mu_n \leq |\mu_m - \mu_n|$ for all $m, n \geq 1$, we have for any F in $\mathcal{F}$,

$$|\mu_m(F) - \mu_n(F)| \leq |\mu_m - \mu_n|(F) \leq |\mu_m - \mu_n|(\Omega)$$
$$= \|\mu_m - \mu_n\|.$$

This shows that $\mu_m(F)$, $m \geq 1$ is a uniform Cauchy sequence of real numbers over $\mathcal{F}$. Let $\mu(F) = \lim_{m \to \infty} \mu_m(F)$, $F \in \mathcal{F}$. Thus μ_m, $m \geq 1$ converges to μ uniformly over $\mathcal{F}$. μ is obviously a charge on $\mathcal{F}$. Since the convergence is uniform, μ is bounded. Since $\|\cdot\|$ is a norm, $|\|\mu_n - \mu\| - \|\mu_m - \mu\|| \leq \|(\mu_n - \mu) - (\mu_m - \mu)\| = \|\mu_n - \mu_m\|$, it follows that $\lim_{m \to \infty} \|\mu_m - \mu\| = 0$. Hence $(\text{ba}(\Omega, \mathcal{F}), \leq, \|\cdot\|)$ is a Banach lattice. □

The above theorem in conjunction with Theorem 1.5.4 brings into focus various aspects of bounded charges. Of particular interest, are the charges μ^+, μ^-, and $|\mu|$ associated with $\mu \in \text{ba}(\Omega, \mathcal{F})$. First, we isolate these entities and write formally their computational equivalents.

Let $\mu \in \text{ba}(\Omega, \mathcal{F})$. Then $\mu^+ = \mu \vee 0$ and $\mu^- = (-\mu) \vee 0$. By Theorem 2.2.1(5),

$$\mu^+(F) = \text{Sup}\{\mu(B); B \subset F, B \in \mathcal{F}\}, F \in \mathcal{F},$$

and

$$\mu^-(F) = -\text{Inf}\{\mu(B); B \subset F, B \in \mathcal{F}\}, F \in \mathcal{F}.$$

μ^+ and $\mu^- \in \text{ba}(\Omega, \mathcal{F})$ and they are called the positive and negative variations of μ respectively. The charge $|\mu| = \mu^+ + \mu^-$ is called the total variation of μ.

Combining Theorem 1.5.4 and Theorem 2.2.1, we chronicle some salient features of charges in the following theorem for future reference.

2.2.2 Theorem. *Let $\mu \in \text{ba}(\Omega, \mathscr{F})$. Then the following are true.*
(1). $\mu = \mu^+ - \mu^-$ and $\mu^+ \wedge \mu^- = 0$.
(2). $|\mu| = \mu^+ + \mu^- = \mu^+ \vee \mu^- = \mu \vee (-\mu)$. *An important consequence of this is* $|\mu|(F) = \text{Sup}\{\mu(B) - \mu(F-B); B \subset F, B \in \mathscr{F}\}$, $F \in \mathscr{F}$.
(3). *If* $\mu = \lambda - \nu$ *with* $\lambda, \nu \in \text{ba}(\Omega, \mathscr{F})$, $\lambda \geq 0$ *and* $\nu \geq 0$, *then* $\lambda \geq \mu^+$ *and* $\nu \geq \mu^-$.
(4). *If* $\mu = \lambda - \nu$ *with* $\lambda, \nu \in \text{ba}(\Omega, \mathscr{F})$ *and* $\lambda \wedge \nu = 0$, *then* $\lambda = \mu^+$ *and* $\nu = \mu^-$.
(5). $\mu^+ = \frac{1}{2}(|\mu| + \mu)$ *and* $\mu^- = \frac{1}{2}(|\mu| - \mu)$.
(6). *If* $\nu \in \text{ba}(\Omega, \mathscr{F})$ *and* $\mu \leq \nu$, *then* $\mu^+ \leq \nu^+$ *and* $\mu^- \geq \nu^-$.
(7). *If* $\nu \in \text{ba}(\Omega, \mathscr{F})$ *and* $\mu \wedge \nu = 0$, *then* $(\mu + \nu)^+ = \mu^+ + \nu^+$ *and* $|\mu + \nu| = |\mu| + |\nu|$.

2.2.3 Remark. Theorem 2.2.2(1) is precisely Jordan Decomposition theorem for bounded charges. Any bounded charge μ can be expressed as the difference of two positive charges with a certain minimality property in the following sense. If $\mu = \lambda - \nu$ also, where $\lambda, \nu \geq 0$, then $\lambda \geq \mu^+$ and $\nu \geq \mu^-$. We will take up the issue of writing a given charge as a difference of two positive charges in Section 2.5.

In the following theorem, we give an alternative description of $|\mu|$, in addition to the one described in Theorem 2.2.2(2).

2.2.4 Theorem. *Let $\mu \in \text{ba}(\Omega, \mathscr{F})$. Then for any F in $\mathscr{F}$,*

$$|\mu|(F) = \text{Sup} \sum_{i=1}^{n} |\mu(F_i)|,$$

where the supremum is taken over all finite partitions $F_1, F_2, \ldots, F_n$ *of F in $\mathscr{F}$. Further,* $|\mu|(\Omega) \leq 2 \text{ Sup}\{|\mu(F)|; F \in \mathscr{F}\}$.

Proof. Let $F_1, F_2, \ldots, F_n$ be any partition of F in $\mathscr{F}$. Let $I = \{1 \leq i \leq n; \mu(F_i) \geq 0\}$ and $J = \{1 \leq i \leq n; \mu(F_i) < 0\}$. Then

$$\sum_{i=1}^{n} |\mu(F_i)| = \sum_{i \in I} \mu(F_i) - \sum_{j \in J} \mu(F_j)$$

$$= \mu\left(\bigcup_{i \in I} F_i\right) - \mu\left(\bigcup_{j \in J} F_j\right)$$

$$= \mu\left(\bigcup_{i \in I} F_i\right) - \mu\left(F - \bigcup_{i \in I} F_i\right)$$

$$\leq |\mu|(F),$$

by Theorem 2.2.2(2). Consequently, $\text{Sup} \sum_{i=1}^{n} |\mu(F_i)| \leq |\mu|(F)$, where the supremum is taken as described above. On the other hand, let $B \in \mathscr{F}$ and

$B \subset F$. Then $\{B, F-B\}$ is a partition of F in $\mathscr{F}$ and $\mu(B) - \mu(F-B) \leq |\mu(B)| + |\mu(F-B)| \leq \text{Sup} \sum_{i=1}^{n} |\mu(F_i)|$, where the supremum is taken as described above. Therefore, $|\mu|(F) = \text{Sup}\{\mu(B) - \mu(F-B); B \subset F, B \in \mathscr{F}\} \leq \text{Sup} \sum_{i=1}^{n} |\mu(F_i)|$. This completes the proof. □

2.3 MEASURES

In this section, we introduce measures and establish some properties of measures. We also give a set of necessary and sufficient conditions under which a charge is a measure.

2.3.1 Definition. Let $\mathscr{F}$ be a field of subsets of a set Ω. A *measure* on $\mathscr{F}$ is any map μ from $\mathscr{F}$ to $[-\infty, \infty]$ having the following properties.
 (i). $\mu(\emptyset) = 0$.
 (ii). If F_n, $n \geq 1$ is a sequence of pairwise disjoint sets in $\mathscr{F}$ with $\bigcup_{n \geq 1} F_n$ in $\mathscr{F}$, then

$$\mu\left(\bigcup_{n \geq 1} F_n\right) = \sum_{n \geq 1} \mu(F_n).$$

Any measure is, obviously, a charge. If μ is real valued, we call μ a real measure. If μ is bounded on $\mathscr{F}$, we call μ a bounded measure. If $\mu \geq 0$, we call μ a positive measure.

2.3.2 Proposition. *Let μ be a charge on a field $\mathscr{F}$ of subsets of a set Ω. Then the following statements are true.*
(1). *μ is a measure on $\mathscr{F}$ if and only if $\lim_{n \to \infty} \mu(A_n) = \mu(A)$ whenever A_n, $n \geq 1$ is an increasing sequence of sets in $\mathscr{F}$ with $A = \bigcup_{n \geq 1} A_n$ in $\mathscr{F}$.*
(2). *Let μ be real. The following are equivalent.*
 (i). *μ is a measure on $\mathscr{F}$.*
 (ii). *$\lim_{n \to \infty} \mu(A_n) = \mu(A)$ whenever A_n, $n \geq 1$ is a decreasing sequence of sets in $\mathscr{F}$ with $\bigcap_{n \geq 1} A_n = A$ in $\mathscr{F}$.*
 (iii). *$\lim_{n \to \infty} \mu(A_n) = 0$ whenever A_n, $n \geq 1$ is a decreasing sequence of sets in $\mathscr{F}$ with $\bigcap_{n \geq 1} A_n = \emptyset$.*

Proof. (1). Suppose μ is a measure on $\mathscr{F}$. Let A_n, $n \geq 1$ be an increasing sequence of sets in $\mathscr{F}$ with $A = \bigcup_{n \geq 1} A_n$ in $\mathscr{F}$. Let $B_1 = A_1$, $B_2 = A_2 - A_1, \ldots, B_n = A_n - A_{n-1}, \ldots$. Then B_n, $n \geq 1$ is a sequence of pairwise disjoint sets in $\mathscr{F}$ and

$$\bigcup_{n \geq 1} A_n = \bigcup_{n \geq 1} B_n.$$

So,
$$\mu(A) = \mu\left(\bigcup_{n\geq 1} A_n\right) = \mu\left(\bigcup_{n\geq 1} B_n\right)$$
$$= \sum_{n\geq 1} \mu(B_n) = \lim_{m\to\infty} \sum_{n=1}^{m} \mu(B_n)$$
$$= \lim_{m\to\infty} \mu\left(\bigcup_{n=1}^{m} B_n\right) = \lim_{m\to\infty} \mu(A_m).$$

The converse is simple to prove.

(2). (i)$\Rightarrow$(ii). Let A_n, $n \geq 1$ be a decreasing sequence of sets in $\mathcal{F}$ with $\bigcap_{n\geq 1} A_n = A$ in $\mathcal{F}$. Then A_n^c, $n \geq 1$ is an increasing sequence of sets in $\mathcal{F}$ with $\bigcup_{n\geq 1} A_n^c = A^c$. Let $B_1 = A_1^c$, $B_2 = A_2^c - A_1^c, \ldots, B_n = A_n^c - A_{n-1}^c, \ldots$. Then B_n, $n \geq 1$ is a sequence of pairwise disjoint sets in $\mathcal{F}$ with
$$\bigcup_{n\geq 1} B_n = \bigcup_{n\geq 1} A_n^c.$$

Consequently,
$$\mu(\Omega) - \mu(A) = \mu(A^c) = \mu\left(\bigcup_{n\geq 1} A_n^c\right) = \mu\left(\bigcup_{n\geq 1} B_n\right)$$
$$= \sum_{n\geq 1} \mu(B_n) = \lim_{n\to\infty} \sum_{i=1}^{n} \mu(B_i)$$
$$= \lim_{n\to\infty} \mu(A_n^c) = \lim_{n\to\infty} [\mu(\Omega) - \mu(A_n)].$$

Hence $\lim_{n\to\infty} \mu(A_n) = \mu(A)$. (ii)$\Rightarrow$(iii). This is obvious. (iii)$\Rightarrow$(i). Let B_n, $n \geq 1$ be a sequence of pairwise disjoint sets in $\mathcal{F}$ such that $\bigcup_{n\geq 1} B_n \in \mathcal{F}$. Let $A_n = \bigcup_{m\geq n} B_m$, $n \geq 1$. Then A_n, $n \geq 1$ is a decreasing sequence of sets in $\mathcal{F}$ with $\bigcap_{n\geq 1} A_n = \varnothing$. Therefore,
$$0 = \lim_{n\to\infty} \mu(A_n) = \lim_{n\to\infty} \mu\left(\bigcup_{m\geq n} B_m\right) = \lim_{n\to\infty} \mu\left(\bigcup_{k\geq 1} B_k - \bigcup_{m=1}^{n-1} B_m\right)$$
$$= \lim_{n\to\infty} \left[\mu\left(\bigcup_{k\geq 1} B_k\right) - \sum_{m=1}^{n-1} \mu(B_m)\right].$$

Hence
$$\mu\left(\bigcup_{k\geq 1} B_k\right) = \sum_{m\geq 1} \mu(B_m). \qquad \square$$

Note that Theorem 2.3.2(2) is not valid if μ is not real.

We now give a useful sufficient condition under which a charge is a measure. We begin with a definition.

2.3.3 Definition. A collection $\mathscr{C}$ of subsets of a set Ω is called a *compact class* if it has the following property: if C_n, $n \geq 1$ is a sequence of sets in $\mathscr{C}$ with $\bigcap_{n \geq 1} C_n = \varnothing$, then there exists $m \geq 1$ such that $\bigcap_{n=1}^{m} C_n = \varnothing$.

2.3.4 Theorem. *Let $\mathscr{F}$ be a field of subsets of a set Ω and $\mathscr{C}$ a compact class contained in $\mathscr{F}$. Let μ be a positive bounded charge on $\mathscr{F}$ having the following approximation property:*

$$\mu(F) = \mathrm{Sup}\{\mu(C); C \subset F, C \in \mathscr{C}\}, F \in \mathscr{F}.$$

Then μ is a measure on $\mathscr{F}$.

Proof. In view of Proposition 2.3.2(2), it suffices to show that $\lim_{n \to \infty} \mu(A_n) = 0$ for any decreasing sequence A_n, $n \geq 1$ of sets in $\mathscr{F}$ with $\bigcap_{n \geq 1} A_n = \varnothing$. Let $\varepsilon > 0$. For each $n \geq 1$, there exists C_n in $\mathscr{C}$, $C_n \subset A_n$ such that $\mu(A_n) \leq \mu(C_n) + \varepsilon/2^n$. Since $\bigcap_{n \geq 1} A_n = \varnothing$, $\bigcap_{n \geq 1} C_n = \varnothing$. So, there exists $m \geq 1$ such that $\bigcap_{n=1}^{m} C_n = \varnothing$. Consequently,

$$\mu(A_m) = \mu\left(\bigcap_{n=1}^{m} A_n\right) \leq \mu\left(\bigcup_{n=1}^{m}(A_n - C_n)\right)$$

$$\leq \sum_{n=1}^{m} \mu(A_n - C_n) = \sum_{n=1}^{m}[\mu(A_n) - \mu(C_n)] < \varepsilon.$$

So, for every $n \geq m$, $\mu(A_n) < \varepsilon$. Hence $\lim_{n \to \infty} \mu(A_n) = 0$. □

2.3.5 Examples

(1). Let Ω be any infinite set and $\mathscr{F}$ the finite-cofinite field on Ω. Let μ on $\mathscr{F}$ be defined by

$$\mu(A) = 0, \quad \text{if A is finite,}$$
$$= 1, \quad \text{if A is cofinite.}$$

If Ω is countable, then μ is a charge on $\mathscr{F}$ but not a measure. If Ω is uncountable, then μ is a measure on $\mathscr{F}$. In fact, in this case, every charge on $\mathscr{F}$ is a measure.

(2). A standard example of a measure is the *Lebesgue measure* λ on the Borel σ-field $\mathscr{B}$ of the real line $\mathbf{R}$. This is the measure on $\mathscr{B}$ such that $\lambda\{(a, b)\} = b - a$ for every $-\infty < a < b < \infty$.

(3). Let X be a compact Hausdorff totally disconnected space and $\mathscr{C}$ the field of all clopen subsets of X. If A_n, $n \geq 1$ is a sequence of pairwise disjoint sets in $\mathscr{C}$ such that $\bigcup_{n \geq 1} A_n \in \mathscr{C}$, then $A_n = \varnothing$ for every $n \geq m$ for some $m \geq 1$. Consequently, every charge on $\mathscr{C}$ is a measure.

2.4 THE SPACE OF ALL BOUNDED MEASURES, ca($\Omega, \mathscr{F}$)

In this section, we take up the study of the space ca($\Omega, \mathscr{F}$) of all bounded measures on a field $\mathscr{F}$ of subsets of a set Ω. The main result of this section is that ca($\Omega, \mathscr{F}$) is a normal vector sublattice of ba($\Omega, \mathscr{F}$). First, we need a proposition.

2.4.1 Proposition. *Let* $\mu, \nu \in$ ca($\Omega, \mathscr{F}$). *Then* $\mu^+, \mu^-, |\mu|, \mu \vee \nu$ *and* $\mu \wedge \nu \in$ ca($\Omega, \mathscr{F}$).

Proof. First, we show that μ^+ is a measure. Let A_n, $n \geq 1$ be a sequence of pairwise disjoint sets in $\mathscr{F}$ such that $\bigcup_{n \geq 1} A_n \in \mathscr{F}$. Let $B \subset \bigcup_{n \geq 1} A_n$ and $B \in \mathscr{F}$. Let $B_n = B \cap A_n$, $n \geq 1$. Then

$$\mu(B) = \mu\left(\bigcup_{n \geq 1} B_n\right) = \sum_{n \geq 1} \mu(B_n) \leq \sum_{n \geq 1} \mu^+(A_n).$$

Taking supremum over all $B \in \mathscr{F}$ with $B \subset \bigcup_{n \geq 1} A_n$, we obtain $\mu^+(\bigcup_{n \geq 1} A_n) \leq \sum_{n \geq 1} \mu^+(A_n)$. Since μ^+ is a positive charge on $\mathscr{F}$, $\sum_{n \geq 1} \mu^+(A_n) \leq \mu^+(\bigcup_{n \geq 1} A_n)$. See Proposition 2.1.2(viii). Hence $\mu^+(\bigcup_{n \geq 1} A_n) = \sum_{n \geq 1} \mu^+(A_n)$. Thus $\mu^+ \in$ ca($\Omega, \mathscr{F}$). By a similar argument, we can show that $\mu^- \in$ ca($\Omega, \mathscr{F}$). So, $|\mu| = \mu^+ + \mu^- \in$ ca($\Omega, \mathscr{F}$). Now, we show that $\mu \vee \nu$ and $\mu \wedge \nu$ are measures. By Proposition 2.3.2(2), it suffices to show that $\lim_{m \to \infty} (\mu \vee \nu)(A_m) = 0 = \lim_{n \to \infty} (\mu \wedge \nu)(A_n)$ whenever A_n, $n \geq 1$ is a decreasing sequence of sets in $\mathscr{F}$ with $\bigcap_{n \geq 1} A_n = \varnothing$. By Theorem 1.5.4(14),

$$-|\mu| - |\nu| \leq \mu \wedge \nu \leq \mu \vee \nu \leq |\mu| + |\nu|.$$

Since $|\mu|, |\nu| \in$ ca($\Omega, \mathscr{F}$), we have $\lim_{n \to \infty} |\mu|(A_n) = 0 = \lim_{n \to \infty} |\nu|(A_n)$. From this, it follows that the desired limits above are each equal to zero. Hence $\mu \vee \nu$ and $\mu \wedge \nu$ are measures. □

2.4.2 Theorem. ca($\Omega, \mathscr{F}$) *is a normal vector sublattice of* ba($\Omega, \mathscr{F}$). *In particular,* ca($\Omega, \mathscr{F}$) *is a closed subspace of* ba($\Omega, \mathscr{F}$).

Proof. By Proposition 2.4.1, it follows that ca($\Omega, \mathscr{F}$) is a vector sublattice of ba($\Omega, \mathscr{F}$). We now show that it is normal. Let $\mu \in$ ca($\Omega, \mathscr{F}$) and $\nu \in$ ba($\Omega, \mathscr{F}$) be such that $0 \leq |\nu| \leq |\mu|$. Since $|\mu|$ is a measure and $|\nu(F)| \leq |\nu|(F)$ for every F in $\mathscr{F}$, ν is a measure by Proposition 2.3.2(2). Hence $\nu \in$ ca($\Omega, \mathscr{F}$).

Let $\{\mu_\alpha; \alpha \in \Gamma\} \subset$ ca($\Omega, \mathscr{F}$) be such that $\bigvee_{\alpha \in \Gamma} \mu_\alpha$ exists in ba($\Omega, \mathscr{F}$). Let $\tau = \bigvee_{\alpha \in \Gamma} \mu_\alpha$. We show that $\tau \in$ ca($\Omega, \mathscr{F}$). Let $\mathscr{A}$ be the collection of all finite subsets of Γ. For C, D in $\mathscr{A}$, say $C \geq^* D$ if $C \supset D$. For C in $\mathscr{A}$, let $\mu_C = \bigvee_{\alpha \in C} \mu_\alpha$. Then μ_C, $C \in \mathscr{A}$ is an increasing net of measures whose pointwise limit is τ, i.e.

$$\tau(F) = \lim_{C \in \mathscr{A}} \mu_C(F), \quad F \in \mathscr{F}.$$

2. CHARGES

See the proof of Theorem 2.2.1(9). We show that τ is a measure. Since $\bigvee_{\alpha \in \Gamma} \mu_\alpha = \bigvee_{C \in \mathcal{A}} \mu_C = \tau$, let us assume that Γ itself is a directed set and μ_α, $\alpha \in \Gamma$ is an increasing net of measures. Let $d = \sup_{\alpha \in \Gamma} \mu_\alpha(\Omega)$. Since $\mu_\alpha \leq \tau$ for every α in Γ, $d < \infty$. Let α_n, $n \geq 1$ be a sequence in Γ such that $\lim_{n \to \infty} \mu_{\alpha_n}(\Omega) = d$. We can assume, without loss of generality, that $\mu_{\alpha_1} \leq \mu_{\alpha_2} \leq \cdots$. Let $\tau^*(F) = \lim_{n \to \infty} \mu_{\alpha_n}(F)$, $F \in \mathcal{F}$. We show that τ^* is a measure on $\mathcal{F}$. τ^* is obviously a bounded charge on $\mathcal{F}$. Let A_n, $n \geq 1$ be any increasing sequence of sets in $\mathcal{F}$ with $\bigcup_{n \geq 1} A_n = A \in \mathcal{F}$. In view of Proposition 2.3.2(1), it suffices to show that $\lim_{k \to \infty} \tau^*(A_k) = \tau^*(A)$. At this juncture we assume that each μ_α is a positive measure. Note that

$$\lim_{k \to \infty} \tau^*(A_k) = \sup_{k \geq 1} \tau^*(A_k) = \sup_{k \geq 1} \sup_{n \geq 1} \mu_{\alpha_n}(A_k)$$

$$= \sup_{n \geq 1} \sup_{k \geq 1} \mu_{\alpha_n}(A_k) = \sup_{n \geq 1} \mu_{\alpha_n}(A) = \tau^*(A).$$

This shows that τ^* is a measure on $\mathcal{F}$. Next, we claim that $\tau^* = \tau$. Let α in Γ be arbitrary. We show that $\mu_\alpha \leq \tau^*$. Note that

$$\mu_\alpha \vee \tau^* = \mu_\alpha \vee \left(\bigvee_{n \geq 1} \mu_{\alpha_n} \right) = \bigvee_{n \geq 1} (\mu_\alpha \vee \mu_{\alpha_n}).$$

If $(\mu_\alpha \vee \tau^*)(\Omega) > d$, then $(\mu_\alpha \vee \mu_{\alpha_n})(\Omega) > d$ for some $n \geq 1$. This contradicts the definition of d. So, $(\mu_\alpha \vee \tau^*)(\Omega) \leq d = \tau^*(\Omega) \leq (\mu_\alpha \vee \tau^*)(\Omega)$. So, $(\mu_\alpha \vee \tau^*)(\Omega) = \tau^*(\Omega)$. Since $\tau^* \leq \mu_\alpha \vee \tau^*$ and $\tau^*(\Omega) = (\mu_\alpha \vee \tau^*)(\Omega)$, it follows that $\tau^* = \mu_\alpha \vee \tau^*$! (Check this.) This implies that $\mu_\alpha \leq \tau^*$. Since this is true for every α in Γ, $\bigvee_{\alpha \in \Gamma} \mu_\alpha = \tau \leq \tau^*$. On the other hand, since $\mu_{\alpha_n} \leq \tau$ for every $n \geq 1$, $\bigvee_{n \geq 1} \mu_{\alpha_n} = \tau^* \leq \tau$. Hence $\tau = \tau^*$. This shows that τ is a measure.

In the above, we have shown that τ is a measure under the assumption that each μ_α is a positive measure. Now, we treat the general case. Choose and fix α_0 in Γ. Let $\nu_\alpha = \mu_\alpha \vee \mu_{\alpha_0} + |\mu_{\alpha_0}|$, $\alpha \in \Gamma$. Each ν_α is a positive measure since $\nu_\alpha \geq \mu_{\alpha_0} + |\mu_{\alpha_0}| = 2\mu_{\alpha_0}^+ \geq 0$. Observe also that

$$\bigvee_{\alpha \in \Gamma} \nu_\alpha = \bigvee_{\alpha \in \Gamma} [(\mu_\alpha \vee \mu_{\alpha_0}) + |\mu_{\alpha_0}|]$$

$$= \left[\bigvee_{\alpha \in \Gamma} (\mu_\alpha \vee \mu_{\alpha_0}) \right] + |\mu_{\alpha_0}|, \qquad \text{by Theorem 1.5.4(1)}$$

$$= \tau + |\mu_{\alpha_0}|.$$

By what we have proved above, $\tau + |\mu_{\alpha_0}|$ is a measure. Since $|\mu_{\alpha_0}|$ is a measure, it follows that τ is a measure. Hence $\mathrm{ca}(\Omega, \mathcal{F})$ is a normal sublattice of $\mathrm{ba}(\Omega, \mathcal{F})$.

It follows from Theorem 1.5.19 that $\mathrm{ca}(\Omega, \mathcal{F})$ is a closed sublattice of $\mathrm{ba}(\Omega, \mathcal{F})$. □

2.4.3 Corollary. $ca(\Omega, \mathcal{F})$ is a boundedly complete vector lattice and also a Banach lattice in the usual norm.

2.4.4 Remark. Given any field $\mathcal{F}$ of subsets of a set Ω, one can find a set X and a σ-field $\mathcal{A}$ on X such that $ba(\Omega, \mathcal{F})$ and $ca(X, \mathcal{A})$ are isometrically isomorphic as Banach lattices, i.e. there is a linear map T from $ba(\Omega, \mathcal{F})$ onto $ca(X, \mathcal{A})$ such that $\|T(\mu)\| = \|\mu\|$, $\mu \in ba(\Omega, \mathcal{F})$ and $T(\mu) \geq 0$ whenever $\mu \geq 0$. This can be proved using the Stone Representation theorem for Boolean algebras given in Section 1.4.

2.5 JORDAN DECOMPOSITION THEOREM

Jordan Decomposition theorem for bounded charges has been discussed in Remark 2.2.3. It essentially says that every bounded charge can be written as a difference of two positive charges in a minimal way. The main theorem of this section gives a simple necessary and sufficient condition for such a decomposition to prevail for charges not necessarily bounded. First, we introduce the lattice operations $\vee$ and $\wedge$ for general charges.

2.5.1 Definitions. Let μ and ν be two charges on a field $\mathcal{F}$ of subsets of a set Ω such that either both μ and ν avoid the value $-\infty$ or both avoid the value $+\infty$. Define the set functions λ and τ on $\mathcal{F}$ by

$$\lambda(F) = \mathrm{Sup}\{\mu(E) + \nu(F - E); E \subset F, E \in \mathcal{F}\}, \quad F \in \mathcal{F},$$

and

$$\tau(F) = \mathrm{Inf}\{\mu(E) + \nu(F - E); E \subset F, E \in \mathcal{F}\}, \quad F \in \mathcal{F}.$$

The above set functions were defined for bounded charges in Theorem 2.2.1(4) and (6). We used the same formula for general charges in the above definition. We denote λ by $\mu \vee \nu$ and τ by $\mu \wedge \nu$. This is consistent with the notation used for bounded charges.

2.5.2 Proposition. *Let μ and ν be two charges on a field $\mathcal{F}$ of subsets of a set Ω such that either both μ and ν avoid the value $-\infty$ or both avoid the value $+\infty$. Then the set functions λ $(=\mu \vee \nu)$ and τ $(=\mu \wedge \nu)$ defined above are charges on $\mathcal{F}$.*

Proof. The proof given for Theorem 2.2.1(4) and (6) carries through essentially here. □

The following theorem is the main result of this section.

2.5.3 General Jordan Decomposition Theorem. *Let $\mathscr{F}$ be a field of subsets of a set Ω. Let μ be a charge on $\mathscr{F}$. Define μ^+ and μ^- by*

$$\mu^+(F) = \text{Sup}\{\mu(E); E \subset F, E \in \mathscr{F}\}, \quad F \in \mathscr{F},$$

and

$$\mu^-(F) = -\text{Inf}\{\mu(E); E \subset F, E \in \mathscr{F}\}, \quad F \in \mathscr{F}.$$

Then the following statements are true.
(1). *μ^+ and μ^- are positive charges on $\mathscr{F}$.*
(2). *If μ does not take the value $+\infty$, then $\mu^+ - \mu = \mu^-$.*
(3). *If μ does not take the value $-\infty$, then $\mu + \mu^- = \mu^+$.*
(4). *If μ does not take the value $+\infty$ and $\mu_1 - \mu = \mu_2$ for some positive charges μ_1, μ_2 on $\mathscr{F}$, then $\mu_1 \geq \mu^+$ and $\mu_2 \geq \mu^-$.*
(5). *If μ does not take the value $-\infty$ and $\mu + \lambda_1 = \lambda_2$ for some positive charges λ_1, λ_2 on $\mathscr{F}$, then $\lambda_1 \geq \mu^-$ and $\lambda_2 \geq \mu^+$.*
(6). *$\mu = \mu^+ - \mu^-$ if and only if μ is either bounded below or bounded above. More generally, we can write $\mu = \mu_1 - \mu_2$ for some positive charges μ_1, μ_2 on $\mathscr{F}$ if and only if μ is either bounded below or bounded above.*
(7). *$\mu^+ \wedge \mu^- = 0$ if and only if μ is either bounded below or bounded above.*
(8). *If μ_1 and μ_2 are positive charges on $\mathscr{F}$ satisfying $\mu = \mu_1 - \mu_2$ and $\mu_1 \wedge \mu_2 = 0$, then $\mu_1 = \mu^+$ and $\mu_2 = \mu^-$.*
(9). *If μ is a real charge, then $\mu = \mu^+ - \mu^-$ holds if and only if μ is bounded. In such a case, both μ^+ and μ^- are bounded. More generally, if μ is a real charge, then we can write $\mu = \mu_1 - \mu_2$ for some positive charges μ_1 and μ_2 on $\mathscr{F}$ if and only if μ is bounded.*

Proof. (1) follows from Proposition 2.5.2 if we observe that $\mu^+ = \mu \vee 0$ and $\mu^- = (-\mu) \vee 0$ as per Definition 2.5.1.
(2). Let $F \in \mathscr{F}$. Suppose $\mu(F) = -\infty$. Then, from the definition of μ^-, $\mu^-(F) = \infty$. So, $\mu^+(F) - \mu(F) = \infty = \mu^-(F)$. Suppose $\mu(F) > -\infty$. By the given hypothesis, $-\infty < \mu(F) < \infty$. Consequently, $-\infty < \mu(E) < \infty$ for any E in $\mathscr{F}$ such that $E \subset F$. See Proposition 2.1.2(vii). So,

$$\begin{aligned}\mu^+(F) - \mu(F) &= \text{Sup}\{\mu(E); E \subset F, E \in \mathscr{F}\} - \mu(F) \\ &= \text{Sup}\{\mu(E) - \mu(F); E \subset F, E \in \mathscr{F}\} \\ &= \text{Sup}\{-\mu(F - E); E \subset F, E \in \mathscr{F}\} \\ &= -\text{Inf}\{\mu(C); C \subset F, C \in \mathscr{F}\} \\ &= \mu^-(F).\end{aligned}$$

This completes the proof.
(3). This can be proved as above.

(4). Since $\mu_1 - \mu = \mu_2$ and μ_1 and μ_2 are positive, it follows that $\mu_1 \geq \mu$. So, for any F in $\mathcal{F}$, $\mu^+(F) = \text{Sup}\{\mu(E); E \subset F, E \in \mathcal{F}\} \leq \mu_1(F)$. Hence $\mu^+ \leq \mu_1$. For the second part, observe that $\mu^+ - \mu \leq \mu_1 - \mu = \mu_2$. By (2), $\mu^+ - \mu = \mu^-$. This completes the proof.

(5). The proof is analogous to that of (4).

(6). Suppose μ is bounded above. By (2), $\mu^+ - \mu = \mu^-$. Note that μ^+ is a bounded charge. Consequently, $-\mu = \mu^- - \mu^+$ or $\mu = \mu^+ - \mu^-$. A similar argument works when μ is bounded below. Conversely, if $\mu = \mu^+ - \mu^-$, then either μ^+ is bounded or μ^- is bounded. In the former case, μ is bounded above and in the later case, μ is bounded below. The more general version can be established using (4).

(7). Suppose μ is bounded below. We show that $\mu^+ \wedge \mu^- = 0$. Since μ is bounded below, μ^- is a positive bounded charge. So, for any F in $\mathcal{F}$,

$$(\mu^+ \wedge \mu^-)(F) = \text{Inf}\{\mu^+(E) + \mu^-(F-E); E \subset F, E \in \mathcal{F}\}$$

$$= \text{Inf}\{\mu(E) + \mu^-(E) + \mu^-(F-E); E \subset F, E \in \mathcal{F}\},$$

(by (3))

$$= \text{Inf}\{\mu(E) + \mu^-(F); E \subset F, E \in \mathcal{F}\}$$

$$= \text{Inf}\{\mu(E); E \subset F, E \in \mathcal{F}\} + \mu^-(F),$$

(since μ^- is bounded)

$$= -\mu^-(F) + \mu^-(F) = 0.$$

If μ is bounded above, a similar argument shows that $\mu^+ \wedge \mu^- = 0$.

Conversely, suppose μ is neither bounded below nor bounded above. We show that $\mu^+ \wedge \mu^- \neq 0$. Since μ cannot take both the values $+\infty$ and $-\infty$, assume, without loss of generality, that μ does not take the value $-\infty$. Note that $\mu^-(\Omega) = \infty$. By (3), $\mu + \mu^- = \mu^+$. So,

$$(\mu^+ \wedge \mu^-)(\Omega) = [(\mu + \mu^-) \wedge \mu^-](\Omega)$$

$$= \text{Inf}\{(\mu + \mu^-)(F) + \mu^-(F^c); F \in \mathcal{F}\}$$

$$= \text{Inf}\{\mu(F) + \mu^-(\Omega); F \in \mathcal{F}\} = \infty.$$

This completes the proof.

(8). If $\mu = \mu_1 - \mu_2$, where μ_1 and μ_2 are positive charges, then μ is either bounded below or bounded above. See (6). Assume μ is bounded above. This implies that μ_1 is a bounded charge. Since $\mu = \mu_1 - \mu_2 = \mu^+ - \mu^-$, $\mu_1 - \mu^+ - \mu_2 = -\mu^- \leq 0$. Therefore, $\mu_1 - \mu^+ \leq \mu_2$. By (4), $\mu_1 \geq \mu^+$. So, $0 \leq \mu_1 - \mu^+ \leq \mu_2$. Since $\mu_1 \wedge \mu_2 = 0$ and $0 \leq \mu_1 - \mu^+ \leq \mu_1$, it follows that $0 \leq (\mu_1 - \mu^+) \leq \mu_1 \wedge \mu_2 = 0$. Hence $\mu_1 = \mu^+$. The other equality follows easily now.

(9). If μ is a real charge and $\mu = \mu_1 - \mu_2$, where μ_1 and μ_2 are positive charges, then μ_1 and μ_2 are bounded. So, μ is bounded. If μ is bounded, of course, we can write μ as a difference of two positive charges. This completes the proof of the theorem. □

Some comments are in order on the above theorem.

2.5.4 Remarks. (i). The charge μ described in Example 2.1.3(2) is neither bounded above nor bounded below. There is no way we can write μ as a difference of two positive charges. For this charge, $\mu^+ - \mu = \mu^-$ and $\mu + \mu^- = \mu^+$. μ^+ and μ^- work out as follows.

$\mu^+(A) = n$,   if A is finite and has n elements,

$\quad\quad\quad = \infty$,   if A is cofinite.

$\mu^-(A) = 0$,   if A is finite,

$\quad\quad\quad = \infty$,   if A is cofinite.

Observe also that $(\mu^+ \wedge \mu^-)(\Omega) = \infty$.

(ii). Theorem 2.5.3 goes through *in toto* if we replace the word "charge" by "measure".

An interesting result emerges for measures on σ-fields in contrast to the Remark 2.5.4(i), i.e. every measure on a σ-field can be written as a difference of two positive measures. First, we need a lemma.

2.5.5 Lemma. *Let μ be a measure on a σ-field $\mathfrak{A}$ of subsets of a set Ω. Then μ is either bounded below or bounded above.*

Proof. First, assume that μ is real valued. In this case, we show that μ is bounded. Suppose μ is unbounded. By Theorem 2.1.6, there exists a sequence B_n, $n \geq 1$ of pairwise disjoint sets in $\mathfrak{A}$ such that $|\mu(B_i)| \geq 1$ for every $i \geq 1$. Then, either there exists a sequence $n_1 < n_2 < \cdots$ such that $\mu(B_{n_i}) \geq 1$ for every $i \geq 1$ or there exists a sequence $k_1 < k_2 < \cdots$ such that $\mu(B_{k_i}) \leq -1$ for every $i \geq 1$. Assume that the former holds. Then $\mu(\bigcup_{i \geq 1} B_{n_i}) = \infty$ contradicting the fact that μ is real valued. If the latter holds, we do still get a contradiction. Hence, if μ is a real valued measure, then it is bounded.

In the general case, observe that μ takes values either in $(-\infty, \infty]$ or in $[-\infty, \infty)$. Assume that μ takes values in $(-\infty, \infty]$. Then we show that μ is bounded below. Suppose μ is unbounded below. We find A_1 in $\mathfrak{A}$ such that $\mu(A_1) \leq -1$. Since μ is real valued on $A_1 \cap \mathfrak{A}$, by what we have proved above, μ is bounded on A_1. Consequently, μ is unbounded below on A_1^c. So, we can find A_2 in $\mathfrak{A}$ such that $A_2 \subset A_1^c$ and $\mu(A_2) \leq -1$. By the same reasoning given above, we can show that μ is unbounded below on $A_1^c - A_2$.

Continuing this way, we obtain a sequence A_n, $n \geq 1$ of pairwise disjoint sets in $\mathfrak{A}$ such that $\mu(A_n) \leq -1$ for every $n \geq 1$. So, $\mu(\bigcup_{n \geq 1} A_n) = -\infty$. This is a contradiction. Hence μ is bounded below. This completes the proof. $\square$

2.5.6 Jordan Decomposition Theorem for Measures on σ-fields. *Let μ be a measure on a σ-field $\mathfrak{A}$ of subsets of a set Ω. Then we can write*

$$\mu = \mu^+ - \mu^-$$

with the property that $\mu^+ \wedge \mu^- = 0$. Further, the above decomposition has the following optimality property: if $\mu = \mu_1 - \mu_2$, where μ_1 and μ_2 are positive measures on $\mathfrak{A}$, then $\mu_1 \geq \mu^+$ and $\mu_2 \geq \mu^-$.

Proof. By Lemma 2.5.5, μ is either bounded below or bounded above. Remark 2.5.4(ii) completes the proof. $\square$

2.5.7 Remark. The above theorem is not valid for charges on σ-fields.

2.6 HAHN DECOMPOSITION THEOREM

In this section, we prove Hahn Decomposition theorem for charges. Hahn Decomposition theorem for measures on σ-fields will be proved in Section 6.1.

2.6.1 Definition. Let $\mathcal{F}$ be a field of subsets of a set Ω and μ a charge on $\mathcal{F}$. Let $\varepsilon > 0$. A partition $\{D, D^c\}$ of Ω with D in $\mathcal{F}$ is said to be a ε-*Hahn decomposition* of μ if the following are satisfied.

$$B \in \mathcal{F}, \quad B \subset D \Rightarrow \mu(B) \leq \varepsilon.$$
$$C \in \mathcal{F}, \quad C \subset D^c \Rightarrow \mu(C) \geq -\varepsilon.$$

2.6.2 Hahn Decomposition Theorem for Charges. *Let μ be a charge on a field $\mathcal{F}$ of subsets of a set Ω which is either bounded below or bounded above. Then for any $\varepsilon > 0$, there exists a ε-Hahn decomposition for μ. If μ is neither bounded below nor bounded above, there exists $\varepsilon > 0$ for which there is no ε-Hahn decomposition of μ.*

Proof. We give a direct proof of this result. This can also be proved using Jordan Decomposition theorem. Assume that μ is bounded below. Let $d = \text{Inf}\{\mu(A); A \in \mathcal{F}\}$. Then d is a finite number. Let $\varepsilon > 0$. We can find D in $\mathcal{F}$ such that $d \leq \mu(D) \leq d + \varepsilon$. This implies that $-\infty < \mu(D) < \infty$ and for any B in $\mathcal{F}$, $B \subset D$, we have $-\infty < \mu(B) < \infty$. So, $d \leq \mu(D - B) = \mu(D) - \mu(B) \leq d + \varepsilon - \mu(B)$. Or, $\mu(B) \leq \varepsilon$. Now, let $C \in \mathcal{F}$ and $C \subset D^c$. Then $C \cap D = \emptyset$. So, $d \leq \mu(C \cup D) = \mu(C) + \mu(D) \leq \mu(C) + d + \varepsilon$. Hence $\mu(C) \geq -\varepsilon$. Thus the first part is proved.

The statement that μ admits ε-Hahn decomposition for every $\varepsilon > 0$ is equivalent to the statement that $\mu^+ \wedge \mu^- = 0$. But this is equivalent to the statement that μ is either bounded below or bounded above by Theorem 2.5.3(7). This proves the second part of the thereom. □

2.6.3 Remark. Exact Hahn-decomposition, i.e. ε-Hahn decomposition for $\varepsilon = 0$, need not exist for every charge. For example, let $\Omega = \{1, 2, 3, \ldots\}$, $\mathscr{F}$ the finite–cofinite field on Ω and

$$\mu(A) = \sum_{n \in A} (-1)^n/2^n, \qquad A \in \mathscr{F}.$$

There is no set D in $\mathscr{F}$ such that $\mu(B) \leq 0$ for every B in $\mathscr{F}$, $B \subset D$ and $\mu(C) \geq 0$ for every C in $\mathscr{F}$, $C \subset D^c$.

CHAPTER 3

Extensions of Charges

One way of obtaining charges on a given field $\mathcal{F}$ of subsets of a set Ω is to look at subcollections $\mathscr{C}$ of $\mathcal{F}$ and set functions ν defined already on $\mathscr{C}$ and see if ν can be extended to $\mathcal{F}$ as a charge. For example, if ν is a 0–1 valued charge on a given sub-field $\mathscr{C}$ of $\mathcal{F}$, we can extend ν from $\mathscr{C}$ to $\mathcal{F}$ as a 0–1 valued charge. The underlying theme of this chapter is to seek extensions in the spirit of the above example. In Section 3.1, we associate a natural functional T on a suitable vector space with a given set function ν on a given class $\mathscr{C}$ of sets and study the interplay between ν and T. In Section 3.2, we introduce the notion of a real partial charge on a collection $\mathscr{C}$ of sets contained in $\mathcal{F}$ and show that these are precisely the set functions on $\mathscr{C}$ for which extensions are possible. In Section 3.3, we study the extension procedure developed by Łos and Marczewski. Miscellaneous useful extensions are dealt with in Section 3.5. Finally, in Section 3.6, we look for a common extension to $\mathcal{F}$ of two set functions defined respectively on two classes of sets $\mathscr{C}$ and $\mathscr{D}$ contained in $\mathcal{F}$.

3.1 REAL VALUED SET FUNCTIONS AND INDUCED FUNCTIONALS

Any real valued function μ on a given class $\mathcal{F}$ of subsets of a set Ω, under certain conditions, gives rise to a natural linear functional on a suitable vector space. We study this correspondence in this section.

3.1.1 Definition. Let $\mathcal{F}$ be a collection of subsets of a set Ω. Let

$$\mathscr{L}(\mathcal{F}) = \left\{ f : \Omega \to R \, ; \, f = \sum_{i=1}^{n} r_i I_{A_i} \text{ for some } A_1, A_2, \ldots, A_n \text{ in } \mathcal{F}, \, r_1, r_2, \ldots, r_n \text{ rational numbers}, n \geq 1 \right\}.$$

It is obvious that $\mathscr{L}(\mathcal{F})$ is a linear space over the field of all rational numbers.

3. EXTENSIONS OF CHARGES

3.1.2 Definition. Let $\mathscr{F}$ be a collection of subsets of a set Ω and μ a real valued function on $\mathscr{F}$. We set

$$T\left(\sum_{i=1}^{n} r_i I_{A_i}\right) = \sum_{i=1}^{n} r_i \mu(A_i),$$

for $A_1, A_2, \ldots, A_n$ in $\mathscr{F}$, $r_1, r_2, \ldots, r_n$ rationals and $n \geq 1$.

We study the interplay between μ and T. First of all, we examine the unambiguity of the definition of T. More precisely, if $\sum_{i=1}^{n} r_i I_{A_i} = \sum_{j=1}^{m} s_j I_{B_j}$ for some $A_1, A_2, \ldots, A_n, B_1, B_2, \ldots, B_m$ in $\mathscr{F}$, $r_1, r_2, \ldots, r_n, s_1, s_2, \ldots, s_m$ rationals, is $\sum_{i=1}^{n} r_i \mu(A_i) = \sum_{j=1}^{m} s_j \mu(B_j)$? If T is well defined, we call T the functional on $\mathscr{L}(\mathscr{F})$ induced by μ. The following proposition provides an answer to the above question.

3.1.3 Proposition. *T is well defined if and only if*

$$\sum_{i=1}^{n} \mu(A_i) = \sum_{j=1}^{m} \mu(B_j)$$

holds for any two finite sequences $A_1, A_2, \ldots, A_n$ and $B_1, B_2, \ldots, B_m$ of not necessarily distinct sets from $\mathscr{F}$ satisfying

$$\sum_{i=1}^{n} I_{A_i} = \sum_{j=1}^{m} I_{B_j}.$$

Proof. In order to show that T is well defined, we have to establish

$$\sum_{i=1}^{k} r_i \mu(C_i) = \sum_{j=1}^{t} s_j \mu(D_j)$$

whenever

$$\sum_{i=1}^{k} r_i I_{C_i} = \sum_{j=1}^{t} s_j I_{D_j}$$

for $C_1, C_2, \ldots, C_k, D_1, D_2, \ldots, D_t$ in $\mathscr{F}$ and $r_1, r_2, \ldots, r_k, s_1, s_2, \ldots, s_t$ rationals. Since $r_1, r_2, \ldots, r_k, s_1, s_2, \ldots, s_t$ are rationals, we can rewrite the equality

$$\sum_{i=1}^{k} r_i I_{C_i} = \sum_{j=1}^{t} s_j I_{D_j} \quad \text{as} \quad \sum_{i=1}^{k} m_i I_{C_i} = \sum_{j=1}^{t} n_j I_{D_j}$$

by cancelling the common denominator of all the given rationals leading to the situation that $m_1, m_2, \ldots, m_k, n_1, n_2, \ldots, n_t$ are integers. This later equality can be rewritten in the form

$$\sum_{i=1}^{n} I_{A_i} = \sum_{j=1}^{m} I_{B_j}. \quad \text{(Why?)}$$

Retracing these steps from

$$\sum_{i=1}^{n} \mu(A_i) = \sum_{j=1}^{m} \mu(B_j),$$

we obtain

$$\sum_{i=1}^{k} r_i \mu(C_i) = \sum_{j=1}^{t} s_j \mu(D_j).$$

The converse is trivial. $\square$

Thus, if T is well defined on the linear space $\mathscr{L}(\mathscr{F})$, then it is a linear functional on $\mathscr{L}(\mathscr{F})$.

Now, we aim to show that T is a well defined linear functional on $\mathscr{L}(\mathscr{F})$ when $\mathscr{F}$ is a field of subsets of a set Ω and μ a real charge on $\mathscr{F}$. Moreover, there is a one-to-one and onto correspondence between linear functionals on the linear space $\mathscr{L}(\mathscr{F})$ and real valued charges on $\mathscr{F}$. For this, we need the following results. (For sets $C_1, C_2, \ldots, C_n$ and $k > n$, we use the convention that $\bigcup_{1 \leq i_1 < i_2 < \cdots < i_k \leq n} \bigcap_{j=1}^{k} C_{i_j} = \varnothing$.)

3.1.4 Lemma. *For any two finite sequences* $A_1, A_2, \ldots, A_n$ *and* $B_1, B_2, \ldots, B_m$ *of sets, the following statements are true.*
(a). $\sum_{i=1}^{n} I_{A_i} \leq \sum_{j=1}^{m} I_{B_j}$ *if and only if*

$$\bigcup_{1 \leq i_1 < i_2 < \cdots < i_k \leq n} \bigcap_{j=1}^{k} A_{i_j} \subset \bigcup_{1 \leq i_1 < i_2 < \cdots < i_k \leq m} \bigcap_{j=1}^{k} B_{i_j}$$

for every $1 \leq k \leq n$.
(b). $\sum_{i=1}^{n} I_{A_i} = \sum_{j=1}^{m} I_{B_j}$ *if and only if*

$$\bigcup_{1 \leq i_1 < i_2 < \cdots < i_k \leq n} \bigcap_{j=1}^{k} A_{i_j} = \bigcup_{1 \leq i_1 < i_2 < \cdots < i_k \leq m} \bigcap_{j=1}^{k} B_{i_j}$$

for every $1 \leq k \leq \max\{m, n\}$.

Moreover, one can always write $\sum_{i=1}^{n} I_{A_i} = \sum_{i=1}^{n} I_{C_i}$, *where*

$$C_k = \bigcup_{1 \leq i_1 < i_2 < \cdots < i_k \leq n} \bigcap_{j=1}^{k} A_{i_j}, \quad 1 \leq k \leq n.$$

Proof. It is obvious that $\sum_{i=1}^{n} I_{A_i}(\omega) \geq k$, where k is a positive integer, if and only if

$$\omega \in \bigcup_{1 \leq i_1 < i_2 < \cdots < i_k \leq n} \bigcap_{j=1}^{k} A_{i_j}.$$

From this, (a) follows.
(b) is a simple consequence of (a).

3. EXTENSIONS OF CHARGES

To prove the last part, we note that $C_1 \supset C_2 \supset \cdots \supset C_n$ and

$$\bigcup_{1 \le i_1 < i_2 < \cdots < i_k \le n} \bigcap_{j=1}^{k} C_{i_j} = C_k = \bigcup_{1 \le i_1 < i_2 < \cdots < i_k \le n} \bigcap_{j=1}^{k} A_{i_j}. \qquad \square$$

Recall that a collection $\mathscr{F}$ of subsets of a set Ω is said to be a lattice if $A \cup B$ and $A \cap B \in \mathscr{F}$ whenever $A, B \in \mathscr{F}$. If $\mathscr{F}$ is a lattice of subsets of a set Ω, then a real valued function μ on $\mathscr{F}$ is said to be a *real modular function* on $\mathscr{F}$ if $\mu(A) + \mu(B) = \mu(A \cup B) + \mu(A \cap B)$ for every A, B in $\mathscr{F}$.

3.1.5 Proposition. *Let $\mathscr{F}$ be a lattice of subsets of a set Ω and μ a real modular function on $\mathscr{F}$. Then for any finite number of sets $A_1, A_2, \ldots, A_n$ from $\mathscr{F}$, we have*

$$\sum_{i=1}^{n} \mu(A_i) = \sum_{k=1}^{n} \mu\left(\bigcup_{1 \le i_1 < i_2 < \cdots < i_k \le n} \bigcap_{j=1}^{k} A_{i_j} \right).$$

In particular, if $\mathscr{F}$ is a field and μ a real charge on $\mathscr{F}$, the above assertion is true.

Proof. The proof given for Proposition 2.1.2(x) is a proof exactly for this assertion. $\qquad \square$

The following theorem shows that T is well defined in some special cases.

3.1.6 Theorem. (1). *Let $\mathscr{F}$ be a lattice of subsets of a set Ω with $\varnothing$ in $\mathscr{F}$. Let μ be a strongly additive real function on $\mathscr{F}$, i.e. μ is a real modular function on $\mathscr{F}$ satisfying the condition that $\mu(\varnothing) = 0$. Then the induced functional T on $\mathscr{L}(\mathscr{F})$ is well defined.*
(2). *Let $\mathscr{F}$ be a field of subsets of a set Ω and μ a real charge on $\mathscr{F}$. Then the induced functional T on $\mathscr{L}(\mathscr{F})$ is defined unambiguously.*

Proof. (1). In view of Proposition 3.1.3, it suffices to prove that $\sum_{i=1}^{n} \mu(A_i) = \sum_{j=1}^{m} \mu(B_j)$ whenever $\sum_{i=1}^{n} I_{A_i} = \sum_{j=1}^{m} I_{B_j}$ for $A_1, A_2, \ldots, A_n$, $B_1, B_2, \ldots, B_m$ in $\mathscr{F}$. By Proposition 3.1.5,

$$\sum_{i=1}^{n} \mu(A_i) = \sum_{k=1}^{n} \mu\left(\bigcup_{1 \le i_1 < i_2 < \cdots < i_k \le n} \bigcap_{j=1}^{k} A_{i_j} \right)$$

and

$$\sum_{j=1}^{m} \mu(B_j) = \sum_{k=1}^{m} \mu\left(\bigcup_{1 \le i_1 < i_2 < \cdots < i_k \le m} \bigcap_{j=1}^{k} B_{i_j} \right).$$

Since $\sum_{i=1}^{n} I_{A_i} = \sum_{j=1}^{m} I_{B_j}$, by Lemma 3.1.4(b),

$$\bigcup_{1 \le i_1 < i_2 < \cdots < i_k \le n} \bigcap_{j=1}^{k} A_{i_j} = \bigcup_{1 \le i_1 < i_2 < \cdots < i_k \le m} \bigcap_{j=1}^{k} B_{i_j}$$

for every $1 \le k \le \max\{m, n\}$. Assume, without loss of generality, that $m \le n$. If $m < k \le n$, it is obvious that

$$\bigcup_{1 \le i_1 < i_2 < \cdots < i_k \le n} \bigcap_{j=1}^{k} A_{i_j} = \emptyset.$$

Since μ is a strongly additive real function on $\mathscr{F}$, it follows that

$$\sum_{i=1}^{n} \mu(A_i) = \sum_{k=1}^{m} \mu\left(\bigcup_{1 \le i_1 < i_2 < \cdots < i_k \le n} \bigcap_{j=1}^{k} A_{i_j}\right) + \sum_{k=m+1}^{n} \mu(\emptyset)$$

$$= \sum_{k=1}^{m} \mu\left(\bigcup_{1 \le i_1 < i_2 < \cdots < i_k \le m} \bigcap_{j=1}^{k} B_{i_j}\right) + 0$$

$$= \sum_{k=1}^{m} \mu(B_k).$$

This proves (1). (2) is a special case of (1). $\square$

The following theorem looks at the converse implication of the above theorem.

3.1.7 Theorem. *For a given collection $\mathscr{F}$ of subsets of a set Ω, let T be a linear functional defined on the linear space $\mathscr{L}(\mathscr{F})$ over the field of rational numbers. For A in $\mathscr{F}$, set $\mu(A) = T(I_A)$. Then the following statements are true.*
(1). *If $\mathscr{F}$ is a field of subsets of Ω, μ is a real charge on $\mathscr{F}$.*
(2). *If $\mathscr{F}$ is a lattice on Ω, μ is a real modular function on $\mathscr{F}$.*
(3). *If $\mathscr{F}$ is an arbitrary class of subsets of Ω, the set function μ on $\mathscr{F}$ satisfies the following condition: $\sum_{i=1}^{n} \mu(A_i) = \sum_{j=1}^{m} \mu(B_j)$ whenever $\sum_{i=1}^{n} I_{A_i} = \sum_{j=1}^{m} I_{B_j}$ for $A_1, A_2, \ldots, A_n, B_1, B_2, \ldots, B_m$ in $\mathscr{F}$.*

Proof. We prove (2). Let $A, B \in \mathscr{F}$. Since $I_A + I_B = I_{A \cup B} + I_{A \cap B}$, we have

$$\mu(A) + \mu(B) = T(I_A) + T(I_B) = T(I_A + I_B) = T(I_{A \cup B} + I_{A \cap B})$$

$$= T(I_{A \cup B}) + T(I_{A \cap B}) = \mu(A \cup B) + \mu(A \cap B).$$

Hence μ is a real modular function on $\mathscr{F}$. Note that if $\emptyset \in \mathscr{F}$, then $\mu(\emptyset) = T(I_\emptyset) = T(0) = 0$. Now (1) is a special case of (2). (3). This is obvious. $\square$

3.1.8 Corollary. (1). *Let $\mathscr{F}$ be a lattice of subsets of a set Ω such that $\emptyset \in \mathscr{F}$. Then there is a one-to-one and onto correspondence between the collection of all strongly additive real functions on $\mathscr{F}$ and linear functionals on $\mathscr{L}(\mathscr{F})$.*
(2). *Let $\mathscr{F}$ be a field of subsets of a set Ω. Then there is a one-to-one and onto correspondence between the collection of all real charges on $\mathscr{F}$ and linear functionals on $\mathscr{L}(\mathscr{F})$.* $\square$

3. EXTENSIONS OF CHARGES

Now, we look at the relationship between positive bounded charges μ on a field $\mathscr{F}$ of subsets of a set Ω and the induced linear functionals T on $\mathscr{L}(\mathscr{F})$.

3.1.9 Theorem. *Let $\mathscr{F}$ be any collection of subsets of a set Ω and μ a real valued function on $\mathscr{F}$. Let T be defined on $\mathscr{L}(\mathscr{F})$ as in Definition 3.1.2. Consider the following statements. (a). μ is positive. (b). $\sum_{i=1}^{n} \mu(A_i) \geq \sum_{j=1}^{m} \mu(B_j)$ whenever $\sum_{i=1}^{n} I_{A_i} \geq \sum_{j=1}^{m} I_{B_j}$ for $A_1, A_2, \ldots, A_n$, $B_1, B_2, \ldots, B_m$ in $\mathscr{F}$. (c). T is well defined and $T(f) \geq 0$ whenever $f \in \mathscr{L}(\mathscr{F})$ and $f \geq 0$. Then the following statements are true.*

(i). *(b) and (c) are equivalent.*

(ii). *(b) implies (a).*

(iii). *(a), (b) and (c) are equivalent if $\mathscr{F}$ is a field on Ω and μ a charge on $\mathscr{F}$.*

Proof. (i). Suppose (b) holds. By Proposition 3.1.3, T is well defined. Let $f \in \mathscr{L}(\mathscr{F})$ and $f \geq 0$. Then $f = \sum_{i=1}^{k} r_i I_{C_i}$ for some $C_1, C_2, \ldots, C_k$ in $\mathscr{F}$ and $r_1, r_2, \ldots, r_k$ rationals. Writing $r_i = m_i/m$, $i = 1, 2, \ldots, k$, where $m_1, m_2, \ldots, m_k$ are integers and m a positive integer, we observe that $mf = \sum_{i=1}^{k} m_i I_{C_i}$. We can rewrite this equality as $mf = \sum_{i=1}^{p} I_{A_i} - \sum_{j=1}^{q} I_{B_j}$ with A_i's and B_j's coming from $\{C_1, C_2, \ldots, C_k\}$ and $p, q \geq 1$. (Why?) If $f \geq 0$, then $\sum_{i=1}^{p} I_{A_i} \geq \sum_{j=1}^{q} I_{B_j}$. Since (b) holds, $\sum_{i=1}^{p} \mu(A_i) \geq \sum_{j=1}^{q} \mu(B_j)$. Consequently,

$$T(mf) = mT(f) = T\left(\sum_{i=1}^{p} I_{A_i} - \sum_{j=1}^{q} I_{B_j}\right)$$

$$= \sum_{i=1}^{p} \mu(A_i) - \sum_{j=1}^{q} \mu(B_j) \geq 0.$$

This proves (c). If (c) holds, it is obvious that (b) holds.

(ii). Let $A \in \mathscr{F}$. Then $I_A + I_A \geq I_A$. Consequently, $\mu(A) + \mu(A) \geq \mu(A)$ which implies that $\mu(A) \geq 0$.

(iii). Using Lemma 3.1.4(a), Proposition 3.1.5 and the monotonicity of the positive charge μ on the field $\mathscr{F}$, one can show that (b) holds whenever (a) holds. Thus (a), (b) and (c) are equivalent when $\mathscr{F}$ is a field on Ω. □

3.1.10 Remark. The implication (a) $\Rightarrow$ (b) is not valid if $\mathscr{F}$ is a lattice on Ω, $\varnothing \in \mathscr{F}$ and μ a strongly additive real function on $\mathscr{F}$. Here is an example. Let $\Omega = \{1, 2, 3\}$ and $\mathscr{F} = \{\varnothing, \{1\}, \{1, 2\}, \{1, 2, 3\}\}$. $\mathscr{F}$ is a lattice on Ω. Any function μ on $\mathscr{F}$ is a modular function. One can take the following function μ on $\mathscr{F}$. $\mu(\varnothing) = 0$, $\mu(\{1\}) = 1$, $\mu(\{1, 2\}) = \frac{1}{2}$ and $\mu(\{1, 2, 3\}) = \frac{1}{4}$. Then μ is a positive strongly additive real function on $\mathscr{F}$. For this μ, (b) of Theorem 3.1.9 fails to hold, even though the associated functional is well defined.

3.2 REAL PARTIAL CHARGES AND THEIR EXTENSIONS

In this section, we show that under some natural conditions we can extend a real valued set function on a collection $\mathscr{C}$ of subsets of a set Ω to any field $\mathscr{F}$ on Ω containing $\mathscr{C}$, as a charge.

3.2.1 Definition. Let $\mathscr{C}$ be a collection of subsets of a set Ω. A real valued function μ on $\mathscr{C}$ is called a *real partial charge* if $\sum_{i=1}^{n} \mu(A_i) = \sum_{j=1}^{m} \mu(B_j)$ whenever $\sum_{i=1}^{n} I_{A_i} = \sum_{j=1}^{m} I_{B_j}$ for $A_1, A_2, \ldots, A_n, B_1, B_2, \ldots, B_m$ in $\mathscr{C}$, i.e. the functional T defined on $\mathscr{L}(\mathscr{C})$ is well defined.

3.2.2 Definition. Let $\mathscr{C}$ be a collection of subsets of a set Ω and μ a positive real valued function on $\mathscr{C}$. μ is said to be a *positive real partial charge* if $\sum_{i=1}^{n} \mu(A_i) \leq \sum_{j=1}^{m} \mu(B_j)$ whenever $\sum_{i=1}^{n} I_{A_i} \leq \sum_{j=1}^{m} I_{B_j}$ for $A_1, A_2, \ldots, A_n, B_1, B_2, \ldots, B_m$ in $\mathscr{C}$.

If μ is a real partial charge on $\mathscr{C}$ and $\mu(C) \geq 0$ for every C in $\mathscr{C}$, it does not follow necessarily that μ is a positive real partial charge on $\mathscr{C}$.

If $\mathscr{C}$ is a field or a ring of sets on Ω, then a real valued function on $\mathscr{C}$ is a real partial charge on $\mathscr{C}$ if and only if it is a real charge on $\mathscr{C}$. If $\mathscr{C}$ is a lattice on Ω with $\varnothing$ in $\mathscr{C}$, then a real valued function on $\mathscr{C}$ is a real partial charge on $\mathscr{C}$ if and only if it is strongly additive.

The following proposition demonstrates that for the extension problem of charges, the given set function has to be at least a partial charge.

3.2.3 Proposition. (a). *Let μ be a real charge on a field $\mathscr{F}$ of subsets of a set Ω. Let $\mathscr{C} \subset \mathscr{F}$. Then the restriction $\bar{\mu}$ of μ to $\mathscr{C}$ is a real partial charge on $\mathscr{C}$.*
(b). *Let μ be a positive bounded charge on a field $\mathscr{F}$ of subsets of a set Ω. Let $\mathscr{C} \subset \mathscr{F}$. Then the restriction $\bar{\mu}$ of μ to $\mathscr{C}$ is a positive real partial charge on $\mathscr{C}$.*

Proof. (a) is a consequence of Proposition 3.1.3 and Theorem 3.1.6(2).
(b) is a consequence of Theorem 3.1.9(iii). □

Now, we come to the extension problem for charges.

3.2.4 Theorem. *Let μ be a real partial charge on a collection $\mathscr{C}$ of subsets of a set Ω. Let $A \subset \Omega$ be such that $A \notin \mathscr{C}$. Then there exists a real partial charge $\bar{\mu}$ on $\mathscr{C} \cup \{A\}$ which is an extension of μ from $\mathscr{C}$ to $\mathscr{C} \cup \{A\}$.*

Proof. Let T be the linear functional on $\mathscr{L}(\mathscr{C})$ induced by μ on $\mathscr{C}$. If $I_A \in \mathscr{L}(\mathscr{C})$, define $\bar{\mu}(A) = T(I_A)$. If $I_A \notin \mathscr{L}(\mathscr{C})$, set $\bar{\mu}(A) =$ any arbitrary but fixed real number. For C in $\mathscr{C}$, set $\bar{\mu}(C) = \mu(C)$. Now, we claim that $\bar{\mu}$ is a real partial charge on $\mathscr{C} \cup \{A\}$. It is obvious that $\bar{\mu}$ is an extension of μ from $\mathscr{C}$ to $\mathscr{C} \cup \{A\}$.

3. EXTENSIONS OF CHARGES   65

If $I_A \in \mathscr{L}(\mathscr{C})$, then $\mathscr{L}(\mathscr{C} \cup \{A\}) = \mathscr{L}(\mathscr{C})$. Consequently, the map $T: \mathscr{L}(\mathscr{C} \cup \{A\}) \to R$ is well defined and the set function $\bar{\mu}$ on $\mathscr{C} \cup \{A\}$ defined by $\bar{\mu}(B) = T(I_B)$ for B in $\mathscr{C} \cup \{A\}$ is a real partial charge on $\mathscr{C} \cup \{A\}$. See Theorem 3.1.7(3). $\bar{\mu}$ is the desired extension in this case.

If $I_A \notin \mathscr{L}(\mathscr{C})$, then $\mathscr{L}(\mathscr{C} \cup \{A\}) = \{f + rI_A; f \in \mathscr{L}(\mathscr{C}), r \text{ rational}\}$. On $\mathscr{L}(\mathscr{C} \cup \{A\})$, we define a functional $\bar{T}$ as follows. For $f + rI_A$ in $\mathscr{L}(\mathscr{C} \cup \{A\})$, $\bar{T}(f + rI_A) = T(f) + rd$, where d is a fixed real number. We claim that $\bar{T}$ is a well defined functional on $\mathscr{L}(\mathscr{C} \cup \{A\})$. Let $f_1 + r_1 I_A = f_2 + r_2 I_A$ for some f_1, f_2 in $\mathscr{L}(\mathscr{C})$ and r_1, r_2 rationals. This implies that $f_1 - f_2 = (r_2 - r_1)I_A$. Since $f_1, f_2 \in \mathscr{L}(\mathscr{C})$, $f_1 - f_2 \in \mathscr{L}(\mathscr{C})$. Since $I_A \notin \mathscr{L}(\mathscr{C})$, it follows that $r_2 - r_1 = 0$. Consequently, $f_1 = f_2$. From this, it follows that $\bar{T}$ is well defined on $\mathscr{L}(\mathscr{C} \cup \{A\})$. It is indeed a linear functional on $\mathscr{L}(\mathscr{C} \cup \{A\})$. Consequently, the set function $\bar{\mu}$ on $\mathscr{C} \cup \{A\}$ defined by

$$\bar{\mu}(B) = \bar{T}(I_B) \quad \text{for B in } \mathscr{C} \cup \{A\}$$

is a real partial charge on $\mathscr{C} \cup \{A\}$. See Theorem 3.1.7(3). This completes the proof. □

The following is the desired extension theorem.

3.2.5 Theorem. *Let μ be a real partial charge on a collection $\mathscr{C}$ of subsets of a set Ω. Let $\mathscr{F}$ be any field on Ω containing $\mathscr{C}$. Then there exists a real charge $\bar{\mu}$ on $\mathscr{F}$ which is an extension of μ from $\mathscr{C}$ to $\mathscr{F}$.*

Proof. We give a proof based on Theorem 3.2.4 and transfinite induction. Let α_0 be the least ordinal corresponding to the cardinal number of the collection $\mathscr{F} - \mathscr{C}$. Write $\mathscr{F} - \mathscr{C} = \{A_\alpha; \alpha \text{ ordinal}, \alpha < \alpha_0\}$. For each ordinal $\alpha < \alpha_0$, let $\mathscr{C}_{\alpha+1} = \mathscr{C}_\alpha \cup \{A_\alpha\}$, where $\mathscr{C}_0 = \mathscr{C}$. If α is a limit ordinal, let $\mathscr{C}_\alpha = \bigcup_{\beta < \alpha} \mathscr{C}_\beta$. If μ_α is a real partial charge on $\mathscr{C}_\alpha$, let $\mu_{\alpha+1}$ be a real partial charge on $\mathscr{C}_{\alpha+1}$ which is an extension of μ_α from $\mathscr{C}_\alpha$ to $\mathscr{C}_{\alpha+1}$. See Theorem 3.2.4. If α is a limit ordinal, μ_α on $\mathscr{C}_\alpha$ is defined as follows. If $A \in \mathscr{C}_\alpha$, then $A \in \mathscr{C}_\beta$ for some $\beta < \alpha$, and we define $\mu_\alpha(A) = \mu_\beta(A)$. It is obvious that if $\alpha < \beta < \alpha_0$, then $\mu_\beta = \mu_\alpha$ on $\mathscr{C}_\alpha$. Note that $\bigcup_{\alpha < \alpha_0} \mathscr{C}_\alpha = \mathscr{F}$. $\bar{\mu}$ on $\mathscr{F}$ is defined as follows. Let $A \in \mathscr{F}$. Then $A \in \mathscr{C}_\alpha$ for some $\alpha < \alpha_0$. Let $\bar{\mu}(A) = \mu_\alpha(A)$. Then $\bar{\mu}$ is the desired extension of μ from $\mathscr{C}$ to $\mathscr{F}$. It is now clear that $\bar{\mu}$ is a real charge on $\mathscr{F}$. □

Now, we come to the problem of extending positive real partial charges defined on a given collection $\mathscr{C}$ of subsets of a set Ω to any field $\mathscr{F}$ on Ω containing $\mathscr{C}$ as a positive real partial charge. This may not be possible always. Successful extension depends, to some extent, on whether $\Omega \in \mathscr{C}$ or not. If $\Omega \in \mathscr{C}$, we show that an extension is always possible. If $\Omega \notin \mathscr{C}$, the given positive real partial charge on $\mathscr{C}$ can be extended to $\mathscr{F}$ as a positive partial charge but this extension might take the value ∞ on some sets. First, we consider the case $\Omega \in \mathscr{C}$.

3.2.6 Definitions. Let $\mathscr{C}$ be any collection of subsets of a set Ω with $\Omega \in \mathscr{C}$. Let μ be a positive real partial charge on $\mathscr{C}$. For any subset A of Ω, let

$$\mu_i(A) = \sup \frac{\sum_{j=1}^{n} \mu(A_j) - \sum_{j=1}^{m} \mu(B_j)}{k},$$

where the supremum is taken over all finite sequences $A_1, A_2, \ldots, A_n$ and $B_1, B_2, \ldots, B_m$ from $\mathscr{C}$ satisfying the condition that

$$kI_A + \sum_{j=1}^{m} I_{B_j} \geq \sum_{j=1}^{n} I_{A_j}$$

for some positive integer k.

Also, for any subset A of Ω, let

$$\mu_e(A) = \operatorname{Inf} \frac{\sum_{j=1}^{n} \mu(A_j) - \sum_{j=1}^{m} \mu(B_j)}{k},$$

where the infimum is taken over all finite sequences $A_1, A_2, \ldots, A_n$ and $B_1, B_2, \ldots, B_m$ from $\mathscr{C}$ satisfying the condition that

$$\sum_{j=1}^{n} I_{A_j} \geq \sum_{j=1}^{m} I_{B_j} + kI_A$$

for some positive integer k.

The supremum defined above always makes sense. One can take any set C in $\mathscr{C}$ and note that $I_A + I_C \geq I_C$. The infimum defined above is also meaningful since $\Omega \in \mathscr{C}$.

The following proposition puts μ_i and μ_e into proper perspective in relation to the extension problem.

3.2.7 Proposition. *Let $\mathscr{C}$ be a collection of subsets of a set Ω with $\Omega \in \mathscr{C}$. Let μ be a positive real partial charge on $\mathscr{C}$. Let A be any subset of Ω. If $\bar{\mu}$ is a positive real partial charge on $\mathscr{C} \cup \{A\}$ which is an extension of μ, then*

$$\mu_i(A) \leq \bar{\mu}(A) \leq \mu_e(A).$$

Proof. We prove the inequality $\mu_i(A) \leq \bar{\mu}(A)$. Let $A_1, A_2, \ldots, A_n$, $B_1, B_2, \ldots, B_m$ be sets from $\mathscr{C}$ satisfying the condition that $kI_A + \sum_{j=1}^{m} I_{B_j} \geq \sum_{j=1}^{n} I_{A_j}$ for some positive integer k. Since $\bar{\mu}$ is a positive real partial charge on $\mathscr{C} \cup \{A\}$, we have $k\bar{\mu}(A) + \sum_{j=1}^{m} \bar{\mu}(B_j) \geq \sum_{j=1}^{n} \bar{\mu}(A_j)$. Since $\bar{\mu}$ is an extension of μ from $\mathscr{C}$ to $\mathscr{C} \cup \{A\}$, we have

$$\bar{\mu}(A) \geq \frac{\sum_{j=1}^{n} \mu(A_j) - \sum_{j=1}^{m} \mu(B_j)}{k}.$$

Consequently, we obtain the inequality $\bar{\mu}(A) \geq \mu_i(A)$ by taking supremum over all finite sequences with the property specified above. The proof of the other inequality is similar. □

From the above proposition, it is clear that when seeking an extension $\bar{\mu}$ of μ from $\mathscr{C}$ to $\mathscr{C} \cup \{A\}$, the choice of the number $\bar{\mu}(A)$ should confirm to the inequalities established above. The following proposition gives some properties of the set functions μ_i and μ_e.

3.2.8 Proposition. *Let $\mathscr{C}$ be a collection of subsets of a set Ω containing Ω. Let μ be a positive real partial charge on $\mathscr{C}$. Let μ_i and μ_e be the set functions on the power set $\mathscr{P}(\Omega)$ of Ω as defined in Definition 3.2.6. Then the following statements are true.*

(i). $0 \leq \mu_i(A) \leq \mu_e(A) \leq \mu(\Omega)$ *for every* $A \subset \Omega$.

(ii). *If* $A \in \mathscr{C}$ *or* $I_A \in \mathscr{L}(\mathscr{C})$, *then* $\mu_i(A) = \mu_e(A) = T(I_A)$, *where T is the linear functional on $\mathscr{L}(\mathscr{C})$ induced by μ.*

(iii). *If* $A, B \subset \Omega$ *and* $A \cap B = \varnothing$, *then*

$$\mu_i(A) + \mu_i(B) \leq \mu_i(A \cup B) \leq \mu_i(A) + \mu_e(B) \leq \mu_e(A \cup B) \leq \mu_e(A) + \mu_e(B).$$

(iv). *If* $A \in \mathscr{C}$, $B \subset \Omega$ *and* $A \cap B = \varnothing$, *then*

$$\mu_i(A \cup B) = \mu(A) + \mu_i(B)$$

and

$$\mu_e(A \cup B) = \mu(A) + \mu_e(B).$$

(v). *If* $A, B \subset \Omega$, $A \cap B = \varnothing$ *and* $A \cup B \in \mathscr{C}$, *then*

$$\mu(A \cup B) = \mu_i(A) + \mu_e(B) = \mu_e(A) + \mu_i(B).$$

In particular, for any A contained in Ω, $\mu_i(A) + \mu_e(A^c) = \mu_i(A^c) + \mu_e(A) = \mu(\Omega)$.

Proof. (i). For any B in $\mathscr{C}$, $I_A + I_B \geq I_B$. Consequently,

$$\mu_i(A) \geq \frac{\mu(B) - \mu(B)}{1} = 0.$$

Since Ω is in $\mathscr{C}$ and $I_\Omega + I_\Omega \geq I_\Omega + I_A$, we have

$$\mu_e(A) \leq \frac{\mu(\Omega) + \mu(\Omega) - \mu(\Omega)}{1} = \mu(\Omega).$$

Let $A_1, A_2, \ldots, A_n; B_1, B_2, \ldots, B_m; C_1, C_2, \ldots, C_p;$ and $D_1, D_2, \ldots, D_q$ be sets from $\mathscr{C}$ such that $kI_A + \sum_{j=1}^{m} I_{B_j} \geq \sum_{j=1}^{n} I_{A_j}$ for some positive integer k, and $\sum_{j=1}^{p} I_{C_j} \geq \sum_{j=1}^{q} I_{D_j} + sI_A$ for some positive integer s. From these inequalities, it follows that

$$k\left[\sum_{j=1}^{p} I_{C_j} - \sum_{j=1}^{q} I_{D_j}\right] \geq s\left[\sum_{j=1}^{n} I_{A_j} - \sum_{j=1}^{m} I_{B_j}\right].$$

So,
$$k \sum_{j=1}^{p} I_{C_j} + s \sum_{j=1}^{m} I_{B_j} \geq s \sum_{j=1}^{n} I_{A_j} + k \sum_{j=1}^{q} I_{D_j}.$$

Since μ is a positive real partial charge on $\mathscr{C}$, we obtain
$$k \sum_{j=1}^{p} \mu(C_j) + s \sum_{j=1}^{m} \mu(B_j) \geq s \sum_{j=1}^{n} \mu(A_j) + k \sum_{j=1}^{q} \mu(D_j).$$

Rearranging these terms, we get
$$\frac{\sum_{j=1}^{p} \mu(C_j) - \sum_{j=1}^{q} \mu(D_j)}{s} \geq \frac{\sum_{j=1}^{n} \mu(A_j) - \sum_{j=1}^{m} \mu(B_j)}{k}.$$

Hence $\mu_e(A) \geq \mu_i(A)$.

(ii). If $I_A \in \mathscr{L}(\mathscr{C})$, we can write $I_A = \sum_{i=1}^{m} r_i I_{C_i}$ for some $C_1, C_2, \ldots, C_m$ in $\mathscr{C}$ and $r_1, r_2, \ldots, r_m$ rationals. Writing $r_i = n_i/N$ for $i = 1, 2, \ldots, m$, where $n_1, n_2, \ldots, n_m$ are integers and N is a positive integer, we can rewrite the above as
$$N I_A = \sum_{i=1}^{m} n_i I_{C_i} = \sum_{j=1}^{p} I_{A_j} - \sum_{j=1}^{q} I_{B_j}$$

with $p \geq 1$, $q \geq 1$ and $A_1, A_2, \ldots, A_p$, $B_1, B_2, \ldots, B_q$ coming from $\{C_1, C_2, \ldots, C_m\}$. The above representation gives
$$\mu_i(A) \geq \frac{\sum_{j=1}^{p} \mu(A_j) - \sum_{j=1}^{q} \mu(B_j)}{N}$$

and
$$\mu_e(A) \leq \frac{\sum_{j=1}^{p} \mu(A_j) - \sum_{j=1}^{q} \mu(B_j)}{N}.$$

Hence $\mu_i(A) = \mu_e(A) = T(I_A)$.

(iii). Let $A_1, A_2, \ldots, A_n$; $B_1, B_2, \ldots, B_m$; $C_1, C_2, \ldots, C_p$ and $D_1, D_2, \ldots, D_q$ be sets from $\mathscr{C}$ such that $k I_A + \sum_{j=1}^{m} I_{B_j} \geq \sum_{j=1}^{n} I_{A_j}$ for some positive integer k, and $s I_B + \sum_{j=1}^{q} I_{D_j} \geq \sum_{j=1}^{p} I_{C_j}$ for some positive integer s. Then
$$ks(I_A + I_B) + s \sum_{j=1}^{m} I_{B_j} + k \sum_{j=1}^{q} I_{D_j} \geq s \sum_{j=1}^{n} I_{A_j} + k \sum_{j=1}^{p} I_{C_j}.$$

Consequently, by observing that $I_{A \cup B} = I_A + I_B$, we get
$$\mu_i(A \cup B) \geq \left[s \sum_{j=1}^{n} \mu(A_j) + k \sum_{j=1}^{p} \mu(C_j) - s \sum_{j=1}^{m} \mu(B_j) - k \sum_{j=1}^{q} \mu(D_j) \right] \Big/ ks.$$

The above inequality can be rewritten as

$$\mu_i(A \cup B) \geq \frac{\sum_{j=1}^n \mu(A_j) - \sum_{j=1}^m \mu(B_j)}{k} + \frac{\sum_{j=1}^p \mu(C_j) - \sum_{j=1}^q \mu(D_j)}{s}.$$

From this, it follows that $\mu_i(A \cup B) \geq \mu_i(A) + \mu_i(B)$.

Now, we show that $\mu_i(A \cup B) \leq \mu_i(A) + \mu_e(B)$. Let $A_1, A_2, \ldots, A_n$; $B_1, B_2, \ldots, B_m$; $C_1, C_2, \ldots, C_p$ and $D_1, D_2, \ldots, D_q$ be sets in $\mathscr{C}$ such that $k(I_{A \cup B}) + \sum_{j=1}^m I_{B_j} \geq \sum_{j=1}^n I_{A_j}$ for some positive integer k, and $\sum_{j=1}^q I_{D_j} \geq \sum_{j=1}^p I_{C_j} + s I_B$ for some positive integer s. Since $I_{A \cup B} = I_A + I_B$,

$$\frac{\sum_{j=1}^q I_{D_j} - \sum_{j=1}^p I_{C_j}}{s} \geq I_B \geq \frac{\sum_{j=1}^n I_{A_j} - \sum_{j=1}^m I_{B_j}}{k} - I_A.$$

Multiplying throughout by ks, we obtain

$$k \sum_{j=1}^q I_{D_j} + s \sum_{j=1}^m I_{B_j} + ks I_A \geq s \sum_{j=1}^n I_{A_j} + k \sum_{j=1}^p I_{C_j}.$$

Consequently,

$$\mu_i(A) \geq \frac{\sum_{j=1}^n \mu(A_j) - \sum_{j=1}^m \mu(B_j)}{k} - \frac{\sum_{j=1}^q \mu(D_j) - \sum_{j=1}^p \mu(C_j)}{s}.$$

Taking supremum over $A_1, A_2, \ldots, A_n$; $B_1, B_2, \ldots, B_m$; k, first and then supremum over $C_1, C_2, \ldots, C_p$; $D_1, D_2, \ldots, D_q$; s, we obtain $\mu_i(A) \geq \mu_i(A \cup B) - \mu_e(B)$. From this, the desired inequality follows.

The rest of the inequalities can be proved analogously.

(iv). This follows from (ii) and (iii).

(v). This follows from (iii). The second part follows if we observe that $\Omega \in \mathscr{C}$. □

The following theorem paves the way for the extension of positive real partial charges.

3.2.9 Theorem. *Let $\mathscr{C}$ be any collection of subsets of a set Ω with $\Omega \in \mathscr{C}$. Let μ be a positive real partial charge on $\mathscr{C}$. Let $A \subset \Omega$ be such that $A \notin \mathscr{C}$. Then there exists a positive real partial charge on $\mathscr{C} \cup \{A\}$ which is an extension of μ.*

Proof. Let T be the linear functional on $\mathscr{L}(\mathscr{C})$ induced by μ. *Case (i).* $I_A \in \mathscr{L}(\mathscr{C})$. Then the set function $\bar{\mu}$ on $\mathscr{C} \cup \{A\}$ defined by $\bar{\mu}(B) = T(I_B)$, B in $\mathscr{C} \cup \{A\}$, is a partial charge on $\mathscr{C} \cup \{A\}$ and it is the desired extension of μ from $\mathscr{C}$ to $\mathscr{C} \cup \{A\}$. From Proposition 3.2.8(i) and (ii), it follows that $\bar{\mu}$ is positive. *Case (ii).* $I_A \notin \mathscr{L}(\mathscr{C})$. Observe that $\mathscr{L}(\mathscr{C} \cup \{A\}) = \{f + rI_A; f \in \mathscr{L}(\mathscr{C}) \text{ and } r \text{ rational}\}$. Choose and fix a real number d satisfying $\mu_i(A) \leq d \leq \mu_e(A)$. Obviously, by Proposition 3.2.8(i), $d \geq 0$. Define $\bar{T}$ on $\mathscr{L}(\mathscr{C} \cup \{A\})$ by $\bar{T}(f + rI_A) = T(f) + rd$ for f in $\mathscr{L}(\mathscr{C})$ and r rational. We show that $\bar{T}$ is well defined. Suppose $f_1 + r_1 I_A = f_2 + r_2 I_A$ for some $f_1, f_2 \in \mathscr{L}(\mathscr{C})$ and r_1, r_2 rationals. Consequently, $f_1 - f_2 = (r_2 - r_1)I_A$. Since $I_A \notin \mathscr{L}(\mathscr{C})$ and $\mathscr{L}(\mathscr{C})$ is a linear space, we must have $r_2 = r_1$. Consequently, $f_1 = f_2$. Hence $\bar{T}$ is well defined on $\mathscr{L}(\mathscr{C} \cup \{A\})$. Define $\bar{\mu}$ on $\mathscr{C} \cup \{A\}$ by $\bar{\mu}(B) = \bar{T}(I_B)$ for B in $\mathscr{C} \cup \{A\}$. $\bar{\mu}$ is obviously an extension of μ from $\mathscr{C}$ to $\mathscr{C} \cup \{A\}$. $\bar{\mu}(A) = d \geq 0$. This implies that $\bar{\mu}$ is positive. Since $\bar{T}$ is well defined, $\bar{\mu}$ is a partial charge. From the inequality $\mu_i(A) \leq d \leq \mu_e(A)$, it follows that $\bar{\mu}$ is a positive real partial charge on $\mathscr{C} \cup \{A\}$. □

3.2.10 Theorem. *Let $\mathscr{C}$ be any collection of subsets of a set Ω with $\Omega \in \mathscr{C}$. Let μ be a positive real partial charge on $\mathscr{C}$. Let $\mathscr{F}$ be any field on Ω containing $\mathscr{C}$. Then there is a positive charge $\bar{\mu}$ on $\mathscr{F}$ which is an extension of μ.*

Proof. One can give a proof based on transfinite induction and Theorem 3.2.9. The argument is similar to the one presented in Theorem 3.2.5. □

3.3 EXTENSION PROCEDURE OF ŁOS AND MARCZEWSKI

The extension procedure described in the previous section consists of extending the given set function on a given collection of sets to the collection of sets obtained by adjoining just a single set to the given collection of sets. In this section, we present a procedure due to Łos and Marczewski for the extension problem which can be described as follows. Let $\mathscr{F}$ be a field of subsets of a set Ω and $\mathscr{C}$ a subfield of $\mathscr{F}$. Let μ be a positive bounded charge on $\mathscr{C}$. Let $A \in \mathscr{F}$ be such that $A \notin \mathscr{C}$. Let $\mathscr{F}(\mathscr{C}, A)$ denote the smallest field on Ω containing $\mathscr{C}$ and A. The procedure due to Łos and Marczewski consists of extending μ from $\mathscr{C}$ to $\mathscr{F}(\mathscr{C}, A)$ as a positive charge in one step. Repeating this procedure, one can extend μ from $\mathscr{C}$ to $\mathscr{F}$ as a positive charge.

First, we observe that the expressions for μ_i and μ_e introduced in Section 3.2 simplify if the domain of μ is a field of sets, as the following proposition demonstrates.

3.3.1 Proposition. *Let $\mathscr{C}$ be a field of subsets of a set Ω. Let μ be a positive bounded charge on $\mathscr{C}$. Let μ_i and μ_e be the set functions on the power set $\mathscr{P}(\Omega)$ of Ω as defined in Definition 3.2.6. Then for any subset A of Ω,*
(i). $\mu_i(A) = \text{Sup}\{\mu(B); B \subset A, B \in \mathscr{C}\}$ *and* (ii). $\mu_e(A) = \text{Inf}\{\mu(C); A \subset C, C \in \mathscr{C}\}$.

Proof. We prove (i). The proof of (ii) is similar. Let $A_1, A_2, \ldots, A_n$, $B_1, B_2, \ldots, B_m$ be sets from $\mathscr{C}$ such that
$$k_0 I_A + \sum_{i=1}^{m} I_{B_i} \geq \sum_{i=1}^{n} I_{A_i}$$

for some positive integer k_0. We show that there is a set F_1 in $\mathscr{C}$ such that $F_1 \subset A$ and

$$\frac{\sum_{i=1}^n \mu(A_i) - \sum_{i=1}^m \mu(B_i)}{k_0} \leq \mu(F_1).$$

Since $\varnothing \in \mathscr{C}$, we can assume that $m = n$. Define for each $1 \leq k \leq n$,

$$C_k = \bigcup_{1 \leq i_1 < i_2 < \cdots < i_k \leq n} \bigcap_{j=1}^k A_{i_j}$$

and

$$D_k = \bigcup_{1 \leq i_1 < i_2 < \cdots < i_k \leq n} \bigcap_{j=1}^k B_{i_j}.$$

By Lemma 3.1.4,

$$\sum_{i=1}^n I_{A_i} = \sum_{k=1}^n I_{C_k} \quad \text{and} \quad \sum_{i=1}^n I_{B_i} = \sum_{k=1}^n I_{D_k}.$$

Then

$$k_0 I_A + \sum_{i=1}^n I_{D_i} \geq \sum_{i=1}^n I_{C_i}.$$

This leads to

$$k_0 I_A + \sum_{i=1}^n I_{D_i - C_i} + \sum_{i=1}^n I_{D_i \cap C_i} \geq \sum_{i=1}^n I_{C_i - D_i} + \sum_{i=1}^n I_{C_i \cap D_i}.$$

This can be written in the form

$$k_0 I_A + \sum_{i=1}^n I_{D_i - C_i} \geq \sum_{i=1}^n I_{C_i - D_i}.$$

Since $C_1 \supset C_2 \supset \cdots \supset C_n$ and $D_1 \supset D_2 \supset \cdots \supset D_n$, $(D_i - C_i) \cap (C_j - D_j) = \varnothing$ for all $i, j = 1, 2, \ldots, n$. Now, the inequality reduces to

$$k_0 I_A \geq \sum_{i=1}^n I_{C_i - D_i} = \sum_{i=1}^n I_{E_i}, \quad \text{say.}$$

Let, for each $1 \leq k \leq n$,

$$F_k = \bigcup_{1 \leq i_1 < i_2 < \cdots < i_k \leq n} \bigcap_{j=1}^k E_{i_j}.$$

Note that $F_1 \supset F_2 \supset \cdots \supset F_n$ and by Lemma 3.1.4, $\sum_{i=1}^n I_{E_i} = \sum_{i=1}^n I_{F_i}$. Let k_1 be the largest integer such that $F_{k_1} \neq \varnothing$. (If $F_1 = \varnothing$, clearly $F_1 \subset A$ and

$$\frac{\sum_{i=1}^n \mu(A_i) - \sum_{i=1}^m \mu(B_i)}{k_0} \leq 0 = \mu(F_1).)$$

Obviously, $k_0 \geq k_1$. Now,

$$\frac{\sum_{i=1}^{n}\mu(A_i)-\sum_{i=1}^{n}\mu(B_i)}{k_0} = \frac{\sum_{i=1}^{n}\mu(C_i)-\sum_{i=1}^{n}\mu(D_i)}{k_0}$$

$$= \frac{\sum_{i=1}^{n}\mu(C_i-D_i)-\sum_{i=1}^{n}\mu(D_i-C_i)}{k_0}$$

$$\leq \frac{\sum_{i=1}^{n}\mu(C_i-D_i)}{k_0} = \frac{\sum_{i=1}^{n}\mu(E_i)}{k_0}$$

$$= \frac{\sum_{i=1}^{n}\mu(F_i)}{k_0} = \frac{\sum_{i=1}^{k_1}\mu(F_i)}{k_0}$$

$$\leq \frac{\sum_{i=1}^{k_1}\mu(F_i)}{k_1} \leq \mu(F_1).$$

Since $k_0 I_A \geq \sum_{i=1}^{n} I_{F_i}$ and $F_1 \supset F_2 \supset \cdots \supset F_n$, it follows that $A^c \subset F_1^c$ or $A \supset F_1$. Consequently,

$$\frac{\sum_{i=1}^{n}\mu(A_i)-\sum_{i=1}^{n}\mu(B_i)}{k_0} \leq \mu(F_1) \leq \text{Sup}\{\mu(B); B \subset A, B \in \mathscr{C}\}.$$

Hence $\mu_i(A) \leq \text{Sup}\{\mu(B); B \subset A, B \in \mathscr{C}\}$. On the other hand, for any B in $\mathscr{C}$ and $B \subset A$, observe that $I_A + I_\varnothing \geq I_B$. So, $\mu_i(A) \geq [\mu(B)-\mu(\varnothing)]/1 = \mu(B)$. Therefore,

$$\mu_i(A) \geq \text{Sup}\{\mu(B); B \subset A, B \in \mathscr{C}\}.$$

This proves the desired equality. □

If $\mathscr{C}$ is a field of sets, μ_i and μ_e exhibit some additional properties as the following proposition indicates.

3.3.2 Proposition. *Let $\mathscr{C}$ be a field of subsets of a set Ω and μ a positive bounded charge on $\mathscr{C}$. Let μ_i and μ_e be the set functions defined on $\mathscr{P}(\Omega)$ as in Definition 3.2.6. If A and B are two subsets of Ω satisfying the conditions that $A \subset C$, $B \subset D$, $C \cap D = \varnothing$ and $C, D \in \mathscr{C}$, then*

$$\mu_i(A \cup B) = \mu_i(A) + \mu_i(B)$$

and

$$\mu_e(A \cup B) = \mu_e(A) + \mu_e(B).$$

Proof. We prove the first assertion. The proof of the second is similar. By Proposition 3.2.8(iii), $\mu_i(A) + \mu_i(B) \leq \mu_i(A \cup B)$. By Proposition 3.3.1, for any given $\varepsilon > 0$, we can find $E \subset A \cup B$ such that $E \in \mathscr{C}$ and $\mu(E) \geq$

3. EXTENSIONS OF CHARGES   73

$\mu_i(A \cup B) - \varepsilon$. Note that

$$\mu_i(A \cup B) - \varepsilon \leq \mu(E) = \mu(E \cap (C \cup D)) = \mu((E \cap C) \cup (E \cap D))$$
$$= \mu(E \cap C) + \mu(E \cap D)$$
$$\leq \mu_i(A) + \mu_i(B)$$

as $E \cap C \subset A$ and $E \cap D \subset B$. Since $\varepsilon > 0$ is arbitrary, it follows that $\mu_i(A \cup B) \leq \mu_i(A) + \mu_i(B)$. Hence $\mu_i(A \cup B) = \mu_i(A) + \mu_i(B)$. □

Now, we prove the extension theorem.

3.3.3 Theorem. *Let $\mathscr{C}$ be a field of subsets of a set Ω. Let μ be a positive bounded charge on $\mathscr{C}$. Let $A \subset \Omega$ be such that $A \notin \mathscr{C}$. Let $\mathscr{F}(\mathscr{C}, A)$ be the smallest field on Ω containing $\mathscr{C}$ and A. Then there exists a positive bounded charge $\bar{\mu}$ on $\mathscr{F}(\mathscr{C}, A)$ which is an extension of μ.*

Moreover, if d is any real number between $\mu_i(A)$ and $\mu_e(A)$, then there is a positive charge $\bar{\mu}$ on $\mathscr{F}(\mathscr{C}, A)$ such that $\bar{\mu}(A) = d$ and $\bar{\mu}$ is an extension of μ from $\mathscr{C}$ to $\mathscr{F}(\mathscr{C}, A)$.

Proof. Choose and fix a number $d \in [\mu_i(A), \mu_e(A)]$. Then, for C in $\mathscr{F}(\mathscr{C}, A)$, let

$$\bar{\mu}(C) = a[\mu_i(C \cap A) + \mu_e(C \cap A^c)] + (1-a)[\mu_e(C \cap A) + \mu_i(C \cap A^c)],$$

where a satisfies the equation $d = a\mu_i(A) + (1-a)\mu_e(A)$. Obviously, $0 \leq a \leq 1$. $\bar{\mu}$ is a positive real function on $\mathscr{F}(\mathscr{C}, A)$. We show that $\bar{\mu}$ is an extension of μ. Let $C \in \mathscr{C}$. Since $(C \cap A) \cup (C \cap A^c) = C \in \mathscr{C}$ and $(C \cap A) \cap (C \cap A^c) = \varnothing$, by Proposition 3.2.8(v), $\mu_i(C \cap A) + \mu_e(C \cap A^c) = \mu(C) = \mu_i(C \cap A^c) + \mu_e(C \cap A)$. Consequently,

$$\bar{\mu}(C) = a\mu(C) + (1-a)\mu(C) = \mu(C).$$

This proves that $\bar{\mu}$ is an extension of μ from $\mathscr{C}$ to $\mathscr{F}(\mathscr{C}, A)$. It remains to be shown that $\bar{\mu}$ is a charge on $\mathscr{F}(\mathscr{C}, A)$. Let μ_1, μ_2 be the set functions defined on $\mathscr{F}(\mathscr{C}, A)$ by

$$\mu_1(C) = \mu_i(C \cap A) + \mu_e(C \cap A^c), \text{ C in } \mathscr{F}(\mathscr{C}, A)$$

and

$$\mu_2(C) = \mu_i(C \cap A^c) + \mu_e(C \cap A), \text{ C in } \mathscr{F}(\mathscr{C}, A).$$

It suffices to show that μ_1 and μ_2 are charges on $\mathscr{F}(\mathscr{C}, A)$. Now, we look at μ_1. Let $C, D \in \mathscr{F}(\mathscr{C}, A)$ be such that $C \cap D = \varnothing$. Then we can write $C = (C_1 \cap A) \cup (C_2 \cap A^c)$ and $D = (D_1 \cap A) \cup (D_2 \cap A^c)$ for some C_1, C_2, D_1, D_2 in $\mathscr{C}$. See Proposition 1.1.13. By replacing C_1 by $C_1 - D_1$, D_1 by $D_1 - C_1$, C_2 by $C_2 - D_2$ and D_2 by $D_2 - C_2$, if necessary, we can assume that $C_1 \cap D_1 =$

$\varnothing$ and $C_2 \cap D_2 = \varnothing$. This is possible since $C \cap D = \varnothing$. We note that

$$\mu_1(C \cup D) = \mu_i((C \cup D) \cap A) + \mu_e((C \cup D) \cap A^c)$$
$$= \mu_i((C \cap A) \cup (D \cap A)) + \mu_e((C \cap A^c) \cup (D \cap A^c))$$
$$= \mu_i((C_1 \cap A) \cup (D_1 \cap A)) + \mu_e((C_2 \cap A^c) \cup (D_2 \cap A^c))$$
$$= \mu_i(C_1 \cap A) + \mu_i(D_1 \cap A) + \mu_e(C_2 \cap A^c) + \mu_e(D_2 \cap A^c)$$

(See Proposition 3.3.2.)

$$= \mu_i(C_1 \cap A) + \mu_e(C_2 \cap A^c) + \mu_i(D_1 \cap A) + \mu_e(D_2 \cap A^c)$$
$$= \mu_i(C \cap A) + \mu_e(C \cap A^c) + \mu_i(D \cap A) + \mu_e(D \cap A^c)$$
$$= \mu_1(C) + \mu_1(D).$$

This shows that μ_1 is a charge. By a similar argument, one can show that μ_2 is a charge. Consequently, $\bar{\mu} = a\mu_1 + (1-a)\mu_2$ is a charge. $\square$

The above theorem leads to the general extension result.

3.3.4 Corollary. *Let $\mathscr{C}$ be a field of subsets of a set Ω and μ a positive bounded charge on $\mathscr{C}$. Let $\mathscr{F}$ be a field on Ω containing $\mathscr{C}$. Then there exists a positive bounded charge $\bar{\mu}$ on $\mathscr{F}$ such that $\bar{\mu}$ is an extension of μ from $\mathscr{C}$ to $\mathscr{F}$ and that the range of $\bar{\mu}$ is a subset of the closure of the range of μ on $\mathscr{C}$.*

Proof. Let $\bar{R}(\mu)$ be the closure of the range of μ on $\mathscr{C}$. Let

$$X = \{(\mathscr{B}, \nu, \mathscr{D}, \lambda); \mathscr{C} \subset \mathscr{B} \subset \mathscr{D} \subset \mathscr{F}, \mathscr{B} \text{ and } \mathscr{D} \text{ are}$$
fields on Ω, ν is a positive charge on $\mathscr{B}$,
λ is a positive charge on $\mathscr{D}$, $\lambda/\mathscr{B} = \nu$,
$\nu/\mathscr{C} = \mu$ and the range of λ is contained in $\bar{R}(\mu)\}$.

($\nu/\mathscr{C}$ is the charge ν restricted to the subfield $\mathscr{C}$.) On X, we introduce a partial order $\leq$ as follows. For $(\mathscr{B}_1, \nu_1, \mathscr{D}_1, \lambda_1)$ and $(\mathscr{B}_2, \nu_2, \mathscr{D}_2, \lambda_2)$ in X, say $(\mathscr{B}_1, \nu_1, \mathscr{D}_1, \lambda_1) \leq (\mathscr{B}_2, \nu_2, \mathscr{D}_2, \lambda_2)$ if the following are satisfied.
  (i) $\mathscr{B}_2 \subset \mathscr{B}_1$.
  (ii) $\nu_1/\mathscr{B}_2 = \nu_2$.
  (iii) $\mathscr{D}_1 \subset \mathscr{D}_2$.
  (iv) $\lambda_2/\mathscr{D}_1 = \lambda_1$.

Let $\{(\mathscr{B}_\alpha, \nu_\alpha, \mathscr{D}_\alpha, \lambda_\alpha); \alpha \in \Gamma\}$ be a chain in X. Let $\mathscr{B} = \bigcap_{\alpha \in \Gamma} \mathscr{B}_\alpha$, $\mathscr{D} = \bigcup_{\alpha \in \Gamma} \mathscr{D}_\alpha$, ν the restriction of ν_{α_0} from $\mathscr{B}_{\alpha_0}$ to $\mathscr{B}$ for any α_0 fixed and λ on $\mathscr{D}$ be defined by $\lambda(D) = \lambda_\alpha(D)$ if $D \in \mathscr{D}_\alpha$, $\alpha \in \Gamma$. λ is well defined on $\mathscr{D}$ and is a positive charge on $\mathscr{D}$. The range of λ is contained in $\bar{R}(\mu)$. It follows that $(\mathscr{B}, \nu, \mathscr{D}, \lambda) \in X$ and is indeed an upperbound of the chain $\{(\mathscr{B}_\alpha, \nu_\alpha, \mathscr{D}_\alpha, \lambda_\alpha); \alpha \in \Gamma\}$. By Zorn's lemma, there is a maximal element

$(\mathcal{B}_0, \nu_0, \mathcal{D}_0, \lambda_0)$ in X. It obviously follows that $\mathcal{B}_0 = \mathcal{C}$ and $\nu_0 = \mu$. We claim that $\mathcal{D}_0 = \mathcal{F}$. Suppose $\mathcal{D}_0$ is properly contained in $\mathcal{F}$. Let $A \in \mathcal{F} - \mathcal{D}_0$ and $\mathcal{C}^* = \mathcal{F}(\mathcal{D}_0, A)$. Let λ_{0i} and λ_{0e} be as defined in Definition 3.2.6 for the charge λ_0 on $\mathcal{D}_0$. Define λ^* on $\mathcal{C}^*$ by

$$\lambda^*(C) = \lambda_{0i}(C \cap A) + \lambda_{0e}(C \cap A^c)$$

for C in $\mathcal{C}^*$. We have already seen in Theorem 3.3.3 that λ^* is a positive charge on $\mathcal{C}^*$ and is indeed an extension of λ_0 from $\mathcal{D}_0$ to $\mathcal{C}^*$. We now show that the range of λ^* is contained in $\bar{R}(\mu)$. Let B_n, $n \geq 1$ be a sequence in $\mathcal{D}_0$ such that $B_n \subset C \cap A$ for every $n \geq 1$ and $\lim_{n \to \infty} \lambda_0(B_n) = \lambda_{0i}(C \cap A)$. Let D_n, $n \geq 1$ be a sequence in $\mathcal{D}_0$ such that $C \cap A^c \subset D_n$ for every $n \geq 1$ and $\lim_{n \to \infty} \lambda_0(D_n) = \lambda_{0e}(C \cap A^c)$. Then $C \cap A^c \subset D_n - B_n$ for every $n \geq 1$. Consequently, $\lim_{n \to \infty} \lambda_0(D_n - B_n) = \lambda_{0e}(C \cap A^c)$ since $\lambda_{0e}(C \cap A^c) \leq \lambda_0(D_n - B_n) \leq \lambda_0(D_n)$ for every $n \geq 1$. Therefore,

$$\lim_{n \to \infty} \lambda_0(B_n \cup D_n) = \lim_{n \to \infty} \lambda_0(B_n) + \lim_{n \to \infty} \lambda_0(D_n - B_n)$$

$$= \lambda_{0i}(C \cap A) + \lambda_{0e}(C \cap A^c) = \lambda^*(C).$$

Hence, the range of λ^* is a subset of the closure of the range of λ_0. But the range of λ_0 is contained in $\bar{R}(\mu)$.

It now follows that $(\mathcal{B}_0, \nu_0, \mathcal{C}^*, \lambda^*) \in X$ and $(\mathcal{B}_0, \nu_0, \mathcal{D}_0, \lambda_0) < (\mathcal{B}_0, \nu_0, \mathcal{C}^*, \lambda^*)$. This is a contradiction to the maximality of $(\mathcal{B}_0, \nu_0, \mathcal{D}_0, \lambda_0)$. Hence $\mathcal{D}_0 = \mathcal{F}$. λ_0 is the desired extension of $\bar{\mu}$ from $\mathcal{C}$ to $\mathcal{F}$. This completes the proof. □

This result can also be proved using Transfinite induction. As a special case of the above theorem, we can prove that every filter in a field $\mathcal{F}$ on Ω is contained in a maximal filter in $\mathcal{F}$.

3.3.5 Corollary. *Let μ be a 0–1 valued charge on a field $\mathcal{C}$ of subsets of a set Ω. Let $\mathcal{F}$ be any field on Ω containing $\mathcal{C}$. Then there is a 0–1 valued charge $\bar{\mu}$ on $\mathcal{F}$ which is an extension of μ from $\mathcal{C}$ to $\mathcal{F}$. Equivalently, every filter in $\mathcal{F}$ is contained in a maximal filter in $\mathcal{F}$.*

Corollary 3.3.4 can also be used to extend bounded charges from $\mathcal{C}$ to $\mathcal{F}$.

3.3.6 Corollary. *Let $\mathcal{C}$ be a field of subsets of a set Ω and μ a bounded charge on $\mathcal{C}$. Let $\mathcal{F}$ be any field on Ω containing $\mathcal{C}$. Then there exists a bounded charge $\bar{\mu}$ on $\mathcal{F}$ which is an extension of μ.*

Proof. Apply Jordan Decomposition theorem to μ on $\mathcal{C}$ and then apply Corollary 3.3.4 for positive and negative variations of μ separately. See Theorem 2.2.2(1). □

3.4 EXTENSION OF PARTIAL CHARGES IN THE GENERAL CASE

In Section 3.2, we dealt with the extension of positive real partial charges defined on a given collection $\mathscr{C}$ of subsets of a set Ω when $\Omega \in \mathscr{C}$. In this section, we examine the situation when $\Omega \notin \mathscr{C}$ and also the extension of partial charges on $\mathscr{C}$ taking infinite values.

3.4.1 Lemma. *Let $\mathscr{C}$ be a ring of subsets of a set Ω, where $\Omega \notin \mathscr{C}$. Let μ be a positive real charge on $\mathscr{C}$. Let $\mathscr{F}$ be the field on Ω generated by $\mathscr{C}$. Then there exists a positive charge $\bar{\mu}$ on $\mathscr{F}$ possibly taking the value infinity which is an extension of μ from $\mathscr{C}$ to $\mathscr{F}$.*

Proof. Let $\mathscr{C}_1 = \{A \subset \Omega;\ A^c \in \mathscr{C}\}$. By Theorem 1.1.9(4), $\mathscr{F} = \mathscr{C} \cup \mathscr{C}_1$. Let $d = \sup\{\mu(C);\ C \in \mathscr{C}\}$.
Case (1). $d = \infty$. In this case, define $\bar{\mu}$ on $\mathscr{F}$ as follows.

$$\bar{\mu}(A) = \mu(A), \quad \text{if } A \in \mathscr{C},$$
$$= \infty, \quad \text{if } A \in \mathscr{C}_1.$$

We claim that $\bar{\mu}$ is a positive charge on $\mathscr{F}$. (It is obvious that $\bar{\mu}$ is an extension of μ from $\mathscr{C}$ to $\mathscr{F}$.) Let $A, B \in \mathscr{F}$ and $A \cap B = \emptyset$. *Case (i)*. $A, B \in \mathscr{C}$. Obviously, $\bar{\mu}(A \cup B) = \mu(A \cup B) = \mu(A) + \mu(B) = \bar{\mu}(A) + \bar{\mu}(B)$. *Case (ii)*. $A \in \mathscr{C}$ and $B \in \mathscr{C}_1$. We claim that $A \cup B \in \mathscr{C}_1$. For, $(A \cup B)^c = A^c \cap B^c = B^c - A \in \mathscr{C}$ as $B^c \in \mathscr{C}$, $A \in \mathscr{C}$ and $\mathscr{C}$ is a ring. Consequently, $\bar{\mu}(A \cup B) = \infty = \mu(A) + \infty = \mu(A) + \bar{\mu}(B) = \bar{\mu}(A) + \bar{\mu}(B)$. *Case (iii)*. $A \in \mathscr{C}_1$ and $B \in \mathscr{C}$. This is analogous to Case (ii). *Case (iv)*. $A, B \in \mathscr{C}_1$. Then $\Omega = A^c \cup B^c \in \mathscr{C}$ which is not possible since $\Omega \notin \mathscr{C}$. So, case (iv) does not arise. Hence $\bar{\mu}$ is a charge on $\mathscr{F}$.
Case (2). $d < \infty$. In this case, we define $\bar{\mu}$ on $\mathscr{F}$ as follows.

$$\bar{\mu}(A) = \mu(A), \quad \text{if } A \in \mathscr{C},$$
$$= d - \mu(A^c), \quad \text{if } A \in \mathscr{C}_1.$$

We claim that $\bar{\mu}$ is a positive bounded charge on $\mathscr{F}$. The positivity of $\bar{\mu}$ is obvious. Let $A, B \in \mathscr{F}$ and $A \cap B = \emptyset$. *Case (i)*. $A, B \in \mathscr{C}$. Then $\bar{\mu}(A \cup B) = \mu(A \cup B) = \mu(A) + \mu(B) = \bar{\mu}(A) + \bar{\mu}(B)$. *Case (ii)*. $A \in \mathscr{C}$ and $B \in \mathscr{C}_1$. As proved above, $A \cup B \in \mathscr{C}_1$. Therefore,

$$\bar{\mu}(A \cup B) = d - \mu((A \cup B)^c) = d - \mu(A^c \cap B^c)$$
$$= d - \mu(B^c - A) = d - \mu(B^c) + \mu(A)$$
$$= \bar{\mu}(B) + \mu(A) = \bar{\mu}(A) + \bar{\mu}(B).$$

Case (iii). $A \in \mathscr{C}_1$ and $B \in \mathscr{C}$. This case is similar to Case (ii). *Case (iv)*.

A, B ∈ $\mathscr{C}_1$. As in Case (iv) under Case (1) above, this is not possible. Hence $\bar{\mu}$ is a charge on $\mathscr{F}$ under Case (2) too. □

Now, we come to the result which complements Theorem 3.2.10.

3.4.2 Theorem. *Let $\mathscr{C}$ be a collection of subsets of a set Ω with $\Omega \notin \mathscr{C}$. Let μ be a positive real partial charge on $\mathscr{C}$. Let $\mathscr{F}$ be any field on Ω containing $\mathscr{C}$. Then there exists a positive charge $\bar{\mu}$ on $\mathscr{F}$ (possibly taking the value ∞) which is an extension of μ from $\mathscr{C}$ to $\mathscr{F}$.*

Proof. Let $\mathscr{D} = \{A \in \mathscr{F}; A \subset \bigcup_{i=1}^{n} C_i$ for some $C_1, C_2, \ldots, C_n$ in $\mathscr{C}\}$. It is clear that $\mathscr{D}$ is a ring on Ω. First, we show that μ can be extended from $\mathscr{C}$ to $\mathscr{D}$ as a positive real partial charge on $\mathscr{D}$. Observe that $\mu_i(B)$ is well defined and is a real number for any $B \subset \Omega$. $\mu_e(A)$ is also well defined and is a real number for any A in $\mathscr{D}$. See Definition 3.2.6 and the remarks following this definition. Using the argument given in the proofs of Theorems 3.2.9 and 3.2.10, we can find a positive real partial charge $\bar{\mu}$ on $\mathscr{D}$ which is an extension of μ on $\mathscr{C}$. Note that every partial charge on a ring is a charge. Now, we extend $\bar{\mu}$ from $\mathscr{D}$ to $\mathscr{F}$ as a positive charge. This can be done as follows. Let $d = \text{Sup}\{\bar{\mu}(D); D \in \mathscr{D}\}$.
Case (i). $d = \infty$. Define $\bar{\bar{\mu}}$ on $\mathscr{F}$ as follows.

$$\bar{\bar{\mu}}(A) = \bar{\mu}(A), \quad \text{if } A \in \mathscr{D},$$
$$= \infty, \quad \text{if } A \in \mathscr{F} - \mathscr{D}.$$

As in the proof of the previous lemma, one can show that $\bar{\bar{\mu}}$ is a positive charge on $\mathscr{F}$ which is obviously an extension of μ from $\mathscr{C}$ to $\mathscr{F}$.
Case (ii). $d < \infty$. Let $\mathscr{F}_1$ be the smallest field on Ω containing $\mathscr{D}$. Define $\tilde{\bar{\mu}}$ on $\mathscr{F}_1$ as follows.

$$\tilde{\bar{\mu}}(A) = \bar{\mu}(A), \quad \text{if } A \in \mathscr{D},$$
$$= d - \bar{\mu}(A^c), \quad \text{if } A \in \mathscr{F}_1 - \mathscr{D}.$$

Using the argument given in the proof of Lemma 3.4.1, one can show that $\tilde{\bar{\mu}}$ is a positive bounded charge on $\mathscr{F}_1$ which is obviously an extension of μ from $\mathscr{C}$. By Corollary 3.3.4, there exists a positive bounded charge $\bar{\mu}$ on $\mathscr{F}$ which is an extension of μ from $\mathscr{C}$. This completes the proof. □

Now, we give a condition under which a positive real partial charge on a collection $\mathscr{C}$ of subsets of a set Ω admits an extension $\bar{\mu}$ on $\mathscr{F}$ which is a positive bounded charge, where $\mathscr{F}$ is any given field on Ω containing $\mathscr{C}$.

3.4.3 Theorem. *Let μ be a positive real partial charge on a collection $\mathscr{C}$ of subsets of a set Ω. Let $\mathscr{F}$ be a field on Ω containing $\mathscr{C}$. Let $\mathscr{D} = \{A \in \mathscr{F}; A \subset \bigcup_{i=1}^{n} C_i$ for some $C_1, C_2, \ldots, C_n$ in $\mathscr{C}\}$. Let μ_e be the set function* defined on $\mathcal{D}$ as in Definition 3.2.6. Then there exists a positive bounded charge $\bar{\mu}$ on $\mathcal{F}$ which is an extension of μ from $\mathscr{C}$ if

$$\text{Sup } \{\mu_e(D); D \in \mathcal{D}\} < \infty.$$

Proof. The proof of this result is essentially contained in the proofs of Theorems 3.4.2 and 3.2.9. □

Finally, in this section, we take up the case of charges taking infinite values.

3.4.4 Theorem. *Let $\mathscr{C}$ be a field of subsets of a set Ω and $\mathcal{F}$ be any field on Ω containing $\mathscr{C}$. Let μ be a positive charge on $\mathscr{C}$ possibly taking the value ∞. Then there exists a positive charge $\bar{\mu}$ on $\mathcal{F}$ which is an extension of μ from $\mathscr{C}$ to $\mathcal{F}$.*

Proof. Let $\mathcal{D} = \{A \in \mathscr{C}; \mu(A) < \infty\}$. $\mathcal{D}$ is obviously a ring on Ω. Let λ be the restriction of μ to $\mathcal{D}$. λ is a partial charge on $\mathcal{D}$. Then by Theorem 3.4.2, we can find a positive charge $\bar{\lambda}$ on $\mathcal{F}$ which is an extension of λ from $\mathcal{D}$ to $\mathcal{F}$. Define $\bar{\mu}$ on $\mathcal{F}$ as follows. Let $A \in \mathcal{F}$.

$$\bar{\mu}(A) = \bar{\lambda}(A), \quad \text{if } A \subset \bigcup_{i=1}^{n} A_i \quad \text{for some finite number of sets}$$
$$A_1, A_2, \ldots, A_n \text{ from } \mathcal{D},$$

$$= \infty, \quad \text{otherwise.}$$

$\bar{\mu}$ is the desired extension of μ from $\mathscr{C}$ to $\mathcal{F}$. One can easily check that $\bar{\mu}$ is a charge on $\mathcal{F}$. □

Next, we deal with general charges.

3.4.5 Theorem. *Let $\mathscr{C}$ be a field of subsets of a set Ω and $\mathcal{F}$ any field on Ω containing $\mathscr{C}$. Let μ be a charge on $\mathscr{C}$ not necessarily real valued. Then there exists a charge $\bar{\mu}$ on $\mathcal{F}$ which is an extension of μ from $\mathscr{C}$ to $\mathcal{F}$. If μ is real valued, one can choose $\bar{\mu}$ to be real valued.*

Proof. One can give a proof along the lines of the ideas given in the proof of Theorem 3.4.4, using Theorem 3.2.5. □

3.5 MISCELLANEOUS EXTENSIONS

In this section, we give various classical extension theorems in some special cases, some of them without proofs.

3. EXTENSIONS OF CHARGES

3.5.1 Theorem. (i). *Let $\mathscr{C}$ be a semi-ring of subsets of a set Ω. Let μ be a real charge on $\mathscr{C}$. Let $\mathscr{F}$ be the smallest ring on Ω containing $\mathscr{C}$. Then there exists a unique real charge $\bar{\mu}$ on $\mathscr{F}$ which is an extension of μ from $\mathscr{C}$ to $\mathscr{F}$. If μ is positive on $\mathscr{C}$, so is $\bar{\mu}$ on $\mathscr{F}$.*

(ii). *Let $\mathscr{C}$ be a semi-field of subsets of a set Ω. Let μ be a real charge on $\mathscr{C}$. Let $\mathscr{F}$ be the smallest field on Ω containing $\mathscr{C}$. Then there exists a unique real charge $\bar{\mu}$ on $\mathscr{F}$ which is an extension of μ from $\mathscr{C}$ to $\mathscr{F}$. If μ is positive on $\mathscr{C}$, so is $\bar{\mu}$ on $\mathscr{F}$.*

(iii). *Let $\mathscr{C}$ be a lattice of subsets of a set Ω with $\varnothing \in \mathscr{C}$ and μ a strongly additive real function on $\mathscr{C}$. Let $\mathscr{F}$ be the smallest semi-ring on Ω containing $\mathscr{C}$. Then there exists a unique real charge $\bar{\mu}$ on $\mathscr{F}$ which is an extension of μ from $\mathscr{C}$ to $\mathscr{F}$. If μ is positive on $\mathscr{C}$ and satisfies the condition that $\mu(A) \leq \mu(B)$ whenever $A, B \in \mathscr{C}$ and $A \subset B$, then $\bar{\mu}$ is positive on $\mathscr{F}$.*

(iv). *Let $\mathscr{C}$ be a ring of subsets of a set Ω and μ a real charge on $\mathscr{C}$. Let $\mathscr{F}$ be the smallest field on Ω containing $\mathscr{C}$. Then there exists a unique real charge $\bar{\mu}$ on $\mathscr{F}$ such that $\bar{\mu}(\Omega)$ is a prescribed number. If μ is a positive charge on $\mathscr{C}$, there is a positive charge $\bar{\mu}$ on $\mathscr{F}$ which is an extension of μ. Such a $\bar{\mu}$ is unique if $\bar{\mu}(\Omega) = \mathrm{Sup}\{\mu(B); B \in \mathscr{C}\}$ and μ is positive real valued.*

(v). *Let μ be a positive bounded charge on an additive-class $\mathscr{C}$ of subsets of a set Ω. Let $\mathscr{F}$ be the smallest field on Ω containing $\mathscr{C}$. There need not exist a positive bounded charge on $\mathscr{F}$ which is an extension of μ.*

Proof. Even though the results of this theorem can be proved using some of the results of the previous sections, we give direct proofs.

(i). By Theorem 1.1.9(2), $\mathscr{F} = \{A \subset \Omega; A \text{ is a finite disjoint union of sets from } \mathscr{C}\}$. Define $\bar{\mu}$ on $\mathscr{F}$ as follows. Let $A \in \mathscr{F}$. Then $A = \bigcup_{i=1}^{m} C_i$ for some pairwise disjoint sets in $\mathscr{C}$. Set $\bar{\mu}(A) = \sum_{i=1}^{m} \mu(C_i)$. We show that $\bar{\mu}$ is unambiguously defined. Suppose $A = \bigcup_{i=1}^{m} C_i = \bigcup_{j=1}^{n} D_j$, where $C_1, C_2, \ldots, C_m$ are pairwise disjoint sets in $\mathscr{C}$ and $D_1, D_2, \ldots, D_n$ are pairwise disjoint sets in $\mathscr{C}$. Then

$$A = \bigcup_{i=1}^{m} \bigcup_{j=1}^{n} (C_i \cap D_j).$$

Note that $C_i = \bigcup_{j=1}^{n} (C_i \cap D_j)$ for every $1 \leq i \leq m$ and $C_i \cap D_j \in \mathscr{C}$ for all i and j. Similarly, $D_j = \bigcup_{i=1}^{m} (C_i \cap D_j)$ for every $1 \leq j \leq n$. Since μ is a charge on $\mathscr{C}$, we have

$$\sum_{i=1}^{m} \mu(C_i) = \sum_{i=1}^{m} \sum_{j=1}^{n} \mu(C_i \cap D_j)$$

and

$$\sum_{j=1}^{n} \mu(D_j) = \sum_{j=1}^{n} \sum_{i=1}^{m} \mu(C_i \cap D_j).$$

This proves the unambiguity of the definition of $\bar{\mu}$. It is obvious that $\bar{\mu}$ is a real charge on $\mathscr{F}$ and is an extension of μ from $\mathscr{C}$ to $\mathscr{F}$. The uniqueness of $\bar{\mu}$ is also clear. If μ is positive on $\mathscr{C}$, so is $\bar{\mu}$ on $\mathscr{F}$. This completes the proof of (i).

(ii). This is similar to (i).

(iii). By Theorem 1.1.9(1), $\mathscr{F} = \{F - E; E, F \in \mathscr{C}, E \subset F\}$. Define $\bar{\mu}$ on $\mathscr{F}$ by $\bar{\mu}(F - E) = \mu(F) - \mu(E)$ for $E, F \in \mathscr{C}$ and $E \subset F$. We prove the unambiguity of the definition of $\bar{\mu}$ on $\mathscr{F}$. Suppose $F_1 - E_1 = F_2 - E_2$ for some E_1, F_1, E_2, F_2 in $\mathscr{C}$ with $E_1 \subset F_1$ and $E_2 \subset F_2$. So, $F_2 \cup E_1 = F_1 \cup E_2$ and $F_2 \cap E_1 = F_1 \cap E_2$. Since μ is a modular function on $\mathscr{C}$, $\mu(F_2) + \mu(E_1) = \mu(F_2 \cup E_1) + \mu(F_2 \cap E_1) = \mu(F_1 \cup E_2) + \mu(F_1 \cap E_2) = \mu(F_1) + \mu(E_2)$. Hence $\mu(F_2) - \mu(E_2) = \mu(F_1) - \mu(E_1)$. This shows that $\bar{\mu}$ is unambiguously defined on $\mathscr{F}$. We now show that $\bar{\mu}$ is a charge on $\mathscr{F}$. Note that $\bar{\mu}(\varnothing) = 0$. Let $F_1 - E_1$ and $F_2 - E_2$ be two disjoint sets in $\mathscr{F}$ such that $(F_1 - E_1) \cup (F_2 - E_2) = F_3 - E_3 \in \mathscr{F}$, where $E_1, F_1, E_2, F_2, E_3, F_3 \in \mathscr{C}$, $E_1 \subset F_1$, $E_2 \subset F_2$ and $E_3 \subset F_3$. In order to show that $\bar{\mu}(F_1 - E_1) + \bar{\mu}(F_2 - E_2) = \bar{\mu}(F_3 - E_3)$, it suffices to show that $\mu(F_1) + \mu(F_2) + \mu(E_3) = \mu(F_3) + \mu(E_1) + \mu(E_2)$. Clearly,

$$F_1 \cup F_2 \cup E_3 = F_3 \cup E_1 \cup E_2,$$

$$(F_1 \cap F_2) \cup (F_1 \cap E_3) \cup (F_2 \cap E_3) = (E_1 \cap E_2) \cup (E_1 \cap F_3) \cup (E_2 \cap F_3),$$

and

$$F_1 \cap F_2 \cap E_3 = F_3 \cap E_1 \cap E_2.$$

By Proposition 3.1.5, the desired equality follows. Since $\mu(\varnothing) = 0$, it follows that $\bar{\mu}$ is an extension of μ. It is obvious that $\bar{\mu}$ is unique. If μ is positive and monotone on $\mathscr{C}$, then $\bar{\mu}$ is positive on $\mathscr{F}$.

(iv). Assume that $\Omega \notin \mathscr{C}$. (If $\Omega \in \mathscr{C}$, then $\mathscr{F} = \mathscr{C}$.) Let $\mathscr{C}_1 = \{A; A^c \in \mathscr{C}\}$. By Theorem 1.1.9(4), $\mathscr{F} = \mathscr{C} \cup \mathscr{C}_1$. Let d be any prescribed number. Define $\bar{\mu}$ on $\mathscr{F}$ as follows.

$$\bar{\mu}(A) = \mu(A), \quad \text{if } A \in \mathscr{C},$$
$$= d - \mu(A^c), \quad \text{if } A \in \mathscr{C}_1.$$

As in the proof of Lemma 3.4.1, one can show that $\bar{\mu}$ is a charge on $\mathscr{F}$. $\bar{\mu}$ is, obviously, an extension of μ from $\mathscr{C}$ to $\mathscr{F}$. Uniqueness of $\bar{\mu}$ with the property that $\bar{\mu}(\Omega) = d$ also follows easily.

If μ is a positive charge on $\mathscr{C}$, define $\bar{\mu}$ on $\mathscr{F}$ by

$$\bar{\mu}(A) = \mu(A), \quad \text{if } A \in \mathscr{C},$$
$$= \infty, \quad \text{if } A \in \mathscr{C}_1.$$

$\bar{\mu}$ is a positive charge on $\mathscr{F}$ and is an extension of μ from $\mathscr{C}$ to $\mathscr{F}$.

3. EXTENSIONS OF CHARGES

If μ is a real positive charge on $\mathscr{C}$, define $\bar{\mu}$ on $\mathscr{F}$ by

$$\bar{\mu}(A) = \mu(A), \quad \text{if } A \in \mathscr{C},$$
$$= d - \mu(A^c), \quad \text{if } A \in \mathscr{C}_1,$$

where $d = \text{Sup}\{\mu(B); B \in \mathscr{C}\}$. $\bar{\mu}$ is the desired extension of μ.
(v). Here is an example. Let $\Omega = \{1, 2, 3, 4\}$ and

$$\mathscr{C} = \{\varnothing, \{1, 2\}, \{1, 3\}, \{1, 4\}, \{2, 3\}, \{2, 4\}, \{3, 4\}, \Omega\}.$$

$\mathscr{C}$ is an additive-class on Ω. The smallest field $\mathscr{F}$ on Ω containing $\mathscr{C}$ is the power set $\mathscr{P}(\Omega)$ of Ω. Let μ on $\mathscr{C}$ be defined by $\mu(\varnothing) = 0$, $\mu(\{1, 2\}) = \frac{1}{2} = \mu(\{3, 4\})$, $\mu(\{1, 3\}) = 1$, $\mu(\{2, 4\}) = 0$, $\mu(\{1, 4\}) = \frac{1}{4}$, $\mu(\{2, 3\}) = \frac{3}{4}$, $\mu(\Omega) = 1$. μ is a positive bounded charge on $\mathscr{C}$. But there is no positive bounded charge on $\mathscr{F}$ which is an extension of μ. □

Another important extension theorem is the following theorem of Caratheodory.

3.5.2 Theorem. *Let μ be a positive bounded measure on a field $\mathscr{F}$ of subsets of a set Ω. Let $\mathfrak{A}$ be the smallest σ-field on Ω containing $\mathscr{F}$. Then there exists a unique positive measure on $\mathfrak{A}$ which is an extension of μ.*

Proof. In any standard text book on Measure theory. □

Related to the above theorem is the following approximation theorem.

3.5.3 Theorem. *Let μ be a positive bounded measure on a σ-field $\mathfrak{A}$ of subsets of a set Ω. Let $\mathscr{F}$ be any field on Ω generating $\mathfrak{A}$. Then for any A in $\mathfrak{A}$ and $\varepsilon > 0$, there exists F in $\mathscr{F}$ such that $\mu(F \triangle A) < \varepsilon$.*

Proof. Let $\mathscr{D} = \{A \in \mathfrak{A};$ for every $\varepsilon > 0$ there exists F in $\mathscr{F}$ such that $\mu(A \triangle F) < \varepsilon\}$. Clearly, $\mathscr{F} \subset \mathscr{D}$ and $\mathscr{D}$ is closed under complementation. We shall show that $\mathscr{D}$ is closed under countable unions which will complete the proof. Let A_n, $n \geq 1$ be a sequence in $\mathscr{D}$ and $\varepsilon > 0$. For each $n \geq 1$, let $F_n \in \mathscr{F}$ be such that $\mu(A_n \triangle F_n) < \varepsilon/2^{n+1}$. Let $m \geq 1$ be such that $\mu(\bigcup_{n \geq 1} F_n - \bigcup_{i=1}^m F_i) < \varepsilon/2$. Now,

$$\mu\left(\left(\bigcup_{n \geq 1} A_n\right) \triangle \left(\bigcup_{n=1}^m F_n\right)\right) \leq \mu\left(\left(\bigcup_{n \geq 1} A_n\right) \triangle \left(\bigcup_{n \geq 1} F_n\right)\right)$$
$$+ \mu\left(\left(\bigcup_{n \geq 1} F_n\right) \triangle \left(\bigcup_{n=1}^m F_n\right)\right)$$
$$< \mu\left(\bigcup_{n \geq 1} (A_n \triangle F_n)\right) + \varepsilon/2$$
$$\leq \sum_{n \geq 1} \mu(A_n \triangle F_n) + \varepsilon/2 < \varepsilon. \quad \square$$

We need some results about measures on topological spaces which we shall state here.

3.5.4 Definition. Let X be a compact Hausdorff space and $\mathscr{B}$ its Borel σ-field. A positive bounded measure μ on $\mathscr{B}$ is said to be regular if

$$\mu(A) = \text{Sup}\{\mu(C); C \text{ compact}, C \subset A\}$$

for every A in $\mathscr{B}$.

3.5.5 Theorem. *Let X be a compact Hausdorff space, $\mathscr{B}_0$ its Baire σ-field and $\mathscr{B}$ its Borel σ-field. Then any positive bounded measure on $\mathscr{B}_0$ can be extended uniquely as a regular measure on $\mathscr{B}$.*

Proof. See Theorem D on p. 239 of Halmos' Measure theory. □

3.6 COMMON EXTENSIONS

The problem considered in this section is the following one. Let $\mathscr{F}$ be a field of subsets of a set Ω and $\mathscr{C}$ and $\mathscr{D}$ be two subfields of $\mathscr{F}$. Let μ_1 and μ_2 be two positive charges on $\mathscr{C}$ and $\mathscr{D}$ respectively. Under what conditions can we find a positive charge μ on $\mathscr{F}$ which is an extension of both μ_1 and μ_2 to $\mathscr{F}$? The condition needed, surprisingly, turns out to be a natural one.

3.6.1 Theorem. *Let $\mathscr{C}$ and $\mathscr{D}$ be two fields on a set Ω and μ_1 and μ_2 positive bounded charges on $\mathscr{C}$ and $\mathscr{D}$ respectively. Let $\mathscr{F}$ be a field on Ω containing both $\mathscr{C}$ and $\mathscr{D}$. Then a necessary and sufficient condition for the existence of a positive bounded charge on $\mathscr{F}$ which is a common extension of both μ_1 and μ_2 is that*

$$\mu_1(C) \geq \mu_2(D) \quad \text{whenever } C \in \mathscr{C}, D \in \mathscr{D} \text{ and } C \supset D,$$

and

$$\mu_1(E) \leq \mu_2(F) \quad \text{whenever } E \in \mathscr{C}, F \in \mathscr{D} \text{ and } E \subset F.$$

Proof. The necessity of the condition is obvious. We will prove the sufficiency. From the given condition, it follows that $\mu_1(A) = \mu_2(A)$ whenever $A \in \mathscr{C} \cap \mathscr{D}$. We define a function $\bar{\mu}$ on $\mathscr{C} \cup \mathscr{D}$ unambiguously as follows.

$$\bar{\mu}(A) = \mu_1(A), \quad \text{if } A \in \mathscr{C},$$
$$= \mu_2(A), \quad \text{if } A \in \mathscr{D}.$$

We now show that $\bar{\mu}$ is a positive real partial charge on $\mathscr{C} \cup \mathscr{D}$. Let $A_1, A_2, \ldots, A_{m+n}$ and $B_1, B_2, \ldots, B_{p+q}$ be two finite sequences of sets from

$\mathscr{C} \cup \mathscr{D}$ such that

$$\sum_{i=1}^{m+n} I_{A_i} \geq \sum_{i=1}^{p+q} I_{B_i}.$$

Assume, without loss of generality, that $A_1, A_2, \ldots, A_m \in \mathscr{C}$, $A_{m+1}, A_{m+2}, \ldots, A_{m+n} \in \mathscr{D}$, $B_1, B_2, \ldots, B_p \in \mathscr{C}$ and $B_{p+1}, B_{p+2}, \ldots, B_{p+q} \in \mathscr{D}$. The above inequality can be written in the form

$$\sum_{i=1}^{m} I_{A_i} + \sum_{i=1}^{n} I_{A_{m+i}} \geq \sum_{i=1}^{p} I_{B_i} + \sum_{i=1}^{q} I_{B_{p+i}}.$$

Since $\varnothing \in \mathscr{C} \cap \mathscr{D}$, we can assume, without loss of generality, that $m = p$ and $n = q$. By Lemma 3.1.4, we can assume $A_1 \supset A_2 \supset \cdots \supset A_m$; $A_{m+1} \supset A_{m+2} \supset \cdots \supset A_{m+n}$; $B_1 \supset B_2 \supset \cdots \supset B_p$; $B_{p+1} \supset B_{p+2} \supset \cdots \supset B_{p+q}$. The above inequality can be rewritten in the form

$$\sum_{i=1}^{m} I_{A_i - B_i} + \sum_{i=1}^{n} I_{A_{m+i} - B_{m+i}} \geq \sum_{i=1}^{m} I_{B_i - A_i} + \sum_{i=1}^{n} I_{B_{m+i} - A_{m+i}}.$$

Let $C_i = A_i - B_i$, $i = 1, 2, \ldots, m$; $D_i = B_i - A_i$, $i = 1, 2, \ldots, m$; $E_i = A_{m+i} - B_{m+i}$, $i = 1, 2, \ldots, n$; $F_i = B_{m+i} - A_{m+i}$, $i = 1, 2, \ldots, n$. The above inequality then becomes

$$\sum_{i=1}^{m} I_{C_i} + \sum_{i=1}^{n} I_{E_i} \geq \sum_{i=1}^{m} I_{D_i} + \sum_{i=1}^{n} I_{F_i}.$$

Observe that $C_i \cap D_j = \varnothing$ for all i and j and $E_i \cap F_j = \varnothing$ for all i and j. By Lemma 3.1.4, we can assume that $C_1 \supset C_2 \supset \cdots \supset C_m$; $D_1 \supset D_2 \supset \cdots \supset D_m$; $E_1 \supset E_2 \supset \cdots \supset E_n$; $F_1 \supset F_2 \supset \cdots \supset F_n$. The above inequality leads to two inequalities:

$$\sum_{i=1}^{m} I_{C_i} \geq \sum_{i=1}^{n} I_{F_i} \quad \text{and} \quad \sum_{i=1}^{n} I_{E_i} \geq \sum_{i=1}^{m} I_{D_i}.$$

Note that $C_1, C_2, \ldots, C_m$, $D_1, D_2, \ldots, D_m \in \mathscr{C}$ and $E_1, E_2, \ldots, E_n$, $F_1, F_2, \ldots, F_n \in \mathscr{D}$. Now, observe that $F_i \subset C_i$ and $D_i \subset E_i$ for all i. Consequently, by the given condition,

$$\sum_{i=1}^{m} \mu_1(C_i) \geq \sum_{i=1}^{n} \mu_2(F_i) \quad \text{and} \quad \sum_{i=1}^{n} \mu_2(E_i) \geq \sum_{i=1}^{m} \mu_1(D_i).$$

Hence

$$\sum_{i=1}^{m} \bar{\mu}(C_i) + \sum_{i=1}^{n} \bar{\mu}(E_i) \geq \sum_{i=1}^{m} \bar{\mu}(D_i) + \sum_{i=1}^{n} \bar{\mu}(F_i).$$

From this, it follows that (using Proposition 3.1.5),

$$\sum_{i=1}^{m+n} \bar{\mu}(A_i) \geq \sum_{i=1}^{p+q} \bar{\mu}(B_i).$$

Hence $\bar{\mu}$ is a positive real partial charge on $\mathscr{C} \cup \mathscr{D}$. By Theorem 3.2.10, we can find a positive bounded charge μ on $\mathscr{F}$ which is an extension of $\bar{\mu}$ from $\mathscr{C} \cup \mathscr{D}$ to $\mathscr{F}$. This completes the proof. □

Now, we take up the case of real charges.

3.6.2 Theorem. *Let $\mathscr{C}$ and $\mathscr{D}$ be two fields on a set Ω and μ_1 and μ_2 two real charges on $\mathscr{C}$ and $\mathscr{D}$ respectively. Let $\mathscr{F}$ be a field on Ω containing both $\mathscr{C}$ and $\mathscr{D}$. A necessary and sufficient condition for the existence of a real charge $\bar{\mu}$ on $\mathscr{F}$ which is a common extension of both μ_1 and μ_2 is that $\mu_1(A) = \mu_2(A)$ for every A in $\mathscr{C} \cap \mathscr{D}$.*

Proof. The proof of this theorem is similar to the one given for the previous theorem. □

It may be remarked that these two theorems remain valid if μ_1 and μ_2 take infinite values.

The two theorems proved above are not extendable to the case when there are more than two subfields of $\mathscr{F}$. Here are the relevant examples.

3.6.3 Example. Let

$$\Omega = \{1, 2, 3, 4\}; \qquad \mathscr{F} = \mathscr{P}(\Omega);$$

$$\mathscr{C}_1 = \{\varnothing, \{1, 2\}, \{3, 4\}, \Omega\}; \qquad \mathscr{C}_2 = \{\varnothing, \{1, 3\}, \{2, 4\}, \Omega\};$$

$$\mathscr{C}_3 = \{\varnothing, \{1, 4\}, \{2, 3\}, \Omega\};$$

$$\mu_1(\{1, 2\}) = \tfrac{1}{4}, \quad \mu_1(\{3, 4\}) = \tfrac{3}{4}; \quad \mu_2(\{1, 3\}) = \tfrac{1}{4},$$

$$\mu_2(\{2, 4\}) = \tfrac{3}{4}; \quad \mu_3(\{1, 4\}) = \tfrac{1}{4}, \quad \mu_3(\{2, 3\}) = \tfrac{3}{4}.$$

Each pair of μ_1 and μ_2, μ_1 and μ_3, μ_2 and μ_3 satisfies the condition of Theorem 3.6.1. But there is no positive charge on $\mathscr{F}$ which is a common extension of all the three charges μ_1, μ_2 and μ_3.

3.6.4 Example. Let

$$\Omega = \{1, 2, 3\}; \qquad \mathscr{F} = \mathscr{P}(\Omega);$$

$$\mathscr{C}_1 = \{\varnothing, \{1\}, \{2, 3\}, \Omega\}; \qquad \mathscr{C}_2 = \{\varnothing, \{2\}, \{1, 3\}, \Omega\};$$

$$\mathscr{C}_3 = \{\varnothing, \{3\}, \{1, 2\}, \Omega\};$$

$$\mu_1(\{1\}) = \tfrac{1}{2} = \mu_1(\{2, 3\}); \quad \mu_2(\{2\}) = \tfrac{2}{3}, \mu_2(\{1, 3\}) = \tfrac{1}{3};$$

$$\mu_3(\{3\}) = \tfrac{1}{4}, \quad \mu_3(\{1, 2\}) = \tfrac{3}{4}.$$

Note that $\mu_1 = \mu_2$ on $\mathscr{C}_1 \cap \mathscr{C}_2$, $\mu_1 = \mu_3$ on $\mathscr{C}_1 \cap \mathscr{C}_3$ and $\mu_2 = \mu_3$ on $\mathscr{C}_2 \cap \mathscr{C}_3$. But there is no real charge μ on $\mathscr{F}$ which is a common extension of μ_1, μ_2 and μ_3.

CHAPTER 4

Integration

In this chapter, we develop the theory of integration for real valued functions with respect to charges. Integration with respect to charges requires a good deal of tact, patience and circumspection to get around measurability problems. The treatment of this topic given here is fairly comprehensive. After presenting the preliminaries in the first three sections, we develop D-integral as presented by Dunford and Schwartz in Section 4.4. In Section 4.5, we introduce S-integrals which are of Stieltjes type and make comparisons with D-integrals. L_p-spaces are introduced and studied in Section 4.6. Finally, in Section 4.7, ba$(\Omega, \mathscr{F})$ is realized as a dual space.

4.1 TOTAL VARIATION AND OUTER CHARGES

Let μ be a charge defined on a field $\mathscr{F}$ of subsets of a set Ω. Recall the definitions of positive and negative variations, μ^+ and μ^-, of μ as expostulated in Section 2.5.

$$\mu^+(A) = \text{Sup}\{\mu(B); B \subset A, B \in \mathscr{F}\}, \quad A \in \mathscr{F},$$

and

$$\mu^-(A) = -\text{Inf}\{\mu(B); B \subset A, B \in \mathscr{F}\}, \quad A \in \mathscr{F}.$$

As has been noted in Theorem 2.5.3, μ^+ and μ^- are positive charges on $\mathscr{F}$. The total variation $|\mu|$ of μ has been defined for bounded charges μ on $\mathscr{F}$. This notion can be introduced for any charge μ.

4.1.1 Definition. For any charge μ on a field $\mathscr{F}$ of subsets of a set Ω, the *total variation* $|\mu|$ of μ is defined by

$$|\mu|(A) = \mu^+(A) + \mu^-(A), \quad A \in \mathscr{F}.$$

Clearly, $|\mu|$ is a positive charge on $\mathscr{F}$. $|\mu|$ can also be described in the following way as in Theorem 2.2.4.

4.1.2 Theorem. *For any charge μ on a field $\mathscr{F}$ of subsets of a set Ω and A in $\mathscr{F}$,*

$$|\mu|(A) = \text{Sup} \sum_{i=1}^{n} |\mu(B_i)|$$

holds true, where the supremum is taken over all partitions $\{B_1, B_2, \ldots, B_n\}$ of A in $\mathscr{F}$.

Proof. If $|\mu|(A) < \infty$, then μ is a bounded charge on the field $A \cap \mathscr{F} = \{A \cap B; B \in \mathscr{F}\}$ on A and the above equality follows from Theorem 2.2.4. If $|\mu|(A) = \infty$, then either $\mu^+(A) = \infty$ or $\mu^-(A) = \infty$. From the definitions of μ^+ and μ^-, it follows that $\text{Sup} \sum_{i=1}^{n} |\mu(B_i)| = \infty$, where the supremum is taken over all partitions $\{B_1, B_2, \ldots, B_n\}$ of A in $\mathscr{F}$. This proves the theorem. □

Now, we introduce the concept of an outer charge.

4.1.3 Definition. Let $\mathscr{F}$ be a field of subsets of a set Ω and μ a positive charge on $\mathscr{F}$. The set function $\mu^* : \mathscr{P}(\Omega) \to [0, \infty]$ defined by

$$\mu^*(A) = \text{Inf} \{\mu(B); A \subset B, B \in \mathscr{F}\}, \quad A \subset \Omega$$

is called the *outer charge* induced by μ.

The following proposition chronicles some of the properties of outer charges.

4.1.4 Proposition. *Let $\mathscr{F}$ be a field of subsets of a set Ω and μ a positive charge on $\mathscr{F}$. Then the following are true.*
  (i). $\mu^*(\varnothing) = 0$.
  (ii). $\mu^*(A) \leq \mu^*(B)$ *if* $A \subset B \subset \Omega$.
  (iii). $\mu^*(A) = \mu(A)$ *if* $A \in \mathscr{F}$.
  (iv). $\mu^*(A \cup B) \leq \mu^*(A) + \mu^*(B)$ *if* $A, B \subset \Omega$.
  (v). *If $\mathscr{F}$ is a σ-field and μ is a measure on $\mathscr{F}$,* $\mu^*(\bigcup_{n \geq 1} A_n) \leq \sum_{n \geq 1} \mu^*(A_n)$ *for any sequence A_n, $n \geq 1$ of subsets of Ω.*

Proof. Properties (i), (ii) and (iii) are obvious. To prove (iv), we proceed as follows. If either $\mu^*(A) = \infty$ or $\mu^*(B) = \infty$, the inequality obviously follows. Suppose $\mu^*(A) < \infty$ and $\mu^*(B) < \infty$. Let $\varepsilon > 0$. There exist A_1, B_1 in $\mathscr{F}$ such that $A \subset A_1$, $B \subset B_1$, $\mu(A_1) \leq \mu^*(A) + \varepsilon/2$ and $\mu(B_1) \leq \mu^*(B) + \varepsilon/2$. Consequently, $\mu^*(A \cup B) \leq \mu^*(A_1 \cup B_1) = \mu(A_1 \cup B_1) \leq \mu(A_1) + \mu(B_1) \leq \mu^*(A) + \mu^*(B) + \varepsilon$. Since $\varepsilon > 0$ is arbitrary, the desired inequality follows. The proof of (v) is analogous to that of (iv). □

4.1.5 Remark. Proposition 4.1.4(v) is not valid if $\mathscr{F}$ is only a field. Let $\Omega = \{1, 2, 3, \ldots, \infty\}$ and $\mathscr{F} = \{A$; either A or A^c is a finite subset of

$\{1, 2, 3, \ldots\}\}$. Define μ on $\mathscr{F}$ by

$$\mu(A) = 0, \quad \text{if A is a finite subset of } \{1, 2, 3, \ldots\}.$$
$$= 1, \quad \text{otherwise.}$$

μ is a measure on the field $\mathscr{F}$. Note that $\mu^*(\{1, 2, 3, \ldots\}) = 1$. If we let $A_n = \{n\}$, $n \geq 1$, then $\mu^*(\bigcup_{n \geq 1} A_n) = 1$ and $\sum_{n \geq 1} \mu^*(A_n) = 0$.

Finally, we end this section with a result on the outer charge induced by the sum of two charges.

4.1.6 Proposition. *Let μ_1 and μ_2 be two positive charges on a field $\mathscr{F}$ of subsets of a set Ω. Then*

$$(\mu_1 + \mu_2)^* = \mu_1^* + \mu_2^*.$$

Proof. It is obvious that $(\mu_1 + \mu_2)^*(A) \geq \mu_1^*(A) + \mu_2^*(A)$ for every $A \subset \Omega$. If either $\mu_1^*(A) = \infty$ or $\mu_2^*(A) = \infty$, then $(\mu_1 + \mu_2)^*(A) \leq \mu_1^*(A) + \mu_2^*(A)$ is true. Assume that $\mu_1^*(A) < \infty$ and $\mu_2^*(A) < \infty$. Let $\varepsilon > 0$. There exist B_1, B_2 in $\mathscr{F}$ such that $A \subset B_1$, $A \subset B_2$,

$$\mu_1(B_1) \leq \mu_1^*(A) + \varepsilon/2$$

and

$$\mu_2(B_2) \leq \mu_2^*(A) + \varepsilon/2.$$

Hence $(\mu_1 + \mu_2)^*(A) \leq (\mu_1 + \mu_2)(B_1 \cap B_2) \leq \mu_1^*(A) + \mu_2^*(A) + \varepsilon$. Since $\varepsilon > 0$ is arbitrary, the result follows. $\square$

4.2 NULL SETS AND NULL FUNCTIONS

In this section, we formalize the notions of a null set and a null function. We first introduce the notion of a charge space.

4.2.1 Definition. *A charge space is a triple $(\Omega, \mathscr{F}, \mu)$, where Ω is a set, $\mathscr{F}$ is a field on Ω and μ a charge on $\mathscr{F}$.*

4.2.2 Definition. *Let $(\Omega, \mathscr{F}, \mu)$ be a charge space. A subset A of Ω is said to be a μ-null set, or simply, a null set if μ is understood, if $|\mu|^*(A) = 0$.*

The following properties of null sets follow from Proposition 4.1.4.

4.2.3 Proposition. *Let $(\Omega, \mathscr{F}, \mu)$ be a charge space. Then the following statements are true.*
   (i). *$\varnothing$ is a null set.*
   (ii). *B is a null set if $B \subset A$ and A is a null set.*

(iii). $\bigcup_{i=1}^{n} A_i$ is a null set if $A_1, A_2, \ldots, A_n$ are null sets, where n is any positive integer.

(iv). $\bigcup_{n \geq 1} A_n$ is a null set if A_n, $n \geq 1$ is a sequence of null sets, μ is a measure and $\mathscr{F}$ is a σ-field.

The concept of a null set leads to the concept of a null function.

4.2.4 Definitions. Let $(\Omega, \mathscr{F}, \mu)$ be a charge space.

(i). A real valued function f on Ω is said to be a μ-*null function*, or simply a *null function* if μ is understood, if

$$|\mu|^*(\{\omega \in \Omega; |f(\omega)| > \varepsilon\}) = 0$$

for every $\varepsilon > 0$.

(ii). Two real valued functions f and g defined on Ω are said to be *equal almost everywhere* if $f - g$ is a null function. In this case, we use the notation $f = g$ a.e.$[\mu]$.

(iii). A real valued function f on Ω is said to be *dominated almost everywhere* by a real valued function g on Ω, if there exists a null function h on Ω such that $f \leq g + h$. In such a case, we use the notation $f \leq g$ a.e.$[\mu]$.

A sufficient condition for $f \leq g$ a.e. $[\mu]$ to hold is that $|\mu|^*(\{\omega \in \Omega; f(\omega) > g(\omega)\}) = 0$, though not necessary.

The following proposition follows easily from Proposition 4.1.4.

4.2.5 Proposition. *Let $(\Omega, \mathscr{F}, \mu)$ be a charge space. The following statements are true.*

(i). $cf + dg$, $|f|^p (p > 0)$ *and fg are null functions whenever f and g are null functions on Ω and c and d are real numbers.*

(ii). *g is a null function if $|g| \leq |f|$ a.e. $[\mu]$ and f is a null function.*

(iii). *For real functions f, g, h on Ω, $f = h$ a.e. $[\mu]$ whenever $f = g$ a.e. $[\mu]$ and $g = h$ a.e. $[\mu]$.*

(iv). *For real functions f_1, f_2, g_1, g_2 on Ω and c, d real numbers with $f_1 = g_1$ a.e. $[\mu]$ and $f_2 = g_2$ a.e. $[\mu]$, $cf_1 + df_2 = cg_1 + dg_2$ a.e. $[\mu]$ and $|f_1| = |g_1|$ a.e. $[\mu]$.*

(v). *On the space $C(\Omega, \mathscr{F}, \mu)$ of all real functions on Ω, the binary relation $\sim$ defined by $f \sim g$ if $f = g$ a.e. $[\mu]$ is an equivalence relation.*

(vi). *For real functions f, g, h on Ω, $f \leq h$ a.e. $[\mu]$ whenever $f \leq g$ a.e. $[\mu]$ and $g \leq h$ a.e. $[\mu]$.*

(vii). *A subset A of Ω is a null set if and only if I_A is a null function.*

(viii). *For real functions f, g on Ω, $f = g$ a.e. $[\mu]$ iff $f \leq g$ a.e. $[\mu]$ and $g \leq f$ a.e. $[\mu]$.*

4.2.6 Remark. If $f_1 = g_1$ a.e. $[\mu]$ and $f_2 = g_2$ a.e. $[\mu]$, it is not true that $f_1 f_2 = g_1 g_2$ a.e. $[\mu]$. It is not even true that $f^2 = g^2$ a.e. $[\mu]$ if $f = g$ a.e. $[\mu]$. The following example explains this.

Example. Let $\Omega = \{1, 2, 3, \ldots\}$, $\mathscr{F}$ the finite-cofinite field on Ω and μ the charge on $\mathscr{F}$ defined by

$$\mu(A) = 0, \quad \text{if A is finite,}$$
$$= 1, \quad \text{if A is cofinite.}$$

Let f and g on Ω be defined by $f(n) = n + (1/n)$, $n \geq 1$ and $g(n) = n - (1/n)$, $n \geq 1$. Then $f = g$ a.e. $[\mu]$. But $f^2 = g^2$ a.e. $[\mu]$ does not hold.

A sufficient condition for a real function f on Ω to be a null function is that $|\mu|^*(\{\omega \in \Omega; f(\omega) \neq 0\}) = 0$, though not necessary. The following proposition amplifies this point.

4.2.7 Proposition. *Let $(\Omega, \mathscr{F}, \mu)$ be a charge space. Let f be a real valued function defined on Ω.*

(i). *If $|\mu|^*(\{\omega \in \Omega; f(\omega) \neq 0\}) = 0$, then f is a null function.*

(ii). *The converse of (i) is not true.*

(iii). *If $\mathscr{F}$ is a σ-field and μ is a measure on $\mathscr{F}$, then f is a null function if and only if $|\mu|^*(\{\omega \in \Omega; f(\omega) \neq 0\}) = 0$.*

Proof. (i). Note that for any $\varepsilon > 0$,

$$\{\omega \in \Omega; |f(\omega)| > \varepsilon\} \subset \{\omega \in \Omega; f(\omega) \neq 0\}.$$

The monotonicity of $|\mu|^*$ completes the proof.

(ii). *Example.* Let $(\Omega, \mathscr{F}, \mu)$ be as in Remark 4.2.6. Let the function f on Ω be defined by $f(n) = 1/n$, $n \geq 1$. Then f is a null function. For, $\{\omega \in \Omega; |f(\omega)| > \varepsilon\}$ is a finite set for any $\varepsilon > 0$. On the other hand, $|\mu|^*(\{\omega \in \Omega; f(\omega) \neq 0\}) = 1$.

(iii). Observe that

$$\bigcup_{\varepsilon > 0} \{\omega \in \Omega; |f(\omega)| > \varepsilon\} = \bigcup_{n \geq 1} \{\omega \in \Omega; |f(\omega)| > 1/n\}$$
$$= \{\omega \in \Omega; f(\omega) \neq 0\}.$$

By Proposition 4.2.3, the result follows. □

Now, we come to the concept of essential boundedness.

4.2.8 Definition. Let $(\Omega, \mathscr{F}, \mu)$ be a charge space and f a real valued function on Ω. f is said to be *essentially bounded* if there exists a null set A contained in Ω such that f is bounded on A^c. If f is essentially bounded, the *essential supremum* of f is denoted by $\|f\|_\infty$ and is defined by

$$\|f\|_\infty = \text{Inf Sup}\{|f(\omega)|; \omega \in A^c\},$$

where the infimum is taken over all null sets A contained in Ω.

The following proposition gives equivalent conditions for essential boundedness of a function.

4.2.9 Proposition. *Let $(\Omega, \mathscr{F}, \mu)$ be a charge space and f a real valued function on Ω. Then the following statements are equivalent.*

(i). *f is essentially bounded.*
(ii). *There exists $k > 0$ such that $|\mu|^*(\{\omega \in \Omega; |f(\omega)| > k\}) = 0$.*
(iii). *There exists $k > 0$ such that $|f| \le k$ a.e. $[\mu]$.*
(iv). *There exists a bounded function g on Ω such that $f = g$ a.e. $[\mu]$.*

The following proposition gives the properties of essentially bounded functions which are easily established.

4.2.10 Proposition. *Let $(\Omega, \mathscr{F}, \mu)$ be a charge space. (All functions considered below are real valued functions on Ω.)*

(i). *If f and g are essentially bounded and c and d are real numbers, then $cf + dg$ is essentially bounded. Further,*

$$\|cf + dg\|_\infty \le |c|\|f\|_\infty + |d|\|g\|_\infty.$$

(ii). *$\|f\|_\infty = 0$ if and only if f is a null function.*
(iii). *If f is essentially bounded and $f = g$ a.e. $[\mu]$, then g is essentially bounded and $\|f\|_\infty = \|g\|_\infty$.*
(iv). *If $B(\Omega, \mathscr{F}, \mu)$ is the collection of all essentially bounded functions on Ω, then the map $\|\cdot\|_\infty : B(\Omega, \mathscr{F}, \mu) \to [0, \infty)$ is a pseudo-norm on $B(\Omega, \mathscr{F}, \mu)$.*

4.2.11 Remark. Introduce an equivalence relation $\sim$ on $B(\Omega, \mathscr{F}, \mu)$ by $f \sim g$ for f, g in $B(\Omega, \mathscr{F}, \mu)$ if $f = g$ a.e. $[\mu]$. Let $C_\infty(\Omega, \mathscr{F}, \mu)$ be the collection of all equivalence classes of $B(\Omega, \mathscr{F}, \mu)$ under $\sim$. For f in $B(\Omega, \mathscr{F}, \mu)$, let $[f]$ denote the equivalence class containing f. The map $\|\cdot\|_\infty : C_\infty(\Omega, \mathscr{F}, \mu) \to [0, \infty)$ defined by $\|[f]\|_\infty = \|f\|_\infty$ for $[f]$ in $C_\infty(\Omega, \mathscr{F}, \mu)$ is a norm on $C_\infty(\Omega, \mathscr{F}, \mu)$.

The following concept of a simple function is important for the development of D-integral.

4.2.12 Definition. *Let $\mathscr{F}$ be a field of subsets of a set Ω. A real valued function f on Ω is said to be a simple function if it can be written in the form*

$$f = \sum_{i=1}^{n} c_i I_{F_i}$$

for some real numbers $c_1, c_2, \ldots, c_n$; $F_1, F_2, \ldots, F_n$ in $\mathscr{F}$ with $F_i \cap F_j = \emptyset$ for every $i \ne j$ and $\bigcup_{i=1}^{n} F_i = \Omega$.

The following properties of simple functions are clear.

4.2.13 Proposition. *Let $\mathscr{F}$ be a field of subsets of a set Ω. If f and g are simple functions on Ω and c and d are real numbers, then $cf + dg$, fg and $|f|^p (p > 0)$ are all simple functions.*

Now, we introduce smooth functions.

4.2.14 Definition. Let $(\Omega, \mathscr{F}, \mu)$ be a charge space. A real valued function f on Ω is said to be *smooth* if for every $\varepsilon > 0$ there exists $k > 0$ such that

$$|\mu|^*(\{\omega \in \Omega; |f(\omega)| > k\}) < \varepsilon.$$

Not every function is a smooth function as the following example demonstrates.

4.2.15 Example. Let $(\Omega, \mathscr{F}, \mu)$ be as in Remark 4.2.6. The function f on Ω defined by $f(n) = n$, $n \in \Omega$ is not smooth.

However, in some cases, many natural functions are smooth. In what follows, let $\mathscr{B}$ denote the Borel σ-field on the real line R, i.e. $\mathscr{B}$ is the smallest σ-field on R containing all subintervals of R.

4.2.16 Definition. Let $(\Omega, \mathscr{F}, \mu)$ be a charge space in which $\mathscr{F}$ is a σ-field on Ω and μ a measure on $\mathscr{F}$. A real valued function f on Ω is said to be *μ-measurable* if there exists a null set $N \subset \Omega$ such that $f^{-1}(B) \cap N^c \in \mathscr{F}$ for every B in $\mathscr{B}$.

4.2.17 Proposition. *Let $(\Omega, \mathscr{F}, \mu)$ be a charge space in which $\mathscr{F}$ is a σ-field and μ a real measure on $\mathscr{F}$. Then every μ-measurable function f on Ω is smooth.*

Proof. Let N be a null set such that $f^{-1}(B) \cap N^c \in \mathscr{F}$ for every B in $\mathscr{B}$. Let $B_n = \{\omega \in \Omega; |f(\omega)| \geq n\}$, $n \geq 1$. Note that B_n, $n \geq 1 \downarrow \emptyset$. Since μ is a real measure on the σ-field $\mathscr{F}$, $|\mu|$ is a bounded measure. Consequently, $\lim_{n \to \infty} |\mu|(B_n \cap N^c) = 0$. For any $\varepsilon > 0$, we can find a positive integer k such that $|\mu|(B_k \cap N^c) < \varepsilon$. Now, observe that

$$|\mu|^*(B_k) = |\mu|^*((B_k \cap N) \cup (B_k \cap N^c))$$
$$\leq |\mu|^*(B_k \cap N) + |\mu|^*(B_k \cap N^c)$$
$$\leq |\mu|^*(N) + |\mu|^*(B_k \cap N^c) < 0 + \varepsilon = \varepsilon.$$

This proves the result. □

The following proposition records some of the properties of smooth functions.

4.2.18 Proposition. *Let $(\Omega, \mathscr{F}, \mu)$ be a charge space.*
(i). *If f and g are smooth functions on Ω and c and d are real numbers, then $cf + dg$, fg and $|f|^p$ ($p > 0$) are all smooth functions.*
(ii). *If f is essentially bounded, then f is smooth.*

4.3 HAZY CONVERGENCE

In this section, we introduce the notion of hazy convergence in the space of all real valued functions on Ω of a charge space $(\Omega, \mathcal{F}, \mu)$. This notion is commonly known as "convergence in measure" which is a misnomer in the present context of charges.

4.3.1 Definition. Let $(\Omega, \mathcal{F}, \mu)$ be a charge space. A sequence f_n, $n \geq 1$ of real valued functions on Ω is said to converge to a real valued function f on Ω *hazily* if

$$\lim_{n \to \infty} |\mu|^*(\{\omega \in \Omega; |f_n(\omega) - f(\omega)| > \varepsilon\}) = 0$$

for every $\varepsilon > 0$.

The following proposition establishes that the limit function in hazy convergence is essentially unique.

4.3.2 Proposition. *Let $(\Omega, \mathcal{F}, \mu)$ be a charge space. If a sequence f_n, $n \geq 1$ of real valued functions on Ω converges to a real valued function f on Ω hazily and $f = g$ a.e. $[\mu]$, then f_n, $n \geq 1$ converges to g hazily. Conversely, if f_n, $n \geq 1$ converges to both f and g hazily, then $f = g$ a.e. $[\mu]$.*

Proof. For the first part, observe that for any $\varepsilon > 0$,

$$\{\omega \in \Omega; |f_n(\omega) - g(\omega)| > \varepsilon\} \subset \{\omega \in \Omega; |f_n(\omega) - f(\omega)| > \varepsilon/2\}$$
$$\cup \{\omega \in \Omega; |f(\omega) - g(\omega)| > \varepsilon/2\}.$$

For the second part, observe that for any $\varepsilon > 0$,

$$\{\omega \in \Omega; |f(\omega) - g(\omega)| > \varepsilon\} \subset \{\omega \in \Omega; |f_n(\omega) - f(\omega)| > \varepsilon/2\}$$
$$\cup \{\omega \in \Omega; |f_n(\omega) - g(\omega)| > \varepsilon/2\}.$$

The monotonicity and the sub-additivity properties of $|\mu|^*$ complete the proof. □

The following theorem gives algebraic properties of hazy convergence.

4.3.3 Theorem. *Let $(\Omega, \mathcal{F}, \mu)$ be a charge space. Let f_n, $n \geq 1$ and g_n, $n \geq 1$ be two sequences of real valued functions on Ω converging hazily to real valued functions f and g on Ω respectively. Then the following statements are true.*

(i). *$cf_n + dg_n$, $n \geq 1$ converges to $cf + dg$ hazily for any two real numbers c and d.*

(ii). *$|f_n|$, $n \geq 1$ converges to $|f|$ hazily.*

4. INTEGRATION

(iii). f_n^2, $n \geq 1$ converges to f^2 hazily, if f is a null function.
(iv). $f_n h$, $n \geq 1$ converges to fh hazily, if h is a smooth function on Ω.
(v). If f is smooth and ψ is a real valued continuous function defined on the real line, then $\psi(f_n)$, $n \geq 1$ converges to $\psi(f)$ hazily. In particular, $|f_n|^p$, $n \geq 1$ converges to $|f|^p$ hazily for any $p > 0$.
(vi). $f_n g_n$, $n \geq 1$ converges to fg hazily if f and g are smooth functions.
(vii). f_n^+, $n \geq 1$ converges to f^+ and f_n^-, $n \geq 1$ converges to f^- hazily.
(viii). $f_n \vee g_n$, $n \geq 1$ converges to $f \vee g$ and $f_n \wedge g_n$, $n \geq 1$ converges to $f \wedge g$ hazily.

Proof. (i), (ii) and (iii) are easy to prove.
(iv). Let $\varepsilon > 0$. Since h is a smooth function, there exists a real number $k > 0$ such that $|\mu|^*(\{\omega \in \Omega; |h(\omega)| > k\}) < \varepsilon/2$. Since f_n, $n \geq 1$ converges to f hazily, there exists an integer $m \geq 1$ such that $|\mu|^*(\{\omega \in \Omega; |f_n(\omega) - f(\omega)| > \delta/k\}) < \varepsilon/2$ whenever $n \geq m$, where δ is a given positive number. So, if $n \geq m$,

$$|\mu|^*(\{\omega \in \Omega; |f_n(\omega)h(\omega) - f(\omega)h(\omega)| > \delta\})$$
$$\leq |\mu|^*(\{\omega \in \Omega; |f_n(\omega) - f(\omega)||h(\omega)| > \delta, |h(\omega)| \leq k\})$$
$$+ |\mu|^*(\{\omega \in \Omega; |f_n(\omega) - f(\omega)||h(\omega)| > \delta, |h(\omega)| > k\})$$
$$< |\mu|^*(\{\omega \in \Omega; |f_n(\omega) - f(\omega)| > \delta/k\}) + \varepsilon/2$$
$$< \varepsilon/2 + \varepsilon/2 = \varepsilon.$$

This proves (iv).
(v). Let $\varepsilon_1 > 0$ and $\varepsilon_2 > 0$. Since f is smooth, there exists a real number $k > 0$ such that $|\mu|^*(\{\omega \in \Omega; |f(\omega)| > k\}) < \varepsilon_1/2$. Since ψ is uniformly continuous on $[-2k, 2k]$, there exists $\delta > 0$ such that $|\psi(x) - \psi(y)| < \varepsilon_2$ whenever $|x|, |y| \leq 2k$ and $|x - y| < \delta$. Without loss of generality, assume that $\delta < k$. Since f_n, $n \geq 1$ converges to f hazily, there exists $m \geq 1$ such that $|\mu|^*(\{\omega \in \Omega; |f_n(\omega) - f(\omega)| > \delta\}) < \varepsilon_1/2$ whenever $n \geq m$. Now, if $n \geq m$,

$$|\mu|^*(\{\omega \in \Omega; |\psi(f_n(\omega)) - \psi(f(\omega))| > \varepsilon_2\})$$
$$\leq |\mu|^*(\{\omega \in \Omega; |\psi(f_n(\omega)) - \psi(f(\omega))| > \varepsilon_2, |f(\omega)| > k\})$$
$$+ |\mu|^*(\{\omega \in \Omega; |\psi(f_n(\omega)) - \psi(f(\omega))| > \varepsilon_2, |f(\omega)| \leq k, |f_n(\omega) - f(\omega)| > \delta\})$$
$$+ |\mu|^*(\{\omega \in \Omega; |\psi(f_n(\omega)) - \psi(f(\omega))| > \varepsilon_2, |f(\omega)| \leq k, |f_n(\omega) - f(\omega)| \leq \delta\})$$
$$< \varepsilon_1/2 + \varepsilon_1/2 + 0 = \varepsilon_1.$$

Thus $\psi(f_n)$, $n \geq 1$ converges to $\psi(f)$ hazily.
(vi). Observe that for any $n \geq 1$,

$$f_n g_n = \tfrac{1}{4}[(f_n + g_n)^2 - (f_n - g_n)^2].$$

Since f and g are smooth, $f+g$ and $f-g$ are smooth. See Proposition 4.2.18(i). By (i) and (v), $(f_n+g_n)^2$, $n \geq 1$ and $(f_n-g_n)^2$, $n \geq 1$ converge hazily to $(f+g)^2$ and $(f-g)^2$ respectively. Again, by (i), $f_n g_n$, $n \geq 1$ converges to fg hazily.

(vii). Observe that $f^+ = \frac{1}{2}(f+|f|)$ and $f^- = \frac{1}{2}(f-|f|)$. The assertion now follows from (i) and (ii).

(viii). Observe that $f_n \vee g_n = \frac{1}{2}(f_n+g_n+|f_n-g_n|)$ and $f_n \wedge g_n = \frac{1}{2}(f_n+g_n-|f_n-g_n|)$ for all $n \geq 1$. □

4.3.4 Remark. The assertion of Theorem 4.3.3(iv) is not valid unconditionally. We need to impose some conditions on h to ensure the hazy convergence of $f_n h$, $n \geq 1$ to fh. The following is a relevant example.

Let $(\Omega, \mathcal{F}, \mu)$ be as in Remark 4.2.6. For $n \geq 1$, let

$$f_n(k) = k, \quad \text{if } 1 \leq k \leq n,$$
$$= 1 - 1/n, \quad \text{if } k > n;$$
$$f(k) = 1 \quad \text{for all } k \text{ in } \Omega;$$
$$h(k) = k \quad \text{for all } k \text{ in } \Omega.$$

It can be checked easily that f_n, $n \geq 1$ converges to f hazily. But $f_n h$, $n \geq 1$ fails to converge to $fh = h$ hazily.

Let $(\Omega, \mathcal{F}, \mu)$ be a charge space and $C(\Omega, \mathcal{F}, \mu)$ the collection of all real valued functions on Ω. One can introduce a pseudo-metric ρ on $C(\Omega, \mathcal{F}, \mu)$ such that convergence in the pseudo-metric space $C(\Omega, \mathcal{F}, \mu)$ coincides with hazy convergence.

For f in $C(\Omega, \mathcal{F}, \mu)$, Let

$$\psi(f, c) = c + |\mu|^*(\{\omega \in \Omega; |f(\omega)| > c\}), \quad c > 0.$$

$\psi(f, c)$ is a nonnegative number and could be equal to ∞. If μ is bounded, then $\psi(f, c)$ is a real number for all $c > 0$. Now, define

$$\|f\| = \inf_{c > 0} \frac{\psi(f, c)}{1 + \psi(f, c)}.$$

If $\psi(f, c) = \infty$ for every $c > 0$, we define $\|f\| = 1$. Now, we give the properties of the function $\|\cdot\|$.

4.3.5 Proposition. *The function $\|\cdot\|$ defined on $C(\Omega, \mathcal{F}, \mu)$ above has the following properties.*

(i). $\|f\| = 0$ *if and only if f is a null function.*
(ii). $\|f+g\| \leq \|f\| + \|g\|$.
(iii). *The function ρ defined by $\rho(f, g) = \|f-g\|$ for f, g in $C(\Omega, \mathcal{F}, \mu)$ is a pseudo-metric on $C(\Omega, \mathcal{F}, \mu)$.*

(iv). f_n, $n \geq 1$ converges to f hazily if and only if $\rho(f_n, f)$, $n \geq 1$ converges to zero.

Proof. (i). If f is a null function, then $\psi(f, c) = c$ for every $c > 0$. Consequently,
$$\|f\| = \operatorname*{Inf}_{c>0} \frac{c}{1+c} = 0.$$
Conversely, let $\|f\| = 0$. Let k be any positive number. In order to show that $|\mu|^*(\{\omega \in \Omega; |f(\omega)| > k\}) = 0$, it suffices to show that $|\mu|^*(\{\omega \in \Omega; |f(\omega)| > k\}) < \varepsilon$ for any $0 < \varepsilon < k$. Let $0 < \varepsilon < k$. Since $\|f\| = 0$, there exists $c > 0$ such that $\psi(f, c)/(1 + \psi(f, c)) < \varepsilon/(1 + \varepsilon)$. This implies that $\psi(f, c) < \varepsilon$. So, $c < \varepsilon$ and $|\mu|^*(\{\omega \in \Omega; |f(\omega)| > c\}) < \varepsilon$. Consequently, $|\mu|^*(\{\omega \in \Omega; |f(\omega)| > k\}) \leq |\mu|^*(\{\omega \in \Omega; |f(\omega)| > \varepsilon\}) \leq |\mu|^*(\{\omega \in \Omega; |f(\omega)| > c\}) < \varepsilon$. This shows that f is a null function.

(ii). Observe that
$$\|f + g\| = \operatorname*{Inf}_{c>0} \frac{\psi(f+g, c)}{1 + \psi(f+g, c)}$$
$$= \operatorname*{Inf}_{r>0, s>0} \frac{\psi(f+g, r+s)}{1 + \psi(f+g, r+s)}$$
$$\leq \operatorname*{Inf}_{r>0, s>0} \frac{\psi(f, r)}{1 + \psi(f, r)} + \frac{\psi(g, s)}{1 + \psi(g, s)}$$
$$\leq \operatorname*{Inf}_{r>0} \frac{\psi(f, r)}{1 + \psi(f, r)} + \operatorname*{Inf}_{s>0} \frac{\psi(g, s)}{1 + \psi(g, s)}$$
$$\leq \|f\| + \|g\|.$$

(The first of the above inequalities can be proved as follows. The function $\gamma(x) = x/(1+x)$, $0 \leq x \leq \infty$ is an increasing function having the additional property $\gamma(x + y) \leq \gamma(x) + \gamma(y)$ for all $0 \leq x, y \leq \infty$. We use the convention that $\gamma(\infty) = 1$. If
$$c_1 = |\mu|^*(\{\omega \in \Omega; |f(\omega) + g(\omega)| > r + s\}),$$
$$c_2 = |\mu|^*(\{\omega \in \Omega; |f(\omega)| > r\})$$
and
$$c_3 = |\mu|^*(\{\omega \in \Omega; |g(\omega)| > s\}),$$
then $c_1 \leq c_2 + c_3$. Further,
$$\frac{\psi(f+g, r+s)}{1 + \psi(f+g, r+s)} = \gamma(r + s + c_1) \leq \gamma(r + s + c_2 + c_3)$$
$$= \gamma(r + c_2 + s + c_3) \leq \gamma(r + c_2) + \gamma(s + c_3)$$
$$= \frac{\psi(f, r)}{1 + \psi(f, r)} + \frac{\psi(g, s)}{1 + \psi(g, s)}.$$

From this, the first of the above inequalities follows. The rest of the equalities and inequalities are obvious.)

(iii). This follows from (i) and (ii).

(iv). Suppose f_n, $n \geq 1$ converges to f hazily. Then for any $c > 0$ and $\varepsilon > 0$, there exists $m \geq 1$ such that

$$|\mu|^*(\{\omega \in \Omega; |f_n(\omega) - f(\omega)| > c\}) < \varepsilon$$

whenever $n \geq m$. Consequently, for $n \geq m$,

$$\rho(f_n, f) = \|f_n - f\| = \operatorname*{Inf}_{d > 0} \frac{\psi(f_n - f, d)}{1 + \psi(f_n - f, d)} \leq \frac{\psi(f_n - f, c)}{1 + \psi(f_n - f, c)} < \frac{c + \varepsilon}{1 + c + \varepsilon}$$

$$\leq \frac{c}{1+c} + \frac{\varepsilon}{1+\varepsilon} < \frac{c}{1+c} + \varepsilon.$$

Consequently, $\limsup_{n \to \infty} \rho(f_n, f) \leq c/(1+c) + \varepsilon$. Since $c > 0$ is arbitrary, we have $\limsup_{n \to \infty} \rho(f_n, f) \leq \varepsilon$. Since $\varepsilon > 0$ is arbitrary, it follows that f_n, $n \geq 1$ converges to f in the pseudo-metric space $(C(\Omega, \mathscr{F}, \mu), \rho)$.

Conversely, let $\rho(f_n, f)$, $n \geq 1$ converge to zero. Let k be any positive number. Let $0 < \varepsilon < k$ be arbitrary. There exists $m \geq 1$ such that $\rho(f_n, f) < \varepsilon/(1+\varepsilon)$ whenever $n \geq m$. Now, let $n \geq m$ be given. Since

$$\operatorname*{Inf}_{d > 0} \frac{\psi(f_n - f, d)}{1 + \psi(f_n - f, d)} < \frac{\varepsilon}{1+\varepsilon},$$

there exists $c > 0$ such that $\psi(f_n - f, c)/(1 + \psi(f_n - f, c)) < \varepsilon/(1+\varepsilon)$. This implies that $\psi(f_n - f, c) < \varepsilon$. So, $c < \varepsilon$ and $|\mu|^*(\{\omega \in \Omega; |f_n(\omega) - f(\omega)| > c\}) < \varepsilon$. Consequently,

$$|\mu|^*(\{\omega \in \Omega; |f_n(\omega) - f(\omega)| > k\}) \leq |\mu|^*(\{\omega \in \Omega; |f_n(\omega) - f(\omega)| > \varepsilon\})$$

$$\leq |\mu|^*(\{\omega \in \Omega; |f_n(\omega) - f(\omega)| > c\}) < \varepsilon.$$

This shows that f_n, $n \geq 1$ converges to f hazily. □

4.4 D-INTEGRAL

In this section, we develop the basic ideas concerning D-integrals. We start with simple functions.

4.4.1 Definition. Let $(\Omega, \mathscr{F}, \mu)$ be a charge space and f a simple function on Ω with a representation $f = \sum_{i=1}^{n} c_i I_{F_i}$ for some real numbers $c_1, c_2, \ldots, c_n$ and partition $\{F_1, F_2 \ldots, F_n\}$ of Ω in $\mathscr{F}$. f is said to be *D-integrable* if $|\mu|(F_i) < \infty$ whenever $c_i \neq 0$, and the D-integral of f, denoted by $\mathrm{D} \int f \, d\mu$, is defined to be the real number $\sum_{i=1}^{n} c_i \mu(F_i)$. (We adopt the convention that $0 \cdot (\pm \infty) = 0$.)

4. INTEGRATION

We settle the question of unambiguity of the above definition in the following proposition.

4.4.2 Proposition. *Let $(\Omega, \mathscr{F}, \mu)$ be a charge space.*
(i). *Let f be a simple function with a representation $f = \sum_{i=1}^{m} c_i I_{E_i}$ for some real numbers $c_1, c_2, \ldots, c_m$ and partition $\{E_1, E_2, \ldots, E_m\}$ of Ω in $\mathscr{F}$ such that $|\mu|(E_i) < \infty$ whenever $c_i \neq 0$. If f has another representation $\sum_{j=1}^{n} d_j I_{F_j}$ for some real numbers $d_1, d_2, \ldots, d_n$ and for some partition $\{F_1, F_2, \ldots, F_n\}$ of Ω in $\mathscr{F}$, then $|\mu|(F_j) < \infty$ whenever $d_j \neq 0$, and*

$$\sum_{i=1}^{m} c_i \mu(E_i) = \sum_{j=1}^{n} d_j \mu(F_j).$$

(ii). *If f and g are D-integrable simple functions, then $f + g$ is D-integrable and $D \int (f+g) \, d\mu = D \int f \, d\mu + D \int g \, d\mu$.*
(iii). *If f is a D-integrable simple function and g is a simple function on Ω such that $f = g$ a.e. $[\mu]$, then g is D-integrable and $D \int f \, d\mu = D \int g \, d\mu$.*
(iv). *If f is a D-integrable simple function and E is in $\mathscr{F}$, then $I_E f$ is also a D-integrable simple function.*

Proof. (i). Note that

$$0 = f - f = \sum_{i=1}^{m} c_i I_{E_i} - \sum_{j=1}^{n} d_j I_{F_j} = \sum_{i=1}^{m} \sum_{j=1}^{n} (c_i - d_j) I_{E_i \cap F_j}.$$

This implies that $c_i = d_j$ whenever $E_i \cap F_j \neq 0$. Suppose $d_j \neq 0$. Let $J = \{1 \leq i \leq m; E_i \cap F_j \neq \varnothing\}$. If $i \in J$, then $c_i \neq 0$ and so $|\mu|(E_i) < \infty$. Further, $\bigcup_{i \in J}(E_i \cap F_j) = F_j$. Consequently, $|\mu|(F_j) < \infty$. Also,

$$\sum_{i=1}^{m} c_i \mu(E_i) = \sum_{i=1}^{m} \sum_{j=1}^{n} c_i \mu(E_i \cap F_j)$$

$$= \sum_{j=1}^{n} \sum_{i=1}^{m} c_i \mu(E_i \cap F_j)$$

$$= \sum_{j=1}^{n} \sum_{i=1}^{m} d_j \mu(E_i \cap F_j)$$

$$= \sum_{j=1}^{n} d_j \mu(F_j).$$

(ii). This is simple to prove.
(iii). To prove this, it suffices to show that $\sum_{i=1}^{m} c_i \mu(F_i) = 0$ whenever $h = \sum_{i=1}^{m} c_i I_{F_i}$ is a simple and null function. We claim that for every $1 \leq i \leq m$ either $c_i = 0$ or $\mu(F_i) = 0$. Suppose $c_i \neq 0$. Let $\varepsilon > 0$ be any number such that $\varepsilon < |c_i|$. Then $\{\omega \in \Omega : |h(\omega)| > \varepsilon\} = \bigcup_{j \in J} F_j$, where $J = \{j; 1 \leq j \leq m$ and

$|c_i| > \varepsilon\}$. Obviously, $i \in J$. Since h is a null function, $|\mu|^*(\bigcup_{j \in J} F_j) = |\mu|(\bigcup_{j \in J} F_j) = 0$. Consequently, $|\mu|(F_i) = 0$. This proves (iii).

(iv). The condition of Definition 4.4.1 for $I_E f$ is easily verified from the hypothesis that f is D-integrable and simple. □

Part (iv) of the above proposition enables us to define $D \int_E f \, d\mu$ for any D-integrable simple function f and E in $\mathcal{F}$.

4.4.3 Definition. If f is a D-integrable simple function and $E \in \mathcal{F}$, then $D \int_E f \, d\mu$ stands for $D \int I_E f \, d\mu$.

The following theorem gives the properties of integrals of simple functions.

4.4.4 Theorem. *Let $(\Omega, \mathcal{F}, \mu)$ be a charge space.*

(i). *If f is a simple function on Ω and D-integrable with respect to μ, then f is D-integrable with respect to μ^+ and μ^- also. Further, for any E in $\mathcal{F}$,*

$$D \int_E f \, d\mu = D \int_E f \, d\mu^+ - D \int_E f \, d\mu^-.$$

(ii). *If f and g are D-integrable simple functions on Ω, c and d are real numbers and $E \in \mathcal{F}$, then $cf + dg$ is a D-integrable simple function and*

$$D \int_E (cf + dg) \, d\mu = c \left(D \int_E f \, d\mu \right) + d \left(D \int_E g \, d\mu \right).$$

(iii). *If f is a simple function on Ω and D-integrable with respect to μ, then $|f|$ is a simple function on Ω which is D-integrable with respect to $|\mu|$ and for any E in $\mathcal{F}$*

$$\left| D \int_E f \, d\mu \right| \leq D \int_E |f| \, d|\mu|.$$

(iv). *If f is a D-integrable simple function on Ω and $f \geq 0$ a.e. $[\mu]$ on E in $\mathcal{F}$, i.e. $I_E f \geq 0$ a.e. $[\mu]$, then*

$$D \int_E f \, d|\mu| \geq 0.$$

(v). *If f and g are D-integrable simple functions on Ω and $f \leq g$ a.e. $[\mu]$ on E in $\mathcal{F}$, i.e. $I_E f \leq I_E g$ a.e. $[\mu]$, then*

$$D \int_E f \, d|\mu| \leq D \int_E g \, d|\mu|.$$

(vi). *If f and g are D-integrable simple functions on Ω, then $|f|, |g|, |f+g|,$*

4. INTEGRATION

$\|f|-|g\|$ are all D-integrable simple functions and for any E in $\mathscr{F}$,

$$\left| D\int_E |f| \, d|\mu| - D\int_E |g| \, d|\mu| \right| \leq D\int_E \left| |f|-|g| \right| d|\mu|$$

$$\leq D\int_E |f+g| \, d|\mu| \leq D\int_E |f| \, d|\mu| + D\int_E |g| \, d|\mu|.$$

(vii). *If f is a D-integrable simple function on Ω and $c \leq f \leq d$ a.e. $[\mu]$ on E in $\mathscr{F}$, i.e. $cI_E \leq I_E f \leq dI_E$ a.e. $[\mu]$ for some real numbers c and d, then*

$$c|\mu|(E) \leq D\int_E f \, d|\mu| \leq d|\mu|(E).$$

(viii). *If f is a D-integrable simple function on Ω, then the set function λ on $\mathscr{F}$ defined by*

$$\lambda(F) = D\int_F f \, d\mu, \qquad F \in \mathscr{F}$$

is a bounded charge on $\mathscr{F}$. Also, $|\lambda|(F) = D\int_F |f| \, d|\mu|$, $F \in \mathscr{F}$. Further, λ is absolutely continuous with respect to μ in the following sense. Given $\varepsilon > 0$, there exists $\delta > 0$ such that $|\lambda(E)| < \varepsilon$ whenever $E \in \mathscr{F}$ and $|\mu|(E) < \delta$.

Proof. Note that for simple functions f and g on Ω, $f \leq g$ a.e. $[\mu]$ if and only if $|\mu|(\{\omega \in \Omega; f(\omega) > g(\omega)\}) = 0$. (The set $\{\omega \in \Omega; f(\omega) > g(\omega)\}$ does belong to $\mathscr{F}$.) With this observation, we proceed as follows.

(i). Let $f = \sum_{i=1}^m c_i I_{E_i}$ be a representation of f. Obviously, $\mu^+(E_i) < \infty$ and $\mu^-(E_i) < \infty$ whenever $c_i \neq 0$. Hence f is D-integrable with respect to μ^+ as well as with respect to μ^-. Further, if we let $J = \{1 \leq i \leq m; c_i \neq 0\}$, then

$$D\int f \, d\mu = \sum_{i=1}^m c_i \mu(E_i) = \sum_{i \in J} c_i \mu(E_i)$$

$$= \sum_{i \in J} c_i [\mu^+(E_i) - \mu^-(E_i)]$$

$$= \sum_{i=1}^m c_i \mu^+(E_i) - \sum_{i=1}^m c_i \mu^-(E_i)$$

$$= D\int f \, d\mu^+ - D\int f \, d\mu^-.$$

(ii). Let $f = \sum_{i=1}^m c_i I_{E_i}$ and $g = \sum_{j=1}^n d_j I_{F_j}$ be representations of f and g respectively such that $|\mu|(E_i) < \infty$ whenever $c_i \neq 0$ and $|\mu|(F_j) < \infty$ whenever $d_j \neq 0$.

Then

$$cf + dg = \sum_{i=1}^{m} \sum_{j=1}^{n} (cc_i + dd_j) I_{E_i \cap F_j}.$$

Suppose $cc_i + dd_j \neq 0$. Then, either $cc_i \neq 0$ or $dd_j \neq 0$. This implies that either $c_i \neq 0$ or $d_j \neq 0$. This means that either $|\mu|(E_i) < \infty$ or $|\mu|(F_j) < \infty$. In any case, $|\mu|(E_i \cap F_j) < \infty$. Hence $cf + dg$ is D-integrable. The desired equality follows easily.

(iii). It is obvious that $|f|$ is D-integrable with respect to $|\mu|$. If $f = \sum_{i=1}^{m} c_i I_{E_i}$ is a representation of f, then

$$\left| D \int f \, d\mu \right| = \left| \sum_{i=1}^{m} c_i \mu(E_i) \right| \leq \sum_{i=1}^{m} |c_i| |\mu(E_i)|$$

$$\leq \sum_{i=1}^{m} |c_i| |\mu|(E_i) = D \int |f| \, d|\mu|.$$

(iv). We can assume that $E = \Omega$ and f is a simple function having a representation $\sum_{i=1}^{m} c_i I_{E_i}$ where each $c_i \geq 0$. Now, (iv) is easily proved.

(v). This is a consequence of (ii). and (iv).

(vi). This follows from (iii) and (v).

(vii). If $|\mu|(E) < \infty$, the result is clear. If $|\mu|(E) = \infty$, observe that $c \leq 0 \leq d$.

(viii). $\lambda(\varnothing) = 0$ since $I_\varnothing f = 0$. Let E and F be two disjoint sets in $\mathscr{F}$. Since $I_{E \cup F} f = I_E f + I_F f$, it follows that $\lambda(E \cup F) = \lambda(E) + \lambda(F)$. So, λ is a charge on $\mathscr{F}$. Further, for any F in $\mathscr{F}$,

$$|\lambda(F)| = \left| D \int_F f \, d\mu \right| \leq D \int_F |f| \, d|\mu| \leq D \int |f| \, d|\mu|.$$

Thus λ is bounded. Next, we show that $|\lambda|(F) = D \int_F |f| \, d|\mu|$ for any F in $\mathscr{F}$. This, we proceed to establish in stages. Clearly, the result is true for $f = I_\Omega$. Next, the result is clear for cI_Ω also, for any real number c. If $f = \sum_{i=1}^{m} c_i I_{E_i}$ for some real numbers $c_1, c_2, \ldots, c_m$ and $E_1, E_2, \ldots, E_m$ pairwise disjoint sets in $\mathscr{F}$, then $\lambda(F) = \sum_{i=1}^{m} c_i D \int_F I_{E_i} d\mu$. Now, considering each $D \int_F I_{E_i} d\mu$ as a charge with reference to the charge space $(E_i, E_i \cap \mathscr{F}, \mu/E_i \cap \mathscr{F})$ and applying the above argument, we obtain the desired result for f.

For the last part, if we let $c = \max\{|c_i|; 1 \leq i \leq m\}$, then for any F in $\mathscr{F}$, $|\lambda|(F) \leq c |\mu|(F)$. Consequently, λ is absolutely continuous with respect to μ. □

We introduce two important concepts before defining the integrability of a general function.

4. INTEGRATION

4.4.5 Definition. Let $(\Omega, \mathscr{F}, \mu)$ be a charge space. A real valued function on Ω is said to be T_1-*measurable* if there exists a sequence f_n, $n \geq 1$ of simple functions on Ω converging to f hazily.

If we denote the collection of all simple functions on Ω by $S(\Omega, \mathscr{F}, \mu)$, then f is T_1-measurable if and only if $f \in \overline{S(\Omega, \mathscr{F}, \mu)}$, where the closure is taken in the pseudo-metric space $(C(\Omega, \mathscr{F}, \mu), \rho)$. See Proposition 4.3.5.

4.4.6 Definition. Let $(\Omega, \mathscr{F}, \mu)$ be a charge space. A real valued function f on Ω is said to be T_2-*measurable* if for every $\varepsilon > 0$, there exists a partition $\{F_0, F_1, F_2, \ldots, F_n\}$ of Ω in $\mathscr{F}$ such that $|\mu|(F_0) < \varepsilon$ and $|f(\omega) - f(\omega')| < \varepsilon$ for every ω, ω' in F_i for every $i = 1, 2, \ldots, n$.

The following theorem establishes the relation between T_1-measurability and T_2-measurability.

4.4.7 Theorem. *Let $(\Omega, \mathscr{F}, \mu)$ be a charge space. A real valued function f on Ω is T_1-measurable if and only if it is T_2-measurable.*

Proof. Suppose f is T_1-measurable. Given $\varepsilon > 0$, there exists a simple function g on Ω such that $|\mu|^*(\{\omega \in \Omega; |f(\omega) - g(\omega)| > \varepsilon/2\}) < \varepsilon$. Let $g = \sum_{i=1}^{m} c_i I_{E_i}$ be a representation of g. Let $G = \{\omega \in \Omega; |f(\omega) - g(\omega)| > \varepsilon/2\}$. Since $|\mu|^*(G) = \operatorname{Inf}\{|\mu|(F); G \subset F, F \in \mathscr{F}\}$, there exists F_0 in $\mathscr{F}$ such that $G \subset F_0$ and $|\mu|(F_0) < \varepsilon$. Let $F_i = E_i \cap F_0^c$, $i = 1, 2, \ldots, m$. Now, if $1 \leq i \leq m$ and $\omega, \omega' \in F_i$, then $|f(\omega) - g(\omega)| \leq \varepsilon/2$, $|f(\omega') - g(\omega')| \leq \varepsilon/2$ and $g(\omega) = c_i = g(\omega')$. Therefore,

$$|f(\omega) - f(\omega')| \leq |f(\omega) - c_i| + |c_i - f(\omega')|$$
$$\leq |f(\omega) - g(\omega)| + |f(\omega') - g(\omega')| \leq \varepsilon.$$

This shows that f is T_2-measurable.

Conversely, let f be T_2-measurable. For each $n \geq 1$, let $\{F_{n0}, F_{n1}, F_{n2}, \ldots, F_{nk_n}\}$ be a partition of Ω in $\mathscr{F}$ such that $|\mu|(F_{n0}) < 1/n$ and $|f(\omega) - f(\omega')| < 1/n$ for every ω, ω' in F_{ni} and for every $i = 1, 2, \ldots, k_n$. For each $n \geq 1$ and $1 \leq i \leq k_n$, choose and fix ω_{ni} in F_{ni}. For each $n \geq 1$, let

$$f_n = \sum_{i=1}^{k_n} f(\omega_{ni}) I_{F_{ni}} + 0 \cdot I_{F_{n0}}.$$

Each f_n is a simple function and f_n, $n \geq 1$ converges to f hazily. For, let $\varepsilon > 0$ and $m \geq 1$ be such that $1/m < \varepsilon$. If $n \geq m$, $\{\omega \in \Omega; |f_n(\omega) - f(\omega)| > \varepsilon\} \subset F_{n0}$. Consequently, $|\mu|^*(\{\omega \in \Omega; |f_n(\omega) - f(\omega)| > \varepsilon\}) \leq |\mu|^*(F_{n0}) < 1/n$. This establishes the result. □

4.4.8 Corollary. *Every T_1-measurable function is smooth.*

102 THEORY OF CHARGES

Proof. From the definition of T_2-measurability, observe that every T_2-measurable function is smooth. The result now follows from Theorem 4.4.7. □

Relations between T_1-measurable functions, Smooth functions and bounded functions can be described as follows.

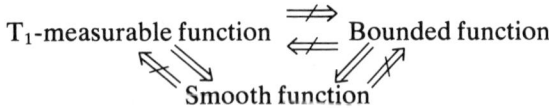

4.4.9 Corollary. *Let $(\Omega, \mathcal{F}, \mu)$ be a charge space.*
(i). *If f and g are T_1-measurable functions on Ω and c and d are real numbers, then $cf + dg$ and fg are all T_1-measurable.*
(ii). *If ψ is a real valued continuous function defined on the real line and f is T_1-measurable, then $\psi(f)$ is T_1-measurable. In particular, $|f|^p$ ($p > 0$) is T_1-measurable if f is T_1-measurable.*
(iii). *If f_n, $n \geq 1$ is a sequence of T_1-measurable functions converging to f hazily, then f is T_1-measurable and $\psi(f_n)$, $n \geq 1$ converges to $\psi(f)$ hazily, where ψ is as in (ii).*

Proof. (i) and (ii) follow from Corollary 4.4.8 and Theorem 4.3.3(i) and (vi). (iii) follows from the Remark following Definition 4.4.5, Theorem 4.3.3(v) and Corollary 4.4.8. □

The following proposition is instrumental in establishing the unambiguity in the general definition of D-integral.

4.4.10 Proposition. *Let $(\Omega, \mathcal{F}, \mu)$ be a charge space. Let f_{n1}, $n \geq 1$ and f_{n2}, $n \geq 1$ be two sequences of D-integrable simple functions on Ω converging to a real valued function f on Ω hazily. Suppose*

$$\lim_{m,n \to \infty} D \int |f_{ni} - f_{mi}| \, d|\mu| = 0$$

for $i = 1, 2$. Then $D \int_E f_{ni} \, d\mu$ converges uniformly over E in $\mathcal{F}$ for each $i = 1, 2$, and the limits coincide.

Proof. By Proposition 4.4.4(iii), for any $m, n \geq 1$ and $i = 1, 2$

$$\left| D \int_E f_{ni} \, d\mu - D \int_E f_{mi} \, d\mu \right| \leq D \int_\Omega |f_{ni} - f_{mi}| \, d|\mu|$$

for every E in $\mathcal{F}$. Consequently, $D \int_E f_{ni} \, d\mu$, $n \geq 1$ converges uniformly over E in $\mathcal{F}$ for each $i = 1, 2$. It remains to be shown that for each E in $\mathcal{F}$,

$$\lim_{n \to \infty} D \int_E f_{n1} \, d\mu = \lim_{n \to \infty} D \int_E f_{n2} \, d\mu.$$

This is carried out in the following steps.

1°. Let $g_n = |f_{n1} - f_{n2}|$, $n \geq 1$, and $\mu_n(F) = D \int_F g_n \, d|\mu|$ for F in $\mathcal{F}$ and $n \geq 1$. Each μ_n is a positive bounded charge on $\mathcal{F}$.

2°. We claim that μ_n, $n \geq 1$ converges uniformly over $\mathcal{F}$. For any F in $\mathcal{F}$ and $n, m \geq 1$, by Theorem 4.4.4(vi),

$$|\mu_n(F) - \mu_m(F)| = \left| D \int_F |f_{n1} - f_{n2}| \, d|\mu| - D \int_F |f_{m1} - f_{m2}| \, d|\mu| \right|$$

$$\leq D \int_F |f_{n1} - f_{m1} - f_{n2} + f_{m2}| \, d|\mu|$$

$$\leq D \int_F |f_{n1} - f_{m1}| \, d|\mu| + D \int_F |f_{n2} - f_{m2}| \, d|\mu|$$

$$\leq D \int |f_{n1} - f_{m1}| \, d|\mu| + D \int |f_{n2} - f_{m2}| \, d|\mu|.$$

From this, it follows that μ_n, $n \geq 1$ is uniformly Cauchy over $\mathcal{F}$. Thus the claim is established.

3°. Let $\lambda(F) = \lim_{n \to \infty} \mu_n(F)$, $F \in \mathcal{F}$. It suffices to show that $\lambda = 0$. For, for any E in $\mathcal{F}$,

$$\left| D \int_E f_{n1} \, d\mu - D \int_E f_{n2} \, d\mu \right| \leq D \int |f_{n1} - f_{n2}| \, d|\mu| = \mu_n(E).$$

If $\lim_{n \to \infty} \mu_n(E) = 0$, then

$$\lim_{n \to \infty} D \int f_{n1} \, d\mu = \lim_{n \to \infty} D \int f_{n2} \, d\mu.$$

4°. Now, we claim that λ is absolutely continuous with respect to μ. (See Theorem 4.4.4(viii).) Let $\varepsilon > 0$. There exists $N \geq 1$ such that $|\mu_n(E) - \lambda(E)| < \varepsilon/2$ for every E in $\mathcal{F}$ and $n \geq N$. Since μ_N is absolutely continuous with respect to μ, there exists $\delta > 0$ such that $\mu_N(E) < \varepsilon/2$ whenever $E \in \mathcal{F}$ and $|\mu|(E) < \delta$. Consequently, if $E \in \mathcal{F}$ and $|\mu|(E) < \delta$, then $\lambda(E) \leq |\lambda(E) - \mu_N(E)| + \mu_N(E) < \varepsilon/2 + \varepsilon/2 = \varepsilon$.

5°. The above deliberations can be summarized as follows. Let $\varepsilon > 0$. There exists $\delta > 0$ and $N \geq 1$ such that $\lambda(E) < \varepsilon$ whenever $E \in \mathcal{F}$ and $|\mu|(E) < \delta$, and $|\lambda(F) - \mu_n(F)| < \varepsilon$ for every F in $\mathcal{F}$ and for every $n \geq N$.

6°. We show that $\lambda(\Omega) < 4\varepsilon$. Let $A^c = \{\omega \in \Omega; g_N(\omega) = 0\}$. Since g_N is a simple function, $A^c \in \mathcal{F}$. Since g_N is D-integrable, $|\mu|(A) < \infty$. Also $\mu_N(A^c) = 0$. From 5°, we conclude that $\lambda(A^c) < \varepsilon$. If $|\mu|(A) = 0$, then $\lambda(A) = 0$ and $\lambda(\Omega) = \lambda(A) + \lambda(A^c) < \varepsilon < 4\varepsilon$. The next step covers the case when $|\mu|(A) > 0$.

7°. Suppose $|\mu|(A) > 0$. Since g_n, $n \geq 1$ converges to 0 hazily, there exists an integer $N_1 > N$ such that $|\mu|^*(\{\omega \in \Omega; |g_n(\omega)| > \varepsilon/|\mu|(A)\}) < \delta$ whenever $n \geq N_1$. Let $B^c = \{\omega \in \Omega; |g_{N_1}(\omega)| > \varepsilon/|\mu|(A)\}$. Since g_{N_1} is a simple function, $B^c \in \mathscr{F}$. Further, if $\omega \in B$, then $|g_{N_1}(\omega)| \leq \varepsilon/|\mu|(A)$.

8°.
$$\lambda(\Omega) = \lambda(A \cap B) + \lambda(A \cap B^c) + \lambda(A^c)$$
$$< [\mu_{N_1}(A \cap B) + \varepsilon] + \varepsilon + \varepsilon.$$

The inequality $\lambda(A \cap B^c) < \varepsilon$ follows from the following argument. By 7°, $|\mu|(A \cap B^c) \leq |\mu|(B^c) = |\mu|^*(B^c) < \delta$. So, by 5°, the above inequality follows. Again, by 5°, the inequality $\lambda(A \cap B) < \mu_{N_1}(A \cap B) + \varepsilon$ follows. Now,

$$\mu_{N_1}(A \cap B) = D \int_{A \cap B} g_{N_1} \, d|\mu| \leq \frac{\varepsilon}{|\mu|(A)} |\mu|(A \cap B) \leq \varepsilon.$$

Consequently, $\lambda(\Omega) < 4\varepsilon$. Since $\varepsilon > 0$ is arbitrary, it follows that $\lambda(\Omega) = 0$. Since λ is a positive charge, we have that $\lambda = 0$. This completes the proof. □

Now, we are in a position to define D-integrability of a general function.

4.4.11 Definition. Let $(\Omega, \mathscr{F}, \mu)$ be a charge space. A real valued function f on Ω is said to be *D-integrable* if there exists a sequence f_n, $n \geq 1$ of D-integrable simple functions on Ω such that

(i). f_n, $n \geq 1$ converges to f hazily.

(ii). $\lim_{m,n \to \infty} D \int |f_n - f_m| \, d|\mu| = 0$.

If f is D-integrable, the D-integral of f is denoted by $D \int f \, d\mu$ and is defined to be the number $\lim_{n \to \infty} D \int f_n \, d\mu$. The sequence f_n, $n \geq 1$ is called a *determining sequence of D-integrable simple functions for f*, or simply a *determining sequence for f*.

From Proposition 4.4.10, it is clear that $\lim_{n \to \infty} D \int f_n \, d\mu$ exists and is finite, and is independent of the choice of the determining sequence f_n, $n \geq 1$. Also, observe that every D-integrable function is T_1-measurable.

If f is D-integrable and $E \in \mathscr{F}$, then the function $I_E f$ is also D-integrable. If f_n, $n \geq 1$ is a determining sequence of D-integrable simple functions for f, it is easily checked that $I_E f_n$, $n \geq 1$ is a determining sequence of D-integrable simple functions for $I_E f$. Consequently, we define $D \int_E f \, d\mu = D \int I_E f \, d\mu$.

We need a lemma which will be useful in the proof of Theorem 4.4.13(xi).

4.4.12 Lemma. *Let $(\Omega, \mathscr{F}, \mu)$ be a charge space and f a D-integrable function. Let f_n, $n \geq 1$ be a determining sequence for f. Then $f_n - f$ is D-integrable for every $n \geq 1$ and*

$$\lim_{n \to \infty} D \int |f_n - f| \, d|\mu| = 0.$$

4. INTEGRATION

Proof. Fix $n \geq 1$. Then $f_n - f_m$, $m \geq 1$ converges to $f_n - f$ hazily. Further, each $f_n - f_m$ is a D-integrable simple function. Also, $\lim_{m,p \to \infty} D \int |(f_n - f_m) - (f_n - f_p)| \, d|\mu| = 0$. Hence $f_n - f$ is D-integrable, and $|f_n - f_m|$, $m \geq 1$ is a determining sequence for $|f_n - f|$. Thus

$$\lim_{n \to \infty} D \int |f_n - f| \, d|\mu| = \lim_{n \to \infty} \left[\lim_{m \to \infty} D \int |f_n - f_m| \, d|\mu| \right]$$

$$= \lim_{m,n \to \infty} D \int |f_n - f_m| \, d|\mu| = 0.$$

This proves the lemma. □

Now, we give a comprehensive list of properties of D-integrable functions.

4.4.13 Theorem. *Let $(\Omega, \mathcal{F}, \mu)$ be a charge space.*

(i). *If a real valued function f on Ω is D-integrable with respect to μ, then f is D-integrable with respect to μ^+ as well as with respect to μ^-. Further, for every E in $\mathcal{F}$,*

$$D \int_E f \, d\mu = D \int_E f \, d\mu^+ - D \int_E f \, d\mu^-.$$

(ii). *If f and g are D-integrable and c and d are real numbers, then $cf + dg$ is D-integrable and for every E in $\mathcal{F}$,*

$$D \int_E (cf + dg) \, d\mu = c \left(D \int_E f \, d\mu \right) + d \left(D \int_E g \, d\mu \right).$$

(iii). *If f is D-integrable with respect to μ, then $|f|$ is D-integrable with respect to μ as well as with respect to $|\mu|$ and for any E in $\mathcal{F}$,*

$$\left| D \int_E f \, d\mu \right| \leq D \int_E |f| \, d|\mu|.$$

(iv). *f is D-integrable if and only if f^+ and f^- are D-integrable. Further, if f is D-integrable, then*

$$D \int_E f \, d\mu = D \int_E f^+ \, d\mu - D \int_E f^- \, d\mu$$

for every E in $\mathcal{F}$. More generally, if f and g are D-integrable, then $f \vee g$ and $f \wedge g$ are D-integrable and

$$D \int_E f \, d\mu + D \int_E g \, d\mu = D \int_E (f \vee g) \, d\mu + D \int_E (f \wedge g) \, d\mu$$

for every E in $\mathcal{F}$.

(v). *If f is D-integrable and $f \geq 0$ a.e. $[\mu]$ on E in $\mathscr{F}$, i.e. $I_E f \geq 0$ a.e. $[\mu]$, then*

$$D \int_E f \, d|\mu| \geq 0.$$

(vi). *If f and g are D-integrable and $f \leq g$ a.e. $[\mu]$ on E in $\mathscr{F}$, i.e. $I_E f \leq I_E g$ a.e. $[\mu]$, then*

$$D \int_E f \, d|\mu| \leq D \int_E g \, d|\mu|.$$

(vii). *If f is D-integrable and $c \leq f \leq d$ a.e. $[\mu]$ on E in $\mathscr{F}$ for some real numbers c and d, then*

$$c|\mu|(E) \leq D \int_E f \, d|\mu| \leq d|\mu|(E).$$

In fact, if $c > 0$, then $|\mu|(E) < \infty$.

(viii). *If f and g are D-integrable and $E \in \mathscr{F}$, then*

$$\left| D \int_E |f| \, d|\mu| - D \int_E |g| \, d|\mu| \right| \leq D \int_E ||f| - |g|| \, d|\mu|$$

$$\leq D \int_E |f + g| \, d|\mu|$$

$$\leq D \int_E |f| \, d|\mu| + D \int_E |g| \, d|\mu|.$$

(ix). *f is a null function if and only if f is D-integrable and $\int |f| \, d|\mu| = 0$.*

(x). *If f is D-integrable and $g = f$ a.e. $[\mu]$, then g is D-integrable and $D \int_E f \, d\mu = D \int_E g \, d\mu$ for all E in $\mathscr{F}$.*

(xi). *If f is D-integrable, then the set function λ on $\mathscr{F}$ defined by*

$$\lambda(F) = D \int_F f \, d\mu, \qquad F \in \mathscr{F}$$

is a bounded charge on $\mathscr{F}$ satisfying

$$|\lambda|(F) = D \int_F |f| \, d|\mu|, \qquad F \in \mathscr{F}.$$

Further, λ is absolutely continuous with respect to μ in the following sense. Given $\varepsilon > 0$, there exists $\delta > 0$ such that $|\lambda(E)| < \varepsilon$ whenever $E \in \mathscr{F}$ and $|\mu|(E) < \delta$.

4. INTEGRATION

(xii). *If μ is positive, f is D-integrable and λ is the charge on $\mathscr{F}$ defined by* $\lambda(F) = D \int_F f \, d\mu$, $F \in \mathscr{F}$, *then*

$$\lambda^+(F) = D \int_F f^+ \, d\mu \quad \text{and} \quad \lambda^-(F) = D \int_F f^- \, d\mu$$

for all F in $\mathscr{F}$.

(xiii). *For a D-integrable function f, $D \int |f| \, d|\mu| = 0$ if and only if $D \int_E f \, d\mu = 0$ for every E in $\mathscr{F}$. If f and g are D-integrable, then $f = g$ a.e. $[\mu]$ if and only if*

$$D \int_E f \, d\mu = D \int_E g \, d\mu$$

for every E in $\mathscr{F}$.

Proof. (i). If f_n, $n \geq 1$ is a determining sequence for f with respect to μ, then it has the same property with respect to $|\mu|$, μ^+ and μ^-. This follows from the inequalities that $\mu^+ \leq |\mu|$ and $\mu^- \leq |\mu|$. Consequently, f is D-integrable with respect to μ^+ as well as with respect to μ^-. By Theorem 4.4.4(i), we have for any E in $\mathscr{F}$,

$$D \int_E f \, d\mu = \lim_{n \to \infty} D \int_E f_n \, d\mu = \lim_{n \to \infty} \left[D \int_E f_n \, d\mu^+ - D \int_E f_n \, d\mu^- \right]$$

$$= \lim_{n \to \infty} D \int_E f_n \, d\mu^+ - \lim_{n \to \infty} D \int_E f_n \, d\mu^-$$

$$= D \int_E f \, d\mu^+ - D \int_E f \, d\mu^-.$$

(ii). This follows from Theorem 4.3.3(i) and Theorem 4.4.4(ii).

(iii). If f_n, $n \geq 1$ is a determining sequence for f with respect to μ, then $|f_n|$, $n \geq 1$ is a determining sequence for $|f|$ with respect to μ as well as with respect to $|\mu|$. This follows from Theorem 4.3.3(ii) and Theorem 4.4.4(vi). Further, from Theorem 4.4.4(iii),

$$\left| D \int_E f \, d\mu \right| = \lim_{n \to \infty} \left| D \int_E f_n \, d\mu \right| \leq \lim_{n \to \infty} D \int_E |f_n| \, d|\mu|$$

$$= D \int_E |f| \, d|\mu|$$

for any E in $\mathscr{F}$.

(iv). If f_n, $n \geq 1$ is a determining sequence for f, g_n, $n \geq 1$ is a determining sequence for g, then $f_n \vee g_n$, $n \geq 1$ converges to $f \vee g$ hazily, by Theorem

4.3.3(viii). Further,

$$D \int |(f_n \vee g_n) - (f_m \vee g_m)| \, d|\mu| \leq D \int |f_n - f_m| \, d|\mu| + D \int |g_n - g_m| \, d|\mu|.$$

(Note that $|(a \vee b) - (c \vee d)| \leq |a-c| + |b-d|$ and $|(a \wedge b) - (c \wedge d)| \leq |a-c| + |b-d|$ for any numbers a, b, c and d.) Hence $f_n \vee g_n$, $n \geq 1$ is a determining sequence for $f \vee g$. Similarly, one can show that $f_n \wedge g_n$, $n \geq 1$ is a determining sequence for $f \wedge g$. Since $f + g = (f \vee g) + (f \wedge g)$, by (ii), we have $D \int_E (f + g) \, d\mu = D \int_E (f \vee g) \, d\mu + D \int_E (f \wedge g) \, d\mu$ for any E in $\mathscr{F}$. Taking $g = 0$, we get the first part of (iv).

(v). Since $I_E f \geq 0$ a.e. $[\mu]$, there exists a null function h such that $h + I_E f \geq 0$. If f_n, $n \geq 1$ is a determining sequence for f, then $I_E f_n$, $n \geq 1$ is a determining sequence for $h + I_E f$. See Proposition 4.3.2. Hence $h + I_E f$ is D-integrable, and $(I_E f_n)^+$, $n \geq 1$ is a determining sequence for $(h + I_E f)^+ = h + I_E f$ with respect to $|\mu|$. Since $f_n^+ \geq 0$ for every $n \geq 1$, by Theorem 4.4.4(iv), and (ii) and (iii) above, we have

$$D \int_E f \, d|\mu| = \lim_{n \to \infty} D \int I_E f_n \, d|\mu| = D \int (h + I_E f) \, d|\mu|$$

$$= \lim_{n \to \infty} D \int (I_E f_n)^+ \, d|\mu| \geq 0.$$

This proves (v).

(vi). This is a consequence of (v) if one looks at the function $I_E(g - f)$.

(vii). If $|\mu|(E) < \infty$, then cI_E and dI_E are D-integrable and the desired inequalities follow from (vi). Suppose $|\mu|(E) = \infty$. By adding a null function if necessary, we can assume that $c \leq f$ on E. Without loss of generality, assume that $E = \Omega$. We show that $c \leq 0$. Suppose not. Then $0 < c \leq f$. Let f_n, $n \geq 1$ be a determining sequence for f. We can assume that $f_n \geq 0$ for all n. Then $f_n \wedge c$, $n \geq 1$ converges hazily to the constant function c. Note that each $f_n \wedge c$ is D-integrable because $0 \leq f_n \wedge c \leq f_n$ and $f_n \wedge c$ is a simple function. Also,

$$D \int |(f_n \wedge c) - (f_m \wedge c)| \, d|\mu| \leq D \int |f_n - f_m| \, d|\mu|$$

for all m, $n \geq 1$. Hence $f_n \wedge c$, $n \geq 1$ is a determining sequence for the constant function c. Hence the constant function c is D-integrable. But this is not possible since $|\mu|(\Omega) = \infty$. This shows that $c \leq 0$. Hence, it follows that $c|\mu|(E) \leq D \int f \, d|\mu|$. If $|\mu|(E) = \infty$, we can show that $d \geq 0$ by an analogous argument. Thus the desired inequalities are established for all cases.

4. INTEGRATION

(viii). This is a consequence of Theorem 4.4.4(vi).

(ix). If f is a null function, it is easily verified that $0, 0, \ldots$ is a determining sequence for f and hence $D \int |f| \, d|\mu| = 0$.

Conversely, let f be D-integrable and $D \int |f| \, d|\mu| = 0$. Let f_n, $n \geq 1$ be a determining sequence for f. Then $\lim_{n \to \infty} D \int |f_n| \, d|\mu| = 0$. For any $\delta > 0$ and $n \geq 1$, let $E_n(\delta) = \{\omega \in \Omega; \ |f_n(\omega)| > \delta\}$ and $F_n(\delta) = \{\omega \in \Omega; |f_n(\omega) - f(\omega)| > \delta\}$. Since f_n is a simple function, $E_n(\delta) \in \mathcal{F}$. Since f_n is D-integrable, $|\mu|(E_n(\delta)) < \infty$. Further, for any $n \geq 1$, by (vi) and (vii),

$$D \int |f_n| \, d|\mu| \geq D \int_{E_n(\delta)} |f_n| \, d|\mu| \geq \delta |\mu|(E_n(\delta)).$$

Consequently, $\lim_{n \to \infty} |\mu|(E_n(\delta)) = 0$. Since f_n, $n \geq 1$ converges to f hazily, $\lim_{n \to \infty} |\mu|^*(F_n(\delta)) = 0$. Thus, for any $n \geq 1$,

$$|\mu|^*(\{\omega \in \Omega; |f(\omega)| > 2\delta\}) \leq |\mu|^*(\{\omega \in \Omega; |f_n(\omega)| > \delta\})$$
$$+ |\mu|^*(\{\omega \in \Omega; |f_n(\omega) - f(\omega)| > \delta\})$$

which converges to zero as n tends to ∞. Consequently, $|\mu|^*(\{\omega \in \Omega; |f(\omega)| > 2\delta\}) = 0$ for any $\delta > 0$. Hence f is a null function.

(x). This follows from (ix) and (ii).

(xi). It is now obvious that λ is a charge. The boundedness of λ follows from $|\lambda(F)| = |D \int_F f \, d\mu| \leq D \int_F |f| \, d|\mu| \leq D \int |f| \, d|\mu|$ for any F in $\mathcal{F}$.

Now, we show that $|\lambda|(F) = D \int_F |f| \, d|\mu|$ for every F in $\mathcal{F}$. Let f_n, $n \geq 1$ be a determining sequence for f. Let λ_n, $n \geq 1$ be defined by $\lambda_n(E) = D \int_E f_n \, d\mu$, $E \in \mathcal{F}$. We show that $\lim_{n \to \infty} |\lambda_n|(E) = |\lambda|(E)$ for every E in $\mathcal{F}$. Then, by Theorem 4.4.4(viii), and (ii) above, we would have

$$\lim_{n \to \infty} |\lambda_n|(E) = \lim_{n \to \infty} D \int_E |f_n| \, d|\mu| = D \int_E |f| \, d|\mu| = |\lambda|(E)$$

for every E in $\mathcal{F}$. Let $F \in \mathcal{F}$ be fixed and $\varepsilon > 0$. Since f_n, $n \geq 1$ is a determining sequence for f, by Lemma 4.4.12, there exists $N \geq 1$ such that $D \int_F |f - f_n| \, d|\mu| < \varepsilon$ whenever $n \geq N$. Let $\{F_1, F_2, \ldots, F_k\}$ be any partition of F in $\mathcal{F}$. Then for any $n \geq N$,

$$\left| \sum_{i=1}^{k} \left| D \int_{F_i} f \, d\mu \right| - \sum_{i=1}^{k} \left| D \int_{F_i} f_n \, d\mu \right| \right| \leq \sum_{i=1}^{k} \left| D \int_{F_i} f \, d\mu - D \int_{F_i} f_n \, d\mu \right|$$

$$\leq \sum_{i=1}^{k} D \int_{F_i} |f - f_n| \, d|\mu|$$

$$= D \int_F |f - f_n| \, d|\mu| < \varepsilon.$$

We show that for any $n \geq N$, $||\lambda_n|(F) - |\lambda|(F)| < 3\varepsilon$. Let $n \geq N$ be fixed. From the definition of $|\lambda_n|(F)$ and $|\lambda|(F)$, we can find a partition $\{A_1, A_2, \ldots, A_m\}$ of F in $\mathscr{F}$ and a partition $\{B_1, B_2, \ldots, B_k\}$ of F in $\mathscr{F}$ such that

$$\left||\lambda|(F) - \sum_{i=1}^{m} |\lambda(A_i)|\right| = \left||\lambda|(F) - \sum_{i=1}^{m} \left|D\int_{A_i} f\,d\mu\right|\right| < \varepsilon$$

and

$$\left||\lambda_n|(F) - \sum_{i=1}^{k} |\lambda_n(B_i)|\right| = \left||\lambda_n|(F) - \sum_{i=1}^{k} \left|D\int_{B_i} f_n\,d\mu\right|\right| < \varepsilon.$$

Let $F_{ij} = A_i \cap B_j$, $1 \leq i \leq m$ and $1 \leq j \leq k$. Then $\{F_{ij}; 1 \leq i \leq m \text{ and } 1 \leq j \leq k\}$ is a partition of F in $\mathscr{F}$. Further,

$$0 \leq |\lambda|(F) - \sum_{i=1}^{m}\sum_{j=1}^{k} \left|D\int_{F_{ij}} f\,d\mu\right| \leq |\lambda|(F) - \sum_{i=1}^{m} \left|D\int_{A_i} f\,d\mu\right| < \varepsilon$$

and

$$0 \leq |\lambda_n|(F) - \sum_{i=1}^{m}\sum_{j=1}^{k} \left|D\int_{F_{ij}} f_n\,d\mu\right| \leq |\lambda_n|(F) - \sum_{j=1}^{k} \left|D\int_{B_j} f_n\,d\mu\right| < \varepsilon.$$

Therefore, for $n \geq N$,

$$||\lambda|(F) - |\lambda_n|(F)| \leq \left||\lambda|(F) - \sum_{i=1}^{m}\sum_{j=1}^{k}\left|D\int_{F_{ij}} f\,d\mu\right|\right|$$

$$+ \left|\sum_{i=1}^{m}\sum_{j=1}^{k}\left|D\int_{F_{ij}} f\,d\mu\right| - \sum_{i=1}^{m}\sum_{j=1}^{k}\left|D\int_{F_{ij}} f_n\,d\mu\right|\right|$$

$$+ \left|\sum_{i=1}^{m}\sum_{j=1}^{k}\left|D\int_{F_{ij}} f_n\,d\mu\right| - |\lambda_n|(F)\right|$$

$$< \varepsilon + \varepsilon + \varepsilon = 3\varepsilon.$$

This completes the proof of the desired assertion.

Now, we prove that λ is absolutely continuous with respect to μ. Let $\varepsilon > 0$. By Lemma 4.4.12, there exists a D-integrable simple function g such that $D\int |f-g|\,d|\mu| < \varepsilon$. Let M be a positive number such that $|g(\omega)| \leq M$ for every ω in Ω. Take $\delta = \varepsilon/M$. Let $E \in \mathscr{F}$ and $|\mu|(E) < \delta$. Then

$$|\lambda|(E) = D\int_E |f|\,d|\mu| \leq D\int_E |f-g|\,d|\mu| + D\int_E |g|\,d|\mu|$$

$$\leq D\int |f-g|\,d|\mu| + M|\mu|(E) < \varepsilon + M\frac{\varepsilon}{M} = 2\varepsilon.$$

This shows that λ is absolutely continuous with respect to μ.
(xii). From Theorem 2.2.2(5), $\lambda^+ = \frac{1}{2}(\lambda + |\lambda|)$ and $\lambda^- = \frac{1}{2}(\lambda - |\lambda|)$ for the bounded charge λ. Then,

$$\lambda^+(F) = \frac{1}{2}\left(D\int_F f \, d\mu + D\int_F |f| \, d\mu\right) = D\int_F \frac{1}{2}(f + |f|) \, d\mu$$

$$= D\int_F f^+ \, d\mu.$$

Similarly, the result about λ^- is proved.
(xiii). Let $\lambda(F) = D\int_F f \, d\mu$. Then $\lambda(F) = 0$ for all F in $\mathcal{F}$, if $D\int_F f \, d\mu = 0$ for all F in $\mathcal{F}$. Hence $|\lambda|(\Omega) = 0$. So, by (xii) above, we have $D\int |f| \, d|\mu| = 0$. The converse is clear. Now, the second part follows from the first part and (ix).

This completes the proof of the theorem. □

Now, we enumerate all the D-integrable functions for the simple example given below.

4.4.14 Example. *Let $(\Omega, \mathcal{F}, \mu)$ be a charge space in which μ is a 0–1 valued charge on $\mathcal{F}$. A real valued function f on Ω is D-integrable if and only if $f = c$ a.e. $[\mu]$, where c is a constant, and if $f = c$ a.e. $[\mu]$, then $D\int f \, d\mu = c$.*

Proof. Since μ is bounded, every constant function c is D-integrable. By Theorem 4.4.13(x), any function f which satisfies $f = c$ a.e. $[\mu]$ for some constant c is D-integrable and $D\int f \, d\mu = c$.

Conversely, suppose f is D-integrable. Let f_n, $n \geq 1$ be a determining sequence for f. Since each f_n is a simple function and μ is a 0–1 valued charge, $f_n = c_n$ a.e. $[\mu]$ for some constant c_n. Without loss of generality, we can assume that $f_n \equiv c_n$ for every $n \geq 1$. Note that c_n, $n \geq 1$ is a Cauchy sequence of real numbers since $D\int |f_n - f_m| \, d\mu = |c_n - c_m|$ which converges to zero as $n, m \to \infty$. Let $c = \lim_{n \to \infty} c_n$. Then

$$D\int f \, d\mu = \lim_{n \to \infty} D\int f_n \, d\mu = \lim_{n \to \infty} c_n = c.$$

We show that $f = c$ a.e. $[\mu]$. Let $\varepsilon > 0$. Let $N_1 \geq 1$ be such that $|c_n - c| < \varepsilon/2$ for every $n \geq N_1$. Since f_n, $n \geq 1$ converges to f hazily, there exists $N_2 \geq 1$ such that $\mu^*(\{\omega \in \Omega; |f_n(\omega) - f(\omega)| > \varepsilon/2\}) = 0$ for every $n \geq N_2$. Let $N = \max\{N_1, N_2\}$. If $n \geq N$, then $\mu^*(\{\omega \in \Omega; |f(\omega) - c| > \varepsilon\}) \leq \mu^*(\{\omega \in \Omega; |f(\omega) - f_n(\omega)| > \varepsilon/2\}) + \mu^*(\{\omega \in \Omega; |f_n(\omega) - c| > \varepsilon/2\}) \leq 0 + 0 = 0$. Hence $f = c$ a.e. $[\mu]$. □

The following three lemmas are needed to prove Theorem 4.4.18.

4.4.15 Lemma. *Let $(\Omega, \mathscr{F}, \mu)$ be a charge space and f a D-integrable function on Ω. Let λ on $\mathscr{F}$ be defined by $\lambda(F) = D\int_F f\, d\mu$, $F \in \mathscr{F}$. Then given $\varepsilon > 0$, there exists A in $\mathscr{F}$ such that $|\mu|(A) < \infty$ and $|\lambda|(A^c) < \varepsilon$.*

Proof. By Lemma 4.4.12, there exists a D-integrable simple function f on Ω such that $D\int |g - f|\, d|\mu| < \varepsilon$. Let $A^c = \{\omega \in \Omega; g(\omega) = 0\}$. Since g is a simple function, $A^c \in \mathscr{F}$. Since g is D-integrable $|\mu|(A) < \infty$. Further,

$$|\lambda|(A^c) = D\int_{A^c} |f|\, d|\mu| \leq D\int_{A^c} |f - g|\, d|\mu| + D\int_{A^c} |g|\, d|\mu|$$

$$\leq D\int |f - g|\, d|\mu| + 0 < \varepsilon.$$

This completes the proof. □

4.4.16 Lemma. *Let $(\Omega, \mathscr{F}, \mu)$ be a charge space and f a D-integrable function. Suppose g is a simple function and $|g| \leq |f|$ a.e. $[\mu]$. Then g is D-integrable.*

Proof. This follows from Theorem 4.4.13(vii). □

4.4.17 Lemma. *Let $(\Omega, \mathscr{F}, \mu)$ be a charge space. Let g_n, $n \geq 1$ be a sequence of simple functions converging to g hazily. Then there exists a sequence h_n, $n \geq 1$ of simple functions converging to g hazily such that $|h_n| \leq 2|g|$ for every $n \geq 1$.*

Proof. From the definition of hazy convergence, we can find a subsequence g_{n_k}, $k \geq 1$ of g_n, $n \geq 1$ such that $|\mu|^*(\{\omega \in \Omega; |g_{n_k}(\omega)| > 1/k\}) < 1/k$ for every $k \geq 1$. Consequently, we can find A_k in $\mathscr{F}$ such that $|\mu|(A_k) < 1/k$ and $\{\omega \in \Omega; |g_{n_k}(\omega) - g(\omega)| > 1/k\} \subset A_k$ for every $k \geq 1$. So, if $\omega \in A_k^c$, then $|g_{n_k}(\omega) - g(\omega)| \leq 1/k$ for every $k \geq 1$. Now, for each $k \geq 1$, define

$$h_k(\omega) = g_{n_k}(\omega), \quad \text{if } \omega \in A_k^c \text{ and } |g_{n_k}(\omega)| > 2/k,$$

$$= 0, \qquad \text{otherwise.}$$

Since each g_{n_k} is a simple function, each h_k is a simple function. Let $\omega \in A_k^c$ and satisfy the inequality $|g_{n_k}(\omega)| > 2/k$. Then $|h_k(\omega)| = |g_{n_k}(\omega)| \leq |g_{n_k}(\omega) - g(\omega)| + |g(\omega)| \leq (1/k) + |g(\omega)|$. But $1/k \geq |g_{n_k}(\omega)| - |g(\omega)| > 2/k - |g(\omega)|$. Therefore, $|g(\omega)| > 1/k$. Consequently, $|h_k(\omega)| \leq 1/k + |g(\omega)| < 2|g(\omega)|$. Thus, for any ω in Ω, $|h_k(\omega)| \leq 2|g(\omega)|$ for all $k \geq 1$.

Now, we show that h_k, $k \geq 1$ converges to g hazily. Let $\varepsilon > 0$. Then there exists $N \geq 1$ such that $1/N < \varepsilon$. If $k \geq N$, then

$$|\mu|^*(\{\omega \in \Omega; |h_k(\omega) - g(\omega)| > \varepsilon\}) \leq |\mu|^*(\{\omega \in A_k; |h_k(\omega) - g(\omega)| > \varepsilon\})$$

$$+ |\mu|^*(\{\omega \in A_k^c; |h_k(\omega) - g(\omega)| > \varepsilon\})$$

$$\leq |\mu|(A_k) + 0 < 1/k.$$

Hence h_k, $k \geq 1$ converges to g hazily. □

The following theorem gives a necessary and sufficient condition under which a given function dominated by a D-integrable function is D-integrable.

4.4.18 Theorem. Let $(\Omega, \mathcal{F}, \mu)$ be a charge space. Let f and g be two real valued functions on Ω such that $|g| \leq |f|$ a.e. $[\mu]$ and f is D-integrable. Then g is D-integrable if and only if g is T_1-measurable.

Proof. "Only if" part follows from the definition of D-integrability.

"If" part. Suppose that g is T_1-measurable. By Lemma 4.4.17, there exists a sequence h_n, $n \geq 1$ of simple functions converging to g hazily and such that $|h_n| \leq 2|g|$ for every $n \geq 1$. Since each $|h_n| \leq 2|f|$ and f is D-integrable, by Lemma 4.4.16, we have that h_n is D-integrable.

Now, we show that $\lim_{m,n \to \infty} D \int |h_n - h_m| \, d|\mu| = 0$. Given $\varepsilon > 0$, we exhibit $N \geq 1$ such that $D \int |h_n - h_m| \, d|\mu| < 3\varepsilon$ for every $n, m \geq N$. Observe that for every $n, m \geq 1$ and E in $\mathcal{F}$, $D \int_E |h_n - h_m| \, d|\mu| \leq 4D \int_E |f| \, d|\mu|$.

Let $\varepsilon > 0$. By Lemma 4.4.15, there exists a set A in $\mathcal{F}$ such that $|\mu|(A) < \infty$ and $4D \int_{A^c} |f| \, d|\mu| < \varepsilon$. Hence $D \int_{A^c} |h_n - h_m| \, d|\mu| < \varepsilon$ for every $n, m \geq 1$. If $|\mu|(A) = 0$, then

$$D \int |h_n - h_m| \, d|\mu| = D \int_A |h_n - h_m| \, d|\mu| + D \int_{A^c} |h_n - h_m| \, d|\mu|$$
$$< 0 + \varepsilon < 3\varepsilon$$

for all $m, n \geq 1$. If $|\mu|(A) > 0$, we proceed as follows.

By Theorem 4.4.13(xi), there exists $\delta > 0$ such that $4D \int_E |f| \, d|\mu| < \varepsilon$ whenever $E \in \mathcal{F}$ and $|\mu|(E) < \delta$. This implies that $D \int_E |h_n - h_m| \, d|\mu| < \varepsilon$ whenever $E \in \mathcal{F}$ and $|\mu|(E) < \delta$. Since $\lim_{m,n \to \infty} |\mu|^*(\{\omega \in \Omega; |h_n(\omega) - h_m(\omega)| > \varepsilon/|\mu|(A)\}) = 0$, there exists $N \geq 1$ and sets E_{nm} in $\mathcal{F}$ for $n, m \geq N$ such that $|\mu|(E_{nm}) < \delta$ and $|h_n(\omega) - h_m(\omega)| \leq \varepsilon/|\mu|(A)$ if $\omega \in E_{nm}^c$ for $n, m \geq N$. Now, let $n, m \geq N$. Then

$$D \int |h_n - h_m| \, d|\mu| = D \int_{A^c} |h_n - h_m| \, d|\mu| + D \int_{A \cap E_{nm}} |h_n - h_m| \, d|\mu|$$
$$+ D \int_{A \cap E_{nm}^c} |h_n - h_m| \, d|\mu|$$
$$< \varepsilon + \varepsilon + \frac{\varepsilon}{|\mu|(A)} \cdot |\mu|(A) = 3\varepsilon.$$

This shows that h_n, $n \geq 1$ is a determining sequence for g. Hence g is D-integrable. □

4.4.19 Corollary. Let $(\Omega, \mathcal{F}, \mu)$ be a charge space and f a T_1-measurable function on Ω. Then f is D-integrable if and only if $|f|$ is D-integrable.

The following theorem is an extension of Lemma 4.4.12.

4.4.20 Theorem. *Let $(\Omega, \mathscr{F}, \mu)$ be a charge space and f_n, $n \geq 1$ a sequence of D-integrable functions such that*

$$\lim_{m,n \to \infty} D \int |f_n - f_m| \, d|\mu| = 0.$$

Let f be a real valued function on Ω such that f_n, $n \geq 1$ converges to f hazily. Then f is D-integrable and

$$\lim_{n \to \infty} D \int |f_n - f| \, d|\mu| = 0.$$

Proof. If f_n's are simple functions, the assertion follows from the definition of D-integrability of f and by Lemma 4.4.12. Since f_n is D-integrable, by Lemma 4.4.12, there exists a D-integrable simple function h_n on Ω such that

$$D \int |f_n - h_n| \, d|\mu| < 1/n$$

and

$$|\mu|^*(\{\omega \in \Omega; |f_n(\omega) - h_n(\omega)| > 1/n\}) < 1/n$$

for every $n \geq 1$. We claim that h_n, $n \geq 1$ converges to f hazily. Let $\varepsilon > 0$. Since f_n, $n \geq 1$ converges to f hazily, there exists $N_1 \geq 1$ such that $|\mu|^*(\{\omega \in \Omega; |f_n(\omega) - f(\omega)| > \varepsilon/2\}) < \varepsilon/2$ whenever $n \geq N_1$. Let $N \geq 1$ be such that $1/N < \varepsilon/2$ and $N \geq N_1$. If $n \geq N$, then

$$\{\omega \in \Omega; |h_n(\omega) - f(\omega)| > \varepsilon\} \subset \{\omega \in \Omega; |h_n(\omega) - f_n(\omega)| > \varepsilon/2\}$$
$$\cup \{\omega \in \Omega; |f_n(\omega) - f(\omega)| > \varepsilon/2\}$$
$$\subset \{\omega \in \Omega; |h_n(\omega) - f_n(\omega)| > 1/n\}$$
$$\cup \{\omega \in \Omega; |f_n(\omega) - f(\omega)| > \varepsilon/2\}.$$

From these set inclusions, it follows that

$$|\mu|^*(\{\omega \in \Omega; |h_n(\omega) - f(\omega)| > \varepsilon\}) < \varepsilon/2 + \varepsilon/2 = \varepsilon$$

whenever $n \geq N$. Next, observe that

$$D \int |h_n - h_m| \, d|\mu| \leq D \int |h_n - f_n| \, d|\mu| + D \int |f_n - f_m| \, d|\mu|$$
$$+ D \int |f_m - h_m| \, d|\mu|$$

for all $m, n \geq 1$. From this, it follows that $\lim_{n,m \to \infty} D \int |h_n - h_m| \, d|\mu| = 0$.

Hence h_n, $n \geq 1$ is a determining sequence for f. So, f is D-integrable. Further, for $n \geq 1$,

$$D \int |f_n - f| \, d|\mu| \leq D \int |f_n - h_n| \, d|\mu| + D \int |h_n - f| \, d|\mu|.$$

The second term on the right above converges to zero as $n \to \infty$, by Lemma 4.4.12. Hence

$$\lim_{n \to \infty} D \int |f_n - f| \, d|\mu| = 0.$$

This completes the proof. □

4.5 S-INTEGRAL

In this section, we introduce S-integrals which are of Stieltjes type in the framework of charge spaces. We also show that D-integrals and S-integrals coincide in the case of positive bounded charges and bounded functions.

In what follows, we assume that all the charges are positive bounded unless otherwise specified.

For a given field $\mathscr{F}$ on a set Ω, let $\mathscr{P}$ denote the collection of all finite partitions of Ω in $\mathscr{F}$. On $\mathscr{P}$, we define a partial order by $P_1 \geq P_2$ for P_1, P_2 in $\mathscr{P}$ if P_1 is a refinement of P_2, i.e. every set in P_2 is a union of sets in P_1. Indeed, $(\mathscr{P}, \geq)$ is a directed set.

4.5.1 Definition. Let $(\Omega, \mathscr{F}, \mu)$ be a charge space. Let f be a bounded real valued function on Ω. For $P = \{F_1, F_2, \ldots, F_n\}$ in $\mathscr{P}$, let

$$L(P) = \sum_{i=1}^{n} (\inf_{\omega \in F_i} f(\omega)) \mu(F_i)$$

and

$$U(P) = \sum_{i=1}^{n} (\sup_{\omega \in F_i} f(\omega)) \mu(F_i).$$

L(P) is called the *lower sum* associated with P and U(P) is called the *upper sum* associated with P. (Since f and μ are bounded, L(P) and U(P) are real numbers.)

The following proposition gives some inequalities between these sums.

4.5.2 Proposition. *Let $(\Omega, \mathscr{F}, \mu)$ be a charge space and f a bounded real valued function on Ω. Then for any $P_1 \geq P_2$ in $\mathscr{P}$,*

$$(\inf_{\omega \in \Omega} f(\omega)) \mu(\Omega) \leq L(P_2) \leq L(P_1) \leq U(P_1) \leq U(P_2) \leq (\sup_{\omega \in \Omega} f(\omega)) \mu(\Omega).$$

Proof. Since $P_1 \geq P_2$, every set in P_2 is a union of sets in P_1. Hence the above inequalities easily follow. Of course, we use the fact that μ is positive in proving the above inequalities. □

Thus, we observe that the net $\{U(P); P \in \mathcal{P}\}$ defined on the directed set $(\mathcal{P}, \geq)$ is a decreasing net of real numbers bounded below and hence has a limit. The net $\{L(P); P \in \mathcal{P}\}$ defined over the directed set $(\mathcal{P}, \geq)$ is an increasing net of real numbers bounded above and therefore, has a limit.

4.5.3 Definitions. Let $(\Omega, \mathcal{F}, \mu)$ be a charge space and f a bounded real valued function on Ω. Let

$$\overline{\int} f \, d\mu = \operatorname*{Inf}_{P \in \mathcal{P}} U(P) = \lim_{P \in \mathcal{P}} U(P)$$

and

$$\underline{\int} f \, d\mu = \operatorname*{Sup}_{P \in \mathcal{P}} L(P) = \lim_{P \in \mathcal{P}} L(P).$$

$\overline{\int} f \, d\mu$ is called the *upper integral* of f with respect to μ and $\underline{\int} f \, d\mu$ the *lower integral* of f with respect to μ.

The following proposition is obvious in view of Proposition 4.5.2.

4.5.4 Proposition. *Let $(\Omega, \mathcal{F}, \mu)$ be a charge space and f a bounded real valued function on Ω. Then*

$$\underline{\int} f \, d\mu \leq \overline{\int} f \, d\mu.$$

Now, we define the S-integral.

4.5.5 Definition. Let $(\Omega, \mathcal{F}, \mu)$ be a charge space and f a bounded real valued function on Ω. f is said to be S-*integrable* if

$$\underline{\int} f \, d\mu = \overline{\int} f \, d\mu.$$

If f is S-integrable, the S-integral of f is denoted by $S \int f \, d\mu$ and is defined to be the common number $\underline{\int} f \, d\mu = \overline{\int} f \, d\mu$.

4.5.6 Remark. If f is a simple function, then $S \int f \, d\mu = D \int f \, d\mu$.

4. INTEGRATION

We link S-integrability and D-integrability of a function in the following result.

4.5.7 Theorem. *Let $(\Omega, \mathscr{F}, \mu)$ be a charge space and f a bounded real valued function on Ω. Then the following statements are equivalent.*

(i). *f is T_1-measurable.*
(ii). *f is T_2-measurable.*
(iii). *f is S-integrable.*
(iv). *There exists a real number a with the following property. For every $\varepsilon > 0$, there exists a partition P_0 in $\mathscr{P}$ such that for every partition P in $\mathscr{P}$ with $P = \{F_1, F_2, \ldots, F_n\} \geq P_0$ and for every ω_i in F_i, $i = 1, 2, \ldots, n$,*

$$\left| \sum_{i=1}^{n} f(\omega_i) \mu(F_i) - a \right| < \varepsilon$$

holds.

(v). *For every $\varepsilon > 0$, there exists a partition P_0 in $\mathscr{P}$ such that for every partition P in $\mathscr{P}$ with $P = \{F_1, F_2, \ldots, F_n\} \geq P_0$ and for every ω_{i1}, ω_{i2} in F_i, $i = 1, 2, \ldots, n$,*

$$\left| \sum_{i=1}^{n} (f(\omega_{i1}) - f(\omega_{i2})) \mu(F_i) \right| < \varepsilon$$

holds.

(vi). *For every $\varepsilon > 0$, there exists a partition P_0 in $\mathscr{P}$ such that for every partition P in $\mathscr{P}$ with $P = \{F_1, F_2, \ldots, F_n\} \geq P_0$,*

$$\sum_{i=1}^{n} [\sup_{\omega_{i1}, \omega_{i2} \in F_i} |f(\omega_{i1}) - f(\omega_{i2})|] \mu(F_i) < \varepsilon$$

holds.

(vii). *For every $\varepsilon > 0$, there exists a partition $P_0 = \{E_1, E_2, \ldots, E_m\}$ in $\mathscr{P}$ such that for any partition $\{E_{11}, E_{12}, \ldots, E_{1k_1}, E_{21}, E_{22}, \ldots, E_{2k_2}, \ldots, E_{m1}, E_{m2}, \ldots, E_{mk_m}\}$ in $\mathscr{P}$ with $E_i = \bigcup_{j=1}^{k_i} E_{ij}$, $i = 1, 2, \ldots, m$ and for every choice λ_{ij}, $j = 1, 2, \ldots, k_i$, $i = 1, 2, \ldots, m$, of real numbers satisfying*

$$|\lambda_{ij}| \leq \sup_{\omega_{i1}, \omega_{i2} \in E_i} |f(\omega_{i1}) - f(\omega_{i2})|, \quad j = 1, 2, \ldots, k_i, i = 1, 2, \ldots, m,$$

$$\left| \sum_{i=1}^{m} \sum_{j=1}^{k_i} \lambda_{ij} \mu(E_{ij}) \right| < \varepsilon$$

holds.
(viii). *f is D-integrable.*

Proof. (i)$\Rightarrow$(ii). This follows from Theorem 4.4.7.
(ii)$\Rightarrow$(iii). Let $\varepsilon > 0$. We show that

$$\overline{\int} f \, d\mu - \underline{\int} f \, d\mu < \varepsilon.$$

This then would prove that f is S-integrable. Let $M = \operatorname{Sup}_{\omega \in \Omega} |f(\omega)|$. Since f is T_2-measurable, there exists a partition $P = \{F_0, F_1, F_2, \ldots, F_m\}$ in $\mathcal{P}$ such that

$$\mu(F_0) < \varepsilon/4M \quad \text{and} \quad |f(\omega_{i1}) - f(\omega_{i2})| < \varepsilon/2\mu(\Omega)$$

for every ω_{i1}, ω_{i2} in F_i for $i = 1, 2, \ldots, m$. Then

$$\overline{\int} f \, d\mu - \underline{\int} f \, d\mu \leq U(P) - L(P)$$

$$= \sum_{i=0}^{m} (\operatorname{Sup}_{\omega \in F_i} f(\omega))\mu(F_i) - \sum_{i=0}^{m} (\operatorname{Inf}_{\omega \in F_i} f(\omega))\mu(F_i)$$

$$= \sum_{i=0}^{m} (\operatorname{Sup}_{\omega \in F_i} f(\omega) - \operatorname{Inf}_{\omega \in F_i} f(\omega))\mu(F_i)$$

$$\leq 2M\mu(F_0) + (\varepsilon/2\mu(\Omega)) \sum_{i=1}^{m} \mu(F_i)$$

$$< \varepsilon/2 + \varepsilon/2 = \varepsilon.$$

This shows that f is S-integrable.

(iii)$\Rightarrow$(iv). Let $a = S\int f \, d\mu = \overline{\int} f \, d\mu = \underline{\int} f \, d\mu$. Let $\varepsilon > 0$. There exists a partition P_0 in $\mathcal{P}$ such that for every $P \geq P_0$, $U(P) - a < \varepsilon$ and $a - L(P) < \varepsilon$. Let $P = \{F_1, F_2, \ldots, F_n\}$ be any partition in $\mathcal{P}$ such that $P \geq P_0$. Let ω_i in F_i, $i = 1, 2, \ldots, n$ be arbitrary. Then

$$\sum_{i=1}^{n} f(\omega_i)\mu(F_i) - a \leq \sum_{i=1}^{n} (\operatorname{Sup}_{\omega \in F_i} f(\omega))\mu(F_i) - a$$

$$= U(P) - a < \varepsilon,$$

and

$$a - \sum_{i=1}^{n} f(\omega_i)\mu(F_i) \leq a - \sum_{i=1}^{n} (\operatorname{Inf}_{\omega \in F_i} f(\omega))\mu(F_i)$$

$$= a - L(P) < \varepsilon.$$

Consequently, $|\sum_{i=1}^{n} f(\omega_i)\mu(F_i) - a| < \varepsilon$. This proves (iv).
(iv)⇒(v). This is obvious.
(v)⇒(vi). This is obvious. The modulus sign can be taken inside the summation in (v) and the inequality still remains valid.
(vi)⇒(vii). This is obvious.
(vii)⇒(viii). For each $k \geq 1$, let $P_k = \{E_{k1}, E_{k2}, \ldots, E_{kp_k}\}$ be a partition in $\mathcal{P}$ satisfying (vii) for $\varepsilon = 1/k$. Assume, without loss of generality, that $P_1 \leq P_2 \leq P_3 \leq \cdots$. For each $k \geq 1$ and $i = 1, 2, \ldots, p_k$, choose and fix ω_{ki} in E_{ki}. Let

$$f_k = \sum_{i=1}^{p_k} f(\omega_{ki}) I_{E_{ki}}, \quad k \geq 1.$$

We claim that $\lim_{m,k \to \infty} D \int |f_m - f_k| \, d\mu = 0$. To begin with, we look at the function $|f_k - f_m|$. Suppose $m \geq k$. Since $P_k \leq P_m$, every set in P_k is a union of sets in P_m. Assume, without loss of generality, that the sets in P_k are of the following form.

$$E_{k1} = E_{m1} \cup E_{m2} \cup \cdots \cup E_{mr_1},$$
$$E_{k2} = E_{mr_1+1} \cup E_{mr_1+2} \cup \cdots \cup E_{mr_2},$$
$$\cdots\cdots\cdots\cdots\cdots\cdots\cdots\cdots\cdots\cdots\cdots\cdots\cdots$$
$$E_{kp_k} = E_{mr_{p_k-1}+1} \cup E_{mr_{p_k-1}+2} \cup \cdots \cup E_{mp_m}.$$

Write $r_0 = 0$ and $r_{p_k} = p_m$. Then

$$|f_k - f_m| = \left| \sum_{i=1}^{p_k} f(\omega_{ki}) I_{E_{ki}} - \sum_{i=1}^{p_m} f(\omega_{mi}) I_{E_{mi}} \right|$$
$$= \left| \sum_{i=1}^{p_k} \sum_{j=r_{i-1}+1}^{r_i} f(\omega_{ki}) I_{E_{mj}} - \sum_{i=1}^{p_m} f(\omega_{mi}) I_{E_{mi}} \right|$$
$$= \left| \sum_{i=1}^{p_k} \sum_{j=r_{i-1}+1}^{r_i} (f(\omega_{ki}) - f(\omega_{mj})) I_{E_{mj}} \right|.$$

Therefore,

$$D \int |f_k - f_m| \, d\mu = \sum_{i=1}^{p_k} \sum_{j=r_{i-1}+1}^{r_i} |f(\omega_{ki}) - f(\omega_{mj})| \mu(E_{mj}).$$

Obviously, $|f(\omega_{ki}) - f(\omega_{mj})| = \lambda_{ij} \leq \mathrm{Sup}_{\omega, \omega' \in E_{ki}} |f(\omega) - f(\omega')|$ for every $j = r_{i-1}+1, r_{i-1}+2, \ldots, r_i$ and $i = 1, 2, \ldots, p_k$. Hence $D \int |f_k - f_m| \, d\mu < 1/k$. This shows that $\lim_{m,k \to \infty} D \int |f_k - f_m| \, d\mu = 0$.

Now, we claim that f_k, $k \geq 1$ converges to f hazily. Observe that for any $k \geq 1$,

$$|f_k - f| = \left| \sum_{i=1}^{p_k} f(\omega_{ki}) I_{E_{ki}} - f \sum_{i=1}^{p_k} I_{E_{ki}} \right|$$

$$= \left| \sum_{i=1}^{p_k} (f(\omega_{ki}) - f) I_{E_{ki}} \right|$$

$$\leq \sum_{i=1}^{p_k} \left(\sup_{\omega, \omega' \in E_{ki}} |f(\omega) - f(\omega')| \right) I_{E_{ki}}$$

$$= \sum_{i=1}^{p_k} \lambda_{ki} I_{E_{ki}}, \quad \text{say.}$$

Therefore, $\mu^*(\{\omega \in \Omega; |f_k(\omega) - f(\omega)| > \varepsilon\}) \leq \mu^*\left(\left\{\omega \in \Omega; \sum_{i=1}^{p_k} \lambda_{ki} I_{E_{ki}}(\omega) > \varepsilon\right\}\right)$

$$\leq \mu\left(\bigcup_{i \in J} E_{ki}\right),$$

where $J = \{1 \leq i \leq p_k; \lambda_{ki} > \varepsilon\}$.

$$\leq \sum_{i \in J} \mu(E_{ki})$$

$$\leq \frac{\sum_{i=1}^{p_k} \lambda_{ki} \mu(E_{ki})}{\varepsilon}$$

$$< \frac{1}{k\varepsilon}.$$

Hence $\lim_{k \to \infty} \mu^*(\{\omega \in \Omega; |f_k(\omega) - f(\omega)| > \varepsilon\}) = 0$. This shows that f is D-integrable. □

The definition of S-integral can be extended to general charge spaces and real functions. See Gould (1965).

We link S-integrals and D-integrals in the following theorem.

4.5.8 Theorem. *Let $(\Omega, \mathscr{F}, \mu)$ be a charge space and f a bounded real valued function on Ω. If f is S-integrable, then f is D-integrable and $S \int f \, d\mu = D \int f \, d\mu$.*

Proof. The proof of this theorem can be recovered from the proof of the above theorem as follows. For any partition $P = \{F_1, F_2, \ldots, F_n\}$ in $\mathscr{P}$ and fixed points $\omega_i \in F_i$, $i = 1, 2, \ldots, n$, define

$$a_P = \sum_{i=1}^{n} f(\omega_i) \mu(F_i).$$

Then, by Theorem 4.5.7 (iii) $\Rightarrow$ (iv), $\lim_{P \in \mathscr{P}} a_P = a = S \int f \, d\mu$. Next, observe that the proof of Theorem 4.5.7 (vii) $\Rightarrow$ (viii) can be improved to show that if $P \geq P_k$, $Q \geq P_m$ and $k \leq m$, then $|a_P - a_Q| \leq 1/k$. These two observations together show that

$$D\int f\,d\mu = \lim_{n\to\infty} D\int f_n\,d\mu$$
$$= \lim_{n\to\infty} a_{P_n} = a = S\int f\,d\mu. \qquad \square$$

As a consequence of Theorem 4.5.7, we obtain the following result.

4.5.9 Corollary. *Let $\mathscr{F}$ be a σ-field of subsets of a set Ω and μ a bounded charge on $\mathscr{F}$. Let f be a bounded measurable function on Ω, i.e. $f^{-1}(B) \in \mathscr{F}$ for every Borel set $B \subset R$. Then f is D-integrable.*

Proof. We show that f is the uniform limit of a sequence of simple functions. Let $M = \operatorname{Sup}_{\omega \in \Omega} |f(\omega)|$ and $d = 2M$. For $n \geq 1$, let for ω in Ω,

$$f_n(\omega) = -M + \frac{i-1}{2^n}d, \quad \text{if } -M + \frac{i-1}{2^n} \leq f(\omega) < -M + \frac{i}{2^n},$$

$$\text{for } i = 1, 2, \ldots, 2^n - 1,$$

$$= M - \frac{d}{2^n}, \quad \text{if } M - \frac{d}{2^n} \leq f(\omega) \leq M.$$

It can be checked that f_n, $n \geq 1$ converges to f uniformly. Since f is measurable, each f_n is a simple function. Hence f is T_1-measurable. An application of Theorem 4.5.7 completes the proof. $\qquad \square$

Note that if f is not bounded in the above proof, there is no sequence of simple functions converging to f uniformly.

4.6 L_p-SPACES

Let $(\Omega, \mathscr{F}, \mu)$ be a charge space. In this section, we introduce some function spaces, namely L_p-spaces, associated with $(\Omega, \mathscr{F}, \mu)$ and study some of the properties of these spaces. We establish Hölder's inequality and Minkowski's inequality to aid the study of these spaces. We also prove Lebesgue dominated convergence theorem. First, we introduce L_p-spaces.

4.6.1 Definitions. Let $(\Omega, \mathscr{F}, \mu)$ be a charge space. For $1 \leq p < \infty$, let

$$L_p(\Omega, \mathscr{F}, \mu) = \{f; f \text{ is a } T_1\text{-measurable real valued}$$

$$\text{function on } \Omega \text{ such that } |f|^p \text{ is D-integrable}\},$$

$$\|f\|_p = \left(D\int |f|^p\,d|\mu|\right)^{1/p} \quad \text{for } f \in L_p(\Omega, \mathscr{F}, \mu),$$

$L_\infty(\Omega, \mathscr{F}, \mu) = \{f; f \text{ is essentially bounded and } T_1\text{-measurable}\}$,

and

$$\|f\|_\infty = \text{Essential supremum of } |f|$$
$$= \text{Inf } \{k > 0; |\mu|^*(\{\omega \in \Omega; |f(\omega)| > k\}) = 0\}.$$

(We use the convention that the infimum over an empty set is ∞.)

If f is a null function, obviously, $\|f\|_p = 0$ for any $1 \leq p \leq \infty$. If f and g are such that $f = g$ a.e. $[\mu]$, $f \in L_p(\Omega, \mathscr{F}, \mu)$ for some $1 \leq p \leq \infty$, then $g \in L_p(\Omega, \mathscr{F}, \mu)$ and $\|f\|_p = \|g\|_p$. We want to show that the nonnegative function $\|\cdot\|_p$ defined on $L_p(\Omega, \mathscr{F}, \mu)$ for $1 \leq p \leq \infty$ is a pseudo-norm on $L_p(\Omega, \mathscr{F}, \mu)$. We need the following inequalities for this purpose. The first of these is Hölder's inequality.

4.6.2 Theorem. *Let $(\Omega, \mathscr{F}, \mu)$ be a charge space and p and q be two positive numbers satisfying $1/p + 1/q = 1$. If $f \in L_p(\Omega, \mathscr{F}, \mu)$ and $g \in L_q(\Omega, \mathscr{F}, \mu)$, then $fg \in L_1(\Omega, \mathscr{F}, \mu)$ and*

$$\|fg\|_1 \leq \|f\|_p \|g\|_q.$$

Proof. Assume, first, that $p > 1$ and $q > 1$. The function

$$\psi(t) = \frac{t^p}{p} + \frac{t^{-q}}{q}, t > 0$$

has a global minimum at $t = 1$. Therefore, for every $t > 0$, $\psi(t) \geq \psi(1) = 1/p + 1/q = 1$. Let a and b be any two positive numbers and $t = (a^{1/q})/(b^{1/p})$. Then

$$1 \leq \psi(t) = \frac{a^{p-1}}{pb} + \frac{b^{q-1}}{qa}.$$

This implies that $ab \leq a^p/p + b^q/q$. This inequality is valid even if $a = 0$ or $b = 0$. Now, we turn to the proof of the theorem.

If f or g is a null function, then fg is a null function. This can be proved as follows. Suppose f is a null function. Since any T_1-measurable function is smooth, g is smooth. So, for a given $\varepsilon > 0$, there exists $k > 0$ such that $|\mu|^*(\{\omega \in \Omega; |g(\omega)| > k\}) \leq \varepsilon$. Consequently, for any $\delta > 0$,

$$|\mu|^*(\{\omega \in \Omega; |f(\omega)g(\omega)| > \delta\}) \leq |\mu|^*(\{\omega \in \Omega; |f(\omega)| > \delta/k\})$$
$$+ |\mu|^*(\{\omega \in \Omega; |g(\omega)| > k\})$$
$$\leq 0 + \varepsilon = \varepsilon.$$

This shows that fg is a null function. In this case, the theorem is evidently true.

Let $\|f\|_p > 0$ and $\|g\|_q > 0$. By setting $a = |f(\omega)|/\|f\|_p$ and $b = |g(\omega)|/\|g\|_q$ for any ω in Ω, we obtain

$$\frac{|fg|}{\|f\|_p \|g\|_q} \leq \frac{|f|^p}{p\|f\|_p^p} + \frac{|g|^q}{q\|g\|_q^q}.$$

Since f and g are T_1-measurable, fg is T_1-measurable. See Corollary 4.4.9(i). By Theorem 4.4.18, fg is D-integrable. It is now obvious that

$$\|fg\|_1 \leq (1/p + 1/q)\|f\|_p \|g\|_q = \|f\|_p \|g\|_q.$$

If $p = 1$, then $q = \infty$. Therefore, $|fg| \leq k|f|$ a.e. $[\mu]$ for any number $k > \|g\|_\infty$. Since fg is T_1-measurable, by Theorem 4.4.18, it follows that fg is D-integrable and $\|fg\|_1 \leq k\|f\|_1$ for any $k > \|g\|_\infty$. Consequently, $\|fg\|_1 \leq \|f\|_1 \|g\|_\infty$. The case $p = \infty$ and $q = 1$ can be disposed of in a similar vein. $\square$

A more general version of the above theorem is the following result.

4.6.3 Corollary. *Let $(\Omega, \mathcal{F}, \mu)$ be a charge space and p, q, r be numbers satisfying $1 \leq p, q, r \leq \infty$ and $1/r = 1/p + 1/q$. If $f \in L_p(\Omega, \mathcal{F}, \mu)$ and $g \in L_q(\Omega, \mathcal{F}, \mu)$, then $fg \in L_r(\Omega, \mathcal{F}, \mu)$ and $\|fg\|_r \leq \|f\|_p \|g\|_q$.*

Proof. There are only three possibilities involving ∞. Case (i). $p = 1$, $q = \infty$, $r = 1$. Case (ii). $p = \infty$, $q = 1$, $r = 1$. Case (iii). $p = \infty$, $q = \infty$, $r = \infty$. In Cases (i) and (ii), the result follows from Theorem 4.6.2. For the case (iii), we proceed as follows. For any $k > 0$ and $t > 0$ satisfying $k > \|f\|_\infty$ and $t > \|g\|_\infty$, we have

$$|\mu|^*(\{\omega \in \Omega; |f(\omega)g(\omega)| > kt\}) \leq |\mu|^*(\{\omega \in \Omega; |f(\omega)| > k\})$$
$$+ |\mu|^*(\{\omega \in \Omega; |g(\omega)| > t\})$$
$$= 0.$$

This shows that fg is essentially bounded. Further, $\|fg\|_\infty \leq kt$ for any $k > \|f\|_\infty$ and $t > \|g\|_\infty$. Hence $\|fg\|_\infty \leq \|f\|_\infty \|g\|_\infty$.

Let us look into the case $1 < p < \infty$ and $1 < q < \infty$. In this case, $1 \leq r < \infty$. Since $1/r = 1/p + 1/q$, we have $1 = 1/(p/r) + 1/(q/r)$. Note that $p/r > 1$, $q/r > 1$, $|f|^r \in L_{(p/r)}(\Omega, \mathcal{F}, \mu)$ and $|g|^r \in L_{(q/r)}(\Omega, \mathcal{F}, \mu)$. By Theorem 4.6.2, $|f|^r |g|^r \in L_1(\Omega, \mathcal{F}, \mu)$ and $\||fg|^r\|_1 \leq \||f|^r\|_{(p/r)} \||g|^r\|_{(q/r)}$. Equivalently,

$$D\int |fg|^r \, d|\mu| \leq \left(D\int |f|^p \, d|\mu|\right)^{r/p} \left(D\int |g|^q \, d|\mu|\right)^{r/q}.$$

Consequently, $\|fg\|_r \leq \|f\|_p \|g\|_q$. Since fg is T_1-measurable, $fg \in L_r(\Omega, \mathcal{F}, \mu)$. $\square$

A special case of Hölder's inequality is Cauchy-Schwartz inequality.

4.6.4 Corollary. *Let $(\Omega, \mathscr{F}, \mu)$ be a charge space and $f, g \in L_2(\Omega, \mathscr{F}, \mu)$. Then $fg \in L_1(\Omega, \mathscr{F}, \mu)$ and*

$$\|fg\|_1 \le \|f\|_2 \|g\|_2. \qquad \square$$

A consequence of Corollary 4.6.3 is the inclusion relations among the L_p-spaces.

4.6.5 Corollary. *Let $(\Omega, \mathscr{F}, \mu)$ be a charge space in which μ is bounded. Let r and s be any numbers satisfying $1 \le r \le s \le \infty$. Then*

$$L_1(\Omega, \mathscr{F}, \mu) \supset L_r(\Omega, \mathscr{F}, \mu) \supset L_s(\Omega, \mathscr{F}, \mu) \supset L_\infty(\Omega, \mathscr{F}, \mu).$$

Proof. Since μ is bounded, every constant function (which is obviously T_1-measurable) is D-integrable. Consequently, $|f|^p$ is D-integrable whenever $f \in L_\infty(\Omega, \mathscr{F}, \mu)$ and $1 \le p < \infty$. So, the last inclusion relation in the corollary follows. Now, let $1 \le r < s < \infty$. Let $q = rs/(s-r)$. It is easy to check that $1/r = 1/s + 1/q$. Let $\underline{1}$ denote the constant function on Ω which is identically equal to unity. Then $\underline{1} \in L_q(\Omega, \mathscr{F}, \mu)$. If $f \in L_s(\Omega, \mathscr{F}, \mu)$, by Corollary 4.6.3, $f\underline{1} = f \in L_r(\Omega, \mathscr{F}, \mu)$. This shows that $L_s(\Omega, \mathscr{F}, \mu) \subset L_r(\Omega, \mathscr{F}, \mu)$. This completes the proof. $\qquad \square$

The following inequality is known as Minkowski's inequality.

4.6.6 Theorem. *Let $(\Omega, \mathscr{F}, \mu)$ be a charge space and $1 \le p \le \infty$. If $f, g \in L_p(\Omega, \mathscr{F}, \mu)$, then $f + g \in L_p(\Omega, \mathscr{F}, \mu)$ and*

$$\|f+g\|_p \le \|f\|_p + \|g\|_p.$$

Proof. The case $p = 1$ follows from Theorem 4.4.13(viii). If $p = \infty$, it is easy to check that $\|f+g\|_\infty \le \|f\|_\infty + \|g\|_\infty$ by using an argument similar to the one used in the proof of Case (iii) of Corollary 4.6.3. Assume $1 < p < \infty$. Let q be the positive number satisfying $1/p + 1/q = 1$. For any four real numbers a_1, b_1, a_2, b_2, we have

$$|a_1 b_1 + a_2 b_2| \le (|a_1|^p + |a_2|^p)^{1/p} (|b_1|^q + |b_2|^q)^{1/q}.$$

This can be seen as follows. Let $\Omega = \{\omega_1, \omega_2\}$, $\mathscr{F} = \mathscr{P}(\Omega)$ and $\mu(\{\omega_1\}) = 1 = \mu(\{\omega_2\})$. Let $f(\omega_1) = a_1$, $f(\omega_2) = a_2$, $g(\omega_1) = b_1$ and $g(\omega_2) = b_2$. Applying Hölder's inequality, we have precisely the above inequality. Note that

$$|f+g|^p \le |f||f+g|^{p-1} + |g||f+g|^{p-1}$$
$$\le (|f|^p + |g|^p)^{1/p} (2|f+g|^{q(p-1)})^{1/q}$$
$$\le (|f|^p + |g|^p)^{1/p} (2^{1/q})(|f+g|)^{p/q}.$$

So, $|f+g|^{p-(p/q)} = |f+g| \le 2^{1/q}(|f|^p + |g|^p)^{1/p}$. Thus, we obtain the inequality $|f+g|^p \le 2^{p/q}(|f|^p + |g|^p)$. By Corollary 4.4.9 and Theorem 4.4.18,

$f + g \in L_p(\Omega, \mathscr{F}, \mu)$. Since $|f|, |g| \in L_p(\Omega, \mathscr{F}, \mu)$ and $|f+g|^{p-1} \in L_q(\Omega, \mathscr{F}, \mu)$, by Hölder's inequality, $|f||f+g|^{p-1}$ and $|g||f+g|^{p-1} \in L_1(\Omega, \mathscr{F}, \mu)$. Hence

$$\mathrm{D}\int |f+g|^p \, \mathrm{d}|\mu| \leq \mathrm{D}\int |f||f+g|^{p-1} \, \mathrm{d}|\mu| + \mathrm{D}\int |g||f+g|^{p-1} \, \mathrm{d}|\mu|$$

$$\leq \left(\mathrm{D}\int |f|^p \, \mathrm{d}|\mu|\right)^{1/p} \left(\mathrm{D}\int |f+g|^{(p-1)q} \, \mathrm{d}|\mu|\right)^{1/q}$$

$$+ \left(\mathrm{D}\int |g|^p \, \mathrm{d}|\mu|\right)^{1/p} \left(\mathrm{D}\int |f+g|^{(p-1)q} \, \mathrm{d}|\mu|\right)^{1/q}.$$

This follows from Hölder's inequality. Therefore,

$$\mathrm{D}\int |f+g|^p \, \mathrm{d}|\mu| \leq (\|f\|_p + \|g\|_p)\left(\mathrm{D}\int |f+g|^p \, \mathrm{d}|\mu|\right)^{p/q}.$$

From the above inequality, it follows that

$$\left(\mathrm{D}\int |f+g|^p \, \mathrm{d}|\mu|\right)^{1-(1/q)} = \|f+g\|_p \leq \|f\|_p + \|g\|_p.$$

This completes the proof. □

4.6.7 Theorem. *Let $(\Omega, \mathscr{F}, \mu)$ be a charge space. Then, for each $1 \leq p \leq \infty$, $(L_p(\Omega, \mathscr{F}, \mu), \|\cdot\|_p)$ is a linear space with a pseudo-norm $\|\cdot\|_p$.*

Proof. It is now obvious that each $L_p(\Omega, \mathscr{F}, \mu)$ is a linear space. Further, if $f = 0$, then $\|f\|_p = 0$. For any real number c and f in $L_p(\Omega, \mathscr{F}, \mu)$, it is obvious that $cf \in L_p(\Omega, \mathscr{F}, \mu)$ and that $\|cf\|_p = |c|\|f\|_p$. The inequality $\|f+g\|_p \leq \|f\|_p + \|g\|_p$ for f, g in $L_p(\Omega, \mathscr{F}, \mu)$ follows from Theorem 4.6.6. □

4.6.8 Remark. If μ is a 0–1 valued charge on a field $\mathscr{F}$ of subsets of a set Ω, then $L_p(\Omega, \mathscr{F}, \mu)/\sim$ is isometrically isomorphic to the real line R for any $1 \leq p \leq \infty$, where $L_p(\Omega, \mathscr{F}, \mu)/\sim$ is the collection of all equivalence classes of $L_p(\Omega, \mathscr{F}, \mu)$ under the equivalence relation $\sim$ induced by the notion of a null function. (See Example 4.4.14.) Consequently, $L_p(\Omega, \mathscr{F}, \mu)$ is complete.

In general, $(L_p(\Omega, \mathscr{F}, \mu), \|\cdot\|_p)$ need not be complete. Let $\Omega = \{1, 2, \ldots\}$, $\mathscr{F}$ the finite-cofinite field on Ω and μ the charge on $\mathscr{F}$ defined by

$$\mu(A) = \sum_{n \in A} \frac{1}{2^n}, \quad \text{if } A \text{ is finite},$$

$$= 2 - \sum_{n \in A^c} \frac{1}{2^n}, \quad \text{if } A \text{ is cofinite}.$$

Let $A_n = \{1, 2, 3, \ldots, n\}$, $n \geq 1$. We claim that

$$\lim_{m,n \to \infty} D \int |I_{A_m} - I_{A_n}| \, d\mu = 0.$$

This claim is established if we observe that $D \int |I_{A_m} - I_{A_n}| \, d\mu = \mu(A_n \Delta A_m)$ which converges to zero as $m, n \to \infty$. Suppose I_{A_n}, $n \geq 1$ converges to some function f on Ω hazily. We show that $f \equiv 1$. For every $k \geq 1$, there exists $n_k \geq 1$ such that $\mu^*(\{\omega \in \Omega; |I_{A_{n_k}}(\omega) - f(\omega)| > 1/2^k\}) < 1/2^k$. Let B_k in $\mathscr{F}$ be any set such that $\mu(B_k) < 1/2^k$ and $\{\omega \in \Omega; |I_{A_{n_k}}(\omega) - f(\omega)| > 1/2^k\} \subset B_k$ for $k \geq 1$. Assume, without loss of generality, that $n_1 < n_2 < n_3 < \cdots$. We now give the properties of the sets B_k, $k \geq 1$.
  (i). Each B_k is a finite set. For, for any infinite set A, $\mu^*(A) \geq 1$.
  (ii). $B_k \subset \{k+1, k+2, \ldots\}$, $k \geq 1$.
  (iii). $|I_{A_{n_k}}(\omega) - f(\omega)| \leq 1/2^k$, if $\omega \notin B_k$, for $k \geq 1$.
  (iv). $k \notin B_k$, for each $k \geq 1$.
Now, let k_0 in Ω be fixed. Let $\varepsilon > 0$. We show that $|f(k_0) - 1| < \varepsilon$. This then would imply that $f(k_0) = 1$. Let $N \geq 1$ be such that $1/2^N < \varepsilon$. Let $p \geq \max\{N, k_0\}$. Since $B_p \subset \{p+1, p+2, \ldots\}$ and $p \geq k_0$, $k_0 \notin B_p$. Further, $k_0 \in A_{k_0} \subset A_{n_{k_0}} \subset A_{n_p}$. Therefore, $|f(k_0) - I_{A_{n_p}}(k_0)| = |f(k_0) - 1| \leq 1/2^p \leq 1/2^N < \varepsilon$. This shows that $f \equiv 1$.

Next, we show that I_{A_n}, $n \geq 1$ does not converge to the constant function identically equal to 1 hazily. Let $\varepsilon = \frac{1}{2}$. Then the set $\{\omega \in \Omega; |I_{A_n}(\omega) - 1| > \frac{1}{2}\}$ is a cofinite set and consequently, $\mu(\{\omega \in \Omega; |I_{A_n}(\omega) - 1| > \frac{1}{2}\}) \geq 1$. So, I_{A_n}, $n \geq 1$ fails to converge to $\underline{1}$ hazily. Thus, we have a Cauchy sequence in $L_1(\Omega, \mathscr{F}, \mu)$ not convergent in $L_1(\Omega, \mathscr{F}, \mu)$. Hence $L_1(\Omega, \mathscr{F}, \mu)$ is not complete.

4.6.9 Remark. In $L_p(\Omega, \mathscr{F}, \mu)$, if we introduce the equivalence relation $\sim$ by $f \sim g$ for f, g in $L_p(\Omega, \mathscr{F}, \mu)$ if $f = g$ a.e. $[\mu]$, then the collection of all equivalence classes $L_p(\Omega, \mathscr{F}, \mu)/\sim$ of $L_p(\Omega, \mathscr{F}, \mu)$ equipped with the norm

$$\|[f]\|_p = \|f\|_p$$

for f in $L_p(\Omega, \mathscr{F}, \mu)$ is a normed linear space, where $[f]$ is the equivalence class in $L_p(\Omega, \mathscr{F}, \mu)$ containing f.

Next, we aim at proving Lebesgue dominated convergence theorem. For this, we need the following theorem on convergence in L_p-spaces.

4.6.10 Theorem. *Let $(\Omega, \mathscr{F}, \mu)$ be a charge space and $1 \leq p < \infty$. Let f_n, $n \geq 1$ be a sequence in $L_p(\Omega, \mathscr{F}, \mu)$ and f any real valued function on Ω. Then $f \in L_p(\Omega, \mathscr{F}, \mu)$ and $\lim_{n \to \infty} \|f_n - f\|_p = 0$ if and only if the following three conditions are satisfied.*

(i). f_n, $n \geq 1$ converges to f hazily.

(ii). The charges λ_n on $\mathscr{F}$ defined by $\lambda_n(F) = D \int_F |f_n|^p \, d|\mu|$, F in $\mathscr{F}$, $n \geq 1$ are uniformly absolutely continuous with respect to μ, i.e. given $\varepsilon > 0$, there exists $\delta > 0$ such that $\lambda_n(E) < \varepsilon$ for every $n \geq 1$ whenever $E \in \mathscr{F}$ and $|\mu|(E) < \delta$.

(iii). For each $\varepsilon > 0$, there exists a set $E_\varepsilon \in \mathscr{F}$ such that $|\mu|(E_\varepsilon) < \infty$ and $\lambda_n(E_\varepsilon^c) < \varepsilon$ for every $n \geq 1$.

Proof. The proof is carried out in the following steps.

$1°$. "Only if" part. If h is a nonnegative simple D-integrable function on Ω, the following inequality known as Chebychev's inequality is easy to establish.

$$|\mu|(\{\omega \in \Omega; h(\omega) > r\}) \leq \frac{D \int h \, d|\mu|}{r}$$

for any $r > 0$.

$2°$. In order to show that f_n, $n \geq 1$ converges to f hazily, it suffices to show that for any given $\varepsilon_1 > 0$ and $\varepsilon_2 > 0$, there exist $N \geq 1$ and sets A_n in $\mathscr{F}$ for $n \geq N$ such that $|\mu|(A_n) < \varepsilon_1$ and $|f_n(\omega) - f(\omega)| < \varepsilon_2$ for ω in A_n^c whenever $n \geq N$.

$3°$. Let r and ε be two positive numbers satisfying $(2r)^{1/p} < \varepsilon_2$ and $3\varepsilon/r < \varepsilon_1$.

$4°$. Let $g_n = |f_n - f|^p$, $n \geq 1$. Since $\lim_{n \to \infty} D \int |f_n - f|^p \, d|\mu| = 0$, there exists $N \geq 1$ such that $D \int g_n \, d|\mu| < \varepsilon$ whenever $n \geq N$.

$5°$. Since each g_n is nonnegative and D-integrable, there exists a nonnegative simple function h_n on Ω such that

$$\left| D \int g_n \, d|\mu| - D \int h_n \, d|\mu| \right| < \varepsilon$$

and

$$|\mu|^*(\{\omega \in \Omega; |h_n(\omega) - g_n(\omega)| \geq r\}) < \varepsilon/r.$$

See Lemma 4.4.12. Let B_n in $\mathscr{F}$ be any set such that $|\mu|(B_n) < \varepsilon/r$ and $\{\omega \in \Omega; |h_n(\omega) - g_n(\omega)| \geq r\} \subset B_n$, $n \geq 1$. Let $C_n = \{\omega \in \Omega; h_n(\omega) > r\}$, $n \geq 1$. Then, by $1°$, $|\mu|(C_n) \leq (D \int h_n \, d|\mu|)/r$ for every $n \geq 1$. Let $A_n = B_n \cup C_n$, $n \geq 1$. Then

$$|\mu|(A_n) \leq |\mu|(B_n) + |\mu|(C_n) < \varepsilon/r + \left(D \int h_n \, d|\mu|\right)\bigg/r$$

$$< \varepsilon/r + \left(D \int g_n \, d|\mu| + \varepsilon\right)\bigg/r$$

$$< \varepsilon/r + 2\varepsilon/r < \varepsilon_1, \quad \text{if } n \geq N.$$

If $\omega \in A_n^c = B_n^c \cap C_n^c$, $|f_n(\omega) - f(\omega)|^p = g_n(\omega) \leq h_n(\omega) + r \leq 2r$ which implies that $|f_n(\omega) - f(\omega)| \leq (2r)^{1/p} < \varepsilon_2$ for every $n \geq 1$. This shows that f_n, $n \geq 1$ converges to f hazily and thus (i) is established.

6°. Since $|f|^p$ is D-integrable, for $\varepsilon > 0$, there exists $\delta_1 > 0$ such that $D \int_F |f|^p \, d|\mu| < \varepsilon/2^p$ whenever $F \in \mathscr{F}$ and $|\mu|(F) < \delta_1$. Since $\lim_{n \to \infty} \|f_n - f\|_p = 0$, there exists $N \geq 1$ such that $D \int |f_n - f|^p \, d|\mu| < \varepsilon/2^p$ for all $n > N$. Since $\lambda_1, \lambda_2, \ldots, \lambda_N$ are all absolutely continuous with respect to μ, by Theorem 4.4.13(xi), there exists $\delta_2 > 0$ such that $\lambda_n(F) < \varepsilon/2$ whenever $F \in \mathscr{F}$ and $|\mu|(F) < \delta_2$ for $n = 1, 2, \ldots, N$. Let $\delta = \min\{\delta_1, \delta_2\}$. If $F \in \mathscr{F}$ and $|\mu|(F) < \delta$, then

$$\lambda_n(F) < \varepsilon/2, \qquad \text{for } 1 \leq n \leq N, \text{ and}$$

$$\lambda_n(F) = D \int_F |f_n - f + f|^p \, d|\mu|$$

$$\leq \left[\left(D \int_F |f_n - f|^p \, d|\mu| \right)^{1/p} + \left(D \int_F |f|^p \, d|\mu| \right)^{1/p} \right]^p$$

$$< [(\varepsilon/2^p)^{1/p} + (\varepsilon/2^p)^{1/p}]^p = \varepsilon, \qquad \text{for } n > N.$$

This proves (ii).

7°. We now prove (iii). Let $\varepsilon > 0$. There exists E_0 in $\mathscr{F}$ such that $|\mu|(E_0) < \infty$ and $D \int_{E_0^c} |f|^p \, d|\mu| < \varepsilon/2^p$. See Lemma 4.4.15. Further, there exists $N \geq 1$ such that $D \int |f_n - f|^p \, d|\mu| < \varepsilon/2^p$ for every $n > N$. Also, for each $n = 1, 2, \ldots, N$, there exists E_n in $\mathscr{F}$ such that $|\mu|(E_n) < \infty$ and $\lambda_n(E_n^c) < \varepsilon$. See Lemma 4.4.15. Let $E_\varepsilon = E_0 \cup E_1 \cup \cdots \cup E_N$. Obviously, $|\mu|(E_\varepsilon) < \infty$. If $1 \leq n \leq N$, then $\lambda_n(E_\varepsilon^c) \leq \lambda_n(E_n^c) < \varepsilon$. If $n > N$, then also,

$$\lambda_n(E_\varepsilon^c) = D \int_{E_\varepsilon^c} |f_n|^p \, d|\mu| = D \int_{E_\varepsilon^c} |f_n - f + f|^p \, d|\mu|$$

$$\leq \left[\left(D \int_{E_\varepsilon^c} |f_n - f|^p \, d|\mu| \right)^{1/p} + \left(D \int_{E_\varepsilon^c} |f|^p \, d|\mu| \right)^{1/p} \right]^p$$

$$< [(\varepsilon/2^p)^{1/p} + (\varepsilon/2^p)^{1/p}]^p = \varepsilon.$$

This completes the proof of "only if" part.

8°. Now, we prove "if" part. First, we show that $g = |f|^p$ is D-integrable. Let $g_n = |f_n|^p$, $n \geq 1$. In view of Theorem 4.4.20, it suffices to show that g_n, $n \geq 1$ converges to g hazily and

$$\lim_{m,n \to \infty} D \int |g_n - g_m| \, d|\mu| = 0.$$

Hazy convergence can be proved as follows. Since each f_n is T_1-measurable and f_n, $n \geq 1$ converges to f hazily (by (i)), f is T_1-measurable and so, $|f_n|^p$, $n \geq 1$ converges to $|f|^p$ hazily. See Corollary 4.4.9(iii).

9°. It remains to be shown that the above limit is indeed equal to zero. For this, first, we show that if $E \in \mathscr{F}$ and $|\mu|(E) < \infty$, then

$$\lim_{m,n \to \infty} D \int_E |g_n - g_m| \, d|\mu| = 0.$$

Let $\varepsilon > 0$. By (ii), there exists $\delta > 0$ such that $\lambda_n(F) < \varepsilon/(2 + |\mu|(E))$ for every $n \geq 1$ whenever $F \in \mathscr{F}$ and $|\mu|(F) < \delta$. Since g_n, $n \geq 1$ converges to g hazily, we have

$$\lim_{m,n \to \infty} |\mu|^*(\{\omega \in \Omega; |g_n(\omega) - g_m(\omega)| \geq \varepsilon/(2 + |\mu|(E))\}) = 0.$$

Consequently, there exist $N \geq 1$ and sets E_{nm} in $\mathscr{F}$ for $n, m \geq N$ such that $|\mu|(E_{nm}) < \delta$ and $|g_n(\omega) - g_m(\omega)| < \varepsilon/(2 + |\mu|(E))$ for ω in E_{nm}^c whenever $n, m \geq N$. Now, if $n, m \geq N$,

$$D \int_E |g_n - g_m| \, d|\mu| = D \int_{E_{nm} \cap E} |g_n - g_m| \, d|\mu| + D \int_{E_{nm}^c \cap E} |g_n - g_m| \, d|\mu|$$

$$\leq D \int_{E_{nm}} |g_n| \, d|\mu| + D \int_{E_{nm}} |g_m| \, d|\mu| + \frac{\varepsilon}{2 + |\mu|(E)} |\mu|(E)$$

$$\leq \lambda_n(E_{nm}) + \lambda_m(E_{nm}) + \frac{\varepsilon |\mu|(E)}{2 + |\mu|(E)}$$

$$\leq \frac{\varepsilon}{2 + |\mu|(E)} + \frac{\varepsilon}{2 + |\mu|(E)} + \frac{\varepsilon |\mu|(E)}{2 + |\mu|(E)} = \varepsilon.$$

This establishes the desired assertion.

10°. Next, we show that $\lim_{m,n \to \infty} D \int |g_n - g_m| \, d|\mu| = 0$. Let $\varepsilon > 0$. By (iii), there exists a set E in $\mathscr{F}$ such that $|\mu|(E) < \infty$ and $\lambda_n(E^c) < \varepsilon$ for all $n \geq 1$. Consequently,

$$D \int |g_n - g_m| \, d|\mu| = D \int_E |g_n - g_m| \, d|\mu| + D \int_{E^c} |g_n - g_m| \, d|\mu|$$

$$\leq D \int_E |g_n - g_m| \, d|\mu| + D \int_{E^c} |g_n| \, d|\mu| + D \int_{E^c} |g_m| \, d|\mu|$$

$$\leq D \int_E |g_n - g_m| \, d|\mu| + \varepsilon + \varepsilon$$

for all $n, m \geq 1$. Taking limits as $n, m \to \infty$, we obtain, by 9° as $|\mu|(E) < \infty$, that $\lim_{m,n \to \infty} D \int |g_n - g_m| \, d|\mu| \leq 2\varepsilon$. Since $\varepsilon > 0$ is arbitrary, the above limit is indeed equal to zero. Hence g is D-integrable.

11°. Finally, we show that $\lim_{n \to \infty} \|f_n - f\|_p = 0$. Let λ_0 on $\mathscr{F}$ be defined by $\lambda_0(F) = D \int_F |f|^p \, d|\mu|$, F in $\mathscr{F}$. By Theorem 4.4.13(xi) and Lemma 4.4.15, (ii) and (iii) hold for the sequence $\lambda_0, \lambda_1, \lambda_2, \ldots$ also.

Let $\varepsilon > 0$. By (iii), there exists E_ε in $\mathscr{F}$ such that $|\mu|(E_\varepsilon) < \infty$ and $(\lambda_n(E_\varepsilon^c))^{1/p} < \varepsilon$ for all $n \geq 0$. Let r be any positive number such that $r(|\mu|(E_\varepsilon))^{1/p} < \varepsilon$. By (ii), there exists $\delta > 0$ such that $(\lambda_n(F))^{1/p} < \varepsilon$ for every $n \geq 0$ whenever $F \in \mathscr{F}$ and $|\mu|(F) < \delta$. Since f_n, $n \geq 1$ converges to f hazily, there exist sets A_n in $\mathscr{F}$ such that $\lim_{n \to \infty} |\mu|(A_n) = 0$ and $|f_n(\omega) - f(\omega)| < r$ for every ω in A_n^c and $n \geq 1$. So, there exists $N \geq 1$ such that $|\mu|(A_n) < \delta$ if $n \geq N$. Now, let $n \geq N$. Then

$$\left(D \int |f_n - f|^p \, d|\mu| \right)^{1/p}$$

$$= \left[D \int [(I_{(E_\varepsilon \cup A_n)^c} + I_{E_\varepsilon - A_n} + I_{A_n}) |f_n - f|]^p \, d|\mu| \right]^{1/p}$$

$$\leq \left(D \int_{(E_\varepsilon \cup A_n)^c} |f_n - f|^p \, d|\mu| \right)^{1/p} + \left(D \int_{E_\varepsilon - A_n} |f_n - f|^p \, d|\mu| \right)^{1/p}$$

$$+ \left(D \int_{A_n} |f_n - f|^p \, d|\mu| \right)^{1/p}, \qquad \text{by Minkowski's inequality}$$

$$\leq \left(D \int_{(E_\varepsilon \cup A_n)^c} |f_n|^p \, d|\mu| \right)^{1/p} + \left(D \int_{(E_\varepsilon \cup A_n)^c} |f|^p \, d|\mu| \right)^{1/p}$$

$$+ \left(D \int_{E_\varepsilon - A_n} |f_n - f|^p \, d|\mu| \right)^{1/p} + \left(D \int_{A_n} |f_n|^p \, d|\mu| \right)^{1/p}$$

$$+ \left(D \int_{A_n} |f|^p \, d|\mu| \right)^{1/p}$$

$$< \varepsilon + \varepsilon + r[|\mu|(E_\varepsilon - A_n)]^{1/p} + \varepsilon + \varepsilon$$

$$< \varepsilon + \varepsilon + r[|\mu|(E_\varepsilon)]^{1/p} + \varepsilon + \varepsilon < 5\varepsilon.$$

Hence $\lim_{n \to \infty} D \int |f_n - f|^p \, d|\mu| = 0$.

This completes the proof of the theorem. □

4.6.11 Remark. If μ is bounded, condition (iii) is not required in Theorem 4.6.10.

We need the following notion of hazy convergence for nets and the attendant result in the proof of Lebesgue dominated convergence theorem.

4.6.12 Definition. Let $(\Omega, \mathscr{F}, \mu)$ be a charge space and f_α, $\alpha \in D$ be a net of real valued functions on Ω. f_α, $\alpha \in D$ is said to *converge to f hazily* if for every $\varepsilon > 0$,

$$\lim_{\alpha \in D} |\mu|^*(\{\omega \in \Omega; |f_\alpha(\omega) - f(\omega)| > \varepsilon\}) = 0.$$

4. INTEGRATION   131

The following result says that the limit function f is T_1-measurable if each f_α is T_1-measurable.

4.6.13 Proposition. *Let $(\Omega, \mathscr{F}, \mu)$ be a charge space and f_α, $\alpha \in D$ a net of T_1-measurable functions on Ω converging to a function f hazily. Then f is T_1-measurable.*

Proof. In view of Theorem 4.4.7, it suffices to show that f is T_2-measurable. Let $\varepsilon > 0$. There exists α_0 in D and a set A in $\mathscr{F}$ such that $|\mu|(A) < \varepsilon/2$ and $|f_{\alpha_0}(\omega) - f(\omega)| < \varepsilon/3$ for all ω in A^c. Since f_{α_0} is T_2-measurable, there exists a partition $\{F_0, F_1, F_2, \ldots, F_n\}$ of Ω in $\mathscr{F}$ such that $|\mu|(F_0) < \varepsilon/2$ and $|f_{\alpha_0}(\omega) - f_{\alpha_0}(\omega')| < \varepsilon/2$ for all ω, ω' in F_i for $i = 1, 2, \ldots, n$. Let $E_0 = A \cup F_0$, and $E_i = F_i \cap A^c$, $i = 1, 2, \ldots, n$. $\{E_0, E_1, E_2, \ldots, E_n\}$ is obviously a partition of Ω in $\mathscr{F}$. Further, $|\mu|(E_0) < \varepsilon/2 + \varepsilon/2 = \varepsilon$. If $\omega, \omega' \in E_i$ for any $i = 1, 2, \ldots, n$, then

$$|f(\omega) - f(\omega')| \leq |f(\omega) - f_{\alpha_0}(\omega)| + |f_{\alpha_0}(\omega) - f_{\alpha_0}(\omega')| + |f_{\alpha_0}(\omega') - f(\omega')|$$
$$< \varepsilon/3 + \varepsilon/3 + \varepsilon/3 = \varepsilon.$$

This completes the proof. □

Now, we are ready to prove Lebesgue dominated convergence theorem.

4.6.14 Theorem *Let $(\Omega, \mathscr{F}, \mu)$ be a charge space and $g \in L_p(\Omega, \mathscr{F}, \mu)$ for some $p \geq 1$. Let f_α, $\alpha \in D$ be a net of T_1-measurable functions on Ω such that $|f_\alpha| \leq |g|$ a.e. $[\mu]$. Let f be a real valued function on Ω. Then the following statements are equivalent.*

(a). *f_α, $\alpha \in D$ converges to f hazily.*
(b). *$f \in L_p(\Omega, \mathscr{F}, \mu)$ and $\lim_{\alpha \in D} \|f_\alpha - f\|_p = 0$.*

If $p = 1$, the following statements are equivalent.

(a'). *f_α, $\alpha \in D$ converges to f hazily.*
(b'). *$f \in L_1(\Omega, \mathscr{F}, \mu)$ and $\lim_{\alpha \in D} D \int_F (f_\alpha - f) \, d|\mu| = 0$ uniformly over F in $\mathscr{F}$.*
(c'). *$f \in L_1(\Omega, \mathscr{F}, \mu)$ and $\lim_{\alpha \in D} D \int_F |f_\alpha - f| \, d|\mu| = 0$ uniformly over F in $\mathscr{F}$.*
(d'). *$f \in L_1(\Omega, \mathscr{F}, \mu)$ and $\lim_{\alpha \in D} D \int |f_\alpha - f| \, d|\mu| = 0$.*

Proof. We prove the first part.

By Theorem 4.4.18, each $f_\alpha \in L_p(\Omega, \mathscr{F}, \mu)$. Assume that the net f_α, $\alpha \in D$ is really a sequence f_n, $n \geq 1$. Suppose (a) holds. We verify (i), (ii) and (iii) of Theorem 4.6.10. (i) holds by virtue of the validity of (a). We show that (ii) holds. Let $\varepsilon > 0$. By Theorem 4.4.13(xi), there exists $\delta > 0$ such that $D \int_F |g|^p \, d|\mu| < \varepsilon$ whenever $F \in \mathscr{F}$ and $|\mu|(F) < \delta$. Since $|f_n| \leq |g|$ a.e. $[\mu]$ for every $n \geq 1$, it follows that $D \int_F |f_n|^p \, d|\mu| < \varepsilon$ whenever $F \in \mathscr{F}$ and $|\mu|(F) < \delta$. Thus (ii) holds. Next, we show that (iii) of Theorem 4.6.10 holds. By Lemma

4.4.15, there exists a set F_ε in $\mathscr{F}$ such that $|\mu|(F_\varepsilon)<\infty$ and $D\int_{F_\varepsilon^c}|g|^p\,d|\mu|<\varepsilon$. From this, it follows that $D\int_{F_\varepsilon^c}|f_n|^p\,d|\mu|<\varepsilon$ for every $n\geq 1$. Thus (iii) Theorem 4.6.10 holds. Hence (b) follows. The implication (b)$\Rightarrow$(a) follows from Theorem 4.6.10.

Now, we treat nets. Suppose that f_α, $\alpha\in D$ converges to f hazily. Since $C(\Omega,\mathscr{F},\mu)$ is a pseudo-metric space, there exists a sequence $\alpha_1<\alpha_2<\cdots$ such that f_{α_n}, $n\geq 1$ converges to f hazily. Hence $f\in L_p(\Omega,\mathscr{F},\mu)$. Suppose $\lim_{\alpha\in D}\|f_\alpha-f\|_p\neq 0$. Then there exists $\varepsilon>0$ and $\beta_\alpha>\alpha$ for every α in D such that $\|f_{\beta_\alpha}-f\|_p>\varepsilon$. Then f_{β_α}, $\alpha\in D$, being a subnet of f_α, $\alpha\in D$, also converges to f hazily. So, again, there exists a sequence $\beta_{\alpha_1}<\beta_{\alpha_2}<\cdots$ such that $f_{\beta_{\alpha_n}}$, $n\geq 1$ converges to f hazily. But, by what we have proved for sequences, $\lim_{n\to\infty}\|f_{\beta_{\alpha_n}}-f\|_p=0$. This is a contradiction. Hence $\lim_{\alpha\in D}\|f_\alpha-f\|_p=0$. Thus for nets, (a)$\Rightarrow$(b) holds. Similarly, (b)$\Rightarrow$(a) can be established for nets too.

We come to the second part. The equivalence of (a') and (d') follows from the first part. The implications (d')$\Rightarrow$(c')$\Rightarrow$(b') are clear. We prove (b')$\Rightarrow$(d'). Let $\lambda_\alpha(F)=D\int_F(f_\alpha-f)\,d|\mu|$, $F\in\mathscr{F}$, $\alpha\in D$. Then the uniform convergence of λ_α, $\alpha\in D$ to zero implies that $|\lambda_\alpha|(\Omega)$, $\alpha\in D$ converges to zero, because $|\lambda_\alpha|(\Omega)=\mathrm{Sup}_{F\in\mathscr{F}}|\lambda_\alpha(F)-\lambda_\alpha(F^c)|$. $\square$

We end this section with a result on the denseness of D-integrable simple functions in $L_p(\Omega,\mathscr{F},\mu)$ for $1\leq p<\infty$.

4.6.15 Theorem. *Let $(\Omega,\mathscr{F},\mu)$ be a charge space. Let $\mathrm{Sim}(\Omega,\mathscr{F},\mu)$ be the space of all D-integrable simple functions on Ω. Then $\mathrm{Sim}(\Omega,\mathscr{F},\mu)$ is dense in $L_p(\Omega,\mathscr{F},\mu)$ for every $1\leq p<\infty$.*

Proof. Let $1\leq p<\infty$ be fixed and $f\in L_p(\Omega,\mathscr{F},\mu)$. For a given $\varepsilon>0$, by Lemma 4.4.15, there exists a set A in $\mathscr{F}$ such that $|\mu|(A)<\infty$ and $D\int_{A^c}|f|^p\,d|\mu|<\varepsilon^p$. Then the function $I_A f$ has the property that $\|f-I_A f\|_p<\varepsilon$. For,

$$\|f-I_A f\|_p^p = D\int |f-I_A f|^p\,d|\mu|$$
$$= D\int_A |f-I_A f|^p\,d|\mu| + D\int_{A^c} |f-I_A f|^p\,d|\mu|$$
$$< 0+\varepsilon^p.$$

From this, it follows that $\|f-I_A f\|_p<\varepsilon$.

Now, we look at $I_A f$. Let f_n, $n\geq 1$ be a sequence of simple functions converging to $I_A f$ hazily. By Lemma 4.4.17, we can assume that $|f_n|\leq 2I_A|f|$ for every $n\geq 1$. By Lemma 4.4.16, $|f_n|^p$ is D-integrable for every $n\geq 1$. By Lebesgue dominated convergence theorem, $\lim_{n\to\infty}\|f_n-I_A f\|_p=0$. Consequently, there exists $N\geq 1$ such that $\|f_N-I_A f\|_p<\varepsilon$. From this, it follows that

$$\|f_N - f\|_p \le \|f_N - I_A f\|_p + \|I_A f - f\|_p < 2\varepsilon.$$

This completes the proof. □

4.7 ba(Ω, $\mathscr{F}$) AS A DUAL SPACE

In this section, we introduce a Banach space whose dual is the space of all bounded charges ba(Ω, $\mathscr{F}$) on the field $\mathscr{F}$ of subsets of a set Ω. We also consider some extensions of this result.

4.7.1 Definition. Let $\mathscr{F}$ be a field of subsets of a set Ω. A real valued function f on Ω is said to be $\mathscr{F}$-*continuous* if given $\varepsilon > 0$, there exists a partition $\{F_1, F_2, \ldots, F_n\}$ of Ω in $\mathscr{F}$ such that $|f(\omega) - f(\omega')| < \varepsilon$ for all ω, ω' in F_i and for every $i = 1, 2, \ldots, n$. Let $\mathscr{C}(\Omega, \mathscr{F})$ denote the collection of all $\mathscr{F}$-continuous functions on Ω.

It is obvious that every $\mathscr{F}$-continuous function is bounded. Further, the space $\mathscr{C}(\Omega, \mathscr{F})$ is a linear space. All simple functions are available in $\mathscr{C}(\Omega, \mathscr{F})$. On $\mathscr{C}(\Omega, \mathscr{C})$, we introduce a norm by $\|f\| = \text{Sup}\{|f(\omega)|; \omega \in \Omega\}$ for f in $\mathscr{C}(\Omega, \mathscr{F})$. It is obvious that $\|\cdot\|$ is indeed a norm on $\mathscr{C}(\Omega, \mathscr{F})$. If $A \in \mathscr{F}$ and $A \ne \varnothing$, then $\|I_A\| = 1$.

4.7.2 Proposition. *Let $\mathscr{F}$ be a field of subsets of a set Ω. Then the following statements are true.*
   (i). *$(\mathscr{C}(\Omega, \mathscr{F}), \|\cdot\|)$ is a Banach space.*
   (ii). *The collection of all simple functions is a dense subset of $\mathscr{C}(\Omega, \mathscr{F})$.*
   (iii). *f is $\mathscr{F}$-continuous if and only if f is the uniform limit of a sequence of simple functions on Ω.*

Proof. (i). Let f_n, $n \ge 1$ be a Cauchy sequence in $\mathscr{C}(\Omega, \mathscr{F})$. Let $f(\omega) = \lim_{n \to \infty} f_n(\omega)$, $\omega \in \Omega$. Then f_n, $n \ge 1$ converges to f uniformly over Ω. We show that f is $\mathscr{F}$-continuous. Let $\varepsilon > 0$. There exists $N \ge 1$ such that $\text{Sup}\{|f_N(\omega) - f(\omega)|; \omega \in \Omega\} < \varepsilon/3$. For f_N, there is a partition $\{F_1, F_2, \ldots, F_k\}$ of Ω in $\mathscr{F}$ such that $|f_N(\omega) - f_N(\omega')| < \varepsilon/3$ for all ω, ω' in F_i and for every $i = 1, 2, \ldots, k$. Let $i \in \{1, 2, \ldots, k\}$ and $\omega, \omega' \in F_i$. Then

$$|f(\omega) - f(\omega')| \le |f(\omega) - f_N(\omega)| + |f_N(\omega) - f_N(\omega')| + |f_N(\omega') - f(\omega')|$$
$$< \varepsilon/3 + \varepsilon/3 + \varepsilon/3 = \varepsilon.$$

This shows that f is $\mathscr{F}$-continuous. It is obvious that $\lim_{n \to \infty} \|f_n - f\| = 0$. Hence $(\mathscr{C}(\Omega, \mathscr{F}), \|\cdot\|)$ is a Banach space.

(ii). Let f be $\mathscr{F}$-continuous. For $n \geq 1$, there is a partition $\{F_{n1}, F_{n2}, \ldots, F_{nk_n}\}$ of Ω in $\mathscr{F}$ such that for every $i \in \{1, 2, \ldots, k_n\}$, $|f(\omega) - f(\omega')| < 1/n$ for all ω, ω' in F_{ni}. For each $n \geq 1$ and $i = 1, 2, \ldots, k_n$, choose and fix ω_{ni} in F_{ni}. Let

$$f_n = \sum_{i=1}^{k_n} f(\omega_{ni}) I_{F_{ni}}, \quad n \geq 1.$$

We claim that f_n, $n \geq 1$ converges to f uniformly. If $\omega \in F_{ni}$, $|f(\omega) - f_n(\omega)| = |f(\omega) - f(\omega_{ni})| < 1/n$ for all $n \geq 1$ and $i = 1, 2, \ldots, k_n$. Consequently, $\|f - f_n\| \leq 1/n$ for all $n \geq 1$. Hence $\lim_{n \to \infty} \|f - f_n\| = 0$. This shows that the class of all simple functions on Ω is a dense subset of $\mathscr{C}(\Omega, \mathscr{F})$.

(iii). This is now obvious. □

We now give a characterization of $\mathscr{F}$-continuous functions based on D-integrability.

4.7.3 Theorem. *Let $\mathscr{F}$ be a field of subsets of a set Ω. Then $f \in \mathscr{C}(\Omega, \mathscr{F})$ if and only if f is D-integrable with respect to every bounded charge λ on $\mathscr{F}$.*

Proof. Let $f \in \mathscr{C}(\Omega, \mathscr{F})$. Note that if λ is a bounded charge on $\mathscr{F}$, then f is T_2-measurable with respect to λ and hence is D-integrable. See Theorem 4.5.7. Conversely, suppose f is D-integrable with respect to every bounded charge on $\mathscr{F}$. Suppose f is not $\mathscr{F}$-continuous. There exists $\varepsilon > 0$ such that given any partition $\{F_1, F_2, \ldots, F_n\}$ of Ω in $\mathscr{F}$, there exists $1 \leq i \leq n$ satisfying $O(f, F_i) = \text{Sup}\{|f(\omega) - f(\omega')|; \omega, \omega' \in F_i\} > \varepsilon$. For each partition $P = \{F_1, F_2, \ldots, F_n\}$ of Ω in $\mathscr{F}$, let $A(P) = \bigcup \{F_i; 1 \leq i \leq n, O(f, F_i) > \varepsilon\}$. Let $\mathscr{P}$ denote the collection of all finite partitions of Ω in $\mathscr{F}$. Then $\{A(P); P \in \mathscr{P}\}$ is a filter base in $\mathscr{F}$. To see this, let $P_1, P_2 \in \mathscr{P}$ and P any partition in $\mathscr{P}$ finer than both P_1 and P_2. Suppose for some F in P, $O(f, F) > \varepsilon$. Then F is contained in some G_1 in P_1 and in some G_2 in P_2. Consequently, $O(f, G_1) > \varepsilon$ and $O(f, G_2) > \varepsilon$. Thus $F \subset G_1 \cap G_2$ and from this, it follows that $A(P) \subset A(P_1) \cap A(P_2)$. Further, note that $A(P) \neq \emptyset$ for every P in $\mathscr{P}$. Thus we have proved that $\{A(P); P \in \mathscr{P}\}$ is a filter base in $\mathscr{F}$. Hence there is a maximal filter $\mathscr{E}$ in $\mathscr{F}$ containing $\{A(P); P \in \mathscr{P}\}$. Define λ on $\mathscr{F}$ by

$$\lambda(E) = 1, \quad \text{if } E \in \mathscr{E},$$
$$= 0, \quad \text{if } E \notin \mathscr{E} \text{ and } E \in \mathscr{F}.$$

λ is a 0–1 valued charge on $\mathscr{F}$.

We claim that f is not D-integrable with respect to λ. Suppose f is D-integrable with respect to λ. Then $f = c$ a.e. $[\lambda]$ for some constant c. See Example 4.4.14. Then $\lambda^*(\{\omega \in \Omega; |f(\omega) - c| > \varepsilon/2\}) = 0$. Consequently, there exists B in $\mathscr{F}$ such that $\{\omega \in \Omega; |f(\omega) - c| > \varepsilon/2\} \subset B$ and $\lambda(B) = 0$. Now, we look at the partition $P = \{B, B^c\}$ in $\mathscr{P}$. Then

4. INTEGRATION

$B^c \subset \{\omega \in \Omega; |f(\omega) - c| \le \varepsilon/2\}$. For any ω, ω' in B^c, $|f(\omega) - f(\omega')| \le |f(\omega) - c| + |f(\omega') - c| \le \varepsilon/2 + \varepsilon/2 = \varepsilon$. Therefore, $O(f, B^c) \le \varepsilon$. Hence $A(P) = B \in \mathscr{E}$. By the definition of λ, $\lambda(B) = 1$. This contradiction proves the desired assertion. $\square$

Now, we characterize the continuous linear functionals on $(\mathscr{C}(\Omega, \mathscr{F}), \|\cdot\|)$.

4.7.4 Theorem. *Let $\mathscr{F}$ be a field of subsets of a set Ω. Let T be a continuous linear functional on $\mathscr{C}(\Omega, \mathscr{F})$. Then there exists a unique bounded charge μ on $\mathscr{F}$ such that*

$$T(f) = D \int f \, d\mu$$

for every f in $\mathscr{C}(\Omega, \mathscr{F})$. Further, $\|T\| = \mathrm{Sup}\{|T(f)|; \|f\| \le 1\} = |\mu|(\Omega)$.

Conversely, for any given bounded charge λ on $\mathscr{F}$, the functional T' on $\mathscr{C}(\Omega, \mathscr{F})$ defined by $T'(f) = D \int f \, d\lambda$, f in $\mathscr{C}(\Omega, \mathscr{F})$ is a continuous linear functional on $\mathscr{C}(\Omega, \mathscr{F})$ with $\|T'\| = |\lambda|(\Omega)$.

If T is a nonnegative linear functional on $\mathscr{C}(\Omega, \mathscr{F})$, i.e. $T(f) \ge 0$ if $f \ge 0$, then μ is a positive charge.

Proof. For A in $\mathscr{F}$, $I_A \in \mathscr{C}(\Omega, \mathscr{F})$ and so, define $\mu(A) = T(I_A)$. It is obvious that μ is a charge on $\mathscr{F}$. Also,

$$|\mu(A)| = |T(I_A)| \le \|T\| \|I_A\| \le \|T\|$$

for every A in $\mathscr{F}$. This shows that μ is a bounded charge on $\mathscr{F}$. Now, we claim that for any f in $\mathscr{C}(\Omega, \mathscr{F})$, $T(f) = D \int f \, d\mu$. If $f = \sum_{i=1}^n c_i I_{A_i}$ is a simple function in $\mathscr{C}(\Omega, \mathscr{F})$, then

$$T(f) = \sum_{i=1}^n c_i T(I_{A_i}) = \sum_{i=1}^n c_i \mu(A_i) = D \int f \, d\mu.$$

For any f in $\mathscr{C}(\Omega, \mathscr{F})$, by Theorem 4.7.2, there exists a sequence f_n, $n \ge 1$ of simple functions in $\mathscr{C}(\Omega, \mathscr{F})$ converging to f uniformly. It is obvious that f_n, $n \ge 1$ converges to f hazily. Further, since μ is bounded,

$$\lim_{m,n \to \infty} D \int |f_n - f_m| \, d|\mu| \le \lim_{m,n \to \infty} \|f_n - f_m\| |\mu|(\Omega) = 0.$$

Thus, f_n, $n \ge 1$ is a determining sequence for f with respect to μ. Consequently,

$$D \int f \, d\mu = \lim_{n \to \infty} D \int f_n \, d\mu = \lim_{n \to \infty} T(f_n) = T(f),$$

as T is continuous on $\mathscr{C}(\Omega, \mathscr{F})$. The claim is thus established.

Now, if $f \in \mathscr{C}(\Omega, \mathscr{F})$ and $\|f\| \leq 1$, then $|T(f)| = |D \int f \, d\mu| \leq D \int |f| \, d|\mu| \leq |\mu|(\Omega)$. See Theorem 4.4.13 (iii). Hence $\|T\| \leq |\mu|(\Omega)$. Since $|\mu|(\Omega) = \operatorname{Sup} \sum_{i=1}^{n} |\mu(A_i)|$, where the supremum is taken over all finite partitions $\{A_1, A_2, \ldots, A_n\}$ of Ω in $\mathscr{F}$, for any given $\varepsilon > 0$, there exists a partition $\{F_1, F_2, \ldots, F_m\}$ of Ω in $\mathscr{F}$ such that

$$|\mu|(\Omega) - \varepsilon \leq \sum_{i=1}^{m} |\mu(F_i)| = \sum_{i \in J_1} \mu(F_i) - \sum_{i \in J_2} \mu(F_i),$$

where $J_1 = \{1 \leq i \leq m; \mu(F_i) \geq 0\}$ and $J_2 = \{1 \leq i \leq m; \mu(F_i) < 0\}$. Therefore,

$$|\mu|(\Omega) - \varepsilon \leq \sum_{i \in J_1} T(I_{F_i}) - \sum_{i \in J_2} T(I_{F_i})$$

$$= T\left(\sum_{i \in J_1} I_{F_i} - \sum_{i \in J_2} I_{F_i}\right)$$

$$\leq \|T\| \left\|\sum_{i \in J_1} I_{F_i} - \sum_{i \in J_2} I_{F_i}\right\| = \|T\|.$$

Since $\varepsilon > 0$ is arbitrary, we have $|\mu|(\Omega) \leq \|T\|$. This shows that $\|T\| = |\mu|(\Omega)$. The rest of the theorem is obvious. □

4.7.5 Corollary. *The dual of $\mathscr{C}(\Omega, \mathscr{F}) = \mathscr{C}^*(\Omega, \mathscr{F}) = \operatorname{ba}(\Omega, \mathscr{F})$.*

4.7.6 Corollary. *Let X be a compact Hausdorff totally disconnected space, $\mathscr{F}$ the field of all clopen subsets of X, $\mathscr{B}$ the Borel σ-field on X, $\mathscr{C}(X)$ the space of all real continuous functions on X and $\operatorname{ca}(X, \mathscr{B})$ the space of all bounded regular measures on $\mathscr{B}$. ($\mathscr{C}(X)$ is a Banach space under supremum norm and $\operatorname{ca}(X, \mathscr{B})$ is a Banach space under total variation norm.) Then the following statements are true.*

(i). $\mathscr{C}(X, \mathscr{F}) = $ *The space of all $\mathscr{F}$-continuous real functions on X*
$= \mathscr{C}(X)$.

(ii). *(Riesz Representation Theorem.) If T is a continuous linear functional on $\mathscr{C}(X)$, there exists a unique bounded regular measure μ on $\mathscr{B}$ such that*

$$T(f) = D \int f \, d\mu$$

for f in $\mathscr{C}(X)$ having the property that $\|T\| = |\mu|(X)$.

(iii). *The dual of $\mathscr{C}(X) = \mathscr{C}^*(X) = \operatorname{ca}(X, \mathscr{B})$.*

Proof. (i). Suppose f is a continuous real valued function on X. For $\varepsilon > 0$ and for each x in X, there exists a clopen set C_x containing x such that

$|f(y)-f(x)| < \varepsilon/2$ for all y in C_x. $\{C_x; x \in X\}$ is an open cover for X. Since X is compact, there exists a finite subcover $\{C_{x_1}, C_{x_2}, \ldots, C_{x_n}\}$ of X. For each $1 \le i \le n$ and for every x, y in C_{x_i}, we have $|f(x)-f(y)| \le |f(x)-f(x_i)| + |f(x_i)-f(y)| < \varepsilon/2 + \varepsilon/2 = \varepsilon$. Let $D_1 = C_{x_1}$, $D_2 = C_{x_2} - C_{x_1}, \ldots, D_n = C_{x_n} - (C_{x_1} \cup C_{x_2} \cup \cdots \cup C_{x_{n-1}})$. Then $\{D_1, D_2, \ldots, D_n\}$ is a partition of X in $\mathscr{F}$ and for this partition, we still have $|f(x)-f(y)| < \varepsilon$ for all x, y in D_i and for all $i = 1, 2, \ldots, n$. Hence f is $\mathscr{F}$-continuous.

Conversely, if f is $\mathscr{F}$-continuous, it follows easily that $f \in \mathscr{C}(X)$. This proves (i).

(ii). Let T be a continuous linear functional on $\mathscr{C}(X)$. By Theorem 4.7.5, there exists a bounded charge $\tilde{\mu}$ on $\mathscr{F}$ such that $T(f) = D \int f \, d\tilde{\mu}$ for every f in $\mathscr{C}(X)$. $\tilde{\mu}$ can be extended as a measure $\tilde{\tilde{\mu}}$ on the Baire σ-field $\mathscr{B}_0$ of X. $\tilde{\tilde{\mu}}$ can be extended as a regular measure μ on $\mathscr{B}$. See Theorem 3.5.5. This completes the proof of (ii).

(iii). This is obvious now. □

4.7.7 Remark If $\mathscr{F}$ is a field of subsets of a set Ω, then $\mathscr{C}(\Omega, \mathscr{F})$ can be realized as $\mathscr{C}(X)$ for some compact Hausdorff totally disconnected space X.

Now, we obtain some natural subspaces of $ba(\Omega, \mathscr{F})$ as dual spaces when $\mathscr{F}$ is a σ-field.

We consider the following set-up. Let Ω be a set, $\mathfrak{A}$ a σ-field of subsets of Ω and $\mathscr{I}$ a proper σ-ideal in $\mathfrak{A}$, i.e. (i) $\mathscr{I} \subset \mathfrak{A}$, (ii) $C \in \mathscr{I}$ if $C \subset A$, $C \in \mathfrak{A}$ and $A \in \mathscr{I}$, (iii) $\Omega \notin \mathscr{I}$ and (iv) $\bigcup_{n \ge 1} A_n \in \mathscr{I}$ if A_n, $n \ge 1$ is a sequence in $\mathscr{I}$. A real valued function f on Ω is said to be measurable if $f^{-1}(B) \in \mathfrak{A}$ for every Borel set $B \subset R$. A measurable function f on Ω is said to be essentially bounded if there exists $k > 0$ such that $\{\omega \in \Omega; |f(\omega)| > k\} \in \mathscr{I}$. If f is an essentially bounded measurable function on Ω, define

$$\|f\|_\infty = \operatorname{Inf}\{k; k > 0 \quad \text{and} \quad \{\omega \in \Omega; |f(\omega)| > k\} \in \mathscr{I}\}.$$

Obviously, $0 \le \|f\|_\infty < \infty$. Let $L_\infty(\Omega, \mathfrak{A}, \mathscr{I})$ denote the collection of all essentially bounded measurable functions on Ω.

If $f \in L_\infty(\Omega, \mathfrak{A}, \mathscr{I})$, then

$$A = \{\omega \in \Omega; |f(\omega)| > \|f\|_\infty\}$$
$$= \bigcup_{n \ge 1} \{\omega \in \Omega; |f(\omega)| > \|f\|_\infty + 1/n\} \in \mathscr{I}.$$

Further, we show that A has the following properties.

(i). $\operatorname{Sup}\{|f(\omega)|; \omega \in A^c\} = \|f\|_\infty$.

(ii). $\operatorname{Sup}\{|f(\omega)|; \omega \in A^c - B\} = \|f\|_\infty$ for any B in $\mathscr{I}$.

Obviously, $\operatorname{Sup}\{|f(\omega)|; \omega \in A^c\} \le \|f\|_\infty$. If $\operatorname{Sup}\{|f(\omega)|; \omega \in A^c\} < \|f\|_\infty$, then $\{\omega' \in \Omega; |f(\omega')| > \operatorname{Sup}\{|f(\omega)|; \omega \in A^c\}\} \subset A$ and hence $\{\omega' \in \Omega; |f(\omega')| > \operatorname{Sup}\{|f(\omega)|; \omega \in A^c\}\} \in \mathscr{I}$. But this contradicts the definition of $\|f\|_\infty$. (ii) can be proved analogously.

4.7.8 Proposition. *The function $\|\cdot\|_\infty$ defined on $L_\infty(\Omega, \mathfrak{A}, \mathscr{I})$ above has the following properties.*
  (i). $\|cf\|_\infty = |c|\|f\|_\infty$ *for any real number c and f in* $L_\infty(\Omega, \mathfrak{A}, \mathscr{I})$.
  (ii). $\|f+g\|_\infty \leq \|f\|_\infty + \|g\|_\infty$ *for all f, g in* $L_\infty(\Omega, \mathfrak{A}, \mathscr{I})$.

Proof. (i). This is obvious.
  (ii). Note that
$$\{\omega \in \Omega; |f(\omega) + g(\omega)| > \|f\|_\infty + \|g\|_\infty\}$$
$$\subset \{\omega \in \Omega; |f(\omega)| > \|f\|_\infty\}$$
$$\cup \{\omega \in \Omega; |g(\omega)| > \|g\|_\infty\} \in \mathscr{I}.$$

Consequently, $\|f+g\|_\infty \leq \|f\|_\infty + \|g\|_\infty$. □

4.7.9 Proposition. $(L_\infty(\Omega, \mathfrak{A}, \mathscr{I}), \|\cdot\|_\infty)$ *is a linear space and $\|\cdot\|_\infty$ is a pseudo-norm under which $L_\infty(\Omega, \mathfrak{A}, \mathscr{I})$ is complete.*

Proof. It is clear the $L_\infty(\Omega, \mathfrak{A}, \mathscr{I})$ is a linear space and that $\|\cdot\|_\infty$ is a pseudo-norm on $L_\infty(\Omega, \mathfrak{A}, \mathscr{I})$. We show that $L_\infty(\Omega, \mathfrak{A}, \mathscr{I})$ is complete. Let f_n, $n \geq 1$ be a Cauchy sequence in $L_\infty(\Omega, \mathfrak{A}, \mathscr{I})$. For every $n, m \geq 1$, let $A_{nm} = \{\omega \in \Omega; |f_n(\omega) - f_m(\omega)| > \|f_n - f_m\|_\infty\}$. Then for any B in $\mathscr{I}$,
$$\text{Sup}\{|f_n(\omega) - f_m(\omega)|; \omega \in A_{nm}^c - B\} = \|f_n - f_m\|_\infty.$$

Let $A = \bigcup_{n \geq 1} \bigcup_{m \geq 1} A_{nm}$. Then $A \in \mathscr{I}$ and
$$\lim_{m,n \to \infty} \text{Sup}\{|f_n(\omega) - f_m(\omega)|; \omega \in A^c\} = \lim_{m,n \to \infty} \|f_n - f_m\|_\infty = 0.$$

Let $f(\omega) = \lim_{n \to \infty} f_n(\omega)$ for ω in A^c. Then f_n, $n \geq 1$ converges to f uniformly on A^c. For ω in A, define $f(\omega) = 0$. We claim that $\|f_n - f\|_\infty$, $n \geq 1$ converges to zero.
$$\|f_n - f\|_\infty = \|(f_n - f)I_{A^c} + (f_n - f)I_A\|_\infty$$
$$\leq \|(f_n - f)I_{A^c}\|_\infty + \|(f_n - f)I_A\|_\infty$$
$$\leq \text{Sup}\{|f_n(\omega) - f(\omega)|; \omega \in A^c\} + 0$$

which converges to zero as $n \to \infty$. This completes the proof. □

In the present context, we define a null function as follows. A real valued measurable function f on Ω is said to be a null function if $\{\omega \in \Omega; |f(\omega)| > k\} \in \mathscr{I}$ for every $k > 0$. This is equivalent to the condition that $\|f\|_\infty = 0$. The space of all null functions is a linear subspace of $L_\infty(\Omega, \mathfrak{A}, \mathscr{I})$. We introduce an equivalence relation $\sim$ on $L_\infty(\Omega, \mathfrak{A}, \mathscr{I})$ by $f \sim g$ if $f - g$ is a null function. Let $\mathscr{L}_\infty(\Omega, \mathfrak{A}, \mathscr{I})$ denote the collection of all equivalence classes of $L_\infty(\Omega, \mathfrak{A}, \mathscr{I})$. The pseudo-norm $\|\cdot\|_\infty$ defines unambiguously a norm on $\mathscr{L}_\infty(\Omega, \mathfrak{A}, \mathscr{I})$ which is again denoted by $\|\cdot\|_\infty$. Now, it is apparent that $\mathscr{L}_\infty(\Omega, \mathfrak{A}, \mathscr{I})$ is a Banach space. We work out its dual.

4.7.10 Theorem. *Let T be a continuous linear functional on $(L_\infty(\Omega, \mathfrak{A}, \mathscr{I}), \|\cdot\|_\infty)$. Then there exists a unique bounded charge μ on $\mathfrak{A}$ with the following properties.*
(i) $T(f) = D \int f \, d\mu$ for every f in $L_\infty(\Omega, \mathfrak{A}, \mathscr{I})$.
(ii). $\|T\| = |\mu|(\Omega)$.
(iii). $\mu(A) = 0$ if $A \in \mathscr{I}$.

Proof. For A in $\mathfrak{A}$, let $\mu(A) = T(I_A)$. μ is obviously a charge on $\mathfrak{A}$. Note that $T(f) = 0$ whenever f is a null function. If $A \in \mathscr{I}$, then I_A is a null function and so, $\mu(A) = 0$. This proves (iii). The boundedness of μ follows from the inequalities

$$|\mu(A)| = |T(I_A)| \le \|T\| \|I_A\|_\infty \le \|T\|$$

for any A in $\mathfrak{A}$. If f is a simple function, it is obvious that $T(f) = D \int f \, d\mu$. Let f be any function in $L_\infty(\Omega, \mathfrak{A}, \mathscr{I})$ with $\|f\|_\infty > 0$. We show that $T(f) = D \int f \, d\mu$. Let $B = \{\omega \in \Omega; |f(\omega)| \le \|f\|_\infty\}$. Note that $B^c \in \mathscr{I}$. Let $d = 2\|f\|_\infty$. For each $n \ge 1$, define (similar to the construction given in the proof of Corollary 4.5.9) f_n as follows. For ω in Ω,

$$f_n(\omega) = -\|f\|_\infty + \frac{i-1}{2^n} d, \text{ if } -\|f\|_\infty + \frac{i-1}{2^n} d \le f(\omega) <$$

$$-\|f\|_\infty + \frac{i}{2^n} d,$$

for $i = 1, 2, \ldots, 2^n - 1$,

$$= \|f\|_\infty - \frac{d}{2^n}, \quad \text{if } \|f\|_\infty - \frac{d}{2^n} \le f(\omega) \le \|f\|_\infty,$$

$$= 0, \quad \text{if } |f(\omega)| > \|f\|_\infty.$$

It is easy to check that f_n, $n \ge 1$ converges to f uniformly on B. We claim that f_n, $n \ge 1$ is a determining sequence for f with respect to μ. It is obvious that each f_n is a simple function. For any $\varepsilon > 0$,

$$|\mu|(\{\omega \in \Omega; |f_n(\omega) - f(\omega)| > \varepsilon\}) = |\mu|(\{\omega \in B; |f_n(\omega) - f(\omega)| > \varepsilon\})$$

$$+ |\mu|(\{\omega \in B^c; |f_n(\omega) - f(\omega)| > \varepsilon\})$$

$$= 0,$$

if n is sufficiently large. This assertion follows from $|\mu|(B^c) = 0$ and f_n, $n \ge 1$ converges to f uniformly on B. This shows that f_n, $n \ge 1$ converges to f

hazily. Also,

$$\lim_{m,n\to\infty} D\int |f_n - f_m|\,d|\mu| = \lim_{m,n\to\infty} D\int_B |f_n - f_m|\,d|\mu|$$
$$+ \lim_{m,n\to\infty} D\int_{B^c} |f_n - f_m|\,d|\mu|$$
$$= 0.$$

Hence f_n, $n \geq 1$ is a determining sequence for f.

Since $\|f_n - f\|_\infty \leq \|(f_n - f)I_B\|_\infty + \|(f_n - f)I_{B^c}\|_\infty$, it follows that $\lim_{n\to\infty}\|f_n - f\|_\infty = 0$. Since T is a continuous linear functional on $L_\infty(\Omega, \mathfrak{A}, \mathscr{I})$,

$$T(f) = \lim_{n\to\infty} T(f_n) = \lim_{n\to\infty} D\int f_n\,d\mu = D\int f\,d\mu.$$

Next, we show that $\|T\| = |\mu|(\Omega)$. For any f in $L_\infty(\Omega, \mathfrak{A}, \mathscr{I})$, $|T(f)| = |D\int f\,d\mu| \leq D\int |f|\,d|\mu| \leq |\mu|(\Omega)\|f\|_\infty$. Hence, it follows that $\|T\| \leq |\mu|(\Omega)$. Just as in the case of Theorem 4.7.4, we can show that $|\mu|(\Omega) \leq \|T\|$. Thus $\|T\| = |\mu|(\Omega)$ holds true. This completes the proof. $\square$

For the following corollary, let $ba(\Omega, \mathfrak{A}, \mathscr{I})$ stand for the space of all bounded charges on $\mathfrak{A}$ vanishing on $\mathscr{I}$. We equip $ba(\Omega, \mathfrak{A}, \mathscr{I})$ with the total variation norm.

4.7.11 Corollary. *The dual of $\mathscr{L}_\infty(\Omega, \mathfrak{A}, \mathscr{I})$ is isometrically isomorphic to $ba(\Omega, \mathfrak{A}, \mathscr{I})$.*

CHAPTER 5

Nonatomic Charges

Classification of charges is an obvious pursuit one embarks on for a good understanding of charges. We have already come across 0–1 valued charges in Chapter 2. In this chapter, we examine what could be an antithesis of the notion of a 0–1 valued charge. Section 5.1 develops the relevant classification of charges. The Sobczyk–Hammer decomposition theorem is presented in Section 5.2. We prove some existence theorems for nonatomic charges in Section 5.3. Finally, in Section 5.4, we consider the plentitude of nonatomic charges.

5.1 BASIC CONCEPTS

The notion corresponding to nonatomicity of measures on σ-fields can be introduced for charges in three different ways. We show that these three ways are actually distinct for charges.

5.1.1 Definition. Let $\mathscr{F}$ be a field of subsets of a set Ω. A set F in $\mathscr{F}$ is said to be a μ-*atom* if the following conditions are satisfied.
  (i). $\mu(F) \neq 0$.
  (ii). If $E \in \mathscr{F}$ and $E \subset F$, then either $\mu(E) = 0$ or $\mu(F-E) = 0$.

If F is a μ-atom, then the restriction of μ to $F \cap \mathscr{F}$ is a 0-$\mu(F)$ valued charge on the field $F \cap \mathscr{F}$ of F. However, if μ restricted to the field $F \cap \mathscr{F}$ for an F in $\mathscr{F}$ is two valued, F need not be a μ-atom. Also, it is not difficult to check that an F in $\mathscr{F}$ is a μ-atom if and only if F is a $|\mu|$-atom. For, if F is a μ-atom, $E \in \mathscr{F}$, $E \subset F$ and $\mu(E) = 0$, then $|\mu|(E) = 0$.
Note that if $F \in \mathscr{F}$, F is an atom of $\mathscr{F}$ and $\mu(F) \neq 0$, then F is a μ-atom. However, a μ-atom need not be an atom of $\mathscr{F}$.

Non-existence of μ-atoms is one way of defining a class of charges. More formally, we give the following definition.

5.1.2 Definition. Let $\mathscr{F}$ be a field of subsets of a set Ω and μ a charge on $\mathscr{F}$. μ is said to be *nonatomic* on $\mathscr{F}$ if there are no μ-atoms in $\mathscr{F}$. Equivalently, if $F \in \mathscr{F}$ and $\mu(F) \neq 0$, then there exists E in $\mathscr{F}$ such that $E \subset F$, $\mu(E) \neq 0$ and $\mu(F-E) \neq 0$.

In view of the remarks made after Definition 5.1.1, it follows that μ is nonatomic if and only if $|\mu|$ is nonatomic. If μ is a positive bounded charge on $\mathscr{F}$, then μ is nonatomic on $\mathscr{F}$, if for every F in $\mathscr{F}$ with $\mu(F)>0$, there exists E in $\mathscr{F}$ such that $E \subset F$ and $0<\mu(E)<\mu(F)$. Now, we prove a simple property of nonatomic charges.

5.1.3 Proposition. *Let $\mathscr{F}$ be a field of subsets of a set Ω and μ a positive bounded nonatomic charge on $\mathscr{F}$. Then given F in $\mathscr{F}$ with $\mu(F)>0$ and $\varepsilon > 0$, there exists E in $\mathscr{F}$ such that $E \subset F$ and $0 < \mu(E) < \varepsilon$.*

Proof. Since μ is nonatomic, there exists E_1 in $\mathscr{F}$ such that $E_1 \subset F$ and $0 < \mu(E_1) < \mu(F)$. Considering $F - E_1$ if necessary, we can assume, without loss of generality, that $0 < \mu(E_1) \leq \frac{1}{2}\mu(F)$. Repeating the same step on E_1, we can find E_2 in $\mathscr{F}$ such that $E_2 \subset E_1$ and $0 < \mu(E_2) \leq \frac{1}{2}\mu(E_1)$. Continuing this way, we obtain a sequence $F \supset E_1 \supset E_2 \supset \cdots$ in $\mathscr{F}$ such that $0 < \mu(E_n) \leq \frac{1}{2}\mu(E_{n-1}) \leq (1/2^n)\mu(F)$. If $N \geq 1$ is such that $(1/2^N)\mu(F) < \varepsilon$, then $E = E_N$ serves the purpose. $\square$

The following definition is inspired by a certain property of nonatomic measures on σ-fields.

5.1.4 Definition. Let $\mathscr{F}$ be a field of subsets of a set Ω and μ a charge on $\mathscr{F}$. μ is said to be *strongly continuous* on $\mathscr{F}$, if for every $\varepsilon > 0$, there exists a partition $\{F_1, F_2, \ldots, F_n\}$ of Ω in $\mathscr{F}$ such that $|\mu|(F_i) < \varepsilon$ for every i.

Obviously, if a charge μ on $\mathscr{F}$ is strongly continuous on $\mathscr{F}$, then it is bounded.

One might conjecture that a charge μ on $\mathscr{F}$ is strongly continuous on $\mathscr{F}$ if and only if for every $\varepsilon > 0$, there exists a partition $\{F_1, F_2, \ldots, F_n\}$ of Ω in $\mathscr{F}$ such that $|\mu(F_i)| < \varepsilon$ for $i = 1, 2, \ldots, n$. But this conjecture is not true. Any non-zero charge μ on $\mathscr{F}$ with $\mu(\Omega) = 0$ satisfies the later property.

The following definition is inspired by yet another property of nonatomic measures on σ-fields.

5.1.5 Definition. Let $\mathscr{F}$ be a field of subsets of a set Ω and μ a charge on $\mathscr{F}$. μ is said to be *strongly nonatomic* on $\mathscr{F}$, if for every F in $\mathscr{F}$ and $0 \leq c \leq |\mu|(F)$, there exists E in $\mathscr{F}$ such that $E \subset F$ and $|\mu|(E) = c$.

The following theorem gives the inter-relations between these concepts.

5.1.6 Theorem. *Let μ be a positive bounded charge on a field $\mathscr{F}$ of subsets of a set Ω. Then each of the following conditions implies the succeeding condition.*
  (i). *μ is strongly nonatomic on $\mathscr{F}$.*
  (ii). *μ is strongly continuous on $\mathscr{F}$.*
  (iii). *μ is nonatomic on $\mathscr{F}$.*

If, in addition, μ is a measure and $\mathcal{F}$ is a σ-field, then these conditions are all equivalent.

Proof. The implications (i)$\Rightarrow$(ii)$\Rightarrow$(iii) are immediate. If μ is a measure and $\mathcal{F}$ is a σ-field, we show that (iii)$\Rightarrow$(i) using what is known as the "principle of exhaustion". Let $F \in \mathcal{F}$ and $0 \le c \le \mu(F)$ be given. Let $\mathscr{C} = \{C \in \mathcal{F}; C \subset F \text{ and } \mu(C) \le c\}$. On $\mathscr{C}$, we introduce the following partial order. For C_1, C_2 in $\mathscr{C}$, $C_1 \le C_2$ if $\mu(C_1 - C_2) = 0$. We show that every chain in $\mathscr{C}$ has an upper bound in $\mathscr{C}$. Let $\{C_\beta; \beta \in D\}$ be a chain in $\mathscr{C}$. Let $r = \text{Sup}\{\mu(C_\beta); \beta \in D\}$. We can find a sequence $C_{\beta_1} \le C_{\beta_2} \le \cdots$ such that $r = \lim_{n \to \infty} \mu(C_{\beta_n})$. Let $C = \bigcup_{n \ge 1} C_{\beta_n}$. Obviously, $C \in \mathscr{C}$ and $\mu(C) = r$. It is easy to verify that C is an upper bound of the given chain. By Zorn's lemma, $\mathscr{C}$ has a maximal element, E say, i.e. if $C \in \mathscr{C}$ and $E \le C$, then $\mu(C - E) = 0$. We claim that $\mu(E) = c$. Suppose $\mu(E) < c$. Then $\mu(F - E) = \mu(F) - \mu(E) \ge c - \mu(E) > 0$. Since μ is nonatomic, by Proposition 5.1.3, there exists a set E_0 in $\mathcal{F}$ such that $E_0 \subset F - E$ and $0 < \mu(E_0) < c - \mu(E)$. Let $E_1 = E_0 \cup E$. Then $E_1 \subset F$, $E \le E_1$, $E_1 \in \mathscr{C}$ and $\mu(E_1 - E) = \mu(E_0) > 0$. This contradicts the maximality of E. Hence $\mu(E) = c$. $\square$

5.1.7 Remarks. (i). A nonatomic positive bounded charge on a field $\mathcal{F}$ of subsets of a set Ω need not be strongly continuous. The following is a relevant example. Let $\Omega = [0, 1]$ and $\mathcal{F}$ the field on Ω generated by the collection of all intervals of the form $(a, b] \subset [\frac{1}{4}, \frac{3}{4})$. Let μ be the restriction of the Lebesgue measure to $\mathcal{F}$. Then μ is nonatomic on $\mathcal{F}$ but not strongly continuous on $\mathcal{F}$. For $\varepsilon = \frac{1}{2}$, there is no decomposition of Ω in $\mathcal{F}$ which satisfies the required properties.

(ii). A strongly continuous positive bounded charge on a field $\mathcal{F}$ of subsets of a set Ω need not be strongly nonatomic. The following is an example substantiating this statement. Let $\Omega = [0, 1)$. Let $\mathcal{F}$ be the field of all sets each of which is a finite disjoint union of intervals of the type $[a, b)$ with rational end points and $0 \le a \le b \le 1$. Let μ be the Lebesgue measure restricted to $\mathcal{F}$. Then μ is strongly continuous on $\mathcal{F}$ but not strongly nonatomic on $\mathcal{F}$.

(iii). If μ is a strongly continuous positive bounded charge on a σ-field $\mathcal{F}$ of subsets of a set Ω, then μ is strongly nonatomic on $\mathcal{F}$. In other words, (ii)$\Rightarrow$(i) in Theorem 5.1.6 is valid if $\mathcal{F}$ is a σ-field on Ω. See Theorem 11.4.5 for a proof.

(iv). If μ is a bounded charge on $\mathcal{F}$, Theorem 5.1.6 still remains valid. If μ is a charge on a field $\mathcal{F}$ without necessarily being bounded, then (i)$\Rightarrow$(iii) in Theorem 5.1.6.

(v). If $\mathcal{F}$ is a σ-field on Ω, (iii)$\Rightarrow$(ii) in Theorem 5.1.6 is not true even though (i)$\Leftrightarrow$(ii). Let $\Omega = [0, 1]$, $\mathcal{F}$ = Borel σ-field on Ω, λ = Lebesgue measure on $\mathcal{F}$ and $\mathcal{I} = \{B \in \mathcal{F}; \lambda(B) = 0\}$. Let τ be any 0–1 valued charge on $\mathcal{F}$ such that $\tau(B) = 0$ for every B in $\mathcal{I}$. Let $\mu = \lambda + 2\tau$. Then μ is nonatomic on $\mathcal{F}$ but not strongly nonatomic on $\mathcal{F}$.

Now, we examine the problem of characterizing strong continuity of a charge μ in terms of the positive and negative variations, μ^+ and μ^-, of μ, as strong continuity plays a dominant role in the decomposition theorem to be proved in the next section.

5.1.8 Proposition. *Let $\mathcal{F}$ be a field of subsets of a set Ω.*
(i). *A bounded charge μ on $\mathcal{F}$ is strongly continuous if and only if μ^+ and μ^- are strongly continuous.*
(ii). *If μ_1 and μ_2 are bounded strongly continuous charges on $\mathcal{F}$, then $\mu_1 + \mu_2$ is strongly continuous.*

Proof. (i) follows from the facts that $\mu^+ \leq |\mu|$, $\mu^- \leq |\mu|$ and $|\mu| = \mu^+ + \mu^-$.
(ii) follows from the fact that $|\mu_1 + \mu_2| \leq |\mu_1| + |\mu_2|$. $\square$

The following proposition is useful in establishing the uniqueness part of the decomposition theorem of Section 5.2.

5.1.9 Proposition. *Let $\mathcal{F}$ be a field of subsets of a set Ω and μ_n, $n \geq 1$ a sequence of 0-1 valued charges on $\mathcal{F}$. Let a_n, $n \geq 1$ be a sequence of real numbers such that $\sum_{n \geq 1} |a_n| < \infty$. If the bounded charge $\mu = \sum_{n \geq 1} a_n \mu_n$ is non-zero, then μ is not strongly continuous on $\mathcal{F}$.*

Proof. Assume, without loss of generality, that $|a_1| > 0$. If μ is strongly continuous on $\mathcal{F}$, then $|\mu|$ is also strongly continuous on $\mathcal{F}$. Now, since $|a_1|\mu_1 \leq |\mu| = \sum_{n \geq 1} |a_n| \mu_n$, μ_1 must be strongly continuous. But no two-valued charge is strongly continuous. $\square$

5.2 SOBCZYK–HAMMER DECOMPOSITION THEOREM

In this section, we prove the Sobczyk–Hammer decomposition theorem for charges. According to this theorem, we can write every charge μ as a sum of two charges, one of which is strongly continuous, the other a countable sum of two-valued charges. We need some preliminary results.

5.2.1 Definition. *Let $\mathcal{F}$ be a field of subsets of a set Ω. A sequence μ_n, $n \geq 1$ of 0-1 valued charges on $\mathcal{F}$ is said to be finitely disjoint if for every $n \geq 1$, there exists a partition $\{F_1, F_2, \ldots, F_n\}$ of Ω in $\mathcal{F}$ such that $\mu_i(F_i) = 1$ for every $i = 1, 2, \ldots, n$.*

This notion essentially means that any finite subcollection of $\{\mu_n, n \geq 1\}$ have disjoint supports. This notion is equivalent to the sequence μ_n, $n \geq 1$ being distinct. This is stated in the following proposition.

5. NONATOMIC CHARGES

5.2.2 Proposition. *Let $\mathscr{F}$ be a field of subsets of a set Ω. A sequence μ_n, $n \geq 1$ of 0–1 valued charges on $\mathscr{F}$ is finitely disjoint if and only if the sequence μ_n, $n \geq 1$ is distinct,* i.e. *no two charges in the sequence are the same.*

Proof. "Only if" part is obvious.
"If" part. This can be proved by induction. Since μ_1 and μ_2 are distinct, we can find A_1 in $\mathscr{F}$ such that $\mu_1(A_1) \neq \mu_2(A_2)$. Assume, without loss of generality, that $\mu_1(A_1) = 1$. Then $\{A_1, A_1^c\}$ is the desired partition for μ_1 and μ_2. Suppose, for $\mu_1, \mu_2, \ldots, \mu_n$, there is a partition $\{F_1, F_2, \ldots, F_n\}$ of Ω in $\mathscr{F}$ such that $\mu_i(F_i) = 1$ for $i = 1, 2, \ldots, n$. Since μ_i and μ_{n+1} are distinct, we can find E_i in $\mathscr{F}$ such that $\mu_i(E_i) = 1$ and $\mu_{n+1}(E_i^c) = 1$ for every $i = 1, 2, \ldots, n$. Let $B_{n+1} = \bigcap_{i=1}^{n} E_i^c$ and $B_i = B_{n+1}^c \cap F_i$, $i = 1, 2, \ldots, n$. Then $\{B_1, B_2, \ldots, B_n, B_{n+1}\}$ is a partition of Ω in $\mathscr{F}$ and $\mu_i(B_i) = 1$ for $i = 1, 2, \ldots, n+1$. □

5.2.3 Remark. (i). If μ_n, $n \geq 1$ is a sequence of distinct 0–1 valued charges on a field $\mathscr{F}$ of subsets of a set Ω, it is natural to ask whether the sequence is *infinitely disjoint*, i.e. there exists a partition $\{F_1, F_2, \ldots\}$ of Ω in $\mathscr{F}$ such that $\mu_n(F_n) = 1$ for every $n \geq 1$. But this is not the case as the following example shows.

Let $\Omega = \{1, 2, 3, \ldots\}$ and $\mathscr{F} = \mathscr{P}(\Omega)$, the class of all subsets of Ω. Let μ_1 be any 0–1 valued charge on $\mathscr{P}(\Omega)$ such that $\mu_1(A) = 0$ for every finite subset of Ω. See Example 2.1.3(4). For $n \geq 2$, let μ_n on $\mathscr{F}$ be defined by

$$\mu_n(A) = 1, \quad \text{if } n \in A,$$
$$= 0, \quad \text{if } n \notin A.$$

This sequence μ_n, $n \geq 1$ of distinct charges is not infinitely disjoint.
(ii). If $\mathscr{F}$ is a σ-field on Ω and μ_n, $n \geq 1$ is a distinct sequence of 0–1 valued measures on $\mathscr{F}$, then μ_n, $n \geq 1$ is infinitely disjoint.

The following definition is useful to express the notion of strong continuity of a charge in a form convenient for the development of the subsequent results.

5.2.4 Definition. Let $\mathscr{F}$ be a field of subsets of a set Ω and μ a charge on $\mathscr{F}$. Let $P = \{F_1, F_2, \ldots, F_n\}$ be a partition of Ω in $\mathscr{F}$. Then the number μ_P is defined by

$$\mu_P = \max_{1 \leq i \leq n} \mu(F_i).$$

Let $\mathscr{P}$ be the collection of all finite partitions of Ω in $\mathscr{F}$. A positive bounded charge μ on $\mathscr{F}$ is strongly continuous if and only if $\text{Inf}_{P \in \mathscr{P}} \mu_P = 0$. This assertion is obvious from the definition of strong continuity of μ.
The following two lemmas lead to the decomposition theorem.

5.2.5 Lemma. *Let $\mathcal{F}$ be a field of subsets of a set Ω and μ a positive bounded charge on $\mathcal{F}$ which is not strongly continuous. Let $a = \mathrm{Inf}_{P \in \mathcal{P}} \mu_P$. Then there exists a set F in $\mathcal{F}$ such that*
(i). $a \leq \mu(F) < 2a$, *and*
(ii). $a \leq \mu(F_i)$ *for some F_i in any partition $\{F_1, F_2, \ldots, F_n\}$ of F in $\mathcal{F}$.*

Proof. Since μ is not strongly continuous, $a > 0$. Choose any $0 < \varepsilon < a$. Then we can find a partition $Q = \{E_1, E_2, \ldots, E_m\}$ in $\mathcal{P}$ such that $\mu_Q = \max_{1 \leq i \leq m} \mu(E_i) < a + \varepsilon < 2a$. If none of the sets in Q has property (ii), there exists a partition $\{E_{ij}; j = 1, 2, \ldots, n_i\}$ of E_i in $\mathcal{F}$ such that $\mu(E_{ij}) < a$ for all $j = 1, 2, \ldots, n_i$ and $i = 1, 2, \ldots, m$. Then $Q' = \{E_{ij}; 1 \leq j \leq n_i$ and $1 \leq i \leq m\} \in \mathcal{P}$ and we have that $\mu_{Q'} < a$. This contradiction shows that there exists E_i satisfying (ii). Obviously, this E_i satisfies (i). Take $F = E_i$. □

5.2.6 Lemma. *Let $\mathcal{F}$ be a field of subsets of a set Ω and μ a positive bounded charge on $\mathcal{F}$ which is not strongly continuous. Let $a = \mathrm{Inf}_{P \in \mathcal{P}} \mu_P$ and $F \in \mathcal{F}$ any set having the properties (i) and (ii) of Lemma 5.2.5. Let λ on $\mathcal{F}$ be defined by*

$$\lambda(A) = a, \quad \text{if } \mu(A \cap F) \geq a,$$
$$= 0, \quad \text{if } \mu(A \cap F) < a.$$

Then λ is a charge on $\mathcal{F}$.

Proof. Let A and B be two disjoint sets in $\mathcal{F}$.
Case (i). $\mu((A \cup B) \cap F) < a$. Then $\mu(A \cap F) < a$ and $\mu(B \cap F) < a$. Consequently, $\lambda(A \cup B) = 0 = 0 + 0 = \lambda(A) + \lambda(B)$.
Case (ii). $\mu((A \cup B) \cap F) \geq a$. Then $\mu(A \cap F) + \mu(B \cap F) \geq a$. Since $\mu(F) = \mu(A \cap F) + \mu(B \cap F) + \mu(F - (A \cup B) \cap F) < 2a$, $\mu(F - (A \cup B) \cap F) < a$. Looking at the partition $\{A \cap F, B \cap F, F - (A \cup B) \cap F\}$ of F and in view of property (ii) of Lemma 5.2.5, either $\mu(A \cap F) \geq a$ or $\mu(B \cap F) \geq a$. Note that both these inequalities cannot hold simultaneously. For, if they hold, then $\mu(F) \geq \mu(A \cap F) + \mu(B \cap F) \geq 2a$ contradicting property (i) of Lemma 5.2.5. Consequently, $\lambda(A \cup B) = a = \lambda(A) + \lambda(B)$.
This proves the lemma. □

Finally, we prove the main theorem of this section.

5.2.7 Sobczyk–Hammer Decomposition Theorem. *Let $\mathcal{F}$ be a field of subsets of a set Ω and μ a positive bounded charge on $\mathcal{F}$. Then there exists a sequence μ_n, $n \geq 0$ of distinct positive bounded charges on $\mathcal{F}$ and a sequence a_n, $n \geq 1$ of nonnegative numbers with the following properties.*
(i). μ_0 *is strongly continuous on $\mathcal{F}$.*
(ii). μ_n *is a 0–1 valued charge on $\mathcal{F}$ for every $n \geq 1$.*

(iii). $\sum_{n\geq 1} a_n < \infty$.
(iv). $\mu = \mu_0 + \sum_{n\geq 1} a_n \mu_n$.
Further, the decomposition (iv) is unique.

Proof. First, we establish uniqueness. Let $\mu = \mu_0^* + \sum_{n\geq 1} a_n^* \mu_n^*$ be another decomposition, where μ_0^* is a strongly continuous charge on $\mathscr{F}$, μ_n^*, $n \geq 1$ is a sequence of 0-1 valued charges on $\mathscr{F}$ and a_n^*, $n \geq 1$ is a sequence of nonnegative numbers satisfying $\sum_{n\geq 1} a_n^* < \infty$. Assume, without loss of generality, that μ_n's are distinct, a_n's positive, μ_n^*'s distinct and a_n^*'s positive. (Either sum $\sum_{n\geq 1} a_n \mu_n$ and $\sum_{n\geq 1} a_n^* \mu_n^*$ or both could be a finite sum. The following argument carries through in these cases also.) We show that $\mu_0^* = \mu_0$ and $a_n \mu_n$, $n \geq 1$ is a permutation of $a_n^* \mu_n^*$, $n \geq 1$. Observe that

$$\mu_0^* - \mu_0 = \sum_{n\geq 1} a_n \mu_n - \sum_{n\geq 1} a_n^* \mu_n^*.$$

By Proposition 5.1.8, $\mu_0^* - \mu_0$ is strongly continuous. By Proposition 5.1.9, it follows that $\mu_0^* = \mu_0$ and $\sum_{n\geq 1} a_n \mu_n = \sum_{n\geq 1} a_n^* \mu_n^*$. Next, we claim that $\{a_n \mu_n; n \geq 1\} = \{a_n^* \mu_n^*; n \geq 1\}$. Suppose this is not true. If $\{a_n \mu_n; n \geq 1\}$ is not a subset of $\{a_n^* \mu_n^*; n \geq 1\}$, there is an element $a_1 \mu_1$, say, of $\{a_n \mu_n; n \geq 1\}$ which is not a member of $\{a_n^* \mu_n^*; n \geq 1\}$. Choose $N \geq 1$ such that $\sum_{n\geq N+1} a_n^* < a_1$. Since $\mu_1, \mu_1^*, \mu_2^*, \ldots, \mu_N^*$ are distinct (why?), we can find F in $\mathscr{F}$ such that $\mu_1(F) = 1$ and $\mu_i^*(F) = 0$ for $i = 1, 2, \ldots, N$. See Proposition 5.2.2. Consequently,

$$a_1 \leq \sum_{n\geq 1} a_n \mu_n(F) = \sum_{n\geq 1} a_n^* \mu_n^*(F) = \sum_{n\geq N+1} a_n^* \mu_n^*(F) < a_1.$$

This contradiction shows that $\{a_n \mu_n; n \geq 1\} \subset \{a_n^* \mu_n^*; n \geq 1\}$. By a similar argument, we can show that $\{a_n^* \mu_n^*; n \geq 1\} \subset \{a_n \mu_n; n \geq 1\}$. Thus we have proved that $\{a_n^* \mu_n^*; n \geq 1\} = \{a_n \mu_n; n \geq 1\}$. Since the charges in each set are distinct, it follows that $a_n \mu_n$, $n \geq 1$ is a permutation of $a_n^* \mu_n^*$, $n \geq 1$. This establishes uniqueness.

Now, we prove the main part of the theorem. If μ is strongly continuous, then the conclusion of the theorem is trivially true. So, assume that μ is not strongly continuous. Let $\text{Inf}_{P\in\mathscr{P}} \mu_P = a_1$. Then $a_1 > 0$. Let μ_1' be the 0-a_1 valued charge on $\mathscr{F}$ as defined in Lemma 5.2.6. Note that $\mu_1' \leq \mu$. Let $\lambda_1 = \mu - \mu_1'$. Or, equivalently, $\mu = \mu_1' + \lambda_1$. If λ_1 is strongly continuous, we stop here and we have the desired decomposition. If λ_1 is not strongly continuous, using the argument given above, we can write $\lambda_1 = \mu_2' + \lambda_2$, where μ_2' is a 0-a_2 valued charge with $a_2 > 0$, where $a_2 = \text{Inf}_{P\in\mathscr{P}} (\mu - \mu_1')_P$. Continuing this way, we either stop at a finite number of steps reaching the desired decomposition or obtain a sequence μ_n', $n \geq 1$ of charges such that each μ_n' is 0-a_n valued with $a_n > 0$. In this case, $\sum_{n\geq 1} a_n < \infty$ since μ is bounded. Let $\mu_0 = \mu - \sum_{n\geq 1} \mu_n'$. We claim that μ_0 is strongly continuous.

Observe that

$$0 \le \operatorname*{Inf}_{P \in \mathscr{P}} (\mu_0)_P \le \operatorname*{Inf}_{P \in \mathscr{P}} \left(\mu - \sum_{i=1}^{N} \mu'_i\right)_P = a_N$$

for all N. Since $a_N \to 0$ as $N \to \infty$, we have that $\operatorname{Inf}_{P \in \mathscr{P}} (\mu_0)_P = 0$, i.e. μ_0 is strongly continuous. Writing each $\mu'_n = a_n \mu_n$, where each μ_n is a 0-1 valued charge, we obtain the desired decomposition. □

5.2.8 Remark. The above decomposition theorem is also valid for any bounded charge μ on $\mathscr{F}$. We can work out decompositions for μ^+ and μ^- separately and from these we can obtain a decomposition of μ in the form $\mu_0 + \sum_{n \ge 1} a_n \mu_n$, where μ_0 is strongly continuous, each μ_n is a 0-1 valued charge and $\sum_{n \ge 1} |a_n| < \infty$. Further, this decomposition is unique and this can be proved along the lines of the first part of the proof of Theorem 5.2.7.

5.2.9 Remark. The above decomposition theorem is not valid for unbounded charges even if we allow 0-∞ valued charges. The following is an example.

Let $\Omega = [0, \infty)$ and $\mathscr{F}$ the field on Ω generated by $\{[a, b); 0 \le a \le b < \infty\}$. Let μ be the restriction of the Lebesgue measure to $\mathscr{F}$. Suppose we can write

$$\mu = \mu_0 + \sum_{n \ge 1} \mu_n,$$

where (i) μ_0 is strongly continuous on $\mathscr{F}$, and (ii) μ_n is 0-a_n valued with $a_n > 0$ for every $n \ge 1$. (a_n could be equal to ∞.) (The sum $\sum_{n \ge 1} \mu_n$ could be a finite sum.)

Since μ is unbounded, μ is not strongly continuous. But for any $x > 0$, the charge μ restricted to $[0, x) \cap \mathscr{F}$ is strongly continuous. It follows that $\mu_n([0, x)) = 0$ for every $n \ge 1$ and $\mu([0, x)) = \mu_0([0, x)) = x$. This shows that μ_0 is an unbounded charge on $\mathscr{F}$. Consequently, it cannot be strongly continuous. Hence μ does not admit a decomposition along the lines of Theorem 5.2.7.

The above example worked because every strongly continuous charge is necessarily bounded. By relaxing the definition of strong continuity of charges, we obtain a Sobczyk–Hammer like decomposition theorem for positive unbounded charges μ. For A in $\mathscr{F}$, let μ/A be the charge on $A \cap \mathscr{F}$ defined by $(\mu/A)(B) = \mu(B)$ for B in $\mathscr{F}$ and $B \subset A$.

5.2.10 Definition. Let $\mathscr{F}$ be a field of subsets of a set Ω. A positive charge μ on $\mathscr{F}$ is said to be strongly continuous on $\mathscr{F}$ if μ/A is strongly continuous on $A \cap \mathscr{F}$ for every A in $\mathscr{F}$ with $\mu(A) < \infty$.

5.2.11 Theorem. *Let $\mathscr{F}$ be a field of subsets of a set Ω and μ a positive unbounded charge on $\mathscr{F}$. Then there exists a strongly continuous charge μ_0*

5. NONATOMIC CHARGES

on $\mathscr{F}$ and a family $\{\mu_\alpha; \alpha \in \Gamma\}$ of two valued charges on $\mathscr{F}$ such that

$$\mu = \mu_0 + \sum_{\alpha \in \Gamma} \mu_\alpha$$

and $\{\mu_\alpha; \alpha \in \Gamma\}$ is finitely disjoint, i.e. for any distinct $\alpha_1, \alpha_2, \ldots, \alpha_n$ in Γ, there exists a partition $\{A_{\alpha_1}, A_{\alpha_2}, \ldots, A_{\alpha_n}\}$ of Ω in $\mathscr{F}$ such that $\mu_{\alpha_i}(A_i) = \mu_{\alpha_i}(\Omega)$ for $i = 1, 2, \ldots, n$ and $n \geq 1$. (*Here Γ could be uncountable!*)

Proof. Let $\mathscr{C} = \{A \in \mathscr{F}; \mu(A) < \infty\}$. $\mathscr{C}$ is a ring on Ω. For each A in $\mathscr{C}$, by Theorem 5.2.7, we can write

$$\mu/A = \mu_{0A} + \sum_{i \geq 1} \mu_{iA},$$

where μ_{0A} is a strongly continuous charge on $A \cap \mathscr{F}$ and μ_{iA}, $i \geq 1$ is a sequence of two valued charges on $A \cap \mathscr{F}$ which are finitely disjoint. ($\sum_{i \geq 1} \mu_{iA}$ could be a finite sum.) Because of the uniqueness of the decomposition in Theorem 5.2.7, the charges μ_{0A}, $A \in \mathscr{C}$ are consistent in the following sense: if $A, B \in \mathscr{C}$ and $B \subset A$, then $\mu_{0A}/B = \mu_{0B}$. Define μ_0 on $\mathscr{F}$ as follows.

$$\mu_0(A) = \mu_{0A}(A), \quad \text{if } A \in \mathscr{C},$$
$$= \infty, \quad \text{if } A \notin \mathscr{C} \text{ and } A \in \mathscr{F}.$$

μ_0 is a charge on $\mathscr{F}$ and is, obviously, strongly continuous in the sense of Definition 5.2.10. For $i \geq 1$ and A in $\mathscr{C}$, let $\tilde{\mu}_{iA}(F) = \mu_{iA}(F \cap A)$ for F in $\mathscr{F}$. Each $\tilde{\mu}_{iA}$ is a two-valued charge on $\mathscr{F}$. Let $\bigcup_{A \in \mathscr{C}} \{\tilde{\mu}_{iA}; i \geq 1\} = \{\mu_\alpha; \alpha \in \Gamma\}$. Now, it is not difficult to verify that

$$\mu = \mu_0 + \sum_{\alpha \in \Gamma} \mu_\alpha. \qquad \square$$

5.2.12 Remark. It is possible to show that the above decomposition is essentially unique.

As a corollary of Theorem 5.2.7, we can obtain the following decomposition theorem of Measure theory.

5.2.13 Corollary. *Let $\mathscr{F}$ be a σ-field of subsets of a set Ω and μ a positive bounded measure on $\mathscr{F}$. Then we can write*

$$\Omega = \bigcup_{n \geq 0} A_n \quad \text{or} \quad \bigcup_{n=0}^{N} A_n$$

for some $N \geq 0$ with the following properties.
 (i). A_n's *are pairwise disjoint.*
 (ii). A_n *is a μ-atom for every $n \neq 0$.*
 (iii). *The restriction of μ to the σ-field $A_0 \cap \mathscr{F}$ is a nonatomic measure.*

Proof. By Theorem 5.2.7, we can write

$$\mu = \mu_0 + \sum_{n \geq 1} a_n \mu_n,$$

where (i) μ_0 is strongly continuous on $\mathscr{F}$,
(ii) μ_n, $n \geq 1$ is a sequence of distinct 0–1 valued charges on $\mathscr{F}$, and
(iii) $a_n > 0$ for every $n \geq 1$ with $\sum_{n \geq 1} a_n < \infty$. ($\sum_{n \geq 1} a_n \mu_n$ could be of the form $\sum_{n=1}^{N} a_n \mu_n$.) Since $0 \leq \mu_0 \leq \mu$ and $0 \leq a_n \mu_n \leq \mu$ for every $n \geq 1$, each μ_n is a measure. We now claim that for every $n \geq 1$, there exists a set B_n in $\mathscr{F}$ such that $\mu_n(B_n) = 1$ and $\mu_0(B_n) = 0$. Since μ_0 is strongly continuous, for every $k \geq 1$, we can find a partition $\{C_{k1}, C_{k2}, \ldots, C_{kp_k}\}$ of Ω in $\mathscr{F}$ such that $\mu_0(C_{ki}) < 1/k$ for every i. For fixed $n \geq 1$, one of these sets, say D_k, is such that $\mu_n(D_k) = 1$. Let $B_n = \bigcap_{k \geq 1} D_k$. Then $\mu_0(B_n) = 0$ and $\mu_n(B_n) = 1$. Now, let $B = \bigcup_{n \geq 1} B_n$. Then $\mu_0(B) = 0$. Consequently, μ_0 restricted to $B^c \cap \mathscr{F}$ is nonatomic. The sequence μ_n, $n \geq 1$ restricted to $B \cap \mathscr{F}$ is a sequence of distinct 0–1 valued measures. These restrictions are infinitely disjoint. See Remark 5.2.3(i). So, we can find a partition $\{A_1, A_2, \ldots\}$ of B in $\mathscr{F}$ such that $\mu_n(A_n) = 1$ for every $n \geq 1$. Obviously, A_n is a μ-atom for every $n \geq 1$. Let $A_0 = B^c$. This proves the result. □

5.3 EXISTENCE OF NONATOMIC CHARGES

In this section, we examine the conditions under which there exists a non-zero nonatomic charge on a given field $\mathscr{F}$ of subsets of a set Ω. The following definition is instrumental in providing a solution.

5.3.1 Definition. Let $\mathscr{F}$ be a field of subsets of a set Ω. A collection of non-empty sets $\{F_{i_1,i_2,\ldots,i_k}; i_1, i_2, \ldots, i_k$ is any finite sequence of 0's and 1's, $k \geq 1\}$ in $\mathscr{F}$ is said to be a *tree* in $\mathscr{F}$ if the following conditions are satisfied.
(i). $F_0 \cup F_1 = \Omega$, $F_0 \cap F_1 = \varnothing$.
(ii). $F_{i_1,i_2,\ldots,i_{k-1},0} \cup F_{i_1,i_2,\ldots,i_{k-1},1} = F_{i_1,i_2,\ldots,i_{k-1}}$
and $F_{i_1,i_2,\ldots,i_{k-1},0} \cap F_{i_1,i_2,\ldots,i_{k-1},1} = \varnothing$ for all $i_1, i_2, \ldots, i_{k-1} \in \{0, 1\}$ and $k \geq 2$.

The following theorem provides a set of equivalent conditions for the existence problem.

5.3.2 Theorem. *Let $\mathscr{F}$ be a field of subsets of a set Ω. The following statements are equivalent.*
  (i). *There is a non-zero positive bounded nonatomic charge on $\mathscr{F}$.*
  (ii). *$\mathscr{F}$ contains a tree.*
  (iii). *There is a non-zero positive bounded strongly continuous charge on $\mathscr{F}$.*

5. NONATOMIC CHARGES 151

Proof. (i)$\Rightarrow$(ii). Let μ be a non-zero positive bounded nonatomic charge on $\mathscr{F}$. Since $\mu(\Omega) > 0$, we can find F_0, F_1 in $\mathscr{F}$ such that $F_0 \cup F_1 = \Omega$, $F_0 \cap F_1 = \varnothing$, $0 < \mu(F_0) < \mu(\Omega)$ and $0 < \mu(F_1) < \mu(\Omega)$. Applying this technique to F_0 and F_1 separately, we obtain F_{00}, F_{01}, F_{10}, F_{11}. Continuing this way, we obtain a tree in $\mathscr{F}$.

(ii)$\Rightarrow$(iii). Let $\{F_{i_1,i_2,\ldots,i_k}; i_1, i_2, \ldots, i_k \in \{0, 1\}$ and $k \geq 1\}$ be a tree in $\mathscr{F}$. Let $\mathscr{F}_0$ be the smallest field on Ω containing this tree. In fact, $\mathscr{F}_0$ is precisely the collection of all finite disjoint unions of sets in the tree. Define $\mu_0(F_{i_1,i_2,\ldots,i_k}) = 1/2^k$ for $i_1, i_2, \ldots, i_k$ in $\{0, 1\}$ and $k \geq 1$. μ_0 can be extended in the obvious fashion to $\mathscr{F}_0$ as a charge μ_1. Note that μ_1 is a strongly continuous positive bounded charge on $\mathscr{F}_0$. Let μ_2 be any positive charge on $\mathscr{F}$ which is an extension of μ_1. See Corollary 3.3.4. It is obvious that μ_2 is strongly continuous on $\mathscr{F}$.

(iii)$\Rightarrow$(i). This is part of Theorem 5.1.6. □

The above theorem provides an interesting implication for σ-fields.

5.3.3 Corollary. *Let $\mathscr{F}$ be an infinite σ-field of subsets of a set Ω. Then there exists a non-zero strongly continuous positive charge on $\mathscr{F}$.*

Proof. In every infinite σ-field, one can find a sequence A_n, $n \geq 1$ of pairwise disjoint non-empty sets whose union is Ω. This can be proved as follows. Let B be any non-empty set in $\mathscr{F}$ whose complement, B^c, is also non-empty. Then either $B \cap \mathscr{F}$ or $B^c \cap \mathscr{F}$ is infinite. If $B \cap \mathscr{F}$ is infinite, we can find non-empty sets B_1, B_2 in $\mathscr{F}$ such that $B_1 \cup B_2 = B$ and $B_1 \cap B_2 = \varnothing$. Then either $B_1 \cap \mathscr{F}$ or $B_2 \cap \mathscr{F}$ is infinite. Proceeding this way, we obtain a sequence of pairwise disjoint non-empty sets. From this, we can obtain A_n, $n \geq 1$ with the stated properties. Write the set $N = \{1, 2, \ldots\} = N_1 \cup N_2$, where N_1 and N_2 are disjoint infinite sets. Let $F_0 = \bigcup_{i \in N_1} A_i$ and $F_1 = \bigcup_{i \in N_2} A_i$. Splitting N_1 into two disjoint infinite subsets as above, we obtain F_{00} and F_{01}. Applying the same technique to N_2, we obtain F_{10} and F_{11}. Proceeding this way, we obtain a tree in $\mathscr{F}$. Theorem 5.3.2, now, completes the proof. □

Theorem 5.3.2 also provides useful information on superatomic fields. Keeping in mind the definitions of an atom, atomic field and nonatomic field in Definitions 1.4.9, we define superatomic fields as follows.

5.3.4 Definition. Let $\mathscr{F}$ be a field of subsets of a set Ω. $\mathscr{F}$ is said to be *superatomic* if every sub-field of $\mathscr{F}$ is atomic.

5.3.5 Remarks. (i). If $\Omega = \{1, 2, \ldots\}$ and $\mathscr{F}$ is the finite-cofinite field on Ω, then $\mathscr{F}$ is a superatomic field.
(ii). Let α be any ordinal and $\Omega = [0, \alpha]$ be equipped with the order topology. This is the topology on Ω which has the collection of all subsets

C of Ω of the form (a, b) or $[0, b)$ or $(a, \alpha]$ for $0 \leq a \leq b \leq \alpha$ as a base. Let $\mathcal{F}$ be the field of all clopen subsets of Ω. Then $\mathcal{F}$ is superatomic. See Pierce (1970).

(iii). Let $\mathcal{F}_1$ and $\mathcal{F}_2$ be superatomic fields on Ω_1 and Ω_2 respectively. Let $\mathcal{F}_1 \otimes \mathcal{F}_2$ be the smallest field on $\Omega_1 \times \Omega_2$ containing $\{A \times B; A \in \mathcal{F}_1$ and $B \in \mathcal{F}_2\}$. Then $\mathcal{F}_1 \otimes \mathcal{F}_2$ is a superatomic field.

(iv). If $\mathcal{F}$ is a field on a set Ω, then $\mathcal{F}$ is superatomic if and only if the Stone space X of $\mathcal{F}$ is scattered, i.e. no non-empty subset of X is perfect. See Sikorski (1969) especially the statement on page 35 under D.

The following theorem gives equivalent versions of superatomicity and describes all the bounded charges explicitly on superatomic fields.

5.3.6 Theorem. *Let $\mathcal{F}$ be a field of subsets of a set Ω. The following statements are equivalent.*

(i). *$\mathcal{F}$ is superatomic.*

(ii). *$\mathcal{F}$ does not contain a tree.*

(iii). *There is no non-zero nonatomic positive bounded charge on $\mathcal{F}$.*

(iv). *There is no non-zero strongly continuous positive charge on $\mathcal{F}$.*

(v). *Every positive bounded charge μ on $\mathcal{F}$ has the following representation*

$$\mu = \sum_{i \geq 1} a_i \mu_i,$$

where (a) $a_i \geq 0$ for every i and $\sum_{i \geq 1} a_i < \infty$, and (b) each μ_i is a 0-1 valued charge on $\mathcal{F}$.

Proof. The implications (ii)$\Rightarrow$(iii)$\Rightarrow$(iv)$\Rightarrow$(v) follow from Theorem 5.3.2 and Theorem 5.2.8.

(v)$\Rightarrow$(i). Suppose (v) holds and (i) is not true. We show that $\mathcal{F}$ contains a tree. Since $\mathcal{F}$ is not superatomic, there is a subfield $\mathcal{F}_0$ of $\mathcal{F}$ which is not atomic. This implies that there exists a non-empty set A in $\mathcal{F}_0$ which does not contain any atom of $\mathcal{F}_0$. Then we can find A_0 and A_1 in $\mathcal{F}_0$ such that $A_0 \cup A_1 = A$, $A_0 \cap A_1 = \varnothing$, $A_0 \neq \varnothing$ and $A_1 \neq \varnothing$. Repeating this procedure on A_0 and A_1, we obtain A_{00}, $A_{01} \subset A_0$ and A_{10}, $A_{11} \subset A_1$ having properties similar to the above sets. Continuing this way, we obtain a collection $\{A_{i_1, i_2, \ldots, i_k}; i_1, i_2, \ldots, i_k \in \{0, 1\}$ and $k \geq 1\}$ of non-empty sets in $\mathcal{F}_0$ with the properties

(a) $A_{i_1, i_2, \ldots, i_{k-1}, 0} \cup A_{i_1, i_2, \ldots, i_{k-1}, 1} = A_{i_1, i_2, \ldots, i_{k-1}}$

and

(b) $A_{i_1, i_2, \ldots, i_{k-1}, 0} \cap A_{i_1, i_2, \ldots, i_{k-1}, 1} = \varnothing$

for all $i_1, i_2, \ldots, i_{k-1}$ in $\{0, 1\}$ and $k \geq 2$. From this, it is not difficult to obtain a tree in $\mathcal{F}$. Consequently, by Theorem 5.3.2, there exists a non-zero strongly continuous positive charge on $\mathscr{F}$. Such a charge can never be decomposed in the way condition (v) stipulates. This contradiction proves the desired implication.

(i) $\Rightarrow$ (ii). If $\mathscr{F}$ contains a tree, then the subfield $\mathscr{F}_0$ generated by the tree is nonatomic and hence $\mathscr{F}$ cannot be superatomic. This contradiction proves the desired implication. □

We now proceed to obtain some topological conditions for the existence problem of nonatomic charges. We need some preliminary results for this purpose.

5.3.7 Proposition. *Let $\mathfrak{A}$ be a σ-field of subsets of a set Ω, $\mathscr{F}$ a field on Ω generating $\mathfrak{A}$ and μ a positive bounded measure on $\mathfrak{A}$. Then μ is strongly continuous on $\mathfrak{A}$ if and only if μ is strongly continuous on $\mathscr{F}$.*

Proof. "If" part is trivial.
"Only if" part. Assume that μ is non-zero. Let $\varepsilon > 0$. Let m be a natural number such that $1/m < \varepsilon$. Since μ is strongly continuous on $\mathfrak{A}$, we can find a partition $\{F_1, F_2, \ldots, F_n\}$ of Ω in $\mathfrak{A}$ such that

$$0 < \mu(F_i) < \left[\frac{m\mu(\Omega)}{m\mu(\Omega)+1}\right]\varepsilon$$

for every i. For each $1 \leq i \leq n$, by Theorem 3.5.3, we can find G_i in $\mathscr{F}$ such that

$$\mu(F_i \Delta G_i) < \frac{\mu(F_i)}{m\mu(\Omega)}.$$

Since $G_i \subset F_i \cup (F_i \Delta G_i)$, we have

$$\mu(G_i) \leq \mu(F_i)\left[1 + \frac{1}{m\mu(\Omega)}\right] = \mu(F_i)\left[\frac{m\mu(\Omega)+1}{\mu(\Omega)}\right] < \varepsilon$$

for every i. Further,

$$\mu\left(\Omega - \bigcup_{i=1}^{n} G_i\right) = \mu\left(\bigcup_{i=1}^{n} F_i - \bigcup_{i=1}^{n} G_i\right) \leq \mu\left(\bigcup_{i=1}^{n} (F_i - G_i)\right)$$

$$\leq \sum_{i=1}^{n} \mu(F_i - G_i) \leq \sum_{i=1}^{n} \mu(F_i \Delta G_i) < 1/m < \varepsilon.$$

Defining $D_1 = G_1$, $D_i = G_i - \bigcup_{j=1}^{i-1} G_j$ for $i = 2, 3, \ldots, n$ and $D_{n+1} = \Omega - \bigcup_{i=1}^{n} G_i$, we observe that $\{D_1, D_2, \ldots, D_{n+1}\}$ is a partition of Ω in $\mathscr{F}$ satisfying $\mu(D_i) < \varepsilon$ for every i. This completes the proof. □

The following result provides a simple necessary and sufficient condition for nonatomicity of regular measures on the Borel σ-field of a compact Hausdorff space.

5.3.8 Lemma. *Let X be a compact Hausdorff space, $\mathscr{B}$ its Borel σ-field and $\mathscr{B}_0$ its Baire σ-field. Let μ be a positive bounded regular measure on $\mathscr{B}$. Then the following statements are equivalent.*
   (i). *μ is nonatomic on $\mathscr{B}$.*
   (ii). *$\mu(\{x\}) = 0$ for every x in X.*
   (iii). *μ is nonatomic on $\mathscr{B}_0$.*

Proof. (i)$\Rightarrow$(ii). This is obvious.
(ii)$\Rightarrow$(iii). Let B in $\mathscr{B}_0$ be a μ-atom. Since μ is regular, $\mu(B) = \text{Sup}\{\mu(C); C \subset B, C \in \mathscr{B}_0$ and C is a compact G_δ subset of $X\}$. Since $\mu(B) > 0$, there exists a compact G_δ set C contained in B such that $\mu(C) > 0$. Let $x \in C$. Since μ is regular, $0 = \mu(\{x\}) = \text{Inf}\{\mu(V); x \in V, V$ an open F_σ set $\subset X\}$ and therefore, we can find an open F_σ set V_x containing x such that $\mu(V_x) < \mu(C)$. $\{V_x; x \in C\}$ is an open cover for C. There is a finite sub-cover $\{V_{x_1}, V_{x_2}, \ldots, V_{x_n}\}$ of C. Since B is a μ-atom, C is a μ-atom. Since $\mu(V_x \cap C) \le \mu(V_x) < \mu(C)$, it follows that $\mu(V_x \cap C) = 0$. Consequently, $\mu(C) = \mu((\bigcup_{i=1}^n V_{x_i}) \cap C) = 0$. This contradiction shows that there are no μ-atoms in $\mathscr{B}_0$. Hence μ is nonatomic on $\mathscr{B}_0$.
(iii)$\Rightarrow$(i). Since μ is nonatomic on $\mathscr{B}_0$, μ is strongly continuous on $\mathscr{B}_0$. See Theorem 5.1.6. Hence μ is strongly continuous on the bigger σ-field $\mathscr{B}$. So, μ is nonatomic on $\mathscr{B}$. □

The following theorem gives some more equivalent versions for the existence of non-zero nonatomic charges, mainly topological in nature.

5.3.9 Theorem. *Let $\mathscr{F}$ be a field of subsets of a set Ω and X its Stone space. The following statements are equivalent.*
   (i). *There is a non-zero nonatomic positive bounded charge on $\mathscr{F}$.*
   (ii). *There exists a countable nonatomic subfield $\mathscr{F}_0$ of $\mathscr{F}$.*
   (iii). *X contains a perfect set.*
   (iv). *$\mathscr{F}$ contains an ideal $\mathscr{I}$ such that the quotient Boolean algebra $\mathscr{F}/\mathscr{I}$ is nonatomic.*

Proof. (i)$\Rightarrow$(ii). By Theorem 5.3.2, $\mathscr{F}$ contains a tree. The smallest field $\mathscr{F}_0$ on Ω containing a tree is a countable nonatomic subfield of $\mathscr{F}$. (Note that a tree is obviously a countable collection of sets.) See Corollary 1.1.14.
(ii)$\Rightarrow$(iii). Let $\mathscr{F}_0$ be a countable nonatomic subfield of $\mathscr{F}$. Without loss of generality, assume that $\mathscr{F}_0$ is generated by a tree. Let Y be the Stone space of $\mathscr{F}_0$. See Theorem 1.4.10. We note that Y is perfect. In fact, Y is homeomorphic to the Cantor set $\{0, 1\}^{\aleph_0}$. This follows from the fact that the field $\mathscr{C}$ of all clopen subsets of $\{0, 1\}^{\aleph_0}$ and $\mathscr{F}_0$ are isomorphic. For, since $\mathscr{C}$ is generated by the tree $\{\{i_1\} \times \{i_2\} \times \cdots \times \{i_n\} \times \{0, 1\} \times \cdots ; i_1, i_2, \ldots, i_n \in \{0, 1\}$ and $n \ge 1\}$ and $\mathscr{F}_0$ is generated by a tree, $\mathscr{C}$ and $\mathscr{F}_0$ are isomorphic. It is obvious that $\{0, 1\}^{\aleph_0}$ has no isolated points, i.e. it is

perfect. Since the inclusion map i from $\mathscr{F}_0$ to $\mathscr{F}$ defined by $i(A) = A$, $A \in \mathscr{F}_0$ is a one-to-one homomorphism, there exists a continuous function f from X onto Y. See Theorem 1.4.11. Now, we claim that there exists a minimal closed subset P of X such that $f(P) = Y$. The collection $\mathscr{E} = \{E \subset X; E$ is closed and $f(E) = Y\}$ is non-empty and is partially ordered by set inclusion, i.e. for E_1, E_2 in $\mathscr{E}$, say $E_1 \leq E_2$ if $E_1 \subset E_2$. Let $\{E_\alpha; \alpha \in D\}$ be a chain in $\mathscr{E}$. We claim that $\bigcap_{\alpha \in D} E_\alpha \in \mathscr{E}$. Let $y \in Y$. For each α in D, there exists x_α in E_α such that $f(x_\alpha) = y$. Since X is compact, there exists a subset of x_α, $\alpha \in D$ converging to some element x in X. Further, $x_\beta \in E_\alpha$ whenever $\beta \leq \alpha$, since $\{E_\alpha; \alpha \in D\}$ is a chain. Consequently, since E_α is closed, $x \in E_\alpha$ for every α in D. Thus $x \in \bigcap_{\alpha \in D} E_\alpha$ and $f(x) = y$. Hence $\bigcap_{\alpha \in D} E_\alpha$ is a lower bound of the chain $\{E_\alpha; \alpha \in D\}$. So, we have proved that every chain in $\mathscr{E}$ has a lower bound. By Zorn's lemma, there exists a minimal closed set $P \subset X$ such that $f(P) = Y$.

We also claim that P is perfect. Suppose not. Let x be an isolated point of P. Since $P - \{x\}$ is compact, $f(P - \{x\}) = Y - \{f(x)\}$ is compact. Consequently, $f(x)$ is an isolated point of Y. This implies that Y is not perfect. This is a contradiction. Hence P is perfect.

(iii) $\Rightarrow$ (iv). Let P be a given perfect subset of X. Let $\mathscr{C}$ be the collection of all clopen subsets of X. Let $\mathscr{I}' = \{C \in \mathscr{C}; P \cap C = \varnothing\}$. $\mathscr{I}'$ is clearly non-empty and is an ideal in $\mathscr{C}$. We claim that the quotient Boolean algebra $\mathscr{C}/\mathscr{I}'$ is nonatomic. Let $[A]$ be a non-zero element in $\mathscr{C}/\mathscr{I}'$. Then $A \cap P \neq \varnothing$. Further, $A \cap P$ contains at least two points $x, y \in X$, because P is perfect. Since X is compact and totally disconnected, there exists C, D in $\mathscr{C}$ such that $x \in C \subset A$, $y \in D \subset A$ and $C \cap D = \varnothing$. Note that $0 < [C] < [A]$ and $0 < [D] < [A]$. Consequently, $\mathscr{C}/\mathscr{I}'$ has no atoms. Since $\mathscr{F}$ and $\mathscr{C}$ are isomorphic, (iv) follows.

(iv) $\Rightarrow$ (i). Since $\mathscr{F}/\mathscr{I}$ is nonatomic, $\mathscr{F}/\mathscr{I}$ is not superatomic. By Theorem 5.3.6, there is a non-zero nonatomic positive bounded charge μ on $\mathscr{F}/\mathscr{I}$. Using the quotient homomorphism from $\mathscr{F}$ to $\mathscr{F}/\mathscr{I}$, the charge μ can be lifted as a charge $\tilde{\mu}$ on $\mathscr{F}$. $\tilde{\mu}$ is obviously nonatomic.

This completes the proof. □

Finally, we close this section with a result on the existence of nonatomic measures on the Borel σ-field of a compact Hausdorff totally disconnected space.

5.3.10 Corollary. *Let X be a compact Hausdorff totally disconnected space and $\mathscr{B}$ its Borel σ-field. Then there exists a non-zero nonatomic positive bounded regular measure on $\mathscr{B}$ if and only if X contains a perfect set.*

Proof. Suppose μ is a non-zero nonatomic positive bounded regular measure on $\mathscr{B}$. Then μ is a nonatomic measure on the Baire σ-field $\mathscr{B}_0$ of

X. See Lemma 5.3.8. Consequently, μ is strongly continuous on $\mathscr{B}_0$. See Theorem 5.1.6. Note that the field $\mathscr{C}$ of all clopen subsets of X is a generator of $\mathscr{B}_0$. See Section 1.3. By Proposition 5.3.7, μ is strongly continuous on $\mathscr{C}$. By Theorem 5.3.9, X contains a perfect set.

Conversely, let X contain a perfect set. Let $\mathscr{C}$ be the field of all clopen subsets of X and $\mathscr{B}_0$ its Baire σ-field. By Theorem 5.3.9, there exists a non-zero strongly continuous positive bounded charge μ on $\mathscr{C}$. Every charge on $\mathscr{C}$ is a measure. See Example 2.3.5(3). Since $\mathscr{B}_0$ is generated by $\mathscr{C}$, there is a measure $\tilde{\mu}$ on $\mathscr{B}_0$ which is an extension of μ. See Section 1.3 and Theorem 3.5.2. By Proposition 5.3.7, $\tilde{\mu}$ is a strongly continuous positive measure on $\mathscr{B}_0$. Let λ be the regular measure on $\mathscr{B}$ which is an extension of $\tilde{\mu}$. See Theorem 3.5.5. By Lemma 5.3.8, λ is nonatomic. This completes the proof. □

5.4 DENSENESS

Let $\mathscr{F}$ be a field of subsets of a set Ω. Let $\mathscr{M}$ be the collection of all probability charges on $\mathscr{F}$. $\mathscr{M}$ is equipped with a topology as follows. A net μ_α, $\alpha \in D$ in $\mathscr{M}$ is said to converge to a μ in $\mathscr{M}$ if $\lim_{\alpha \in D} \mu_\alpha(F) = \mu(F)$ for every F in $\mathscr{F}$. Let $\mathscr{M}_1$ be the collection of all nonatomic probability charges on $\mathscr{F}$ and $\mathscr{M}_2$ the collection of all strongly continuous probability charges on $\mathscr{F}$. Obviously, $\mathscr{M}_2 \subset \mathscr{M}_1 \subset \mathscr{M}$. In this section, we examine the conditions under which $\mathscr{M}_2$ is dense in $\mathscr{M}$. We need a preliminary result.

5.4.1 Proposition. *Let $\mathscr{F}$ be a field of subsets of a set Ω. Let X be the Stone space of $\mathscr{F}$. If X is perfect, then for any given non-empty set F in $\mathscr{F}$, there exists a strongly continuous probability charge μ on $\mathscr{F}$ such that $\mu(F) = 1$.*

Proof. The perfectness of X is equivalent to the fact that $\mathscr{F}$ is nonatomic. In the field $F \cap \mathscr{F}$ on F, one can easily construct a tree. By Theorem 5.3.2, there exists a strongly continuous probability charge λ on $F \cap \mathscr{F}$. Define μ on $\mathscr{F}$ by $\mu(E) = \lambda(E \cap F)$, $E \in \mathscr{F}$. μ is the desired charge. □

The following is the main result of this section.

5.4.2 Theorem. *Let $\mathscr{F}$ be a field of subsets of a set Ω and X its Stone space. Then $\mathscr{M}_2$ is dense in $\mathscr{M}$ if and only if X is perfect.*

Proof. "If" part. Let $\mu \in \mathscr{M}$. We construct a net of strongly continuous probability charges converging to μ. By Theorem 5.2.7, we can write

$$\mu = \mu_0 + \sum_{i \geq 1} a_i \mu_i,$$

where (i) μ_0 is a strongly continuous positive charge on $\mathscr{F}$, (ii) $a_i \geq 0$ for every i and $\mu_0(\Omega) + \sum_{i \geq 1} a_i = 1$, and (iii) μ_i is a 0–1 valued charge on $\mathscr{F}$ for every $i \geq 1$. Let $\mathscr{F}_i = \{F \in \mathscr{F}; \mu_i(F) = 1\}$, $i \geq 1$. For every F in $\mathscr{F}_i$, choose and fix a strongly continuous probability charge $\mu_{i,F}$ on $\mathscr{F}$ such that $\mu_{i,F}(F) = 1$. This can be done in view of Proposition 5.4.1. Consider the product set $\mathscr{F}_1 \times \mathscr{F}_2 \times \cdots$ with the following partial order. For $(F_1, F_2, \ldots)$ and $(E_1, E_2, \ldots)$ in $\mathscr{F}_1 \times \mathscr{F}_2 \times \cdots$, say $(F_1, F_2, \ldots) \geq (E_1, E_2, \ldots)$ if $F_i \subset E_i$ for every i. Under this partial order $\geq$, $\mathscr{F}_1 \times \mathscr{F}_2 \times \cdots$ is a directed set. For every $(F_1, F_2, \ldots)$ in $\mathscr{F}_1 \times \mathscr{F}_2 \times \cdots$, let $\mu_{(F_1, F_2, \ldots)} = \mu_0 + \sum_{i \geq 1} a_i \mu_{i,F_i}$. It is easy to check that $\mu_{(F_1, F_2, \ldots)} \in \mathscr{M}_2$. We claim that the net $\mu_{(F_1, F_2, \ldots)}$, $(F_1, F_2, \ldots) \in \mathscr{F}_1 \times \mathscr{F}_2 \times \cdots$ converges to μ in the topology of $\mathscr{M}$. Let E in $\mathscr{F}$ be fixed. Let $N_1 = \{i \geq 1;\ E \in \mathscr{F}_i\} = \{i_1, i_2, \ldots\}$ and $\{1, 2, 3, \ldots\} - \{i_1, i_2, \ldots\} = \{j_1, j_2, \ldots\}$. Since $\mathscr{F}_i$ is a maximal filter, $E^c \in \mathscr{F}_{j_k}$ for every $k \geq 1$. Define E_n^* for each $n \geq 1$ by

$$E_n^* = E, \quad \text{if } n = i_1, i_2, \ldots,$$

$$= E^c, \quad \text{if } n = j_1, j_2, \ldots.$$

Let $(F_1, F_2, \ldots)$ in $\mathscr{F}_1 \times \mathscr{F}_2 \times \cdots$ be such that $(F_1, F_2, \ldots) \geq (E_1^*, E_2^*, \ldots)$. This implies that $F_i \subset E$ for $i = i_1, i_2, \ldots$ and $F_i \subset E^c$ for $i = j_1, j_2, \ldots$. Consequently, $\mu_{(F_1, F_2, \ldots)}(E) = \mu_0(E) + \sum_{k \geq 1} a_{i_k} = \mu(E)$. Hence the net $\mu_{(F_1, F_2, \ldots)}(E)$, $(F_1, F_2, \ldots) \in \mathscr{F}_1 \times \mathscr{F}_2 \times \cdots$ converges to $\mu(E)$. This completes the proof of the "if" part of the theorem.

"Only if" part. Since the perfectness of X is equivalent to the fact that $\mathscr{F}$ is nonatomic, it suffices to show that $\mathscr{F}$ has no atoms. Suppose F in $\mathscr{F}$ is an atom of $\mathscr{F}$. Take the probability charge μ on $\mathscr{F}$ such that $\mu(F) = 1$. By the hypothesis, there exists a net μ_α, $\alpha \in D$ in $\mathscr{M}_2$ converging to μ in the topology of $\mathscr{M}$. In particular, $\lim_{\alpha \in D} \mu_\alpha(F) = \mu(F) = 1$. But $\mu_\alpha(F) = 0$ for every α in D. This contradiction proves the result. □

The following corollaries follow quite easily from the above theorem.

5.4.3 Corollary. *Let $\mathscr{F}$ be a field of subsets of a set Ω whose Stone space X is perfect. Then given any 0–1 valued charge μ on $\mathscr{F}$, there exists a net μ_α, $\alpha \in D$ of strongly continuous probability charges on $\mathscr{F}$ converging to μ in the topology of $\mathscr{M}$.*

5.4.4 Corollary. *Let $\mathscr{F}$ be a field of subsets of a set Ω. Let X be the Stone space of $\mathscr{F}$. Then the following statements are equivalent.*
  (i). *$\mathscr{F}$ is nonatomic.*
  (ii). *X is perfect.*
  (iii). *$\mathscr{M}_2$ is dense in $\mathscr{M}$.*
  (iv). *$\mathscr{M}_1$ is dense in $\mathscr{M}$.*

Finally, we state a result in topological measure theory. Let X be a compact Hausdorff space and $\mathcal{B}$ its Borel σ-field. Let $\mathcal{M}^*$ be the collection of all regular probability measures on $\mathcal{B}$. The weak* topology on $\mathcal{M}^*$ is described as follows. A net μ_α, $\alpha \in D$ in $\mathcal{M}^*$ converges to a μ in $\mathcal{M}^*$ if $\lim_{\alpha \in D} \int f \, d\mu_\alpha = \int f \, d\mu$ for every real valued continuous function f on X. If, in addition, X is totally disconnected, a net μ_α, $\alpha \in D$ in $\mathcal{M}^*$ converges to a μ in $\mathcal{M}^*$ in the weak* topology of $\mathcal{M}^*$ if and only if $\lim_{\alpha \in D} \mu_\alpha(C) = \mu(C)$ for every clopen set $C \subset X$. This follows from the fact that $\{f: X \to R;$ $f = \sum_{i=1}^n c_i I_{C_i}$ for some $c_1, c_2, \ldots, c_n$ real, $C_1, C_2, \ldots, C_n$ clopen sets $\subset X$ and $n \geq 1\}$ is norm dense in the space of all real continuous functions on X equipped with supremum norm. See Section 1.3. Let $\mathcal{M}_1^*$ be the collection of all regular nonatomic probability measures on $\mathcal{B}$.

5.4.5 Corollary. *Let X be a compact Hausdorff totally disconnected space and $\mathcal{M}^*$ and $\mathcal{M}_1^*$ be as defined above. Then $\mathcal{M}_1^*$ is dense in $\mathcal{M}^*$ in the weak* topology of $\mathcal{M}^*$ if and only if X is perfect.*

CHAPTER 6

Absolute Continuity

In this chapter, we formally introduce the notions of absolute continuity and singularity for charges. In Section 6.1, we study various properties of absolute continuity and singularity in the framework of charges and establish the connection with the existing notions of absolute continuity and singularity in Measure theory. In Section 6.2, we obtain Lebesgue Decomposition theorem for charges using Riesz Decomposition theorem in Vector lattices. Finally, in Section 6.3, we prove Radon–Nikodym theorem.

6.1 ABSOLUTE CONTINUITY AND SINGULARITY

The following notion of absolute continuity is the main one we study extensively in this section.

6.1.1 Definition. Let μ and ν be two charges defined on a field $\mathscr{F}$ of subsets of a set Ω. ν is said to be *absolutely continuous* with respect to μ if given $\varepsilon > 0$, there exists $\delta > 0$ such that $|\nu(E)| < \varepsilon$ whenever $E \in \mathscr{F}$ and $|\mu|(E) < \delta$.

If ν is absolutely continuous with respect to μ, we use the notation $\nu \ll \mu$.

6.1.2 Remarks. There are two other notions one could introduce related to absolute continuity.
(i). ν is said to be *weakly absolutely continuous* with respect to μ if $\nu(E) = 0$ whenever $E \in \mathscr{F}$ and $|\mu|(E) = 0$. In this case, we use the notation $\nu \text{w} \ll \mu$.
(ii). ν is said to be *strongly absolutely continuous* with respect to μ if, given $\varepsilon > 0$, there exists $\delta > 0$ such that $|\nu(E)| < \varepsilon$ whenever $E \in \mathscr{F}$ and $|\mu(E)| < \delta$. In this case, we use the notation $\nu \text{s} \ll \mu$.

In the following, we clarify the inter-relations between these three types of absolute continuity.

6.1.3. Remarks. Let μ and ν be two charges on a field $\mathscr{F}$ of subsets of a set Ω.
(i). $\nu \text{w} \ll \mu$ if and only if $\nu \text{w} \ll |\mu|$.
(ii). $\nu \ll \mu$ if and only if $\nu \ll |\mu|$.

(iii). $\nu \ll \mu$ implies $\nu w \ll \mu$. But the converse is not true. Let $\Omega = \{1, 2, 3, \ldots\}$, $\mathscr{F} = \mathscr{P}(\Omega)$, the class of all subsets of Ω and μ on $\mathscr{F}$ be defined by

$$\mu(A) = \sum_{n \in A} \frac{1}{2^n}$$

for A in $\mathscr{F}$. Let ν be any 0-1 valued charge on $\mathscr{F}$ such that $\nu(F) = 0$ for any finite subset F of Ω. It is obvious that $\nu w \ll \mu$. On the other hand, ν is not absolutely continuous with respect to μ. For $\varepsilon = \frac{1}{2}$, there is no $\delta > 0$ such that $\nu(F) < \varepsilon$ whenever $F \in \mathscr{F}$ and $\mu(F) < \delta$. For any given $\delta > 0$, one can always find a cofinite set F such that $\mu(F) < \delta$.

(iv). If μ is a positive charge, then $\nu \ll \mu$ if and only if $\nu s \ll \mu$.

(v). If $\nu s \ll \mu$, then $\nu \ll \mu$. But the converse is not true. Let $\Omega = \{1, 2\}$, $\mathscr{F} = \mathscr{P}(\Omega)$, $\mu(\{1\}) = \frac{1}{2}$, $\mu(\{2\}) = -\frac{1}{2}$, $\nu(\{1\}) = \frac{1}{2}$ and $\nu(\{2\}) = \frac{1}{2}$. Note that $|\mu| = \nu$. Obviously, $\nu \ll \mu$. But ν is not strongly absolutely continuous with respect to μ. For, $\mu(\Omega) = 0$ and $\nu(\Omega) = 1$.

The following proposition gives some properties of absolute continuity in addition to the ones pointed out in Remarks 6.1.3.

6.1.4 Proposition. *Let μ and ν be two charges defined on a field $\mathscr{F}$ of subsets of a set Ω. Then the following statements are equivalent.*

(i). $\nu \ll \mu$.
(ii). $\nu^+ \ll \mu$ and $\nu^- \ll \mu$.
(iii). $|\nu| \ll |\mu|$.

Further, if $\nu \ll \mu^+$ or $\nu \ll \mu^-$, then $\nu \ll \mu$.

Proof. (i) $\Rightarrow$ (ii). We show that $\nu^+ \ll \mu$. Let $\varepsilon > 0$. Since $\nu \ll \mu$, there exists $\delta > 0$ such that $|\nu(F)| < \varepsilon/2$ whenever $F \in \mathscr{F}$ and $|\mu|(F) < \delta$. So, if $F \in \mathscr{F}$ and $|\mu|(F) < \delta$, then $\nu^+(F) = \text{Sup }\{\nu(B); B \subset F, B \in \mathscr{F}\} \leq \varepsilon/2 < \varepsilon$. Hence $\nu^+ \ll \mu$. Similarly, one can show that $\nu^- \ll \mu$.

(ii) $\Rightarrow$ (iii). If (ii) holds, clearly $\nu^+ \ll |\mu|$ and $\nu^- \ll |\mu|$. Hence $|\nu| = \nu^+ + \nu^- \ll |\mu|$.

(iii) $\Rightarrow$ (i). This is obvious.

For the last part, observe that if $\nu \ll \mu^+$, then $\nu \ll \mu^+ + \lambda$ for any positive charge λ on $\mathscr{F}$. In particular, $\nu \ll \mu^+ + \mu^- = |\mu|$. □

Now, some comments are in order on the above proposition.

6.1.5 Remarks.

(i). If $\nu \ll \mu$, neither $\nu \ll \mu^+$ nor $\nu \ll \mu^-$ need hold. As an example, let $\Omega = \{1, 2\}$, $\mathscr{F} = \mathscr{P}(\Omega)$, $\mu(\{1\}) = \frac{1}{2}$, $\mu(\{2\}) = -\frac{1}{2}$, $\nu(\{1\}) = \frac{1}{2}$ and $\nu(\{2\}) = \frac{1}{2}$. Note that $\mu^+(\{1\}) = \frac{1}{2}$, $\mu^+(\{2\}) = 0$, $\mu^-(\{1\}) = 0$, $\mu^-(\{2\}) = \frac{1}{2}$ and $|\mu| = \nu$. Therefore, $\nu \ll \mu$. But neither $\nu \ll \mu^+$ nor $\nu \ll \mu^-$ is valid.

(ii). Proposition 6.1.4 still remains valid if absolute continuity is replaced by weak absolute continuity in the statement.
(iii). Proposition 6.1.4 is not valid if absolute continuity is replaced by strong absolute continuity. Some simple examples based on those in (i) can be provided.

As has been pointed out earlier, absolute continuity and weak absolute continuity are not equivalent in general. However, for bounded measures on σ-fields, these two notions are equivalent.

6.1.6 Theorem. *Let $\mathscr{F}$ be a σ-field of subsets of a set Ω and μ and ν measures on $\mathscr{F}$ such that ν is bounded. Then $\nu\text{w} \ll \mu$ if and only if $\nu \ll \mu$.*

Proof. "If" part is clear even without the assumption of boundedness of ν. We shall prove the "only if" part. Let us assume $\nu\text{w} \ll \mu$. Then $|\nu|\text{w} \ll |\mu|$. See Remark 6.1.5(ii). Suppose $|\nu|$ is not absolutely continuous with respect to $|\mu|$. There exists $\varepsilon > 0$ such that for every $n \geq 1$, there is a set F_n in $\mathscr{F}$ such that $|\mu|(F_n) < 1/2^n$ but $|\nu|(F_n) \geq \varepsilon$. Then the set $F = \limsup_{n \to \infty} F_n$ has the property that $|\mu|(F) = 0$ and $|\nu|(F) \geq \varepsilon$. To prove this, we proceed as follows.

$$|\mu|(\limsup_{n \to \infty} F_n) = |\mu|\left(\bigcap_{n \geq 1} \bigcup_{k \geq n} F_k\right)$$

$$\leq \lim_{n \to \infty} |\mu|\left(\bigcup_{k \geq n} F_k\right)$$

$$\leq \lim_{n \to \infty} \sum_{k \geq n} |\mu|(F_k)$$

$$\leq \lim_{n \to \infty} \sum_{k \geq n} \frac{1}{2^n} = 0.$$

Hence $|\mu|(\limsup_{n \to \infty} F_n) = 0$. On the other hand, since $|\nu|$ is bounded,

$$|\nu|(F) = |\nu|\left(\bigcap_{n \geq 1} \bigcup_{k \geq n} F_k\right) = \lim_{n \to \infty} |\nu|\left(\bigcup_{k \geq n} F_k\right)$$

$$\geq \limsup_{n \to \infty} |\nu|(F_n) \geq \varepsilon.$$

Thus we have $|\mu|(F) = 0$ and $|\nu|(F) \geq \varepsilon$. But this is a contradiction to the assumption that $\nu\text{w} \ll \mu$. This completes the proof. □

6.1.7 Remark. In the above theorem, neither the assumption of boundedness of ν nor the assumption that the measures are defined on a σ-field can be dropped for its validity.
(i). We treat the boundedness part of the above theorem. Let $\Omega = [0, 1]$, $\mathscr{F}$ = Borel σ-field on Ω and μ the Lebesgue measure on $\mathscr{F}$. Define ν on

$\mathscr{F}$ by

$$\nu(A) = 0, \quad \text{if } A \in \mathscr{F} \text{ and } \mu(A) = 0,$$
$$= \infty, \quad \text{if } A \in \mathscr{F} \text{ and } \mu(A) > 0.$$

It is easy to check that ν is a measure on $\mathscr{F}$ and that $\nu w \ll \mu$. But ν is not absolutely continuous with respect to μ.

(ii). Now, we show that the assumption that $\mathscr{F}$ is a σ-field cannot be dropped. Let $\Omega = \{1, 2, 3, \ldots, \infty\}$ and $\mathscr{F} = \{A;\ A \text{ or } A^c \text{ is a finite subset of } \{1, 2, 3, \ldots\}\}$. Let μ and ν be defined on the field $\mathscr{F}$ as follows.

$$\mu(A) = \sum_{n \in A} \frac{1}{2^n}, \quad \text{if } A \in \mathscr{F} \text{ and } A \text{ is finite,}$$

$$= 1 - \sum_{n \in A^c} \frac{1}{2^n}, \quad \text{if } A \in \mathscr{F} \text{ and } A^c \text{ is finite.}$$

$$\nu(A) = \sum_{n \in A} \frac{1}{2^n}, \quad \text{if } A \in \mathscr{F} \text{ and } A \text{ is finite,}$$

$$= 2 - \sum_{n \in A^c} \frac{1}{2^n}, \quad \text{if } A \in \mathscr{F} \text{ and } A^c \text{ is finite.}$$

Note that μ and ν are charges on $\mathscr{F}$. Every charge on $\mathscr{F}$ is a measure. Obviously, $\nu w \ll \mu$. But ν is not absolutely continuous with respect to μ. If for a given $0 < \varepsilon < 1$, there exists $\delta > 0$ such that $\nu(E) < \varepsilon$ whenever $E \in \mathscr{F}$ and $\mu(E) < \delta$, then we can find a set A in $\mathscr{F}$ such that A^c is finite and $\mu(A) < \delta$. For any such set A, $\nu(A) > 1$. This contradiction shows that ν is not absolutely continuous with respect to μ.

Now, we examine the notion of absolute continuity for measures on fields vis-a-vis with that on σ-fields. The following theorem ties up the notion of absolute continuity for charges and the known notion of absolute continuity for measures on σ-fields.

6.1.8 Theorem. *Let $\mathscr{F}$ be a field of subsets of a set Ω and $\mathfrak{A}$ the σ-field on Ω generated by $\mathscr{F}$. Let μ and ν be two positive bounded measures on $\mathfrak{A}$. Then $\nu \ll \mu$ on $\mathscr{F}$ if and only if $\nu \ll \mu$ on $\mathfrak{A}$.*

Proof. If $\nu \ll \mu$ on $\mathfrak{A}$, then it is obvious that $\nu \ll \mu$ on $\mathscr{F}$. Now, we assume that $\nu \ll \mu$ on $\mathscr{F}$. Let $\varepsilon > 0$. There exists $\delta > 0$ such that $\nu(E) < \varepsilon/2$ whenever $E \in \mathscr{F}$ and $\mu(E)/\delta$. Now, let $F \in \mathfrak{A}$ and $\mu(F) < \delta/2$. We show that $\nu(F) < \varepsilon$. By Theorem 3.5.3, there exists a set E in $\mathscr{F}$ such that $\mu(F \Delta E) + \nu(F \Delta E) < \min\{\varepsilon/2, \delta/2\}$. Then $|\mu(F) - \mu(E)| \leq \mu(F \Delta E) < \delta/2$. Consequently, $\mu(E) < \mu(F) + \delta/2 < \delta/2 + \delta/2 = \delta$. Therefore, $\nu(E) < \varepsilon/2$. On the other hand, $\nu(F \Delta E) < \varepsilon/2$ from which it follows that $|\nu(F) - \nu(E)| < \varepsilon/2$. So, $\nu(F) < \nu(E) + \varepsilon/2 < \varepsilon/2 + \varepsilon/2 = \varepsilon$. This completes the proof of the theorem. □

6.1.9 Remark. The above theorem is valid for any two bounded measures μ and ν.

We now study s-boundedness of charges in the light of absolute continuity. See Definition 2.1.4.

6.1.10 Theorem. *Let μ and ν be two charges on a field $\mathscr{F}$ of subsets of a set Ω. Then the following statements are true.*
  (i). *ν is s-bounded if and only if $|\nu|$ is s-bounded.*
  (ii). *If $\nu \ll \mu$ and μ is s-bounded, then ν is s-bounded.*
  (iii). *If $\nu \ll \mu$, μ is s-bounded and ν is a real charge, then ν is bounded.*
  (iv). *If $\nu \ll \mu$, μ is bounded and ν is a real charge, then ν is bounded.*

Proof. (i). Suppose ν is s-bounded and $|\nu|$ is not s-bounded. Then there exist a sequence A_n, $n \geq 1$ of pairwise disjoint sets in $\mathscr{F}$ and $\varepsilon > 0$ such that $|\nu|(A_n) > \varepsilon$ for every $n \geq 1$. Write $|\nu|(A_n) = \nu^+(A_n) + \nu^-(A_n)$ for every $n \geq 1$. Then there exists a sequence $n_1 < n_2 < \cdots$ such that $\nu^+(A_{n_i}) > \varepsilon/2$ for every $i \geq 1$ or there exists a sequence $k_1 < k_2 < \cdots$ such that $\nu^-(A_{k_i}) > \varepsilon/2$ for every $i \geq 1$. Assume that the former holds. Since $\nu^+(A_{n_i}) > \varepsilon/2$, there exists B_i in $\mathscr{F}$ such that $B_i \subset A_{n_i}$ and $\nu(B_i) > \varepsilon/2$ for every $i \geq 1$. B_i, $i \geq 1$ is a sequence of pairwise disjoint sets in $\mathscr{F}$ and $\lim_{i \to \infty} \nu(B_i) \neq 0$. This contradiction proves that $|\nu|$ is s-bounded. The converse is trivial.
(ii). If $\nu \ll \mu$, then $\nu \ll |\mu|$. By (i), $|\mu|$ is s-bounded. Therefore, ν is s-bounded.
(iii). Every s-bounded real charge is bounded. See Corollary 2.1.7.
(iv). This follows from (iii) and the fact that every bounded charge is s-bounded. □

Now, we study the countable additivity property of charges in the presence of absolute continuity.

6.1.11 Theorem. *Let μ and ν be two charges defined on a field $\mathscr{F}$ of subsets of a set Ω. If $\nu \ll \mu$, ν is a real charge and μ a bounded measure, then ν is a measure.*

Proof. Let A_n, $n \geq 1$ be a decreasing sequence of sets in $\mathscr{F}$ with $\bigcap_{n \geq 1} A_n = \varnothing$. It suffices to show that $\lim_{n \to \infty} \nu(A_n) = 0$. See Proposition 2.3.2(2). Since μ is a bounded measure, $\lim_{n \to \infty} |\mu|(A_n) = |\mu|(\varnothing) = 0$. Since $\nu \ll \mu$, $\lim_{n \to \infty} \nu(A_n) = 0$. This completes the proof. □

6.1.12 Remark. Neither the assumption that ν is real valued nor the assumption that μ is bounded can be relaxed in the above theorem. Suitable examples can be provided in the framework of finite-cofinite field $\mathscr{F}$ on $\Omega = \{1, 2, 3, \ldots\}$.

Now, we consider a sequence μ_n, $n \geq 1$ of bounded charges on a field $\mathscr{F}$ of subsets of a set Ω. Let μ on $\mathscr{F}$ be defined by

$$\mu(F) = \sum_{n \geq 1} \frac{1}{2^n} \frac{|\mu_n|(F)}{1 + |\mu_n|(\Omega)}$$

for F in $\mathscr{F}$. It is obvious that μ is a bounded charge on $\mathscr{F}$. Further, $\mu_n \ll \mu$ for every $n \geq 1$. The following theorem indicates that μ is, in a sense, minimal with respect to the above property of absolute continuity.

6.1.13 Theorem. *Let μ_n, $n \geq 1$ be a sequence of bounded charges on a field $\mathscr{F}$ of subsets of a set Ω. Let μ be defined as above. Let ν be any charge on $\mathscr{F}$ such that $\mu_n \ll \nu$ for every $n \geq 1$. Then $\mu \ll \nu$.*

Proof. Let $\varepsilon > 0$. Choose $N \geq 1$ such that $1/2^{N+1} < \varepsilon/2$. Since $\mu_1, \mu_2, \ldots, \mu_N$ are absolutely continuous with respect to ν, there exists $\delta > 0$ such that $|\mu_i(B)| < (\varepsilon/2)(1 + |\mu_i|(\Omega))$ for $i = 1, 2, \ldots, N$ whenever $B \in \mathscr{F}$ and $|\nu|(B) < \delta$. Now, if $B \in \mathscr{F}$ and $|\nu|(B) < \delta$, then

$$\mu(B) = \sum_{n \geq 1} \frac{1}{2^n} \frac{|\mu_n|(B)}{1 + |\mu_n|(\Omega)}$$

$$= \sum_{n=1}^{N} \frac{1}{2^n} \frac{|\mu_n|(B)}{1 + |\mu_n|(\Omega)} + \sum_{n \geq N+1} \frac{1}{2^n} \frac{|\mu_n|(B)}{1 + |\mu_n|(\Omega)}$$

$$< \varepsilon/2 + \varepsilon/2 = \varepsilon.$$

Hence $\mu \ll \nu$. □

Next, we take up the study of singularity for charges.

6.1.14 Definition. Let μ and ν be two charges defined on a field $\mathscr{F}$ of subsets of a set Ω. μ and ν are said to be *singular* if for every $\varepsilon > 0$, there exists a set D in $\mathscr{F}$ such that $|\mu|(D) < \varepsilon$ and $|\nu|(D^c) < \varepsilon$.

If μ and ν are singular, we use the notation $\mu \perp \nu$.

If μ and ν are positive charges, then $\mu \perp \nu$ if and only if $\mu \wedge \nu = 0$. See Definition 2.5.1.

6.1.15 Remark. The following definition might seem natural as a notion of singularity. μ and ν are said to be *strongly singular*, if there exists a set D in $\mathscr{F}$ such that $|\mu|(D) = 0 = |\nu|(D^c)$. If this is the case, we use the notation $\mu s \perp \nu$.

Some comments are in order about these two concepts.

6.1.16 Remarks. (i). $\mu \perp \nu$ if and only if $|\mu| \perp |\nu|$.
(ii). $\mu \perp \nu$ if and only if $\nu \perp \mu$.
(iii). If $\mu s \perp \nu$, then $\mu \perp \nu$. But the converse is not true. Let $\Omega = \{1, 2, \ldots\}$

and $\mathscr{F} = \mathscr{P}(\Omega)$. Let μ on $\mathscr{F}$ be defined by

$$\mu(A) = \sum_{n \in A} \frac{1}{2^n}, \quad A \in \mathscr{F}.$$

Let ν be any 0–1 valued charge on $\mathscr{F}$ such that $\nu(A) = 0$ for every finite subset A of Ω. Then $\mu \perp \nu$. But μ and ν are not strongly singular.

Strong singularity and singularity coincide for measures on σ-fields.

6.1.17 Theorem. *Let $\mathscr{F}$ be a σ-field of subsets of a set Ω and μ and ν two measures on $\mathscr{F}$. Then $\mu \perp \nu$ if and only if $\mu s \perp \nu$.*

Proof. "If" part is clear.

"Only if" part. Let $\mu \perp \nu$. For each $n \geq 1$, there exists a set D_n in $\mathscr{F}$ such that $|\mu|(D_n) < 1/2^n$ and $|\nu|(D_n^c) < 1/2^n$. Let $D = \limsup_{n \to \infty} D_n$. Then $|\mu|(D) = 0$. The proof is exactly similar to the one given in the proof of Theorem 6.1.6. Let $E = \limsup_{n \to \infty} D_n^c$. By a similar argument, it can be shown that $|\nu|(E) = 0$. Note that $D^c = \liminf_{n \to \infty} D_n^c \subset \limsup_{n \to \infty} D_n^c = E$. So, $|\nu|(D^c) = 0$. Hence $\mu s \perp \nu$. □

As a consequence of the above theorem, we obtain Hahn Decomposition theorem for measures.

6.1.18 Corollary. (*Hahn Decomposition Theorem for Measures*). *Let μ be a measure on a σ-field $\mathfrak{A}$ of subsets of a set Ω. Then there exists a set D in $\mathfrak{A}$ such that*

$$\mu(A) \geq 0 \quad \text{whenever } A \in \mathfrak{A} \text{ and } A \subset D,$$

and

$$\mu(B) \leq 0 \quad \text{whenever } B \in \mathfrak{A} \text{ and } B \subset D^c.$$

Proof. By Lemma 2.5.5, μ is either bounded below or bounded above. So, by Remark 2.5.4(ii), we have

$$\mu = \mu^+ - \mu^- \quad \text{and} \quad \mu^+ \wedge \mu^- = 0.$$

Since $\mu^+ \wedge \mu^- = 0$, we have $\mu^+ \perp \mu^-$. By Theorem 6.1.17, there exists a set D in $\mathfrak{A}$ such that $\mu^+(D^c) = 0 = \mu^-(D)$. From the definition of μ^+, it follows that $\mu(B) \leq 0$ whenever $B \in \mathfrak{A}$ and $B \subset D^c$. By a similar argument, it follows that $\mu(A) \geq 0$ whenever $A \in \mathfrak{A}$ and $A \subset D$. This completes the proof. □

The following theorem is analogous to Theorem 6.1.8. This theorem also indicates that Definition 6.1.14 is a natural definition of singularity for charges in harmony with the corresponding notion used in Measure theory.

6.1.19 Theorem. *Let $\mathfrak{A}$ be a σ-field of subsets of a set Ω and μ and ν two bounded measures on $\mathfrak{A}$. Let $\mathcal{F}$ be a field on Ω generating $\mathfrak{A}$. Then $\mu \perp \nu$ on $\mathfrak{A}$ if and only if $\mu \perp \nu$ on $\mathcal{F}$.*

Proof. "If" part is obvious even without the assumption of boundedness of the charges.
"Only if" part. Let $\mu \perp \nu$ on $\mathfrak{A}$. By Theorem 6.1.17, there exists a set D in $\mathfrak{A}$ such that $|\mu|(D) = 0 = |\nu|(D^c)$. Let $\varepsilon > 0$. By Theorem 3.5.3, there exists a set A in $\mathcal{F}$ such that $|\mu|(A\Delta D) + |\nu|(A\Delta D) < \varepsilon$. Now, $||\mu|(A) - |\mu|(D)| \leq |\mu|(A\Delta D) < \varepsilon$. Hence $|\mu|(A) < \varepsilon$. Further, $||\nu|(A^c) - |\nu|(D^c)| \leq |\nu|(A^c \Delta D^c) = |\nu|(A\Delta D) < \varepsilon$. Hence $|\nu|(A^c) < \varepsilon$. This completes the proof. □

We end this section with some remarks.

6.1.20 Remarks. Let μ, ν and τ be charges on a field $\mathcal{F}$ of subsets of a set Ω.
(i). If $\nu \ll \mu$ and $\nu \perp \mu$, then $\nu = 0$. This can be proved as follows. Let $\varepsilon > 0$. There exists $\delta > 0$ such that $|\nu|(A) < \varepsilon$ whenever $A \in \mathcal{F}$ and $|\mu|(A) < \delta$. One can take $\delta < \varepsilon$. Since $\nu \perp \mu$, there exists a set D in $\mathcal{F}$ such that $|\nu|(D) < \delta$ and $|\mu|(D^c) < \delta$. Therefore, $|\nu|(\Omega) = |\nu|(D) + |\nu|(D^c) < \delta + \varepsilon < 2\varepsilon$. Since $\varepsilon > 0$ is arbitrary, $|\nu|(\Omega) = 0$.
(ii). If $\mu \ll \tau$ and $\nu \perp \tau$, then $\mu \perp \nu$. An argument similar to the one above can be presented here.
(iii). Some additional properties of absolute continuity and singularity will be given in Section 8.5.

6.2 LEBESGUE DECOMPOSITION THEOREM

In this section, we embark on proving Lebesgue Decomposition Theorem for charges using Riesz Decomposition Theorem for Vector lattices. Recall that the space $ba(\Omega, \mathcal{F})$ of all bounded charges on a field $\mathcal{F}$ of subsets of a set Ω is a boundedly complete Vector lattice.

First, let us examine how far the notion of singularity introduced for charges in Section 6.1 is related to the notion of orthogonality introduced in Vector lattices.

6.2.1 Theorem. *Let $\mu, \nu \in ba(\Omega, \mathcal{F})$. Then μ and ν are singular if and only if μ and ν are orthogonal in the vector lattice $ba(\Omega, \mathcal{F})$. (Consequently, the notation used for singularity of two charges is consistent with the notation used for orthogonality of two elements in the Vector lattice $ba(\Omega, \mathcal{F})$.)*

Proof. Suppose μ and ν are singular. Let $\varepsilon > 0$. There exists a set D in $\mathcal{F}$ such that $|\mu|(D) < \varepsilon$ and $|\nu|(D^c) < \varepsilon$. Consequently, $(|\mu| \wedge |\nu|)(\Omega) =$

Inf $\{|\mu|(B)+|\nu|(B^c); B \in \mathscr{F}\} < 2\varepsilon$. Since $\varepsilon > 0$ is arbitrary, it follows that $|\mu| \wedge |\nu| = 0$. This implies that μ and ν are orthogonal in the Vector lattice ba$(\Omega, \mathscr{F})$. The converse can be proved similarly. □

Now, we characterize absolute continuity in terms of notions of Vector lattices. For the following, one has to recall the definition of the orthogonal complement of S, $S^\perp$, for any subset S of a Vector lattice. See Definition 1.5.7.

6.2.2 Theorem. *Let $\mu, \nu \in$ ba$(\Omega, \mathscr{F})$. Then $\nu \ll \mu$ if and only if $\nu \in \{\{\mu\}^\perp\}^\perp$.*

Proof. From the definitions

$$\{\mu\}^\perp = \{\tau \in \text{ba}(\Omega, \mathscr{F}); \tau \perp \mu\}$$

and

$$\{\{\mu\}^\perp\}^\perp = \{\lambda \in \text{ba}(\Omega, \mathscr{F}); \lambda \perp \tau \text{ for every } \tau \text{ in } \{\mu\}^\perp\},$$

it follows that $\nu \in \{\{\mu\}^\perp\}^\perp$ if and only if $\nu \perp \tau$ whenever $\tau \perp \mu$.

Suppose $\nu \ll \mu$. Let $\tau \in$ ba$(\Omega, \mathscr{F})$ and $\tau \perp \mu$. We show that $\nu \perp \tau$. But this follows from Remark 6.1.20(ii).

Conversely, let $\nu \in \{\{\mu\}^\perp\}^\perp$. Since $\{\mu\}^\perp = \{|\mu|\}^\perp$, we assume, without loss of generality, that ν and μ are positive. By Theorem 1.5.12,

$$\nu = \bigvee_{n \geq 1} (\nu \wedge n\mu).$$

Since $\nu \wedge n\mu \leq \nu \wedge (n+1)\mu$ for all $n \geq 1$, the set function λ on $\mathscr{F}$ defined by $\lambda(F) = \lim_{n \to \infty} (\nu \wedge n\mu)(F)$, F in $\mathscr{F}$ is a charge on $\mathscr{F}$. Consequently, the lattice supremum ν is precisely λ, i.e. $\nu(F) = \lim_{n \to \infty} (\nu \wedge n\mu)(F)$, $F \in \mathscr{F}$. Since $\nu \wedge n\mu \leq n\mu$, $\nu \wedge n\mu \ll \mu$ for every $n \geq 1$. Now, we show that the sequence $\nu \wedge n\mu$, $n \geq 1$ of charges converges uniformly to ν. For any F in $\mathscr{F}$, $|\nu - \nu \wedge n\mu|(F) \leq |\nu - \nu \wedge n\mu|(\Omega) = (\nu - \nu \wedge n\mu)(\Omega) = \nu(\Omega) - (\nu \wedge n\mu)(\Omega)$ which converges to zero as $n \to \infty$. Hence $\nu \wedge n\mu$, $n \geq 1$ converges to ν uniformly. Since $\nu \wedge n\mu \ll \mu$ for every $n \geq 1$, it follows that $\nu \ll \mu$. This completes the proof. □

The above two theorems provide considerable insight into the concepts of absolute continuity and singularity, and some of the highlights are recorded in the following corollary.

6.2.3 Corollary. (i). *If $\mu_1, \mu_2, \nu \in$ ba$(\Omega, \mathscr{F})$, $\mu_1 \perp \nu$ and $\mu_2 \perp \nu$, then $(\mu_1 + \mu_2) \perp \nu$.*
(ii). *If $\mu, \nu \in$ ba$(\Omega, \mathscr{F})$, then $\mu \perp \nu$ if and only if $\mu^+ \perp \nu$ and $\mu^- \perp \nu$.*
(iii). *If $\mu, \mu_n, n \geq 1$ is a sequence in ba$(\Omega, \mathscr{F})$, $\lim_{n \to \infty} \mu_n = \mu$ in the norm of ba$(\Omega, \mathscr{F})$ and $\mu_n \perp \nu$ for every $n \geq 1$, then $\mu \perp \nu$.*

(iv). If $\mu_1, \mu_2, \mu \in \text{ba}(\Omega, \mathcal{F})$, $\mu_1 \ll \mu$ and $\mu_2 \ll \mu$, then $(\mu_1 + \mu_2) \ll \mu$.
(v). If μ_n, $n \geq 1$ is a sequence in $\text{ba}(\Omega, \mathcal{F})$ converging to a μ in $\text{ba}(\Omega, \mathcal{F})$ in the norm of $\text{ba}(\Omega, \mathcal{F})$ and $\mu_n \ll \lambda$ for every $n \geq 1$ for some λ in $\text{ba}(\Omega, \mathcal{F})$, then $\mu \ll \lambda$.

Proof. (i) and (ii) follow because $\{\nu\}^\perp$ is a vector sublattice of $\text{ba}(\Omega, \mathcal{F})$. (iii) follows from the fact that $\{\nu\}^\perp$ is a closed subspace of $\text{ba}(\Omega, \mathcal{F})$. See Theorem 1.5.19. (iv) and (v) follow from the fact that $\{\{\mu\}^\perp\}^\perp$ is a closed vector sublattice of $\text{ba}(\Omega, \mathcal{F})$. □

Now, we have developed enough machinery to prove Lebesgue Decomposition theorem.

6.2.4 Lebesgue Decomposition Theorem. *Let μ and $\nu \in \text{ba}(\Omega, \mathcal{F})$. Then there exist ν_1, ν_2 in $\text{ba}(\Omega, \mathcal{F})$ having the following properties.*
 (i). $\nu = \nu_1 + \nu_2$.
 (ii). $\nu_1 \ll \mu$.
 (iii). $\nu_2 \perp \mu$.
 (iv). $|\nu| = |\nu_1| + |\nu_2|$.
 (v). *If ν is positive, then ν_1 and ν_2 are positive.*
Further, any decomposition of ν satisfying (ii) and (iii) is unique.

Proof. By Theorem 1.5.8, $\{\mu\}^\perp$ and $\{\{\mu\}^\perp\}^\perp$ are normal vector sublattices of $\text{ba}(\Omega, \mathcal{F})$. By Riesz Decomposition theorem 1.5.10, for a given $\nu \in \text{ba}(\Omega, \mathcal{F})$, there exist ν_1 in $\{\{\mu\}^\perp\}^\perp$ and ν_2 in $\{\mu\}^\perp$ such that $\nu = \nu_1 + \nu_2$. By Theorem 6.2.2, $\nu_1 \ll \mu$. Obviously, $\nu_2 \perp \mu$. Uniqueness of the decomposition follows from Theorem 1.5.10. By Remark 6.1.20(ii), it follows that $\nu_1 \perp \nu_2$. By Theorem 1.5.4(21) and (22), $|\nu| = |\nu_1 + \nu_2| = |\nu_1| + |\nu_2|$. Thus (iv) follows. (v) follows from the uniqueness of the decomposition and (iv).
This completes the proof. □

The above theorem gives a decomposition for bounded charges only. It is natural to enquire about the validity of Lebesgue Decomposition theorem for charges not necessarily bounded. We present some negative results. (See Notes and Comments for some positive results.)

6.2.5 Theorem. *Let $\mathcal{F}$ be a field of subsets of a set Ω and ν an unbounded real charge on $\mathcal{F}$. Then there exists a bounded charge μ on $\mathcal{F}$ such that Lebesgue Decomposition theorem is not valid for ν with respect to μ.*

Proof. Let $\mathcal{I}' = \{F \in \mathcal{F}; |\nu|(F) < \infty\}$. $\mathcal{I}'$ is an ideal in $\mathcal{F}$. Let $\mathcal{I}$ be a maximal ideal in $\mathcal{F}$ containing $\mathcal{I}'$. Define μ on $\mathcal{F}$ by

$$\mu(A) = 0, \quad \text{if } A \in \mathcal{I},$$
$$= 1, \quad \text{if } A \notin \mathcal{I} \text{ and } A \in \mathcal{F}.$$

μ is a charge on $\mathscr{F}$. We claim that Lebesgue Decomposition theorem is not valid for ν with respect to μ. Suppose $\nu = \nu_1 + \nu_2$, where $\nu_1 \ll \mu$ and $\nu_2 \perp \mu$. Since ν_1 is real and μ is bounded, ν_1 is a bounded charge. See Theorem 6.1.10(iv). Since $\nu_2 \perp \mu$, there is a set E in $\mathscr{F}$ such that $|\nu_2|(E^c) < 1$ and $\mu(E) < 1$. Hence $\mu(E) = 0$. Consequently, $E \in \mathscr{I}$. Hence $E^c \notin \mathscr{I}'$. Therefore, $|\nu|(E^c) = \infty$. ν_1 is obviously bounded on E^c and ν_2 is also bounded on E^c. Therefore, ν is bounded on E^c which contradicts $|\nu|(E^c) = \infty$. Thus the claim is established. □

6.2.6 Theorem. *Let $\mathscr{F}$ be a field of subsets of a set Ω. Let ν be any real charge on $\mathscr{F}$ such that $|\nu|(A) = \infty$ for every non-empty set A in $\mathscr{F}$. Then Lebesgue Decomposition theorem for ν is not valid with respect to any non-zero bounded charge μ on $\mathscr{F}$.*

Proof. Suppose there is a non-zero bounded charge μ on $\mathscr{F}$ such that Lebesgue Decomposition theorem is valid for ν with respect to μ. Let $\nu = \nu_1 + \nu_2$, where $\nu_1 \ll \mu$ and $\nu_2 \perp \mu$. By Theorem 6.1.10, it follows that ν_1 is a bounded charge. As $\nu_2 \perp \mu$, there exists a set A in $\mathscr{F}$ such that $|\nu_2|(A) < |\mu|(\Omega)$ and $|\mu|(A^c) < |\mu|(\Omega)$. Then A is a non-empty set on which ν_2 as well as ν_1 is bounded. This is a contradiction. □

6.2.7 Remark. The charge given in Example 2.1.3(6) is a charge satisfying the condition imposed on ν of Theorem 6.2.6.

We end this section by deriving Lebesgue Decomposition theorem for measures on fields from Theorem 6.2.4.

6.2.8 Theorem. *Let μ and $\nu \in \text{ca}(\Omega, \mathscr{F})$. Then there exist ν_1 and ν_2 in $\text{ca}(\Omega, \mathscr{F})$ having the following properties.*
  (i). $\nu = \nu_1 + \nu_2$.
  (ii). $\nu_1 \ll \mu$.
  (iii). $\nu_2 \perp \mu$.
  (iv). $|\nu| = |\nu_1| + |\nu_2|$.
  (v). *If ν is positive, then ν_1 and ν_2 are positive.*
Further, any decomposition of ν satisfying (ii) and (iii) is unique.

Proof. By Theorem 6.2.4, we obtain ν_1 and ν_2 in $\text{ba}(\Omega, \mathscr{F})$ with the above properties. By Theorem 6.1.11, $\nu_1 \in \text{ca}(\Omega, \mathscr{F})$. Hence $\nu_2 \in \text{ca}(\Omega, \mathscr{F})$. □

6.3 RADON–NIKODYM THEOREM

The aim of this section is to prove Radon–Nikodym theorem in the framework of charges. The proof is mainly based on Hahn Decomposition thereom and involves only elementary calculations. In Chapter 7 also, we obtain Radon–Nikodym theorem for charges as a simple consequence of a result in V_1-spaces.

First, we establish some preliminary results.

6.3.1 Theorem. *Let $\mathscr{F}$ be a field of subsets of a set Ω and $\varepsilon > 0$. Suppose μ and ν are two bounded charges on $\mathscr{F}$ satisfying $\mu(F) \geq -\varepsilon/2$ and $\nu(F) \geq -\varepsilon/2$ for every F in $\mathscr{F}$. Then there exists a set A in $\mathscr{F}$ and a simple function f on Ω having the following properties.*

(i). $|\mu(E)| < \varepsilon$ whenever $E \in \mathscr{F}$ and $E \subset A$.

(ii). $|\nu(F) - D \int_F f \, d\mu| \leq \varepsilon$ whenever $F \in \mathscr{F}$ and $F \subset A^c$.

Proof. By Hahn Decomposition theorem (see Theorem 2.6.2), for any τ in ba$(\Omega, \mathscr{F})$ and $\delta > 0$, there exists a set $B(\tau, \delta)$ in $\mathscr{F}$ such that

$$\tau(E) \geq -\delta \quad \text{whenever } E \in \mathscr{F} \text{ and } E \subset B(\tau, \delta),$$

and

$$\tau(E) \leq \delta \quad \text{whenever } E \in \mathscr{F} \text{ and } E \subset (B(\tau, \delta))^c.$$

Let m be any positive integer

$$> 2 \left(\frac{|\mu|(\Omega) + \varepsilon}{\varepsilon} \right) \left(\frac{|\nu|(\Omega) + \varepsilon}{\varepsilon} \right).$$

For $k = 1, 2, \ldots, m$, define

$$B_k = B\left(\nu - \frac{k\varepsilon}{2(|\mu|(\Omega) + \varepsilon)} \mu, \frac{\varepsilon}{2m} \right).$$

Let $A = B_m$, $A_m = B_{m-1} - B_m$, $A_{m-1} = B_{m-2} - (B_{m-1} \cup B_m), \ldots, A_2 = B_1 - (B_2 \cup B_3 \cup \cdots \cup B_m)$ and $A_1 = \Omega - (B_1 \cup B_2 \cup \cdots \cup B_m)$. Note that A, $A_1, A_2, \ldots, A_m$ are pairwise disjoint sets in $\mathscr{F}$ with union $= \Omega$. Let

$$f = \sum_{i=1}^m \frac{(i-1)\varepsilon}{2(|\mu|(\Omega) + \varepsilon)} I_{A_i}.$$

We show that A and f are the desired ones. Let $E \in \mathscr{F}$ and $E \subset A = B_m$. Then

$$\nu(E) - \frac{m\varepsilon}{2(|\mu|(\Omega) + \varepsilon)} \mu(E) \geq -\frac{\varepsilon}{2m} \geq -\frac{\varepsilon}{2}.$$

So,

$$\mu(E) \leq \left(\nu(E) + \frac{\varepsilon}{2} \right) 2 \left(\frac{|\mu|(\Omega) + \varepsilon}{\varepsilon} \right) \frac{1}{m}$$

$$\leq 2 \left(\frac{|\nu|(\Omega) + \varepsilon}{\varepsilon} \right) \left(\frac{|\mu|(\Omega) + \varepsilon}{\varepsilon} \right) \frac{\varepsilon}{m} < \varepsilon.$$

6. ABSOLUTE CONTINUITY

Also, by the given hypothesis, $\mu(E) \geq -\varepsilon/2 > -\varepsilon$. Consequently, $|\mu(E)| < \varepsilon$. Thus (i) is proved.

Now, let $E \in \mathscr{F}$ and $E \subset A^c = A_1 \cup A_2 \cup \cdots \cup A_m$. So,

$$E = \bigcup_{i=1}^{m} (E \cap A_i) = \bigcup_{i=1}^{m} E_i,$$

where $E_i = E \cap A_i$ for $i = 1, 2, \ldots, m$. Note that for each $2 \leq i \leq m$, $E_i \subset A_i \subset B_{i-1}$ and so

$$\nu(E_i) - \frac{(i-1)\varepsilon}{2(|\mu|(\Omega) + \varepsilon)} \mu(E_i) \geq -\frac{\varepsilon}{2m}.$$

On the other hand, for each $1 \leq i \leq m$, $E_i \subset B_i^c$ and so

$$\nu(E_i) - \frac{i\varepsilon}{2(|\mu|(\Omega) + \varepsilon)} \mu(E_i) \leq \frac{\varepsilon}{2m}.$$

Consequently, from the above two inequalities, we obtain

$$\nu(E) - D\int_E f\, d\mu = \sum_{i=1}^{m} \nu(E_i) - \sum_{i=1}^{m} \frac{(i-1)\varepsilon}{2(|\mu|(\Omega) + \varepsilon)} \mu(E_i)$$

$$= \nu(E_1) + \sum_{i=2}^{m} \left[\nu(E_i) - \frac{(i-1)\varepsilon}{2(|\mu|(\Omega) + \varepsilon)} \mu(E_i)) \right]$$

$$\geq -\frac{\varepsilon}{2} - \frac{(m-1)\varepsilon}{2m} \geq -\varepsilon$$

and

$$\nu(E) - D\int_E f\, d\mu = \sum_{i=1}^{m} \nu(E_i) - \sum_{i=1}^{m} \frac{(i-1)\varepsilon}{2(|\mu|(\Omega) + \varepsilon)} \mu(E_i)$$

$$= \sum_{i=1}^{m} \frac{\varepsilon}{2(|\mu|(\Omega) + \varepsilon)} \mu(E_i)$$

$$+ \sum_{i=1}^{m} \left[\nu(E_i) - \frac{i\varepsilon}{2(|\mu|(\Omega) + \varepsilon)} \mu(E_i) \right]$$

$$\leq \frac{\varepsilon}{2(|\mu|(\Omega) + \varepsilon)} \mu(E) + \frac{m\varepsilon}{2m}$$

$$\leq \varepsilon/2 + \varepsilon/2 = \varepsilon.$$

Consequently, $|\nu(E) - D\int_E f\, d\mu| \leq \varepsilon$. This proves (ii). □

The following theorem is a generalization of the above theorem.

6.3.2 Theorem. *Let $\mathscr{F}$ be a field of subsets of a set Ω. Let μ and ν be two bounded charges on $\mathscr{F}$ and $\varepsilon > 0$. Then there exists a set A in $\mathscr{F}$ and a simple function f on Ω such that*
(i). $|\mu(E)| < \varepsilon$ *whenever* $E \in \mathscr{F}$ *and* $E \subset A$, *and*
(ii). $|\nu(E) - D \int_E f \, d\mu| < \varepsilon$ *whenever* $E \in \mathscr{F}$ *and* $E \subset A^c$
hold.

Proof. In the notation of the proof of the above theorem, let $B_1 = B(\mu, \varepsilon/16)$ and $B_2 = B(\nu, \varepsilon/16)$. Let μ_1, μ_2, ν_1 and ν_2 be defined on $\mathscr{F}$ by

$$\mu_1(F) = \mu(B_1 \cap F), \qquad F \in \mathscr{F},$$

$$\mu_2(F) = -\mu(B_1^c \cap F), \qquad F \in \mathscr{F},$$

$$\nu_1(F) = \nu(B_2 \cap F), \qquad F \in \mathscr{F},$$

and

$$\nu_2(F) = -\nu(B_2^c \cap F), \qquad F \in \mathscr{F}.$$

Note that $\mu = \mu_1 - \mu_2$ and $\nu = \nu_1 - \nu_2$. Further, $\mu_1(F) \geq -\varepsilon/16$, $\mu_2(F) \geq -\varepsilon/16$, $\nu_1(F) \geq -\varepsilon/16$ and $\nu_2(F) \geq -\varepsilon/16$ for every F in $\mathscr{F}$. By Theorem 6.3.1, there exist A_{ij} in $\mathscr{F}$ and simple functions f_{ij} on Ω such that

$$|\mu_i(E)| < \varepsilon/8 \quad \text{whenever } E \in \mathscr{F} \text{ and } E \subset A_{ij}$$

and

$$\left|\nu_j(E) - D \int_E f_{ij} \, d\mu_i\right| \leq \varepsilon/8 \quad \text{whenever } E \in \mathscr{F} \text{ and } E \subset A_{ij}^c$$

for $i, j = 1, 2$.

Let $A = A_{11} \cup A_{12} \cup A_{21} \cup A_{22}$ and $f = (f_{11} - f_{12})I_{B_1} + (f_{21} - f_{22})I_{B_1^c}$. We show that the A and f defined above are the desired ones.

Let $E \in \mathscr{F}$ and $E \subset A$. Then

$$E = E \cap A = (E \cap A_{11}) \cup (E \cap A_{12}) \cup (E \cap A_{21}) \cup (E \cap A_{22})$$

$$= E_{11} \cup E_{12} \cup E_{21} \cup E_{22}$$

where $E_{ij} = E \cap A_{ij}$, $i, j = 1, 2$. Disjointizing E_{ij}'s if necessary, we can assume, without loss of generality, that E_{ij}'s are pairwise disjoint. In the following, we use the property that $E_{ij} \subset A_{ij}$ for $i, j = 1, 2$. Now,

$$|\mu(E)| = \left|\sum_{i=1}^{2} \sum_{j=1}^{2} \mu(E_{ij})\right| = \left|\sum_{i=1}^{2} \sum_{j=1}^{2} \mu_1(E_{ij}) - \mu_2(E_{ij})\right|$$

$$\leq \sum_{i=1}^{2} \sum_{j=1}^{2} |\mu_1(E_{ij})| + \sum_{i=1}^{2} \sum_{j=1}^{2} |\mu_2(E_{ij})|$$

$$< 4(\varepsilon/8 + \varepsilon/8) = \varepsilon.$$

Thus (i) is established. Now, let $E \in \mathscr{F}$ and $E \subset A^c$. Then $E \subset A_{ij}^c$ for $i, j = 1, 2$. Observe that $\mu_2(E \cap B_1) = 0 = \mu_1(E \cap B_1^c)$. So,

$$\left|\nu(E) - D\int_E f \, d\mu\right| \leq \left|\nu(E \cap B_1) - D\int_{E \cap B_1} f \, d\mu\right|$$

$$+ \left|\nu(E \cap B_1^c) - D\int_{E \cap B_1^c} f \, d\mu\right|$$

$$\leq \left|\nu(E \cap B_1) - D\int_{E \cap B_1} (f_{11} - f_{12}) \, d\mu\right|$$

$$+ \left|\nu(E \cap B_1^c) - D\int_{E \cap B_1^c} (f_{21} - f_{22}) \, d\mu\right|$$

$$\leq \left|\nu_1(E \cap B_1) - D\int_{E \cap B_1} f_{11} \, d\mu_1\right|$$

$$+ \left|\nu_2(E \cap B_1) - D\int_{E \cap B_1} f_{12} \, d\mu_2\right|$$

$$+ \left|\nu_1(E \cap B_1^c) - D\int_{E \cap B_1^c} f_{21} \, d\mu_1\right|$$

$$+ \left|\nu_2(E \cap B_1^c) - D\int_{E \cap B_1^c} f_{22} \, d\mu_2\right|$$

$$\leq 4(\varepsilon/8) = \varepsilon/2 < \varepsilon.$$

This completes the proof. □

We need a lemma before proving the main theorem of this section.

6.3.3 Lemma. *Let $\mathscr{F}$ be a field of subsets of a set Ω, $\varepsilon > 0$ and $k > 0$. Let $\mu, \nu \in \mathrm{ba}(\Omega, \mathscr{F})$ be such that $\mu \geq 0$ and $-k\mu(F) - \varepsilon < \nu(F) < k\mu(F) + \varepsilon$ for every F in $\mathscr{F}$. Then the following statements are true.*

(i). *For every $\varepsilon' > \varepsilon$, there exists a two-valued simple function f on Ω such that*

$$-\frac{k}{2}\mu(F) - \varepsilon' < \nu(F) - \lambda(F) < \frac{k}{2}\mu(F) + \varepsilon'$$

for every F in $\mathscr{F}$, where $\lambda(F) = D\int_F f \, d\mu$, $F \in \mathscr{F}$.

(ii). *For every $\varepsilon' > \varepsilon$, there exists a simple function g on Ω such that*

$$-\varepsilon' < \nu(F) - \tau(F) < \varepsilon'$$

for every F in $\mathscr{F}$, where $\tau(F) = D\int_F g \, d\mu$, $F \in \mathscr{F}$.

Proof. (i). By Hahn Decomposition theorem, there exists a set A in $\mathscr{F}$ such that

$$\nu(E) \geq -(\varepsilon' - \varepsilon) \quad \text{whenever } E \subset A \text{ and } E \in \mathscr{F},$$

and

$$\nu(E) \leq (\varepsilon' - \varepsilon) \quad \text{whenever } E \subset A^c \text{ and } E \in \mathscr{F}.$$

Let $f = (k/2)I_A - (k/2)I_{A^c}$. Clearly, f is two-valued. Further, for any F in $\mathscr{F}$,

$$\nu(F \cap A) - D \int_{F \cap A} f \, d\mu = \nu(F \cap A) - \frac{k}{2}\mu(F \cap A)$$

$$< \frac{k}{2}\mu(F \cap A) + \varepsilon,$$

and

$$\nu(F \cap A^c) - D \int_{F \cap A^c} f \, d\mu = \nu(F \cap A^c) + \frac{k}{2}\mu(F \cap A^c)$$

$$\leq (\varepsilon' - \varepsilon) + \frac{k}{2}\mu(F \cap A^c).$$

Consequently, by adding the above two inequalities, we obtain

$$\nu(F) - D \int_F f \, d\mu < \frac{k}{2}\mu(F) + \varepsilon'.$$

One can establish, by a similar argument, the other inequality

$$-\frac{k}{2}\mu(F) - \varepsilon' < \nu(F) - D \int_F f \, d\mu.$$

This proves (i).

(ii). This can be proved by induction. Given $\varepsilon^* > \varepsilon$ and $n \geq 1$, there exists a simple function f_n on Ω such that

$$-\frac{k}{2^n}\mu(F) - \varepsilon^* < \nu(F) - D \int_F f_n \, d\mu < \frac{k}{2^n}\mu(F) + \varepsilon^*$$

for every F in $\mathscr{F}$. The avove assertion is true when $n = 1$, by (i). Using (i) again for the charge $\nu(\cdot) - D \int_{(\cdot)} f_1 \, d\mu$, we can show that the above assertion is true for $n = 2$. The general argument is similar.

Now, let $\varepsilon' > \varepsilon$ be given. Find $\varepsilon^* > \varepsilon$ and $n \geq 1$ such that $(k/2^n)\mu(\Omega) + \varepsilon^* < \varepsilon'$. The corresponding f_n is the desired function. □

The following is the main theorem of this section which we call Radon-Nikodym theorem for charges

6.3.4 Theorem. *Let $\mathscr{F}$ be a field of subsets of a set Ω. Let $\mu, \nu \in \text{ba}(\Omega, \mathscr{F})$ be such that μ is positive. Then the following statements are equivalent.*

(i). $\nu \ll \mu$.

(ii). *For each $\varepsilon > 0$, there exists a charge λ in $\text{ba}(\Omega, \mathscr{F})$ and a nonnegative number k such that*

$$-k\mu(F) \leq \lambda(F) \leq k\mu(F)$$

for every F in $\mathscr{F}$, and $\|\nu - \lambda\| \leq \varepsilon$.

(iii). *For each $\varepsilon > 0$, there exists a simple function f on Ω such that*

$$\left| \nu(F) - D\int_F f \, d\mu \right| < \varepsilon$$

for every F in $\mathscr{F}$.

Proof. (i) $\Rightarrow$ (ii). Without loss of generality, we can assume ν to be positive. (We can argue separately for ν^+ and ν^-.) Let $\varepsilon > 0$ be given. Since $\nu \ll \mu$, there exists $\delta > 0$ such that $\nu(E) < \varepsilon$ whenever $E \in \mathscr{F}$ and $\mu(E) < \delta$. Let $k = \nu(\Omega)/\delta$ and $\lambda = \nu \wedge k\mu$. We show that λ has the properties mentioned in (ii). Let $F \in \mathscr{F}$. If $\mu(F) < \delta$, then $\nu(F) - k\mu(F) \leq \nu(F) < \varepsilon$. If $\mu(F) \geq \delta$, then $\nu(F) - k\mu(F) \leq \nu(F) - k\delta = \nu(F) - \nu(\Omega) \leq 0 < \varepsilon$. In any case, we have $\nu(F) - k\mu(F) < \varepsilon$. So,

$$\|\nu - \lambda\| = |\nu - \lambda|(\Omega) = |\nu - \nu \wedge k\mu|(\Omega)$$
$$= (\nu - \nu \wedge k\mu)(\Omega) = \nu(\Omega) - (\nu \wedge k\mu)(\Omega)$$
$$= \nu(\Omega) - \text{Inf}\{\nu(F^c) + k\mu(F); F \in \mathscr{F}\}$$
$$= \text{Sup}\{\nu(F) - k\mu(F); F \in \mathscr{F}\}$$
$$\leq \varepsilon.$$

From the definition of λ, $\lambda \leq k\mu$ and $\lambda \geq 0$. This proves (ii).

(ii) $\Rightarrow$ (iii). Let $\varepsilon > 0$. By (ii), there exists λ in $\text{ba}(\Omega, \mathscr{F})$ and a nonnegative number k such that

$$-k\mu(F) \leq \lambda(F) \leq k\mu(F)$$

for every F in $\mathscr{F}$, and $\|\nu - \lambda\| \leq \varepsilon/3$. Then we have

$$-\varepsilon/3 - k\mu(F) < \lambda(F) < k\mu(F) + \varepsilon/3$$

for every F in $\mathscr{F}$. By applying Lemma 6.3.3(ii) for $\varepsilon' = 2\varepsilon/3$, there is a simple function f on Ω such that

$$\left| \lambda(F) - D\int_F f \, d\mu \right| < 2\varepsilon/3$$

for every F in $\mathscr{F}$. Consequently, for any F in $\mathscr{F}$,

$$\left|\nu(F) - D\int_F f\,d\mu\right| \leq |\nu(F) - \lambda(F)| + \left|\lambda(F) - D\int_F f\,d\mu\right|$$

$$< \varepsilon/3 + 2\varepsilon/3 = \varepsilon.$$

This proves (iii).

(iii) $\Rightarrow$ (i). Let $\varepsilon > 0$. Let f be a simple function on Ω such that

$$\left|\nu(F) - D\int_F f\,d\mu\right| < \varepsilon/2$$

for every F in $\mathscr{F}$. Let $k = \max\{|f(\omega)|; \omega \in \Omega\}$. Take $\delta = \varepsilon/2k$. (Assume, without loss of generality, that $k > 0$.) We show that if $F \in \mathscr{F}$ and $\mu(F) < \delta$, then $|\nu(F)| < \varepsilon$. Note that

$$\varepsilon/2 > \left|\nu(F) - D\int_F f\,d\mu\right| \geq |\nu(F)| - \left|D\int_F f\,d\mu\right|$$

$$\geq |\nu(F)| - D\int_F |f|\,d\mu \geq |\nu(F)| - k\mu(F)$$

$$> |\nu(F)| - k\delta = |\nu(F)| - \varepsilon/2.$$

Consequently, $|\nu(F)| < \varepsilon$. This proves (i). $\square$

6.3.5 Example. Exact Radon–Nikodym derivative may not exist, i.e. there may not exist a D-integrable function f (with respect to μ) such that

$$\nu(F) = D\int_F f\,d\mu$$

for every F in $\mathscr{F}$, in Theorem 6.3.4.

Let $\Omega = \{1, 2, 3, \ldots\}$ and $\mathscr{F}$ the finite-cofinite field on Ω. Let ν and μ be defined on $\mathscr{F}$ by

$$\nu(F) = 0, \quad \text{if F is finite,}$$
$$= 1, \quad \text{if F is cofinite;}$$

$$\mu(F) = \sum_{k \in F} \frac{1}{2^k}, \quad \text{if F is finite,}$$
$$= 1 + \sum_{k \in F} \frac{1}{2^k}, \quad \text{if F is cofinite.}$$

Note that $\nu \ll \mu$. If there is a D-integrable function f on Ω such that

6. ABSOLUTE CONTINUITY

$\nu(F) = D \int_F f \, d\mu$ for every F in $\mathscr{F}$, then

$$\nu(\{n\}) = 0 = D \int_{\{n\}} f \, d\mu$$
$$= f(n) \mu(\{n\})$$
$$= f(n) \frac{1}{2^n} \text{ for every } n \geq 1.$$

So,

$$1 = \nu(\Omega) = D \int f \, d\mu = 0,$$

a contradiction.

For every $k \geq 1$, by Theorem 6.3.4, there is a simple function f_k on Ω, which can be constructed easily, such that $|\nu(F) - D \int_F f_k \, d\mu| < 1/k$ for every F in $\mathscr{F}$. It can be checked that $\lim_{m,n \to \infty} D \int |f_m - f_n| \, d\mu = 0$. But there is no function f on Ω such that f_n, $n \geq 1$ converges to f hazily.

CHAPTER 7

V_p-Spaces

We have come across some function spaces, namely $ba(\Omega, \mathscr{F})$ and $ca(\Omega, \mathscr{F})$, in Chapter 2, where $\mathscr{F}$ is a field of subsets of a set Ω. It seems difficult to describe the duals of these spaces. However, there are some important function spaces, namely V_p-spaces, related to $ba(\Omega, \mathscr{F})$ which are more tractable and which we study in this chapter. These spaces are closely related to L_p-spaces introduced in Section 4.6. In Section 7.1, we introduce V_p-norms for $1 \leq p \leq \infty$ and establish their connection with the corresponding L_p-norms. V_p-spaces are presented in Section 7.2 and are shown to be Banach spaces. The duals of V_p-spaces for $1 \leq p < \infty$ are identified in Section 7.3. In the last two sections, strong convergence and weak convergence in V_p-spaces for $1 \leq p < \infty$ are characterized.

7.1 L_p-SPACES—AN OVERVIEW

Let $\mathscr{F}$ be a field of subsets of a set Ω and μ a probability charge on $\mathscr{F}$, i.e. μ is a positive charge on $\mathscr{F}$ with $\mu(\Omega) = 1$. In Section 4.6, we have defined

$$L_p(\Omega, \mathscr{F}, \mu) = \{f; f \text{ is a } T_1\text{-measurable function on } \Omega$$
$$\text{and } |f|^p \text{ is D-integrable}\}$$

with a pseudo-norm $\|f\|_p = (D \int |f|^p \, d\mu)^{1/p}$ for f in $L_p(\Omega, \mathscr{F}, \mu)$, where $1 \leq p < \infty$, and

$$L_\infty(\Omega, \mathscr{F}, \mu) = \{f; f \text{ is a } T_1\text{-measurable real valued function}$$
$$\text{on } \Omega \text{ and essentially bounded}\}$$

with a pseudo-norm $\|f\|_\infty = $ essential supremum of $|f|$, for f in $L_\infty(\Omega, \mathscr{F}, \mu)$.

By identifying functions which are equivalent under the equivalence relation $f \sim g$ if $f - g$ is a null function, we form $\mathscr{L}_p(\Omega, \mathscr{F}, \mu)$ as the space of all equivalence classes of $L_p(\Omega, \mathscr{F}, \mu)$ with a norm $\|\cdot\|_p$ unambiguously derived from the corresponding pseudo-norm on $L_p(\Omega, \mathscr{F}, \mu)$ for every $1 \leq p \leq \infty$. These normed linear spaces $(\mathscr{L}_p(\Omega, \mathscr{F}, \mu), \|\cdot\|_p)$ are not Banach spaces in general. See Remark 4.6.8. The V_p spaces to be introduced in the next section are precisely the completions of these $\mathscr{L}_p$-spaces (for $1 \le p < \infty$).

In this section, we give an alternative description of $\mathscr{L}_p$-spaces. This description provides a natural setting for the introduction of V_p-spaces.

We establish some preliminary results.

7.1.1 Lemma. *Let a and b be two non-negative real numbers and $1 \le p < \infty$. Then $(a+b)^p \le a^p t^{1-p} + b^p (1-t)^{1-p}$ for any $0 < t < 1$.*

Proof. Let us treat the case $1 < p < \infty$ and $a, b > 0$ first. The function $f(t) = a^p t^{1-p} + b^p (1-t)^{1-p}$, $0 < t < 1$ has the following properties.

(i). $\lim_{t \to 0} f(t) = \infty = \lim_{t \to 1} f(t)$.

(ii). $f'(t) = 0$ admits only one solution $t = a/(a+b)$.

Therefore, f has a global minimum at $t = a/(a+b)$. Hence

$$f(t) \ge f\left(\frac{a}{a+b}\right) = (a+b)^p \quad \text{for any } 0 < t < 1.$$

If $p = 1$ or $a = 0$ or $b = 0$, the desired inequality is obvious. □

For a given field $\mathscr{F}$ on a set Ω, recall that $\mathscr{P}$ stands for the collection of all finite partitions of Ω in $\mathscr{F}$. Under the notion of refinement of partitions, $\mathscr{P}$ is equipped with a natural partial order $\ge$ with respect to which $(\mathscr{P}, \ge)$ becomes a directed set. If $F \in \mathscr{F}$, we denote the collection of all finite partitions of F in $\mathscr{F}$ by $\mathscr{P}_F$.

7.1.2 Proposition. *Let $(\Omega, \mathscr{F}, \mu)$ be a probability charge space, i.e. $(\Omega, \mathscr{F}, \mu)$ is a charge space and μ is a probability charge on $\mathscr{F}$. Let λ be any real charge on $\mathscr{F}$ such that $\lambda \ll \mu$. Then the net of real numbers*

$$\sum_{i=1}^n \left|\frac{\lambda(F_i)}{\mu(F_i)}\right|^p \mu(F_i), \quad P = \{F_1, F_2, \ldots, F_n\} \in \mathscr{P}$$

is an increasing net for any $1 \le p < \infty$. (If $\mu(F_i) = 0$, then $\lambda(F_i) = 0$ and we interpret $0/0 = 0$.)

Proof. It suffices to show that for any two disjoint sets A and B in $\mathscr{F}$,

$$\left|\frac{\lambda(A \cup B)}{\mu(A \cup B)}\right|^p \mu(A \cup B) \le \left|\frac{\lambda(A)}{\mu(A)}\right|^p \mu(A) + \left|\frac{\lambda(B)}{\mu(B)}\right|^p \mu(B).$$

If $\mu(A) = 0$ or $\mu(B) = 0$, this inequality is obvious. Let $\mu(A) > 0$ and $\mu(B) > 0$. In Lemma 7.1.1, let $a = |\lambda(A)|$, $b = |\lambda(B)|$ and $t = [\mu(A)/\mu(A \cup B)]$. Note that λ is bounded. See Theorem 6.1.10(iv). Then $1 - t = [\mu(B)/\mu(A \cup B)]$ and

$$f(t) = |\lambda(A)|^p \left[\frac{\mu(A)}{\mu(A \cup B)}\right]^{1-p} + |\lambda(B)|^p \left[\frac{\mu(B)}{\mu(A \cup B)}\right]^{1-p}$$

$$\ge [|\lambda(A)| + |\lambda(B)|]^p.$$

Also, $|\lambda(A \cup B)| \leq |\lambda(A)| + |\lambda(B)|$. Therefore,

$$\left|\frac{\lambda(A \cup B)}{\mu(A \cup B)}\right|^p \mu(A \cup B) \leq \left|\frac{|\lambda(A)| + |\lambda(B)|}{\mu(A \cup B)}\right|^p \mu(A \cup B)$$

$$\leq \left|\frac{\lambda(A)}{\mu(A)}\right|^p \mu(A) + \left|\frac{\lambda(B)}{\mu(B)}\right|^p \mu(B).$$

This establishes the desired inequality. □

Now, we introduce some notation. Let $(\Omega, \mathcal{F}, \mu)$ be a probability charge space and λ a real charge on $\mathcal{F}$ such that $\lambda \ll \mu$. For $1 \leq p < \infty$, let

$$\|\lambda\|_p = \left[\operatorname{Sup}\left\{\sum_{i=1}^n \left|\frac{\lambda(E_i)}{\mu(E_i)}\right|^p \mu(E_i); \quad P = \{E_1, E_2, \ldots E_n\} \in \mathcal{P}\right\}\right]^{1/p}$$

$$= \left[\lim_{P \in \mathcal{P}} \sum_{i=1}^n \left|\frac{\lambda(E_i)}{\mu(E_i)}\right|^p \mu(E_i)\right]^{1/p}$$

where the limit is taken over all $P = \{E_1, E_2, \ldots, E_n\} \in \mathcal{P}$.

$$= \operatorname{Sup}\left\{\left[\sum_{i=1}^n \left|\frac{\lambda(E_i)}{\mu(E_i)}\right|^p \mu(E_i)\right]^{1/p}; \quad P = \{E_1, E_2, \ldots, E_n\} \in \mathcal{P}\right\}$$

$$= \lim_{P \in \mathcal{P}} \left[\sum_{i=1}^n \left|\frac{\lambda(E_i)}{\mu(E_i)}\right|^p \mu(E_i)\right]^{1/p},$$

where the limit is taken over all $P = \{E_1, E_2, \ldots, E_n\} \in \mathcal{P}$. The equality of these four numbers is clear from Proposition 7.1.2. Note also that $0 \leq \|\lambda\|_p \leq \infty$.

For B in $\mathcal{F}$, let λ_B be the charge on $\mathcal{F}$ defined by $\lambda_B(A) = \lambda(A \cap B)$, $A \in \mathcal{F}$. For λ_B, note that

$$\|\lambda_B\|_p = \left[\operatorname{Sup}\left\{\sum_{i=1}^n \left|\frac{\lambda(E_i)}{\mu(E_i)}\right|^p \mu(E_i); P = \{E_1, E_2, \ldots, E_n\} \in \mathcal{P}_B\right\}\right]^{1/p}.$$

Clearly, if A and B are disjoint sets in $\mathcal{F}$, then

$$\|\lambda_A\|_p^p + \|\lambda_B\|_p^p = \|\lambda_{(A \cup B)}\|_p^p.$$

7.1.3 Proposition. *Let $(\Omega, \mathcal{F}, \mu)$ be a probability charge space. Let λ be a real charge on $\mathcal{F}$ such that $\lambda \ll \mu$. Then $\|\lambda\|_p = \|\|\lambda\|\|_p$ for any $1 \leq p < \infty$.*

Proof. If $p = 1$, then $\|\lambda\|_p = |\lambda|(\Omega)$ and $\|\|\lambda\|\| = |\lambda|(\Omega)$. So, the desired assertion follows for this case.

Let $1 < p < \infty$. Since for any F in $\mathcal{F}$, $|\lambda(F)| \leq |\lambda|(F)$, it follows that $\|\lambda\|_p \leq \|\|\lambda\|\|_p$. It remains to be shown that $\|\|\lambda\|\|_p \leq \|\lambda\|_p$. Let $B \in \mathcal{F}$. From the definition of $\|\lambda_B\|_p^p$, it follows that $|\lambda(B)/\mu(B)|^p \mu(B) \leq \|\lambda_B\|_p^p$. Consequently, $|\lambda(B)| \leq$

$\|\lambda_B\|_p [\mu(B)]^{(p-1)/p} = \|\lambda_B\|_p [\mu(B)]^{1/q}$, where $1/p + 1/q = 1$. (Note that the convention $0/0 = 0$ is enforced.) Thus for any partition $\{E_1, E_2, \ldots, E_n\} \in \mathcal{P}_B$, we have, by Hölder's inequality for finite sums (See Theroem 4.6.2.),

$$\sum_{i=1}^n |\lambda(E_i)| \leq \sum_{i=1}^n \|\lambda_{E_i}\|_p [\mu(E_i)]^{1/q}$$

$$\leq \left(\sum_{i=1}^n \|\lambda_{E_i}\|_p^p \right)^{1/p} \left(\sum_{i=1}^n \mu(E_i) \right)^{1/q}.$$

Therefore, from the last observation made preceding the statement of this proposition, we obtain the inequality

$$\left(\sum_{i=1}^n |\lambda(E_i)| \right)^p \leq \|\lambda_B\|_p^p [\mu(B)]^{p/q}.$$

Taking supremum over all partitions $\{E_1, E_2, \ldots, E_n\}$ in $\mathcal{P}_B$, we obtain $[|\lambda|(B)]^p \leq \|\lambda_B\|_p^p [\mu(B)]^{p-1}$. Or, equivalently,

$$\left| \frac{|\lambda|(B)}{\mu(B)} \right|^p \mu(B) \leq \|\lambda_B\|_p^p.$$

Now, for any partition $\{F_1, F_2, \ldots, F_m\}$ of Ω in $\mathcal{F}$,

$$\sum_{i=1}^m \left| \frac{|\lambda|(F_i)}{\mu(F_i)} \right|^p \mu(F_i) \leq \sum_{i=1}^m \|\lambda_{F_i}\|_p^p = \|\lambda_\Omega\|_p^p = \|\lambda\|_p^p.$$

By taking supremum over all partitions $\{F_1, F_2, \ldots, F_m\}$ in $\mathcal{P}$, we obtain $\||\lambda|\|_p^p \leq \|\lambda\|_p^p$. From this, the desired equality follows. □

We now give some properties of $\|\cdot\|_p$ introduced above.

7.1.4 Proposition. *Let* $(\Omega, \mathcal{F}, \mu)$ *be a probability charge space. Let* $1 \leq p < \infty$. *Let* $\lambda, \lambda_1, \lambda_2$ *be three real charges on* $\mathcal{F}$ *such that* $\lambda \ll \mu$, $\lambda_1 \ll \mu$ *and* $\lambda_2 \ll \mu$. *Then the following statements are true.*
(i). $\lambda = 0$ *if and only if* $\|\lambda\|_p = 0$.
(ii). $\|c\lambda\|_p = |c| \|\lambda\|_p$ *for any real number c.*
(iii). $\|\lambda_1 + \lambda_2\|_p \leq \|\lambda_1\|_p + \|\lambda_2\|_p$.

Proof. (i) and (ii) are obvious. (iii) can be established using Minkowski's inequality as follows. (See Theorem 4.6.6.) If $\{E_1, E_2, \ldots, E_n\}$ is any partition of Ω in $\mathcal{F}$, then

$$\left[\sum_{i=1}^n \left| \frac{\lambda_1(E_i) + \lambda_2(E_i)}{\mu(E_i)^{(p-1)/p}} \right|^p \right]^{1/p} \leq \left[\sum_{i=1}^n \left| \frac{\lambda_1(E_i)}{\mu(E_i)^{(p-1)/p}} \right|^p \right]^{1/p}$$

$$+ \left[\sum_{i=1}^n \left| \frac{\lambda_2(E_i)}{\mu(E_i)^{(p-1)/p}} \right|^p \right]^{1/p}.$$

Therefore,

$$\left[\sum_{i=1}^{n}\left|\frac{\lambda_1(E_i)+\lambda_2(E_i)}{\mu(E_i)}\right|^p \mu(E_i)\right]^{1/p} \leq \|\lambda_1\|_p + \|\lambda_2\|_p.$$

Hence $\|\lambda_1 + \lambda_2\|_p \leq \|\lambda_1\|_p + \|\lambda_2\|_p$. □

Now, we come to the promised alternative description of $\mathscr{L}_p(\Omega, \mathscr{F}, \mu)$.

7.1.5 Lemma. *Let $(\Omega, \mathscr{F}, \mu)$ be a probability charge space and f a simple function on Ω. Let λ on $\mathscr{F}$ be defined by $\lambda(F) = D\int_F f\,d\mu$, $F \in \mathscr{F}$. Then $\|\lambda\|_p^p = D\int |f|^p\,d\mu$ for every $1 \leq p < \infty$.*

Proof. Let $f = \sum_{i=1}^n a_i I_{E_i}$ for some real numbers $a_1, a_2, \ldots, a_n$ and some partition $\{E_1, E_2, \ldots, E_n\}$ in $\mathscr{P}$. Then

$$D\int |f|^p\,d\mu = \sum_{i=1}^{n} |a_i|^p \mu(E_i).$$

Also,

$$\sum_{i=1}^{n} \left|\frac{\lambda(E_i)}{\mu(E_i)}\right|^p \mu(E_i) = \sum_{i=1}^{n} |a_i|^p \mu(E_i).$$

Further, if we take any partition $\{F_1, F_2, \ldots, F_m\}$ in $\mathscr{P}$ finer than $\{E_1, E_2, \ldots, E_n\}$, then we have

$$\sum_{i=1}^{m} \left|\frac{\lambda(F_i)}{\mu(F_i)}\right|^p \mu(F_i) = \sum_{i=1}^{n} |a_i|^p \mu(E_i).$$

Consequently, $\|\lambda\|_p^p = \sum_{i=1}^n |a_i|^p \mu(E_i)$. This completes the proof. □

7.1.6 Theorem. *Let $1 \leq p < \infty$. Let $(\Omega, \mathscr{F}, \mu)$ be a probability charge space and $f \in L_p(\Omega, \mathscr{F}, \mu)$. Let λ on $\mathscr{F}$ be defined by $\lambda(F) = D\int_F f\,d\mu$, $F \in \mathscr{F}$. Then*

$$D\int |f|^p\,d\mu = \|\lambda\|_p^p.$$

Proof. First, let us treat the case $p = 1$. From the definition of $\|\lambda\|_1$, it is clear that $\|\lambda\|_1 = |\lambda|(\Omega)$. By Theorem 4.4.13(xi), $|\lambda|(\Omega) = D\int |f|\,d\mu$.

Now, we come to the case $1 < p < \infty$. We show that $\|\lambda\|_p^p \leq D\int |f|^p\,d\mu$. Let the positive number q satisfy $1/p + 1/q = 1$. Then for any E in $\mathscr{F}$, by Hölder's inequality,

$$|\lambda|(E) = D\int_E |f|\,d\mu = D\int_E (|f| \cdot 1)\,d\mu \leq \left[D\int_E |f|^p\,d\mu\right]^{1/p} [\mu(E)]^{1/q}.$$

Let $\{E_1, E_2, \ldots, E_n\} \in \mathscr{P}$. Then

$$\sum_{i=1}^{n} \left|\frac{|\lambda|(E_i)}{\mu(E_i)}\right|^p \mu(E_i) = \sum_{i=1}^{n} [|\lambda|(E_i)]^p [\mu(E_i)]^{1-p}$$

$$\leq \sum_{i=1}^{n} \left(D \int_{E_i} |f|^p \, d\mu\right) [\mu(E_i)]^{p/q} [\mu(E_i)]^{1-p}$$

$$= \sum_{i=1}^{n} D \int_{E_i} |f|^p \, d\mu = D \int |f|^p \, d\mu.$$

Consequently, the inequality $\|\lambda\|_p^p \leq D \int |f|^p \, d\mu$ follows.

Now, we show the desired equality. Since simple functions are dense in $L_p(\Omega, \mathscr{F}, \mu)$ (see Theorem 4.6.15), there is a sequence f_n, $n \geq 1$ of simple functions on Ω such that $\lim_{n \to \infty} D \int |f_n - f|^p \, d\mu = 0$. For each $n \geq 1$, let λ_n on $\mathscr{F}$ be defined by $\lambda_n(F) = D \int_F f_n \, d\mu$, $F \in \mathscr{F}$. Then for every $n \geq 1$, $|\|\lambda_n\|_p - \|\lambda\|_p| \leq \|\lambda_n - \lambda\|_p \leq (D \int |f_n - f|^p \, d\mu)^{1/p}$. Hence

$$\|\lambda\|_p^p = \lim_{n \to \infty} \|\lambda_n\|_p^p = \lim_{n \to \infty} D \int |f_n|^p \, d\mu = D \int |f|^p \, d\mu.$$

The second equality above follows from Lemma 7.1.5. This completes the proof. □

7.1.7 Theorem. *Let $(\Omega, \mathscr{F}, \mu)$ be a probability charge space. Let $1 \leq p < \infty$. Then*

$$\mathscr{L}_p(\Omega, \mathscr{F}, \mu) = \{\lambda \in \mathrm{ba}(\Omega, \mathscr{F}); \lambda(F) = D \int_F f \, d\mu, F \in \mathscr{F}$$

for some T_1-measurable function f on Ω

such that $|f|^p$ is D-integrable$\}$.

Proof. This is now obvious from the results established above. □

Now, we take up the case $p = \infty$. Let $(\Omega, \mathscr{F}, \mu)$ be a probability charge space and λ a real charge on $\mathscr{F}$ such that $\lambda \ll \mu$. Let

$$\|\lambda\|_\infty = \mathrm{Sup}\left\{\left|\frac{\lambda(F)}{\mu(F)}\right|; F \in \mathscr{F}\right\}.$$

Obviously, $0 \leq \|\lambda\|_\infty \leq \infty$. We give below some properties of $\|\cdot\|_\infty$.

7.1.8 Proposition. *Let $(\Omega, \mathscr{F}, \mu)$ be a probability charge space and λ, λ_1, λ_2 real charges on $\mathscr{F}$ such that $\lambda \ll \mu$, $\lambda_1 \ll \mu$ and $\lambda_2 \ll \mu$. Then the following statements are true.*

(i). $\|\lambda\|_\infty = 0$ if and only if $\lambda = 0$.
(ii). $\|c\lambda\|_\infty = |c|\|\lambda\|_\infty$ for any real number c.
(iii). $\|\lambda_1 + \lambda_2\|_\infty \le \|\lambda_1\|_\infty + \|\lambda_2\|_\infty$.
(iv). $\|\lambda\|_\infty = \|\|\lambda\|\|_\infty$.

Proof. (i), (ii) and (iii) are obvious. We prove (iv). $\|\lambda\|_\infty \le \|\|\lambda\|\|_\infty$ follows from the fact that $|\lambda(F)| \le |\lambda|(F)$ for any F in $\mathscr{F}$. We show that $\|\|\lambda\|\|_\infty \le \|\lambda\|_\infty$. For any B in $\mathscr{F}$ and $\{B_1, B_2, \ldots, B_m\}$ in $\mathscr{P}_B$, we have

$$\sum_{i=1}^{m} |\lambda(B_i)| \le \sum_{i=1}^{m} \|\lambda\|_\infty \mu(B_i) = \|\lambda\|_\infty \mu(B).$$

Hence $|\lambda|(B) \le \|\lambda\|_\infty \mu(B)$, i.e., $[|\lambda|(B)/\mu(B)] \le \|\lambda\|_\infty$. Taking supremum over all B in $\mathscr{F}$, we obtain $\|\|\lambda\|\|_\infty \le \|\lambda\|_\infty$. This proves (iv). □

7.1.9 Theorem. *Let $(\Omega, \mathscr{F}, \mu)$ be a probability charge space and f a real valued T_1-measurable essentially bounded function on Ω. Let λ on $\mathscr{F}$ be defined by $\lambda(F) = D \int_F f \, d\mu$, $F \in \mathscr{F}$. Then*

$$\|f\|_\infty = \text{Essential supremum of } f = \|\lambda\|_\infty.$$

Proof. Note that f is D-integrable by Theorem 4.5.7 and consequently, λ is well defined.

First, we prove the inequality $\|\lambda\|_\infty \le \|f\|_\infty$. Let k be any positive number such that $\mu^*(\{\omega \in \Omega; |f(\omega)| > k\}) = 0$. Then $|f| \le k$ a.e. $[\mu]$. Then by Theorem 4.4.13(vii),

$$|\lambda|(F) = D \int_F |f| \, d\mu \le k \mu(F)$$

for any F in $\mathscr{F}$. Hence

$$\|\lambda\|_\infty = \text{Sup}\left\{\frac{|\lambda|(F)}{\mu(F)}; F \in \mathscr{F}\right\} \le k.$$

From the definition of $\|f\|_\infty$, it follows that $\|\lambda\|_\infty \le \|f\|_\infty$.

To prove the reverse inequality, we proceed as follows. Let $k = \|\lambda\|_\infty$. Then $|\lambda|(F) \le k\mu(F)$ for every F in $\mathscr{F}$. This implies that $D \int_F (|f| - k) \, d\mu \le 0$ for every F in $\mathscr{F}$. We claim that $|f| \le k$ a.e. $[\mu]$. Define τ on $\mathscr{F}$ by $\tau(F) = D \int_F (|f| - k) \, d\mu$, $F \in \mathscr{F}$. Since $\tau(F) \le 0$ for every F in $\mathscr{F}$, it follows from the definition of τ^+ that $\tau^+(F) = 0$ for every F in $\mathscr{F}$ and so, $\tau^-(F) = -\tau(F)$ for every F in $\mathscr{F}$. But $\tau^+(F) = D \int_F (|f| - k)^+ \, d\mu$ for every F in $\mathscr{F}$. See Theorem 4.4.13(xii). By Theorem 4.4.13(xiii), $(|f| - k)^+$ is a null function. Since $|f| - k = (|f| - k)^+ - (|f| - k)^-$, it follows that $|f| - k \le 0$ a.e. $[\mu]$.

Now, we claim that $\mu^*(\{\omega \in \Omega; |f(\omega)| > k + \varepsilon\}) = 0$ for any $\varepsilon > 0$. Observe that $|f| \le k + (|f| - k)^+$ and that $(|f| - k)^+$ is a null function. Therefore,

$$\mu^*(\{\omega \in \Omega; |f(\omega)| > k + \varepsilon\}) \le \mu^*(\{\omega \in \Omega; (|f| - k)^+ > \varepsilon\}) = 0.$$

7. V_p-SPACES

The claim is thus established. By the definition of $\|f\|_\infty$, it follows that $\|f\|_\infty \le k + \varepsilon$ for any $\varepsilon > 0$. So, $\|f\|_\infty \le k$. Hence $\|f\|_\infty \le \|\lambda\|_\infty$. This completes the proof. $\square$

In the above, we have indeed proved that $|f| \le \|f\|_\infty$ a.e. $[\mu]$ if f is T_1-measurable.

Now, we provide an alternative description of $\mathscr{L}_\infty(\Omega, \mathscr{F}, \mu)$.

7.1.10 Theorem. *Let $(\Omega, \mathscr{F}, \mu)$ be a probability charge space. Then*

$$\mathscr{L}_\infty(\Omega, \mathscr{F}, \mu) = \left\{\lambda \in \mathrm{ba}(\Omega, \mathscr{F}); \lambda(\mathrm{F}) = \mathrm{D}\int_F f\,d\mu, \mathrm{F} \in \mathscr{F}\right.$$

for some essentially bounded T_1-measurable real valued function f on $\Omega\bigg\}$.

Proof. This is now obvious. $\square$

7.2 V_p-SPACES

$\mathscr{L}_p$-spaces, discussed in the previous section, are not Banach spaces in general, or, equivalently, the normed linear spaces $\mathscr{L}_p(\Omega, \mathscr{F}, \mu)$ are not complete in general. In this section, we introduce V_p-spaces and show that they are the completions of the corresponding $\mathscr{L}_p$-spaces for $1 \le p < \infty$.

7.2.1 Definition. Let $(\Omega, \mathscr{F}, \mu)$ be a probability charge space. For each $1 \le p \le \infty$, let

$$V_p(\Omega, \mathscr{F}, \mu) = \{\lambda \in \mathrm{ba}(\Omega, \mathscr{F}); \|\lambda\|_p < \infty\}.$$

The following proposition is a consequence of Propositions 7.1.4 and 7.1.8.

7.2.2 Proposition. *Let $(\Omega, \mathscr{F}, \mu)$ be a probability charge space. Then for each $1 \le p \le \infty$, $(V_p(\Omega, \mathscr{F}, \mu), \|\cdot\|_p)$ is a normed linear space.*

By Theorems 7.1.7 and 7.1.10, we have that $\mathscr{L}_p(\Omega, \mathscr{F}, \mu) \subset V_p(\Omega, \mathscr{F}, \mu)$ for every $1 \le p \le \infty$. Members of $V_p(\Omega, \mathscr{F}, \mu)$ need not admit exact Radon–Nikodym derivatives with respect to μ. Now, we show that each V_p-space is complete.

7.2.3 Theorem *Let $(\Omega, \mathscr{F}, \mu)$ be a probability charge space. Then for each $1 \le p \le \infty$, $(V_p(\Omega, \mathscr{F}, \mu), \|\cdot\|_p)$ is a Banach space.*

Proof. First, we treat the case $1 \leq p < \infty$.
Let λ_n, $n \geq 1$ be a Cauchy sequence in $V_p(\Omega, \mathscr{F}, \mu)$. We claim that $\lim_{m,n \to \infty} \text{Sup}_{E \in \mathscr{F}} |\lambda_n(E) - \lambda_m(E)| = 0$. Observe that for any E in $\mathscr{F}$ and $m, n \geq 1$,

$$\left|\frac{\lambda_n(E) - \lambda_m(E)}{\mu(E)}\right|^p \mu(E) \leq \left|\frac{\lambda_n(E) - \lambda_m(E)}{\mu(E)}\right|^p \mu(E)$$
$$+ \left|\frac{\lambda_n(E^c) - \lambda_m(E^c)}{\mu(E^c)}\right|^p \mu(E^c)$$
$$\leq \|\lambda_n - \lambda_m\|_p^p.$$

Hence

$$|\lambda_n(E) - \lambda_m(E)|^p \leq \|\lambda_n - \lambda_m\|_p^p [\mu(E)]^{p-1} \leq \|\lambda_n - \lambda_m\|_p^p$$

for any E in $\mathscr{F}$. From this inequality, the claim follows. Let

$$\lambda(E) = \lim_{n \to \infty} \lambda_n(E)$$

for E in $\mathscr{F}$. It is obvious that λ is a charge on $\mathscr{F}$. Since λ_n, $n \geq 1$ converges to λ uniformly over $\mathscr{F}$, λ is bounded and $\lambda \ll \mu$. It remains to be shown that $\lambda \in V_p(\Omega, \mathscr{F}, \mu)$ and that $\|\lambda_n - \lambda\|_p$, $n \geq 1$ converges to zero.

Let $\varepsilon > 0$. There exists $N \geq 1$ such that $\|\lambda_n - \lambda_m\|_p^p < \varepsilon$ whenever $n, m \geq N$. Let $P = \{E_1, E_2, \ldots, E_k\} \in \mathscr{P}$. Then for $m \geq N$,

$$\sum_{i=1}^{k} \left|\frac{\lambda(E_i) - \lambda_m(E_i)}{\mu(E_i)}\right|^p \mu(E_i) = \lim_{n \to \infty} \sum_{i=1}^{k} \left|\frac{\lambda_n(E_i) - \lambda_m(E_i)}{\mu(E_i)}\right|^p \mu(E_i) \leq \varepsilon.$$

Taking supremum over all P in $\mathscr{P}$, we obtain $\|\lambda - \lambda_m\|_p^p \leq \varepsilon$ if $m \geq N$. This shows that $\lambda \in V_p(\Omega, \mathscr{F}, \mu)$ and that $\lim_{m \to \infty} \|\lambda_m - \lambda\|_p = 0$. Hence $(V_p(\Omega, \mathscr{F}, \mu), \|\cdot\|_p)$ is complete for $1 \leq p < \infty$.

Now, we look at the case $p = \infty$. Let λ_n, $n \geq 1$ be a Cauchy sequence in $V_\infty(\Omega, \mathscr{F}, \mu)$. As above, we show that λ_n, $n \geq 1$ is a uniform Cauchy sequence over $\mathscr{F}$. For any F in $\mathscr{F}$ and $m, n \geq 1$,

$$|\lambda_n(F) - \lambda_m(F)| \leq \|\lambda_n - \lambda_m\|_\infty \mu(F) \leq \|\lambda_n - \lambda_m\|_\infty.$$

This shows that λ_n, $n \geq 1$ is a uniform Cauchy sequence over $\mathscr{F}$. Let $\lambda(F) = \lim_{n \to \infty} \lambda_n(F)$, $F \in \mathscr{F}$. Then λ is a bounded charge on $\mathscr{F}$ and absolutely continuous with respect to μ. It remains to be shown that $\lambda \in V_\infty(\Omega, \mathscr{F}, \mu)$ and that $\|\lambda_n - \lambda\|_\infty$, $n \geq 1$ converges to zero.

Let $\varepsilon > 0$. There exists $N \geq 1$ such that $\|\lambda_n - \lambda_m\|_\infty < \varepsilon$ whenever $m, n \geq N$. Let $F \in \mathscr{F}$. Then $|\lambda(F) - \lambda_m(F)| = \lim_{n \to \infty} |\lambda_n(F) - \lambda_m(F)| \leq \varepsilon \mu(F)$ if $m \geq N$. Consequently, $\|\lambda - \lambda_m\|_\infty \leq \varepsilon$ if $m \geq N$. This shows that $\lambda \in V_\infty(\Omega, \mathscr{F}, \mu)$ and $\lim_{n \to \infty} \|\lambda_n - \lambda\|_\infty = 0$. Hence $V_\infty(\Omega, \mathscr{F}, \mu)$ is a Banach space. $\square$

Now, we prove inclusion relations among V_p-spaces.

7.2.4 Theorem. *Let $(\Omega, \mathscr{F}, \mu)$ be a probability charge space. Let $1 \leq r \leq s \leq \infty$. Then*

$$V_1(\Omega, \mathscr{F}, \mu) \supset V_r(\Omega, \mathscr{F}, \mu) \supset V_s(\Omega, \mathscr{F}, \mu) \supset V_\infty(\Omega, \mathscr{F}, \mu).$$

In fact, $\|\lambda\|_1 \leq \|\lambda\|_r \leq \|\lambda\|_s \leq \|\lambda\|_\infty$ is valid for any real charge λ on $\mathscr{F}$ absolutely continuous with respect to μ.

Proof. Let $1 \leq r < s < \infty$. Write $s/r = p$. Let $q > 1$ be such that $1/p + 1/q = 1$. Let $P = \{E_1, E_2, \ldots, E_n\} \in \mathscr{P}$. Then

$$\sum_{i=1}^n \left|\frac{\lambda(E_i)}{\mu(E_i)}\right|^r \mu(E_i) = \sum_{i=1}^n \left[\left|\frac{\lambda(E_i)}{\mu(E_i)}\right|^{s/p} [\mu(E_i)]^{1/p}\right][\mu(E_i)]^{1/q}$$

$$\leq \left(\sum_{i=1}^n \left[\left|\frac{\lambda(E_i)}{\mu(E_i)}\right|^{s/p} [\mu(E_i)]^{1/p}\right]^p\right)^{1/p} \left[\sum_{i=1}^n \mu(E_i)\right]^{1/q}$$

(by Hölder's inequality)

$$= \left[\sum_{i=1}^n \left|\frac{\lambda(E_i)}{\mu(E_i)}\right|^s \mu(E_i)\right]^{1/p} \leq [\|\lambda\|_s]^{s/p}.$$

Hence $\|\lambda\|_r^r \leq [\|\lambda\|_s]^{s/p}$, or, equivalently, $\|\lambda\|_r \leq \|\lambda\|_s$.

If $1 \leq s < \infty$, then for any partition $\{E_1, E_2, \ldots, E_n\}$ in $\mathscr{P}$,

$$\sum_{i=1}^n \left|\frac{\lambda(E_i)}{\mu(E_i)}\right|^s \mu(E_i) \leq \sum_{i=1}^n \left[\sup_{F \in \mathscr{F}} \left|\frac{\lambda(F)}{\mu(F)}\right|\right]^s \mu(E_i) = \|\lambda\|_\infty^s.$$

Hence $\|\lambda\|_s \leq \|\lambda\|_\infty$. This completes the proof. □

The following is an interesting consequence of the above theorem.

7.2.5 Corollary. *Let $(\Omega, \mathscr{F}, \mu)$ be a probability charge space. For any real charge λ on $\mathscr{F}$ absolutely continuous with respect to μ,*

$$\lim_{p \to \infty} \|\lambda\|_p = \|\lambda\|_\infty.$$

Proof. By Theorem 7.2.4, $\lim_{p \to \infty} \|\lambda\|_p$ certainly exists (may be equal to ∞) and is less than or equal to $\|\lambda\|_\infty$. On the other hand, for any $p > 1$ and F in $\mathscr{F}$,

$$\left|\frac{\lambda(F)}{\mu(F)}\right|^p \mu(F) \leq \|\lambda\|_p^p.$$

Hence

$$|\lambda(F)| \leq \|\lambda\|_p [\mu(F)]^{1-(1/p)} \leq \|\lambda\|_p \mu(F) \leq \lim_{p \to \infty} \|\lambda\|_p \mu(F).$$

From this, it follows that $\|\lambda\|_\infty \leq \lim_{p \to \infty} \|\lambda\|_p$. This completes the proof. □

Now, we introduce simple charges. Recall that for any charge μ on a field $\mathscr{F}$ of subsets of a set Ω and F in $\mathscr{F}$, μ_F is the charge on $\mathscr{F}$ defined by $\mu_F(E) = \mu(E \cap F)$, $E \in \mathscr{F}$.

7.2.6 Definition. Let $(\Omega, \mathscr{F}, \mu)$ be a probability charge space. A *simple charge* on $\mathscr{F}$ is any charge of the form $\sum_{i=1}^{n} a_i \mu_{E_i}$ for some real numbers $a_1, a_2, \ldots, a_n$ and some partition $\{E_1, E_2, \ldots, E_n\}$ in $\mathscr{P}$. $SC(\Omega, \mathscr{F}, \mu)$ stands for the collection of all simple charges on $\mathscr{F}$.

$SC(\Omega, \mathscr{F}, \mu)$ is a linear space and can be identified with the space of all simple functions on $(\Omega, \mathscr{F})$ as follows. If $f = \sum_{i=1}^{n} a_i I_{E_i}$ is a simple function on Ω for some real numbers $a_1, a_2, \ldots, a_n$ and some partition $\{E_1, E_2, \ldots, E_n\}$ in $\mathscr{P}$, then the charge $\lambda = \sum_{i=1}^{n} a_i \mu_{E_i}$ on $\mathscr{F}$ has the property that $\|f\|_p = \|\lambda\|_p$ for every $1 \leq p \leq \infty$. Conversely, if $\lambda = \sum_{i=1}^{n} a_i \mu_{E_i}$ is a simple charge on $\mathscr{F}$ for some real numbers $a_1, a_2, \ldots, a_n$ and some partition $\{E_1, E_2, \ldots, E_n\}$ in $\mathscr{P}$, then the simple function $f = \sum_{i=1}^{n} a_i I_{E_i}$ on Ω has the property that $\|\lambda\|_p = \|f\|_p$ for every $1 \leq p \leq \infty$.

It is obvious that $SC(\Omega, \mathscr{F}, \mu) \subset V_\infty(\Omega, \mathscr{F}, \mu)$.

Among simple charges, the charges given in the following definition are of special interest.

7.2.7 Definition. Let $(\Omega, \mathscr{F}, \mu)$ be a probability charge space and λ a real charge on $\mathscr{F}$ such that $\lambda \ll \mu$. Let $P = \{E_1, E_2, \ldots, E_n\} \in \mathscr{P}$. Define

$$\lambda_P = \sum_{i=1}^{n} \frac{\lambda(E_i)}{\mu(E_i)} \mu_{E_i}.$$

It is obvious that

$$\|\lambda_P\|_p^p = \sum_{i=1}^{n} \left|\frac{\lambda(E_i)}{\mu(E_i)}\right|^p \mu(E_i)$$

and

$$\lim_{P \in \mathscr{P}} \|\lambda_P\|_p^p = \operatorname*{Sup}_{P \in \mathscr{P}} \|\lambda_P\|_p^p = \|\lambda\|_p^p$$

for any $1 \leq p < \infty$. It is important to note that for any partition $P' = \{F_1, F_2, \ldots, F_m\}$ in $\mathscr{P}$ which is a refinement of P,

$$\sum_{i=1}^{n} \left|\frac{\lambda(E_i)}{\mu(E_i)}\right|^p \mu(E_i) = \sum_{j=1}^{m} \left|\frac{\lambda_P(F_j)}{\mu(F_j)}\right|^p \mu(F_j).$$

Next, we show that $SC(\Omega, \mathscr{F}, \mu)$ is a dense subset of $V_p(\Omega, \mathscr{F}, \mu)$ for any $1 \leq p < \infty$. First, we need the following results.

7.2.8 Lemma. *For any $p > 0$ and $\varepsilon > 0$, there exists a number $k(p, \varepsilon)$ such that*

$$|x - 1|^p \geq k(p, \varepsilon)[|x|^p + p - 1 - px] + \varepsilon |x|^p$$

for any real number x.

Proof. The function f defined on the real line by

$$f(x) = |x|^p + p - 1 - px, \quad x \in \mathbb{R}$$

is continuous and positive for all $x \neq 1$. Further, $\lim_{|x| \to \infty} |x - 1|^p / f(x)$ exists and is finite. For the given $\varepsilon > 0$, we can find $\eta > 0$ such that $|x - 1|^p \leq \varepsilon |x|^p$ whenever $|x - 1| \leq \eta$. This follows from the fact that the function $g(x) = |x - 1|^p / |x|^p$ for $x \in (0, \infty)$ is continuous at $x = 1$. Since $\lim_{|x| \to \infty} |x - 1|^p / f(x)$ is finite, there exists $\eta_1 > \eta + 1$ such that $|x - 1|^p / f(x)$ is bounded on the set $\{x ; |x| > \eta_1\}$. Since $|x - 1|^p / f(x)$ is well defined and continuous on $[-\eta_1, 1 - \eta] \cup [1 + \eta, \eta_1]$, it is bounded on this set. Hence we can find a constant $k(p, \varepsilon)$ such that

$$|x - 1|^p \leq k(p, \varepsilon) f(x)$$

whenever $|x - 1| > \eta$. In any case, we have

$$|x - 1|^p \leq k(p, \varepsilon)[|x|^p + p - 1 - px] + \varepsilon |x|^p$$

for all real numbers x. □

7.2.9 Lemma. *Let $(\Omega, \mathscr{F}, \mu)$ be a probability charge space and $\lambda \in V_p(\Omega, \mathscr{F}, \mu)$ for some $1 < p < \infty$. Then for a given $\varepsilon > 0$, there exists a constant $k(p, \varepsilon)$ such that*

$$\|\lambda - \lambda_P\|_p^p \leq k(p, \varepsilon)[\|\lambda\|_p^p - \|\lambda_P\|_p^p] + \varepsilon \|\lambda\|_p^p$$

for any $P = \{E_1, E_2, \ldots, E_m\}$ in $\mathscr{P}$.

Proof. Let $P' = \{F_1, F_2, \ldots, F_n\}$ be a partition in $\mathscr{P}$ finer than P. Let $\varepsilon > 0$. Let $k(p, \varepsilon)$ be the constant provided by Lemma 7.2.8. Let $1 \leq i \leq n$ be fixed. Then we can find $1 \leq j \leq m$ such that $F_i \subset E_j$. Consequently,

$$\lambda_P(F_i) = \sum_{k=1}^{m} \frac{\lambda(E_k)}{\mu(E_k)} \mu_{E_k}(F_i) = \frac{\lambda(E_j)}{\mu(E_j)} \mu(F_i).$$

In Lemma 7.2.8, if we set $x = \lambda(F_i) / \lambda_P(F_i)$, with the usual convention about $0/0$, we obtain

$$\left| \frac{\lambda(F_i) - \lambda_P(F_i)}{(\lambda_P(F_i))} \right|^p \leq k(p, \varepsilon) \left[\left| \frac{\lambda(F_i)}{\lambda_P(F_i)} \right|^p + p - 1 - p \frac{\lambda(F_i)}{\lambda_P(F_i)} \right] + \varepsilon \left| \frac{\lambda(F_i)}{\lambda_P(F_i)} \right|^p.$$

Using $\lambda_P(F_i) = [\lambda(E_j)/\mu(E_j)]\mu(F_i)$ and multiplying throughout by $|\lambda(E_j)/\mu(E_j)|^p \mu(F_i)$, we obtain

$$\left|\frac{\lambda(F_i) - \lambda_P(F_i)}{\mu(F_i)}\right|^p \mu(F_i) \leq k(p,\varepsilon)\left[\left|\frac{\lambda(F_i)}{\mu(F_i)}\right|^p \mu(F_i) + (p-1)\left|\frac{\lambda(E_j)}{\mu(E_j)}\right|^p \mu(F_i)\right.$$

$$\left. - p\left|\frac{\lambda(E_j)}{\mu(E_j)}\right|^p \frac{\lambda(F_i)}{\mu(F_i)} \frac{\mu(F_i)}{\left(\frac{\lambda(E_j)}{\mu(E_j)}\right)}\right] + \varepsilon\left|\frac{\lambda(F_i)}{\mu(F_i)}\right|^p \mu(F_i).$$

By summing first over all those $i \in \{1, 2, \ldots, n\}$ such that $F_i \subset E_j$, for a given $j \in \{1, 2, \ldots, m\}$ and then summing over all $j \in \{1, 2, \ldots, m\}$, we obtain

$$\sum_{i=1}^{n} \left|\frac{\lambda(F_i) - \lambda_P(F_i)}{\mu(F_i)}\right|^p \mu(F_i) \leq k(p,\varepsilon)[\|\lambda\|_p^p + (p-1)\|\lambda_P\|_p^p - p\|\lambda_P\|_p^p] + \varepsilon\|\lambda\|_p^p.$$

Since the above inequality is true for any partition P′ finer than P, we obtain

$$\|\lambda - \lambda_P\|_p^p \leq k(p,\varepsilon)[\|\lambda\|_p^p - \|\lambda_P\|_p^p] + \varepsilon\|\lambda\|_p^p.$$

This completes the proof. □

7.2.10 Proposition. Let $(\Omega, \mathcal{F}, \mu)$ be a probability charge space and $\lambda = \sum_{i=1}^{m} a_i \mu_{E_i}$ be a simple charge on $\mathcal{F}$, for a partition $P = \{E_1, E_2, \ldots, E_m\}$ in $\mathcal{P}$ and real numbers $a_1, a_2, \ldots, a_m$. Then for any partition $P' \geq P$,

$$\|\lambda - \lambda_{P'}\|_p = 0$$

for any $1 \leq p < \infty$.

Proof. Simply note that $\lambda - \lambda_{P'} = 0$. □

7.2.11 Proposition. Let $(\Omega, \mathcal{F}, \mu)$ be a probability charge space and λ a positive bounded charge on $\mathcal{F}$ such that $\lambda \ll \mu$. Then

$$\lim_{n \to \infty} \|\lambda - (\lambda \wedge n\mu)\|_1 = 0.$$

Proof. Since $ba(\Omega, \mathcal{F})$ is a boundedly complete vector lattice (see Theorem 2.2.1) and $\lambda \ll \mu$, $\lambda = \bigvee_{n \geq 1}(\lambda \wedge n\mu)$. See Theorem 1.5.12 and Theorem 6.2.2. Note that $\lim_{n \to \infty}(\lambda \wedge n\mu)(A) = \lambda(A)$ for every A in $\mathcal{F}$. In particular, $(\lambda - (\lambda \wedge n\mu))(\Omega)$, $n \geq 1$ converges to zero, i.e. $\lim_{n \to \infty}\|\lambda - (\lambda \wedge n\mu)\|_1 = 0$. This proves the result. □

Now, we come to an important result of this section.

7.2.12 Theorem. Let $(\Omega, \mathcal{F}, \mu)$ be a probability charge space. Then, for each $1 \leq p < \infty$, the space $SC(\Omega, \mathcal{F}, \mu)$ of all simple charges is a dense subset of $V_p(\Omega, \mathcal{F}, \mu)$. More precisely, for any $1 \leq p < \infty$ and λ in $V_p(\Omega, \mathcal{F}, \mu)$,

$$\lim_{P \in \mathcal{P}} \|\lambda - \lambda_P\|_p = 0.$$

7. V_p-SPACES

Proof. Let us tackle the case $1 < p < \infty$ first.

For a given $\varepsilon > 0$, by Lemma 7.2.9, there exists a constant $k(p, \varepsilon)$ such that for any P in $\mathscr{P}$,

$$\|\lambda - \lambda_P\|_p^p \leq k(p, \varepsilon)[\|\lambda\|_p^p - \|\lambda_P\|_p^p] + \varepsilon \|\lambda\|_p^p.$$

Since $\lim_{P \in \mathscr{P}} \|\lambda_P\|_p^p = \|\lambda\|_p^p$ (see the remark following Definition 7.2.7) and $\varepsilon > 0$ is arbitrary, it follows that

$$\lim_{P \in \mathscr{P}} \|\lambda - \lambda_P\|_p^p = 0.$$

Now, we come to the case $p = 1$. Let λ be a positive charge in $V_1(\Omega, \mathscr{F}, \mu)$. Then for any P in $\mathscr{P}$ and $n \geq 1$,

$$\|\lambda - \lambda_P\|_1 \leq \|\lambda - (\lambda \wedge n\mu)\|_1 + \|(\lambda \wedge n\mu) - (\lambda \wedge n\mu)_P\|_1 + \|(\lambda \wedge n\mu)_P - \lambda_P\|_1$$

$$\leq 2\|\lambda - (\lambda \wedge n\mu)\|_1 + \|(\lambda \wedge n\mu) - (\lambda \wedge n\mu)_P\|_1$$

(by the remark following Definition 7.2.7)

$$\leq 2\|\lambda - (\lambda \wedge n\mu)\|_1 + \|(\lambda \wedge n\mu) - (\lambda \wedge n\mu)_P\|_2$$

(by Theorem 7.2.4).

Since $0 \leq (\lambda \wedge n\mu) \leq n\mu$ and $n\mu \in V_2(\Omega, \mathscr{F}, \mu)$, it follows that $(\lambda \wedge n\mu) \in V_2(\Omega, \mathscr{F}, \mu)$. Therefore, by what we have proved in the first part, $\lim_{P \in \mathscr{P}} \|(\lambda \wedge n\mu) - (\lambda \wedge n\mu)_P\|_2 = 0$. Hence, by Proposition 7.2.11,

$$\lim_{n \to \infty} \lim_{P \in \mathscr{P}} [2\|\lambda - (\lambda \wedge n\mu)\|_1 + \|(\lambda \wedge n\mu) - (\lambda \wedge n\mu)_P\|_2] = 0.$$

Now, it follows that $\lim_{P \in \mathscr{P}} \|\lambda - \lambda_P\|_1 = 0$. For any λ in $V_1(\Omega, \mathscr{F}, \mu)$, write $\lambda = \lambda^+ - \lambda^-$. Note that $\lambda_P = (\lambda^+)_P - (\lambda^-)_P$. Hence $\lim_{P \in \mathscr{P}} \|\lambda - \lambda_P\|_1 = 0$. This completes the proof. □

For a general bounded charge μ, not necessarily positive, by defining V_p-norms with respect to $|\mu|$, we can, in fact, obtain the above theorem for general bounded charges.

The above theorem for $p = 1$ gives the Radon–Nikodym theorem of Section 6.3 for charges.

7.2.13 Corollary (Radon–Nikodym Theorem). *Let $\mathscr{F}$ be a field of subsets of a set Ω. Let μ and ν be two bounded charges on $\mathscr{F}$ such that $\nu \ll \mu$. Then for each $\varepsilon > 0$, there exists a simple function f on Ω such that*

$$\left| \nu(F) - D \int_F f \, d\mu \right| < \varepsilon$$

for every F in $\mathscr{F}$.

We also obtain from the above theorem that $V_p(\Omega, \mathscr{F}, \mu)$ is the completion of $\mathscr{L}_p(\Omega, \mathscr{F}, \mu)$.

7.2.14 Corollary. *Let $(\Omega, \mathscr{F}, \mu)$ be a probability charge space. Then $V_p(\Omega, \mathscr{F}, \mu)$ is the completion of $\mathscr{L}_p(\Omega, \mathscr{F}, \mu)$ for every $1 \le p < \infty$.*

Proof. This is a consequence of Theorem 4.6.15 and Theorem 7.2.12. □

7.2.15 Remark. We have refrained from making any statement about $\mathscr{L}_\infty(\Omega, \mathscr{F}, \mu)$ in the above corollary. It is not true that $V_\infty(\Omega, \mathscr{F}, \mu)$ is the completion of $\mathscr{L}_\infty(\Omega, \mathscr{F}, \mu)$. The following example justifies this point.

We recall the example given in Remark 4.6.8. Let $\Omega = \{1, 2, 3, \ldots\}$, $\mathscr{F}$ the finite–cofinite field on Ω and μ the charge on $\mathscr{F}$ defined by

$$\mu(A) = \sum_{n \in A} \frac{1}{2^n}, \quad \text{if A is finite,}$$

$$= 2 - \sum_{n \in A^c} \frac{1}{2^n}, \quad \text{if A is cofinite.}$$

We state a number of interesting facts about this $(\Omega, \mathscr{F}, \mu)$.
1. $\mathscr{L}_1(\Omega, \mathscr{F}, \mu)$ is not complete. See Remark 4.6.8.
2. Any bounded charge ν on $\mathscr{F}$ is absolutely continuous with respect to μ. (Let $\varepsilon > 0$. Find $m \ge 1$ such that $\sum_{n \ge m} |\nu|(\{n\}) < \varepsilon$. Take $\delta = \sum_{n \ge m} \mu(\{n\})$.)
3. $V_1(\Omega, \mathscr{F}, \mu) = \text{ba}(\Omega, \mathscr{F})$.
4. A real valued function f on Ω is T_1-measurable (T_2-measurable) if and only if $f(n)$, $n \ge 1$ converges.
5. A real valued function f on Ω is essentially bounded if and only if f is bounded.
6. If a real valued function f on Ω is bounded, then $\|f\|_\infty = \text{Sup}\{|f(n)|; n \in \Omega\}$.
7. $\mathscr{L}_\infty(\Omega, \mathscr{F}, \mu) = c$, the space of all convergent sequences of real numbers equipped with the supremum norm. Thus $\mathscr{L}_\infty(\Omega, \mathscr{F}, \mu)$ is complete.
8. A bounded charge ν on $\mathscr{F}$ is in $V_\infty(\Omega, \mathscr{F}, \mu)$ if and only if $\nu(\{n\})/\mu(\{n\})$, $n \ge 1$ is a bounded sequence of numbers.
9. Let ν on $\mathscr{F}$ be defined by

$$\nu(A) = \sum_{\substack{n \in A \\ n \text{ even}}} \frac{1}{2^n} - \sum_{\substack{n \in A \\ n \text{ odd}}} \frac{1}{2^n}, \quad A \in \mathscr{F}.$$

For this ν, there is no essentially bounded T_1-measurable function f on Ω such that $\nu(F) = D\int_F f \, d\mu$ for every F in $\mathscr{F}$.
10. $\mathscr{L}_\infty(\Omega, \mathscr{F}, \mu)$ is a proper closed subspace of $V_\infty(\Omega, \mathscr{F}, \mu)$. Thus $V_\infty(\Omega, \mathscr{F}, \mu)$ is not the completion of $\mathscr{L}_\infty(\Omega, \mathscr{F}, \mu)$.

We need an analogue of Proposition 7.2.11 for V_p-norms to establish some results on strong convergence in V_p-spaces. This is acheived in the following.

7.2.16 Lemma. *Let $(\Omega, \mathcal{F}, \mu)$ be a probability charge space and ν a simple positive charge on $\mathcal{F}$. Then*

$$\lim_{n \to \infty} \|\nu - (\nu \wedge n\mu)\|_p = 0$$

for any $1 \le p < \infty$.

Proof. If $\nu = \sum_{i=1}^{m} a_i \mu_{E_i}$ for some nonnegative real numbers $a_1, a_2, \ldots, a_m$ and some partition $\{E_1, E_2, \ldots, E_m\}$ in $\mathcal{P}$, then $\nu \le n\mu$ for sufficiently large n. Hence the above limit is zero. □

7.2.17 Theorem. *Let $(\Omega, \mathcal{F}, \mu)$ be a probability charge space and $1 \le p < \infty$. Let ν be a positive charge in $V_p(\Omega, \mathcal{F}, \mu)$. Then*

$$\lim_{n \to \infty} \|\nu - (\nu \wedge n\mu)\|_p = 0.$$

Proof. Let $P = \{E_1, E_2, \ldots, E_n\}$ be any partition in $\mathcal{P}$. Then for any $n \ge 1$,

$$\|\nu - (\nu \wedge n\mu)\|_p \le \|\nu - \nu_P\|_p + \|\nu_P - (\nu_P \wedge n\mu)\|_p + \|(\nu_P \wedge n\mu) - (\nu \wedge n\mu)\|_p.$$

By Theorem 1.5.4(29), $|(\nu_P \wedge n\mu) - (\nu \wedge n\mu)| \le |\nu_P - \nu|$. Consequently, for $n \ge 1$, $\|\nu - (\nu \wedge n\mu)\|_p \le 2\|\nu - \nu_P\|_p + \|\nu_P - (\nu_P \wedge n\mu)\|_p$. By Lemma 7.2.16, $0 \le \limsup_{n \to \infty} \|\nu - (\nu \wedge n\mu)\|_p \le 2\|\nu - \nu_P\|_p + 0$. Since this inequality is true for all P in $\mathcal{P}$, by Theorem 7.2.12, we have $\lim_{n \to \infty} \|\nu - (\nu \wedge n\mu)\|_p = 0$. This completes the proof. □

7.3 DUALS OF V_p-SPACES

In this section, we show that the dual of $V_p(\Omega, \mathcal{F}, \mu)$ for any $1 \le p < \infty$ is $V_q(\Omega, \mathcal{F}, \mu)$, where $1/p + 1/q = 1$.

7.3.1 Theorem. *Let $(\Omega, \mathcal{F}, \mu)$ be a probability charge space. Then the dual of $V_p(\Omega, \mathcal{F}, \mu)$ for any $1 \le p < \infty$ is isometrically ismorphic to $V_q(\Omega, \mathcal{F}, \mu)$, where $1/p + 1/q = 1$.*

Proof. Let $1 < p < \infty$ and $1 < q < \infty$ be such that $1/p + 1/q = 1$. Let $\nu \in V_q(\Omega, \mathcal{F}, \mu)$. We define a linear functional T_ν on $V_p(\Omega, \mathcal{F}, \mu)$ as follows. Let $\lambda \in SC(\Omega, \mathcal{F}, \mu)$. Then $\lambda = \sum_{i=1}^{m} a_i \mu_{E_i}$ for some real numbers $a_1, a_2, \ldots, a_m$ and some partition $\{E_1, E_2, \ldots, E_m\}$ in $\mathcal{P}$. Define $T_\nu(\lambda) = \sum_{i=1}^{m} a_i \nu(E_i)$. T_ν is well defined on $SC(\Omega, \mathcal{F}, \mu)$. Further, T_ν is a linear functional on the linear space SC(Ω, $\mathcal{F}$, μ). Observe also that

$$|T_\nu(\lambda)| = \left|\sum_{i=1}^m a_i \nu(E_i)\right| \le \sum_{i=1}^m \left|\frac{a_i \mu(E_i)}{[\mu(E_i)]^{1/q}}\right| \left|\frac{\nu(E_i)}{[\mu(E_i)]^{1/p}}\right|$$

$$\le \left[\sum_{i=1}^m \left|\frac{a_i \mu(E_i)}{[\mu(E_i)]^{1/q}}\right|^p\right]^{1/p} \left[\sum_{i=1}^m \left|\frac{\nu(E_i)}{[\mu(E_i)]^{1/p}}\right|^q\right]^{1/q}$$

(by Holder's inequality)

$$\le \left[\sum_{i=1}^m |a_i|^p \mu(E_i)\right]^{1/p} \left[\sum_{i=1}^m \left|\frac{\nu(E_i)}{\mu(E_i)}\right|^q \mu(E_i)\right]^{1/q}$$

$$\le \|\lambda\|_p \|\nu\|_q.$$

T_ν can be extended uniquely as a continuous linear functional to $V_p(\Omega, \mathcal{F}, \mu)$ since SC(Ω, $\mathcal{F}$, μ) is dense in $V_p(\Omega, \mathcal{F}, \mu)$. Let us denote this extension again by T_ν. T_ν has the property:

$$|T_\nu(\lambda)| \le \|\nu\|_q \|\lambda\|_p$$

for any λ in $V_p(\Omega, \mathcal{F}, \mu)$.

Let us compute the norm of T_ν. Obviously, $\|T_\nu\| \le \|\nu\|_q$. We shall establish the reverse inequality by showing that for any P in $\mathcal{P}$, there exists a simple charge λ such that $|T_\nu(\lambda)| = \|\nu_P\|_q \|\lambda\|_p$. Then it would follow that

$$\|\nu_P\|_q \|\lambda\|_p = |T_\nu(\lambda)| \le \|T_\nu\| \|\lambda\|_p$$

from which we obtain $\|\nu_P\|_q \le \|T_\nu\|$ for any P in $\mathcal{P}$. Hence $\|\nu\|_q \le \|T_\nu\|$ and so, the equality ensues.

Now, let P = $\{E_1, E_2, \ldots, E_n\} \in \mathcal{P}$. Let

$$c_i = [\text{Sign } \nu(E_i)] \left|\frac{\nu(E_i)}{\mu(E_i)}\right|^{q-1}$$

for $i = 1, 2, \ldots, n$ and $\lambda = \sum_{i=1}^n c_i \mu_{E_i}$. Then

$$|T_\nu(\lambda)| = \sum_{i=1}^n c_i \nu(E_i) = \sum_{i=1}^n \left|\frac{\nu(E_i)}{\mu(E_i)}\right|^q \mu(E_i) = \|\nu_P\|_q^q.$$

Let us calculate $\|\lambda\|_p$.

$$\|\lambda\|_p = \left[\sum_{i=1}^n |c_i|^p \mu(E_i)\right]^{1/p} = \left(\sum_{i=1}^n \left|\frac{\nu(E_i)}{\mu(E_i)}\right|^{p(q-1)} \mu(E_i)\right)^{1/p}$$

$$= \left[\sum_{i=1}^n \left|\frac{\nu(E_i)}{\mu(E_i)}\right|^q \mu(E_i)\right]^{1/p} = \|\nu_P\|_q^{q/p}.$$

Consequently, we have $|T_\nu(\lambda)| = \|\nu_P\|_q \|\lambda\|_p$.

7. V_p-SPACES   195

Thus we have associated a continuous linear functional T_ν on $V_p(\Omega, \mathcal{F}, \mu)$ for any given $\nu \in V_q(\Omega, \mathcal{F}, \mu)$ such that $\|T_\nu\| = \|\nu\|_q$.

The map $\nu \to T_\nu$ from $V_q(\Omega, \mathcal{F}, \mu)$ to $V_p^*(\Omega, \mathcal{F}, \mu)$ is a linear map and is norm preserving. This map is obviously one-to-one. In order to show that this map is onto, we proceed as follows.

Let T be any nonzero continuous linear functional on $V_p(\Omega, \mathcal{F}, \mu)$. Define a set function ν_T on $\mathcal{F}$ by $\nu_T(F) = T(\mu_F)$, $F \in \mathcal{F}$. We enumerate the properties of ν_T.

(i). Since T is a linear functional on $V_p(\Omega, \mathcal{F}, \mu)$, ν_T is a charge on $\mathcal{F}$.

(ii). Since T is a continuous linear functional, ν_T is bounded. For, for any F in $\mathcal{F}$, $|\nu_T(F)| = |T(\mu_F)| \le \|T\| \|\mu_F\|_p \le \|T\| \mu(F) \le \|T\|$.

(iii). $\nu_T \ll \mu$. Let $\varepsilon > 0$. Take $\delta = \varepsilon/\|T\|$. If $F \in \mathcal{F}$ and $\mu(F) < \delta$, then $|\nu_T(F)| < \varepsilon$.

(iv). $\nu_T \in V_q(\Omega, \mathcal{F}, \mu)$. Let $P = \{E_1, E_2, \ldots, E_n\} \in \mathcal{P}$. Then

$$\sum_{i=1}^n \left|\frac{\nu_T(E_i)}{\mu(E_i)}\right|^q \mu(E_i) = \sum_{i=1}^n \left|\frac{T(\mu_{E_i})}{\mu(E_i)}\right|^q \mu(E_i)$$

$$\le \sum_{i=1}^n \|T\|^q [\mu(E_i)]^q [\mu(E_i)]^{1-q} = \|T\|^q.$$

Hence $\|\nu_T\|_q^q \le \|T\|^q$. Thus $\nu_T \in V_q(\Omega, \mathcal{F}, \mu)$. For this ν_T, $T_{\nu_T} = T$. Hence $\|\nu_T\| = \|T\|$.

This completes the proof that $V_p^*(\Omega, \mathcal{F}, \mu)$ and $V_q(\Omega, \mathcal{F}, \mu)$ are isometrically isomorphic for $1 < p < \infty$.

Now, we take up the case $p = 1$. Then $q = \infty$.

If $\nu \in V_\infty(\Omega, \mathcal{F}, \mu)$, we define first a linear functional T_ν on $SC(\Omega, \mathcal{F}, \mu)$ by $T_\nu(\lambda) = \sum_{i=1}^n a_i \nu(E_i)$ for $\lambda = \sum_{i=1}^n a_i \mu_{E_i}$ for some real numbers $a_1, a_2, \ldots, a_n$ and some partition $\{E_1, E_2, \ldots, E_n\}$ in $\mathcal{P}$. T_ν is a well defined linear functional on $SC(\Omega, \mathcal{F}, \mu)$. Further,

$$|T_\nu(\lambda)| \le \sum_{i=1}^n |a_i| |\nu(E_i)| \le \sum_{i=1}^n |a_i| \|\nu\|_\infty \mu(E_i)$$

$$\le \|\nu\|_\infty \sum_{i=1}^n |a_i| \mu(E_i) \le \|\nu\|_\infty \|\lambda\|_1.$$

Thus T_ν is a bounded linear functional on $SC(\Omega, \mathcal{F}, \mu)$ and so can be extended uniquely as a continuous linear functional on $V_1(\Omega, \mathcal{F}, \mu)$. We denote this extension again by T_ν. Then

$$|T_\nu(\lambda)| \le \|\nu\|_\infty \|\lambda\|_1$$

for every λ in $V_1(\Omega, \mathcal{F}, \mu)$. Consequently, $\|T_\nu\| \le \|\nu\|_\infty$. To prove the reverse inequality, we proceed as follows. Let $F \in \mathcal{F}$ and $\mu(F) > 0$. Let $a_1 =$

sign $(\nu(F))/\mu(F)$ and $a_2 = 0$. Let $\lambda = a_1\mu_F + a_2\mu_{F^c}$. Since $\|\lambda\|_1 = 1$, we have

$$|T_\nu(\lambda)| = \left|\frac{\nu(F)}{\mu(F)}\right| \leq \|T_\nu\|\|\lambda\|_1 = \|T_\nu\|.$$

Since this inequality is true for any F in $\mathscr{F}$ with $\mu(F) > 0$, we have

$$\text{Sup}\left\{\left|\frac{\nu(F)}{\mu(F)}\right|; F \in \mathscr{F}\right\} \leq \|T_\nu\|.$$

Hence $\|\nu\|_\infty \leq \|T_\nu\|$. Thus we have proved that $\|T_\nu\| = \|\nu\|_\infty$.

Hence, the map $\nu \to T_\nu$ from $V_\infty(\Omega, \mathscr{F}, \mu)$ to $V_1^*(\Omega, \mathscr{F}, \mu)$ is a linear map such that $\|T_\nu\| = \|\nu\|_\infty$. So, this map is one-to-one. In order to show that this map is an isometric isomorphism, it suffices to show that for any given continuous linear functional T on $L_1(\Omega, \mathscr{F}, \mu)$, there exists ν_T in $V_\infty(\Omega, \mathscr{F}, \mu)$ such that $T = T_{\nu_T}$. The existence of ν_T can be established along the same lines as that of the case $1 < p < \infty$. This completes the proof. □

Finally, we end this section with some remarks on the duality between V_p-spaces.

7.3.2 Remark. Let $(\Omega, \mathscr{F}, \mu)$ be a probability charge space and $p > 1$, $q > 1$ be such that $1/p + 1/q = 1$. For each ν in $V_q(\Omega, \mathscr{F}, \mu)$, T_ν defined above is a continuous linear functional on $V_p(\Omega, \mathscr{F}, \mu)$ and for each λ in $V_p(\Omega, \mathscr{F}, \mu)$, T_λ is a continuous linear functional on $V_q(\Omega, \mathscr{F}, \mu)$. The duality between $V_p(\Omega, \mathscr{F}, \mu)$ and $V_q(\Omega, \mathscr{F}, \mu)$ can be expressed by

$$T_\nu(\lambda) = T_\lambda(\nu)$$

for every λ in $V_p(\Omega, \mathscr{F}, \mu)$ and ν in $V_q(\Omega, \mathscr{F}, \mu)$.

For any parition $P = \{E_1, E_2, \ldots, E_m\}$ in $\mathscr{P}$,

$$T_\nu(\lambda_P) = \sum_{i=1}^m \frac{\lambda(E_i)}{\mu(E_i)} \nu(E_i) = \sum_{i=1}^m \frac{\nu(E_i)}{\mu(E_i)} \lambda(E_i) = T_\lambda(\nu_P).$$

Thus we have shown that $T_\nu(\lambda_P) = T_\lambda(\nu_P)$. Since $\lim_{P \in \mathscr{P}} \|\lambda_P - \lambda\|_p = 0 = \lim_{P \in \mathscr{P}} \|\nu_P - \nu\|_q$, it follows that $T_\nu(\lambda) = T_\lambda(\nu)$ for any λ in $V_p(\Omega, \mathscr{F}, \mu)$ and ν in $V_q(\Omega, \mathscr{F}, \mu)$. This argument also shows that

$$\lim_{P \in \mathscr{P}} \sum_{i=1}^m \frac{\lambda(E_i)\nu(E_i)}{\mu(E_i)}$$

exists and is equal to $T_\nu(\lambda) = T_\lambda(\nu)$ for every λ in $V_p(\Omega, \mathscr{F}, \mu)$ and ν in $V_q(\Omega, \mathscr{F}, \mu)$.

7.4 STRONG CONVERGENCE

In this section, we give some conditions for strong convergence in V_p-spaces for $1 \leq p < \infty$.

First, we need the following result.

7.4.1 Proposition. *Let $(\Omega, \mathcal{F}, \mu)$ be a probability charge space and ν a bounded charge on $\mathcal{F}$ absolutely continuous with respect to μ. Then for any $1 \leq p, p_1, p_2 < \infty$, $t_1, t_2 \geq 0$ with $t_1 + t_2 = 1$ and $p = t_1 p_1 + t_2 p_2$, the inequality*

$$\|\nu\|_p^p \leq \|\nu\|_{p_1}^{p_1 t_1} \|\nu\|_{p_2}^{p_2 t_2}$$

holds.

Proof. Let $P = \{E_1, E_2, \ldots, E_n\} \in \mathcal{P}$ and $t_1, t_2 > 0$. Then

$$\sum_{i=1}^n \left|\frac{\nu(E_i)}{\mu(E_i)}\right|^p \mu(E_i) = \sum_{i=1}^n \left[\left|\frac{\nu(E_i)}{\mu(E_i)}\right|^{p_1} \mu(E_i)\right]^{t_1} \left[\left|\frac{\nu(E_i)}{\mu(E_i)}\right|^{p_2} \mu(E_i)\right]^{t_2}.$$

Applying Hölder's inequality to the numbers $1/t_1$ and $1/t_2$, i.e.,

$$\frac{1}{1/t_1} + \frac{1}{1/t_2} = 1,$$

we obtain

$$\sum_{i=1}^n \left|\frac{\nu(E_i)}{\mu(E_i)}\right|^p \mu(E_i) \leq \left[\sum_{i=1}^n \left|\frac{\nu(E_i)}{\mu(E_i)}\right|^{p_1} \mu(E_i)\right]^{t_1} \left[\sum_{i=1}^n \left|\frac{\nu(E_i)}{\mu(E_i)}\right|^{p_2} \mu(E_i)\right]^{t_2}$$

$$\leq \|\nu\|_{p_1}^{t_1 p_1} \|\nu\|_{p_2}^{t_2 p_2}.$$

Since this inequality is true for every P in $\mathcal{P}$, we obtain

$$\|\nu\|_p^p \leq \|\nu\|_{p_1}^{p_1 t_1} \|\nu\|_{p_2}^{p_2 t_2}.$$

The case $t_1 = 0$ or $t_2 = 0$ is trivial. □

The following theorem provides a sufficient condition for strong convergence in $V_p(\Omega, \mathcal{F}, \mu)$ for $p > 1$ in terms of strong convergence in $V_1(\Omega, \mathcal{F}, \mu)$.

7.4.2 Theorem. *Let $1 < r \leq \infty$. Let $(\Omega, \mathcal{F}, \mu)$ be a probability charge space and ν, ν_n, $n \geq 1$ a sequence in $V_r(\Omega, \mathcal{F}, \mu)$ satisfying the following two conditions.*

(i). ν_n, $n \geq 1$ *converges to ν in* $V_1(\Omega, \mathcal{F}, \mu)$.

(ii). $\sup_{n \geq 1} \|\nu_n\|_r < \infty$.

Then ν_n, $n \geq 1$ converges to ν in $V_p(\Omega, \mathcal{F}, \mu)$ for any $1 \leq p < r$.

Proof. In view of Theorem 7.2.4, it suffices to prove for the case $r<\infty$. Assume, also, without loss of generality, that $\nu=0$. In Proposition 7.4.1, if we take $p_1=1$, $p_2=r$, $t_1=(r-p)/(r-1)$ and $t_2=(p-1)/(r-1)$, then for every $n\geq 1$,

$$\|\nu_n\|_p^p \leq \|\nu_n\|_1^{(r-p)/(r-1)} \|\nu_n\|_r^{r(p-1)/(r-1)}$$

$$\leq \|\nu_n\|_1^{(r-p)/(r-1)} [\sup_{n\geq 1} \|\nu_n\|_r]^{r(p-1)/(r-1)}.$$

By (i), it follows that $\lim_{n\to\infty} \|\nu_n\|_p = 0$. This completes the proof. □

The following therorem is analogous to the Lebesgue Dominated convergence thereorem proved in Section 4.6. This result also gives a sufficient condition for strong convergence in $V_p(\Omega, \mathcal{F}, \mu)$ for $p>1$ in terms of strong convergence in $V_1(\Omega, \mathcal{F}, \mu)$.

7.4.3 Theorem. *Let $(\Omega, \mathcal{F}, \mu)$ be a probability charge space and ν, ν_n, $n\geq 1$ a sequence in $V_1(\Omega, \mathcal{F}, \mu)$ having the following properties.*
   (i). *ν_n, $n\geq 1$ converges to ν in $V_1(\Omega, \mathcal{F}, \mu)$.*
   (ii). *$|\nu_n|\leq \lambda$ for every $n\geq 1$ for some λ in $V_p(\Omega, \mathcal{F}, \mu)$ with $1\leq p<\infty$.*
Then ν_n, $n\geq 1$ converges to ν in $V_p(\Omega, \mathcal{F}, \mu)$.

Proof. For the case $p=1$, there is nothing to be proved. Let us assume $p>1$. By (ii), it follows that each $\nu_n \in V_p(\Omega, \mathcal{F}, \mu)$. By (i), it follows that $\lim_{n\to\infty} \nu_n(F) = \nu(F)$ for every F in $\mathcal{F}$. It also follows that $\lim_{n\to\infty} |\nu_n|(F) = |\nu|(F)$ for every F in $\mathcal{F}$. (Use the fact that $|\nu_n|$, $n\geq 1$ converges to $|\nu|$ in $V_1(\Omega, \mathcal{F}, \mu)$.) Consequently, $|\nu|\leq \lambda$. Hence $\nu \in V_p(\Omega, \mathcal{F}, \mu)$ also.

Assume, without loss of generality, that each ν_n is positive. From the lattice properties of $ba(\Omega, \mathcal{F})$, for any m, $n\geq 1$, we have $(\lambda \wedge m\mu) - (\nu_n \wedge m\mu) \leq \lambda - \nu_n$. See Theorem 1.5.4(29). Therefore, from this inequality, we obtain

$$0\leq \nu_n - (\nu_n \wedge m\mu) \leq \lambda - (\lambda \wedge m\mu).$$

Hence $\|\nu_n - (\nu_n \wedge m\mu)\|_p \leq \|\lambda - (\lambda \wedge m\mu)\|_p$. Thus for any m, $n\geq 1$,

$$\|\nu - \nu_n\|_p \leq \|\nu - (\nu \wedge m\mu)\|_p + \|(\nu \wedge m\mu) - (\nu_n \wedge m\mu)\|_p + \|(\nu_n \wedge m\mu) - \nu_n\|_p$$

$$\leq \|\nu - (\nu \wedge m\mu)\|_p + \|(\nu \wedge m\mu) - (\nu_n \wedge m\mu)\|_p + \|\lambda - (\lambda \wedge m\mu)\|_p$$

Observe that for each fixed $m\geq 1$, $(\nu_n \wedge m\mu)$, $n\geq 1$ converges to $(\nu \wedge m\mu)$ in $V_1(\Omega, \mathcal{F}, \mu)$. This follows from the inequality $\|(\nu \wedge m\mu) - (\nu_n \wedge m\mu)\|_1 \leq \|\nu - \nu_n\|_1$ for every $n\geq 1$ and (i). Further, $(\nu_n \wedge m\mu) \leq m\mu$ for every $n\geq 1$. Consequently, by Theorem 7.4.2, $(\nu_n \wedge m\mu)$, $n\geq 1$ converges to $(\nu \wedge m\mu)$ in $V_p(\Omega, \mathcal{F}, \mu)$ for every fixed $m\geq 1$. Therefore, for every fixed $m\geq 1$,

$$0\leq \limsup_{n\to\infty} \|\nu - \nu_n\|_p \leq \|\nu - (\nu \wedge m\mu)\|_p + 0 + \|\lambda - (\lambda \wedge m\mu)\|_p.$$

Now, by letting $m \to \infty$, by Theorem 7.2.15, we have

$$0 \leq \limsup_{n \to \infty} \|\nu - \nu_n\|_p \leq 0 + 0.$$

Hence $\lim_{n \to \infty} \|\nu - \nu_n\|_p = 0$. This completes the proof. $\square$

The following theorem gives necessary and sufficient conditions for strong convergence in $V_p(\Omega, \mathcal{F}, \mu)$ for $1 < p < \infty$.

7.4.4 Theorem. *Let $1 < p < \infty$. Let $(\Omega, \mathcal{F}, \mu)$ be a probability charge space and ν_n, $n \geq 1$ a sequence in $V_p(\Omega, \mathcal{F}, \mu)$.*
(i) *If ν_n, $n \geq 1$ is strongly convergent to a ν in $V_p(\Omega, \mathcal{F}, \mu)$, then $\nu(F) = \lim_{n \to \infty} \nu_n(F)$ for every F in $\mathcal{F}$ and $\lim_{n \to \infty} \|\nu_n\|_p = \|\nu\|_p$.*
(ii). *If $\nu_n(F)$, $n \geq 1$ converges for every F in $\mathcal{F}$ and $\|\nu_n\|_p$, $n \geq 1$ is convergent, then the set function ν defined by $\nu(F) = \lim_{n \to \infty} \nu_n(F)$, $F \in \mathcal{F}$ belongs to $V_p(\Omega, \mathcal{F}, \mu)$. Further, if $\|\nu_n\|_p$, $n \geq 1$ converges to $\|\nu\|_p$, then ν_n, $n \geq 1$ converges strongly to ν in $V_p(\Omega, \mathcal{F}, \mu)$.*

Proof. (i). This follows from the inequalities

$$\big|\|\nu_n\|_p - \|\nu\|_p\big| \leq \|\nu_n - \nu\|_p$$

and

$$|\nu_n(F) - \nu(F)| \leq \|\nu_n - \nu\|_p [\mu(F)]^{(p-1)/p} \leq \|\nu_n - \nu\|_p$$

for every F in $\mathcal{F}$.
(ii). ν is obviously a charge on $\mathcal{F}$. We show that $\nu \ll \mu$ and that $\nu \in V_p(\Omega, \mathcal{F}, \mu)$. Since $\|\nu_n\|_p$, $n \geq 1$ is convergent, there is a positive constant M such that $\|\nu_n\|_p \leq M$ for every $n \geq 1$. From the definition of $\|\nu_n\|_p$, it follows that for any F in $\mathcal{F}$ and $n \geq 1$,

$$\left|\frac{\nu_n(F)}{\mu(F)}\right|^p \mu(F) \leq \|\nu_n\|_p^p.$$

Equivalently, we have the inequality $|\nu_n(F)| \leq \|\nu_n\|_p [\mu(F)]^{(p-1)/p} \leq M[\mu(F)]^{(p-1)/p}$. Letting $n \to \infty$, we obtain $|\nu(F)| \leq M[\mu(F)]^{(p-1)/p}$. From this, it follows that $\nu \ll \mu$.

For any partition $P = \{E_1, E_2, \ldots, E_m\}$ in $\mathcal{P}$,

$$\sum_{i=1}^{m} \left|\frac{\nu(E_i)}{\mu(E_i)}\right|^p \mu(E_i) = \lim_{n \to \infty} \sum_{i=1}^{m} \left|\frac{\nu_n(E_i)}{\mu(E_i)}\right|^p \mu(E_i)$$

$$\leq \limsup_{n \to \infty} \|\nu_n\|_p^p \leq M^p.$$

Taking supremum over all partitions P in $\mathcal{P}$, we obtain $\|\nu\|_p^p \leq M^p$, or $\nu \in V_p(\Omega, \mathcal{F}, \mu)$.

Now, for any $n \geq 1$ and for any given $\varepsilon > 0$, by Lemma 7.2.9, there exists a constant k such that for any partition $P = \{E_1, E_2, \ldots, E_m\}$ in $\mathscr{P}$, we have

$$\|\nu - \nu_n\|_p \leq \|\nu - \nu_P\|_p + \|\nu_P - (\nu_n)_P\|_p + \|(\nu_n)_P - \nu_n\|_p$$

$$\leq \|\nu - \nu_P\|_p + \|\nu_P - (\nu_n)_P\|_p + [k(\|\nu_n\|_p^p - \|(\nu_n)_P\|_p^p) + \varepsilon \|\nu_n\|_p^p]^{1/p}$$

Observe that

$$\|\nu_P - (\nu_n)_P\|_p^p = \sum_{i=1}^{m} \left|\frac{\nu(E_i) - \nu_n(E_i)}{\mu(E_i)}\right|^p \mu(E_i)$$

and this converges to zero as $n \to \infty$ since $\lim_{n \to \infty} \nu_n(F) = \nu(F)$ for every F in $\mathscr{F}$. Therefore,

$$\limsup_{n \to \infty} \|\nu - \nu_n\|_p \leq \|\nu - \nu_P\|_p + 0 + [k(\|\nu\|_p^p - \|\nu_P\|_p^p) + \varepsilon \|\nu\|_p^p]^{1/p}$$

for every P in $\mathscr{P}$. By taking $\lim_{P \in \mathscr{P}}$, we obtain

$$\limsup_{n \to \infty} \|\nu - \nu_n\|_p \leq 0 + 0 + \varepsilon \|\nu\|_p.$$

Since $\varepsilon > 0$ is arbitrary, we conclude that $\lim_{n \to \infty} \|\nu - \nu_n\|_p = 0$. This completes the proof. □

The following theorem gives a necessary and sufficient condition for strong convergence in $V_1(\Omega, \mathscr{F}, \mu)$.

7.4.5 Theorem. *Let $(\Omega, \mathscr{F}, \mu)$ be a probability charge space and ν_n, $n \geq 1$ a sequence in $V_1(\Omega, \mathscr{F}, \mu)$. Then ν_n, $n \geq 1$ converges strongly in $V_1(\Omega, \mathscr{F}, \mu)$ if and only if $\nu_n(F)$, $n \geq 1$ converges uniformly over F in $\mathscr{F}$.* □

7.5 WEAK CONVERGENCE

Weak convergence in $V_1(\Omega, \mathscr{F}, \mu)$ will be considered quite extensively in Chapter 8. In this section, we prove just one result about weak convergence in $V_p(\Omega, \mathscr{F}, \mu)$ for $1 < p < \infty$.

7.5.1 Theorem. *Let $1 < p < \infty$. Let $(\Omega, \mathscr{F}, \mu)$ be a probability charge space and ν_n, $n \geq 1$ a sequence in $V_p(\Omega, \mathscr{F}, \mu)$.*
(i). If ν_n, $n \geq 1$ is weakly convergent to a ν in $V_p(\Omega, \mathscr{F}, \mu)$, then $\nu(F) = \lim_{n \to \infty} \nu_n(F)$ for F in $\mathscr{F}$ and $\sup_{n \geq 1} \|\nu_n\|_p < \infty$.
(ii). If $\nu_n(F)$, $n \geq 1$ converges for every F in $\mathscr{F}$ and $\sup_{n \geq 1} \|\nu_n\|_p < \infty$, then the set function ν defined by $\nu(F) = \lim_{n \to \infty} \nu_n(F)$, $F \in \mathscr{F}$ belongs to $V_p(\Omega, \mathscr{F}, \mu)$ and ν_n, $n \geq 1$ converges weakly to ν in $V_p(\Omega, \mathscr{F}, \mu)$.

7. V_p-SPACES

Proof. (i) If ν_n, $n \geq 1$ converges to ν weakly in $V_p(\Omega, \mathscr{F}, \mu)$, then $T_\lambda(\nu_n)$, $n \geq 1$ converges to $T_\lambda(\nu)$ for every λ in $V_q(\Omega, \mathscr{F}, \mu)$, where $1/p + 1/q = 1$. For F in $\mathscr{F}$, let $\lambda = \mu_F$ which obviously belongs to $V_q(\Omega, \mathscr{F}, \mu)$. Consequently,

$$\nu(F) = T_\nu(\lambda) = T_\lambda(\nu) = \lim_{n \to \infty} T_\lambda(\nu_n)$$

$$= \lim_{n \to \infty} T_{\nu_n}(\lambda) = \lim_{n \to \infty} \nu_n(F).$$

See Remark 7.3.2. It is well known that every weakly convergent sequence in a Banach space is norm bounded. See Thoerem 1.5.16.

(ii). As in the proof of (ii) of Theorem 7.4.4, it follows that $\nu \in V_p(\Omega, \mathscr{F}, \mu)$. Now, we show that ν_n, $n \geq 1$ converges to ν weakly in $V_p(\Omega, \mathscr{F}, \mu)$. Let $\lambda \in V_q(\Omega, \mathscr{F}, \mu)$ and T_λ be the continuous linear functional on $V_p(\Omega, \mathscr{F}, \mu)$ induced by λ, where $1/p + 1/q = 1$. T_λ is determined on $SC(\Omega, \mathscr{F}, \mu)$ by the following rule. $T_\lambda(\tau) = \sum_{i=1}^k a_i \lambda(E_i)$, where $\tau = \sum_{i=1}^k a_i \mu_{E_i}$ for some real numbers $a_1, a_2, \ldots, a_k$ and some partition $\{E_1, E_2, \ldots, E_k\}$ in $\mathscr{P}$. We show that $T_\lambda(\nu_n)$, $n \geq 1$ converges to $T_\lambda(\nu)$. This would then imply weak convergence of ν_n, $n \geq 1$ to ν in $V_p(\Omega, \mathscr{F}, \mu)$. Observe that for any partition $P = \{F_1, F_2, \ldots, F_m\}$ in $\mathscr{P}$,

$$|T_\lambda(\nu_n) - T_\lambda(\nu)| \leq |T_\lambda(\nu_n) - T_\lambda((\nu_n)_P)|$$
$$+ |T_\lambda((\nu_n)_P) - T_\lambda(\nu_P)| + |T_\lambda(\nu_P) - T_\lambda(\nu)|$$
$$\leq |T_\lambda(\nu_n) - T_\lambda((\nu_n)_P)|$$
$$+ \left| \sum_{i=1}^m \left(\frac{\nu_n(E_i)}{\mu(E_i)}\right) \lambda(E_i) - \sum_{i=1}^m \left(\frac{\nu(E_i)}{\mu(E_i)}\right) \lambda(E_i) \right|$$
$$+ |T_\lambda(\nu_P) - T_\lambda(\nu)|.$$

First, we let $n \to \infty$. Then we obtain

$$\limsup_{n \to \infty} |T_\lambda(\nu_n) - T_\lambda(\nu)| \leq \limsup_{n \to \infty} |T_\lambda(\nu_n) - T_\lambda((\nu_n)_P)|$$
$$+ 0 + |T_\lambda(\nu_P) - T_\lambda(\nu)|$$
$$\leq \limsup_{n \to \infty} |T_{\nu_n}(\lambda - \lambda_P)| + |T_\lambda(\nu_P) - T_\lambda(\nu)|.$$

This follows from the facts that $T_\lambda(\nu_n) = T_{\nu_n}(\lambda)$ and $T_\lambda((\nu_n)_P) = T_{\nu_n}(\lambda_P)$. See Remark 7.3.2. Now,

$$\limsup_{n \to \infty} |T_\lambda(\nu_n) - T_\lambda(\nu)| \leq \sup_{n \geq 1} \|\nu_n\|_p \|\lambda - \lambda_P\|_q + |T_\lambda(\nu_P) - T_\lambda(\nu)|.$$

Since this inequality is true for any P in $\mathscr{P}$, by taking the limit over $P \in \mathscr{P}$ and using Theorem 7.2.12, we obtain

$$\lim_{n \to \infty} T_\lambda(\nu_n) = T_\lambda(\nu).$$

This completes the proof. □

7.5.2 Remark. Given a probability charge space $(\Omega, \mathscr{F}, \mu)$, one can find a charge space $(X, \mathfrak{A}, \lambda)$ in which $\mathfrak{A}$ is a σ-field on X and λ is a probability measure on $\mathfrak{A}$ such that $V_p(\Omega, \mathscr{F}, \mu)$ is isometrically isomorphic to $\mathscr{L}_p(X, \mathfrak{A}, \lambda)$ for every $1 \le p \le \infty$.

CHAPTER 8

Nikodym Theorem, Weak Convergence and Vitali–Hahn–Saks Theorem

Nikodym theorem for measures gives a necessary and sufficient condition for norm boundedness of a set of measures on a σ-field. Vitali–Hahn–Saks theorem for measures characterizes uniform absolute continuity in the space of all bounded measures on a σ-field. These results are useful in the study of weak convergence in the space of all bounded measures on a σ-field. The exact analogues of Nikodym and Vitali–Hahn–Saks theorems are not valid for charges on every field. In this chapter, we seek the exact analogues of these theorems for charges by considering more restrictive fields and yet general enough to include σ-fields as a special case. The presentation of this chapter is somewhat at variance with that of the previous chapters in the following sense. In the earlier chapters, we have obtained weaker versions of some results of Measure theory for charges valid on any field. In this chapter, we obtain exact versions of the above theorems by considering some special class of fields.

Section 8.1 gives the relevant background information about Nikodym and Vitali–Hahn–Saks theorems. Section 8.2 gives some examples to demonstrate that Nikodym and Vitali–Hahn–Saks theorems are not valid for charges on fields. Section 8.3 presents Phillips' lemma which plays an important role in our quest in extending Nikodym and Vitali–Hahn–Saks theorems for charges on some special type of fields. Phillips' lemma is also used in characterizing weak convergence of charges. Section 8.4 presents the desired extension of Nikodym theorem. Section 8.5 studies norm boundedness of a set of bounded charges in the light of uniform absolute continuity. Section 8.6 gives a decomposition theorem for a given set of bounded charges in terms of norm bounded sets and finite dimensional sets. Section 8.7 is mainly concerned with characterizations of weak convergence in the space of bounded charges. Finally, Vitali–Hahn–Saks theorem is extended to special fields in Section 8.8.

8.1 NIKODYM AND VITALI–HAHN–SAKS THEOREMS IN THE CLASSICAL CASE

One of the important problems in the study of the space ca(Ω, $\mathfrak{A}$) of all bounded measures on the σ-field $\mathfrak{A}$ of subsets of a set Ω is the identification of weakly sequentially compact subsets M of ca(Ω, $\mathfrak{A}$). A set of necessary and sufficient conditions for m to be weakly sequentially compact involves norm boundedness of m and uniform absolute continuity of m with respect to some λ in ca(Ω, $\mathfrak{A}$). Nikodym theorem characterizes norm bounded subsets of ca(Ω, $\mathfrak{A}$) and Vitali–Hahn–Saks theorem deals with uniform absolute continuity.

8.1.1 Nikodym Theorem. *Let $\mathfrak{A}$ be a σ-field of subsets of a set Ω. Let M be a subset of* ca(Ω, $\mathfrak{A}$). *Then* $\sup\{|\mu(A)|; \mu \in M\} < \infty$ *for every A in $\mathfrak{A}$ if and only if* $\sup\{\|\mu\|; \mu \in M\} < \infty$.

For a proof of this result, one may refer to Dunford and Schwartz (1964, Theorem IV.9.8, p. 309). We will obtain this result as a corollary of a more general result we prove in Section 8.4 in this direction. It is worth noting that if Nikodym Theorem 8.1.1 is valid for every countable subset of M, then it is also valid for M.

For Vitali–Hahn–Saks theorem, we need the following definitions.

8.1.2 Definition. Let $\mathscr{F}$ be a field of subsets of a set Ω and $M \subset$ ba(Ω, $\mathscr{F}$). Let $\lambda \in$ ba(Ω, $\mathscr{F}$). M is said to be *uniformly absolutely continuous* with respect to λ if for every $\varepsilon > 0$, there exists $\delta > 0$ such that

$$|\nu(A)| < \varepsilon \quad \text{for every } \nu \text{ in M}$$

whenever $A \in \mathscr{F}$ and $|\lambda|(A) < \delta$.

8.1.3 Definition. Let $\mathfrak{A}$ be a σ-field of subsets of a set Ω and $M \subset$ ca(Ω, $\mathfrak{A}$). M is said to be *uniformly countably additive* if for any sequence A_n, $n \geq 1$ of pairwise disjoint sets in $\mathfrak{A}$ and $\varepsilon > 0$, there exists a natural number $N \geq 1$ such that

$$\left|\sum_{k \geq n} \mu(A_k)\right| < \varepsilon$$

for every $n \geq N$ and μ in M.

Now, we state Vitali–Hahn–Saks theorem.

8.1.4 Vitali–Hahn–Saks Theorem. *Let $\mathfrak{A}$ be a σ-field of subsets of a set Ω and μ_n, $n \geq 1$ a sequence in* ca(Ω, $\mathfrak{A}$) *such that $\mu_n(A)$, $n \geq 1$ converges to a real number for every A in $\mathfrak{A}$. Define μ on $\mathfrak{A}$ by*

$$\mu(A) = \lim_{n \to \infty} \mu_n(A), \quad A \in \mathfrak{A}.$$

Then the following statements are valid.

(i). $M = \{\mu_n, n \geq 1\}$ *is uniformly absolutely continuous with respect to* λ *whenever* $\lambda \in ca(\Omega, \mathfrak{A})$ *and* $\mu_n \ll \lambda$ *for every* $n \geq 1$.
(ii). *M is uniformly countably additive.*
(iii). μ *is a bounded measure on* $\mathfrak{A}$.

The non-trivial part of this theorem is the conclusion (i) from which (ii) and (iii) follow easily. For a proof of this result, one may refer to Dunford and Schwartz (1964, Theorem III.7.2, Corollary III.7.3 and Corollary III.7.4, pp. 158, 159 and 160). We also obtain this result as a corollary of a more general result we present in Section 8.8.

8.2 EXAMPLES

In this section, we present two examples which illustrate the point that extensions of Nikodym and Vitali–Hahn–Saks theorems fail in the framework of charges on fields. The first example we give here is concerned with Nikodym theorem.

8.2.1 Example. Let Ω be any infinite set and $\mathscr{F}$ the finite–cofinite field on Ω. Fix a sequence $x_n, n \geq 1$ of distinct points in Ω. For each $n \geq 1$, define μ_n on $\mathscr{F}$ by

$$\mu_n(A) = \begin{cases} n, & \text{if A is finite and } x_n \in A, \\ 0, & \text{if A is finite and } x_n \notin A, \\ -n, & \text{if A is cofinite and } x_n \notin A, \\ 0, & \text{if A is cofinite and } x_n \in A. \end{cases}$$

It is easy to check that each μ_n is a bounded charge on $\mathscr{F}$ and that $\text{Sup}_{n \geq 1} |\mu_n(A)| < \infty$ for every A in $\mathscr{F}$. However, $\text{Sup}_{n \geq 1} \|\mu_n\| = \infty$.

Note also that if Ω is uncountable, each μ_n is a measure on the field $\mathscr{F}$. See Example 2.3.5(1).

The second example is concerned with Vitali–Hahn–Saks theorem.

8.2.2 Example. Let Ω, $\mathscr{F}$, μ_n, $n \geq 1$ be as in Example 8.2.1. Note that $\lim_{n \to \infty} \mu_n(A) = 0$ for every A in $\mathscr{F}$. Define λ on $\mathscr{F}$ by

$$\lambda(A) = \sum_{n \in I(A)} \frac{1}{2^n}, \quad \text{if A is finite,}$$

$$= 1 + \sum_{n \in I(A)} \frac{1}{2^n}, \quad \text{if A is cofinite,}$$

where $I(A) = \{n \geq 1; x_n \in A\}$. λ is a charge on $\mathscr{F}$ and each $\mu_n \ll \lambda$. The later statement can be verified as follows. Fix $n \geq 1$ and let $\varepsilon > 0$. Take $0 < \delta < 1/2^n$. If $A \in \mathscr{F}$ and $\lambda(A) < \delta$, then A is finite and $x_n \notin A$. Consequently, $\mu_n(A) = 0 < \varepsilon$. However, $\{\mu_n, n \geq 1\}$ is not uniformly absolutely continuous with respect to λ.

The above examples also demonstrate that Nikodym and Vitali–Hahn–Saks theorems fail to hold in the space $ca(\Omega, \mathscr{F})$, where $\mathscr{F}$ is a field of subsets of Ω.

8.2.3 Remark. Nikodym theorem for positive charges on any field trivially holds whereas Vitali–Hahn–Saks theorem may not hold. For example, let $\Omega = \{1, 2, 3, \ldots\}$, $\mathscr{F}$ the finite–cofinite field on Ω and

$$\mu_n(A) = 1, \quad \text{if } n \in A,$$
$$= 0, \quad \text{if } n \notin A, A \in \mathscr{F}$$

for every $n \geq 1$. Let $\mu(A) = \sum_{n \in A} 1/2^n$, $A \in \mathscr{F}$. Then $\lim_{n \to \infty} \mu_n(A)$ exists for every A in $\mathscr{F}$ and $\mu_n \ll \mu$ for every $n \geq 1$. But $\{\mu_n, n \geq 1\}$ is not uniformly absolutely continuous with respect to μ.

8.3 PHILLIPS' LEMMA

In this section, we prove a lemma due to Phillips (1940, p. 525) which is basic in our study of Nikodym and Vitali–Hahn–Saks theorems.

First, we need the following concept of semi-variation of a charge.

8.3.1 Definition. Let μ be a charge on a field $\mathscr{F}$ of subsets of a set Ω. For F in $\mathscr{F}$, let

$$\tilde{\mu}(F) = \text{Sup}\{|\mu(E)|; E \subset F, E \in \mathscr{F}\}.$$

$\tilde{\mu}$ is called the *semi-variation* of μ.

We give some properties of $\tilde{\mu}$.

8.3.2 Proposition. *Let μ be a charge on a field $\mathscr{F}$ of subsets of a set Ω and $\tilde{\mu}$ its semi-variation. Then the following statements are true.*
  (i). $\tilde{\mu}(E) \leq \tilde{\mu}(F)$ *whenever* $E, F \in \mathscr{F}$ *and* $E \subset F$.
  (ii). $\tilde{\mu}(E \cup F) \leq \tilde{\mu}(E) + \tilde{\mu}(F)$ *whenever* $E, F \in \mathscr{F}$.
  (iii). $\tilde{\mu}(E) \leq |\mu|(E) \leq 2\tilde{\mu}(E)$ *for any* E *in* $\mathscr{F}$.

8.3.3 Phillips' Lemma. *Let $\Omega = \{1, 2, 3, \ldots\}$ and $\mathscr{F} = \mathscr{P}(\Omega)$, the class of all subsets of Ω. Let μ_n, $n \geq 1$ be a sequence of bounded charges on $\mathscr{F}$. In the following, (i) $\Rightarrow$ (ii) $\Leftrightarrow$ (iii).*

8. NIKODYM AND VITALI-HAHN-SAKS THEOREMS

(i). $\lim_{n\to\infty} \mu_n(A) = 0$ *for every* A *in* $\mathscr{F}$.
(ii). $\lim_{n\to\infty} \text{Sup}\{|\mu_n(A)|; A \subset \Omega, A \text{ finite}\} = 0$, *i.e.* μ_n, $n \geq 1$ *converges to* 0 *uniformly over all finite subsets of* Ω.
(iii). $\lim_{n\to\infty} \sum_{k\geq 1} |\mu_n(\{k\})| = 0$.
In particular, if (i) *holds, then* $\lim_{n\to\infty} \mu_n(\{n\}) = 0$.

Proof. (i)$\Rightarrow$(ii). This is proved in the following steps.

1°. Given any integer $m_0 \geq 1$ and $\varepsilon > 0$, we can find $n_0 \geq 1$ such that $\sum_{k=1}^{m_0} |\mu_n(\{k\})| < \varepsilon$ whenever $n \geq n_0$. For this, use $\lim_{n\to\infty} |\mu_n(\{k\})| = 0$ for each $k = 1, 2, \ldots, m_0$, from (i).

2°. Suppose (ii) is not true. Then there exists $\eta > 0$ such that for any given $m \geq 1$, there is an $m' \geq m$ and a finite set $A \subset \Omega$ such that $|\mu_{m'}(A)| \geq 2\eta$.

3°. We develop some notation. If A is a finite subset of Ω, let min (A) stand for the smallest element in A and max (A) for the largest element in A.

4°. We claim that for any given integer $p \geq 1$, η given above in 2° and any integer $q_1 \geq 1$, there is an integer $q \geq q_1$ and a finite set $A \subset \Omega$ such that $p \leq \min(A)$ and $|\mu_q(A)| \geq \eta$.

The above claim obviously follows when $p = 1$ from 2°. Suppose the above claim is false for some $p > 1$. Then there exists an integer $q_2 \geq 1$ such that for any $q \geq q_2$ and every finite set $A \subset \Omega$ with $p \leq \min(A)$, $|\mu_q(A)| < \eta$ holds true. Since there are finitely many sets $B \subset \Omega$ having the property that max (B) $< p$ and $\lim_{n\to\infty} \mu_n(B) = 0$, we can find $q_3 \geq 1$ such that $|\mu_q(B)| < \eta$ for every finite set $B \subset \Omega$ with max (B) $< p$ and $q \geq q_3$. Now, let q be any integer $\geq \max\{q_2, q_3\}$. Let A be any finite subset of Ω. Let $B = \{i \in A; i < p\}$ and $C = \{i \in A; i \geq p\}$. Then $B \cap C = \varnothing$, $B \cup C = A$, max (B) $< p$ and min (C) $\geq p$. Further, $|\mu_q(A)| = |\mu_q(B) + \mu_q(C)| \leq |\mu_q(B)| + |\mu_q(C)| < \eta + \eta = 2\eta$. This contradicts 2°. Hence the claim is established.

5°. Now, we obtain two sequences $n_1 < n_2 < \cdots$ and $m_1 < m_2 < \cdots$ of positive integers and a sequence A_n, $n \geq 1$ of finite subsets of Ω having the following properties.

(a). $m_i \leq \min(A_i) \leq \max(A_i) < m_{i+1}$, $i = 1, 2, 3, \ldots$.
(b). $|\mu_{n_i}(A_i)| \geq \eta$, $i = 1, 2, 3, \ldots$.
(c). $\sum_{i=1}^{m_i} |\mu_{n_i}(\{k\})| < \eta/8$, $i = 2, 3, \ldots$.

First, we obtain m_1, A_1, n_1 and m_2. Take $m_1 = 1$. In 4°, take $p = 1$ and $q_1 = 1$. We find an integer $n_1 \geq q_1$ and a finite set $A_1 \subset \Omega$ such that $m_1 = p \leq \min(A_1)$ and $|\mu_{n_1}(A_1)| \geq \eta$. Let $m_2 = \max(A_1) + 1$.

Using m_2, we proceed to obtain A_2, n_2 and m_3 satisfying (a), (b) and (c). In 1°, take $m_0 = m_2$ and $\varepsilon = \eta/8$. We can find $n_2' \geq 1$ such that $\sum_{k=1}^{m_2} |\mu_n(\{k\})| < \eta/8$ for every $n \geq n_2'$. In 4°, take $p = m_2$ and $q_1 = \max\{n_2', n_1\} + 1$. We can find $n_2 \geq q_1$ and a finite set $A_2 \subset \Omega$ such that $\min(A_2) \geq m_2$ and $|\mu_{n_2}(A_2)| \geq \eta$. Note also that $\sum_{k=1}^{m_2} |\mu_{n_2}(\{k\})| < \eta/8$.

Using m_3, as in the above, we can obtain A_3, n_3 and m_4 satisfying (a), (b) and (c).

Proceeding this way, we obtain the desired sequences. Observe that A_n, $n \geq 1$ is a sequence of pairwise disjoint sets.

6°. Now, we find a sequence S_n, $n \geq 1$ of infinite subsets of Ω and a sequence $p_1 < p_2 < \cdots$ of positive integers with the following properties.

(a)'. p_k is the smallest element in S_k, $k \geq 1$.
(b)'. $S_{k+1} \subset S_k - \{p_k\}$, $k \geq 1$.
(c)'. $\tilde{\mu}_{n_{p_k}}(\bigcup_{i \in S_{k+1}} A_i) \leq \eta/8$, $k \geq 1$.

First, we obtain S_1, p_1 and S_2. Write $\Omega = \bigcup_{j \geq 1} T_j$, where T_n, $n \geq 1$ is a sequence of pairwise disjoint sets and each T_j is infinite. We claim that there is a $j_0 \geq 1$ such that $\tilde{\mu}_{n_1}(\bigcup_{i \in T_{j_0}} A_i) \leq \eta/8$. Suppose not. For every $j \geq 1$, $\tilde{\mu}_{n_1}(\bigcup_{i \in T_j} A_i) > \eta/8$. We can find, for each $j \geq 1$, a set $B_j \subset \bigcup_{i \in T_j} A_i$ such that $|\mu_{n_1}(B_j)| < \eta/8$. Since μ_{n_1} is a bounded charge on $\mathcal{F}$ and B_j, $j \geq 1$ is a sequence of pairwise disjoint sets in $\mathcal{F}$, $\lim_{j \to \infty} \mu_{n_1}(B_j) = 0$. This contradiction establishes the claim. Let $S_1 = T_{j_0}$. Let p_1 be the smallest element in S_1. Then $p_1 \neq 1$. For, if $p_1 = 1$, then $\eta/8 \geq \tilde{\mu}_{n_1}(\bigcup_{j \in S_1} A_j) \geq |\mu_{n_1}(A_1)| \geq \eta$, by 5°(b). This contradiction shows that $p_1 \neq 1$. By decomposing $S_1 - \{p_1\}$ into a countable number of pairwise disjoint infinite sets, as above, we can find an infinite subset $S_2 \subset S_1 - \{p_1\}$ such that $\tilde{\mu}_{n_{p_1}}(\bigcup_{i \in S_2} A_i) \leq \eta/8$. Hence the desired S_1, p_1 and S_2 are obtained satisfying (a)', (b)' and (c)'. If p_2 is the smallest element in S_2, then $p_1 < p_2$.

Continuing this way, we obtain the desired sequences having properties (a)', (b)' and (c)'.

7°. Let $A = \bigcup_{i \geq 1} A_{p_i}$. For any fixed $i \geq 1$, write

$$A = A_{p_i} \cup \bigcup_{j=1}^{i-1} A_{p_j} \cup \bigcup_{j \geq i+1} A_{p_j}.$$

By 5°(b), $|\mu_{n_{p_i}}(A_{p_i})| \geq \eta$. Note that $\max(\bigcup_{j=1}^{i-1} A_{p_j}) < m_{p_{i-1}+1} \leq m_{p_i}$, by 5°(a). Consequently, by 5°(c),

$$\left|\mu_{n_{p_i}}\left(\bigcup_{j=1}^{i-1} A_{p_j}\right)\right| \leq \sum_{k=1}^{m_{p_i}} |\mu_{n_{p_i}}(\{k\})| < \eta/8.$$

Since

$$\bigcup_{j \geq i+1} A_{p_j} \subset \bigcup_{k \in S_{i+1}} A_k,$$

$$\left|\mu_{n_{p_i}}\left(\bigcup_{j \geq i+1} A_{p_j}\right)\right| \leq \tilde{\mu}_{n_{p_i}}\left(\bigcup_{k \in S_{i+1}} A_k\right) \leq \eta/8,$$

8. NIKODYM AND VITALI-HAHN-SAKS THEOREMS

by 6°(c)'. Consequently,

$$\left|\mu_{n_{p_i}}(A)\right| = \left|\mu_{n_{p_i}}(A_{n_{p_i}}) + \mu_{n_{p_i}}\left(\bigcup_{j=1}^{i-1} A_{n_{p_j}}\right) + \mu_{n_{p_i}}\left(\bigcup_{j\geq i+1} A_{n_{p_j}}\right)\right|$$

$$\geq \left|\mu_{n_{p_i}}(A_{n_{p_i}})\right| - \left|\mu_{n_{p_i}}\left(\bigcup_{j=1}^{i-1} A_{n_{p_j}}\right)\right| - \left|\mu_{n_{p_i}}\left(\bigcup_{j\geq i+1} A_{n_{p_j}}\right)\right|$$

$$> \eta - \eta/8 - \eta/8 = \tfrac{3}{4}\eta.$$

Since $p_1 < p_2 < \cdots$, $n_{p_1} < n_{p_2} < \cdots$. Therefore, $\mu_{n_{p_1}}, \mu_{n_{p_2}}, \ldots$ is an infinite subsequence of $\mu_1, \mu_2, \ldots$. So, by (i), $\lim_{i\to\infty} \mu_{n_{p_i}}(A) = 0$. The above inequality is a contradiction to this limit. Hence (ii) is valid.

(ii)$\Rightarrow$(iii). Note that

$$0 \leq \limsup_{n\to\infty} \sum_{k\geq 1} |\mu_n(\{k\})| = \limsup_{n\to\infty} \lim_{m\to\infty} \sum_{k=1}^{m} |\mu_n(\{k\})|$$

$$\leq \limsup_{n\to\infty} \lim_{m\to\infty} \left[\sum_{k\in J_m} \mu_n(\{k\}) - \sum_{k\in I_m} \mu_n(\{k\})\right]$$

($J_m = \{1 \leq k \leq m; \mu_n(\{k\}) \geq 0\}$ and $I_m = \{1 \leq k \leq m; \mu_n(\{k\}) < 0\}$.)

$$\leq 2 \limsup_{n\to\infty} \mathrm{Sup}\{|\mu_n(A)|; A \subset \Omega \text{ and } A \text{ finite}\} = 0,$$

by (ii). Hence (iii) follows.
(iii)$\Rightarrow$(ii). Note that

$$0 \leq \limsup_{n\to\infty} \mathrm{Sup}\{|\mu_n(A)|; A \subset \Omega \text{ and } A \text{ finite}\}$$

$$\leq \limsup_{n\to\infty} \sum_{k\geq 1} |\mu_n(\{k\})| = 0, \quad \text{by (iii)}.$$

Hence (ii) follows. $\square$

8.3.4 Remark. In the above lemma, (iii)$\Rightarrow$(i) is not true. Take any sequence μ_n, $n \geq 1$ of distinct 0–1 valued charges on $\mathcal{P}(\Omega)$ each of which vanishes on finite sets.

8.4 NIKODYM THEOREM

We develop this theorem in the framework of Boolean algebras. The various notions connected with Boolean algebras are given in Section 1.4.

8.4.1 Definition. A Boolean algebra $\mathbb{B}$ is said to have *Seever property* if for every two sequences a_n, $n \geq 1$ and b_n, $n \geq 1$ in $\mathbb{B}$ satisfying the following conditions
   (i). $a_1 \leq a_2 \leq a_3 \leq \cdots$,
   (ii). $b_1 \geq b_2 \geq b_3 \geq \cdots$ and
   (iii). $a_m \leq b_n$ for every $m, n \geq 1$,
there exists c in $\mathbb{B}$ such that $a_m \leq c \leq b_n$ for every $m, n \geq 1$.

Boolean σ-algebras and σ-fields of subsets of any set constitute an important class of Boolean algebras with Seever property. We give some examples.

8.4.2(i) Example. Let $\Omega = \{1, 2, 3, \ldots\}$ and $\mathscr{F}$ the finite–cofinite field on Ω. The Boolean algebra $\mathscr{F}$ does not have Seever property. Take

$$A_m = \{2, 4, 6, \ldots, 2m\}, \quad m \geq 1,$$

and

$$B_n = \Omega - \{1, 3, 5, \ldots, 2n+1\}, \quad n \geq 1.$$

There is no C in $\mathscr{F}$ such that $A_m \subset C \subset B_n$ for all $m, n \geq 1$.

The following is an example of a Boolean algebra with Seever property which is not a Boolean σ-algebra.

8.4.2(ii) Example. First, we prove the following result. Let $\mathbb{B}_1$ and $\mathbb{B}_2$ be two Boolean algebras and h onto homomorphism from $\mathbb{B}_1$ to $\mathbb{B}_2$. If $\mathbb{B}_1$ is a Boolean algebra with Seever property, then $\mathbb{B}_2$ also has Seever property.

Let a_n, $n \geq 1$ and b_n, $n \geq 1$ be two sequences in $\mathbb{B}_2$ such that $a_n \leq a_{n+1} \leq b_{m+1} \leq b_m$ for all $m, n \geq 1$. Let c_n, $n \geq 1$ and d_n, $n \geq 1$ be two sequences in $\mathbb{B}_1$ such that $h(c_n) = a_n$ for every $n \geq 1$ and $h(d_m) = b_m$ for every $m \geq 1$. Define $e_1 = c_1 \wedge d_1$, $f_1 = c_1 \vee d_1$, $f_{n+1} = f_n \wedge (c_{n+1} \vee d_{n+1} \vee e_n)$ and $e_{n+1} = e_n \vee (c_{n+1} \wedge d_{n+1} \wedge f_n)$, $n \geq 1$. Then $e_1 \leq e_2 \leq \cdots \leq f_2 \leq f_1$, $h(e_n) = a_n$ and $h(f_m) = b_m$ for all $m, n \geq 1$. Since $\mathbb{B}_1$ has Seever property, there is an x in $\mathbb{B}_1$ such that $e_n \leq x \leq f_m$ for all $m, n \geq 1$. Then $a_n \leq h(x) \leq b_m$ for all $m, n \geq 1$. This completes the proof.

Now, we give the desired example. Let $\Omega = \{1, 2, \ldots\}$ and $\mathscr{I}$ the ideal of all finite subsets of Ω. Since the natural homomorphism from $\mathscr{P}(\Omega)$ to the quotient Boolean algebra $\mathscr{P}(\Omega)/\mathscr{I}$ is onto and $\mathscr{P}(\Omega)$ has Seever property, by what we have proved above, $\mathscr{P}(\Omega)/\mathscr{I}$ has Seever property. We show that $\mathscr{P}(\Omega)/\mathscr{I}$ is not a Boolean σ-algebra. Let B_n, $n \geq 1$ be a sequence of pairwise disjoint subsets of Ω each of which is infinite. Then $\bigvee_{n \geq 1} [B_n]$ does not exist. This we prove as follows. Let $[C] \geq [B_n]$ for every $n \geq 1$. Then we exhibit $[D]$ in $\mathscr{P}(\Omega)/\mathscr{I}$ such that $[C] > [D] \geq [B_n]$ for every $n \geq 1$.

Since $B_n - C$ is finite for every $n \geq 1$, we can find a point x_n in $C \cap B_n$ for every $n \geq 1$. Let $A = \{x_n; n \geq 1\}$ and $D = C - A$. This D serves the purpose. Consequently, $\bigvee_{n \geq 1}[B_n]$ does not exist.

The most important result about Boolean algebras with Seever property is the following.

8.4.3 Theorem. *Let $\mathbb{B}$ be a Boolean algebra with Seever property. Let μ be any bounded charge on $\mathbb{B}$. Let $\mathcal{N} = \{b \in \mathbb{B}; |\mu|(b) = 0\}$. Then the quotient Boolean algebra $\mathbb{B}/\mathcal{N}$ is complete.*

Proof. Since $\mathbb{B}/\mathcal{N}$ satisfies the countable chain condition, i.e. any family of pairwise disjoint elements in $\mathbb{B}/\mathcal{N}$ is at most countable (because there cannot be more than n pairwise disjoint elements b in $\mathbb{B}$ such that $|\mu|(b) > |\mu|(1)/n$), it suffices to show that $\mathbb{B}/\mathcal{N}$ is a Boolean σ-algebra. See Theorem 1.4.8. Let $[a_n]$, $n \geq 1$ be a sequence of equivalence classes in $\mathbb{B}/\mathcal{N}$. We show that $\bigvee_{n \geq 1}[a_n]$ exists in $\mathbb{B}/\mathcal{N}$. Without loss of generality, assume that $a_1 \leq a_2 \leq \cdots$. Let $C = \{b \in \mathbb{B}; b \geq a_n \text{ for every } n \geq 1\}$. Let $r = \text{Inf}\{|\mu|(b); b \in C\}$. Then there exists a sequence b_n, $n \geq 1$ in C such that $r = \lim_{n \to \infty} |\mu|(b_n)$. Without loss of generality, assume that $b_1 \geq b_2 \geq \cdots$. Since $\mathbb{B}$ has Seever property, there exists c in $\mathbb{B}$ such that $a_m \leq c \leq b_n$ for every $m, n \geq 1$. It is now obvious that $\bigvee_{n \geq 1}[a_n] = [c]$. Thus $\mathbb{B}/\mathcal{N}$ is a Boolean σ-algebra. □

Now, we are ready to prove Nikodym theorem for charges on complete Boolean algebras.

8.4.4 Theorem. *Let $\mathbb{B}$ be a complete Boolean algebra and μ_n, $n \geq 1$ a sequence of bounded charges on $\mathbb{B}$. Suppose for every b in $\mathbb{B}$, $\text{Sup}_{n \geq 1}|\mu_n(b)| < \infty$. Then $\text{Sup}_{n \geq 1}\|\mu_n\| < \infty$.*

Proof. Suppose $\text{Sup}_{n \geq 1}\|\mu_n\| = \infty$, i.e. $\text{Sup}_{n \geq 1}\text{Sup}_{b \in \mathbb{B}}|\mu_n(b)| = \infty$. First, we find a sequence c_n, $n \geq 1$ of pairwise disjoint elements in $\mathbb{B}$ and an increasing sequence m_k, $k \geq 1$ of positive integers such that $\lim_{k \to \infty}|\mu_{m_k}(c_k)| = \infty$.

For each a in $\mathbb{B}$, define $t_a = \text{Sup}_{n \geq 1}\text{Sup}_{b \leq a}|\mu_n(b)|$. Note that for each a in $\mathbb{B}$, either $t_a = \infty$ or $t_{1-a} = \infty$, where 1 is the unit element in $\mathbb{B}$. For, if $t_a < \infty$ and $t_{1-a} < \infty$, then $t_1 = \text{Sup}_{n \geq 1}\text{Sup}_{b \in \mathbb{B}}|\mu_n(b)| < \infty$. Similarly, if $t_a = \infty$ for a in $\mathbb{B}$, then either $t_b = \infty$ or $t_{a-b} = \infty$ for any b in $\mathbb{B}$ satisfying $b \leq a$.

From the supposition that $\text{Sup}_{n \geq 1}\|\mu_n\| = \infty$, we can find $m_1 \geq 1$ and d_1 in $\mathbb{B}$ such that $|\mu_{m_1}(d_1)| > \text{Sup}_{n \geq 1}|\mu_n(1)| + 2$. If $t_{d_1} = \infty$, take $c_1 = 1 - d_1$. We find that

$$|\mu_{m_1}(c_1)| = |\mu_{m_1}(1 - d_1)| \geq |\mu_{m_1}(d_1)| - |\mu_{m_1}(1)|$$
$$\geq |\mu_{m_1}(d_1)| - \text{Sup}_{n \geq 1}|\mu_n(1)| > 2.$$

If $t_{1-d_1} = \infty$, take $c_1 = d_1$. In this case, we find that $|\mu_{m_1}(c_1)| = |\mu_{m_1}(d_1)| > 2$. In any case, we have $|\mu_{m_1}(c_1)| > 2$ and $t_{1-c_1} = \infty$.

Since $t_{1-c_1} = \infty$, we can find $d_2 \leq 1 - c_1$ and $m_2 > m_1$ such that $|\mu_{m_2}(d_2)| > \text{Sup}_{n \geq 1} |\mu_n(1 - c_1)| + 3$. If $t_{d_2} = \infty$, take $c_2 = (1 - c_1) - d_2$. Note that

$$|\mu_{m_2}(c_2)| = |\mu_{m_2}((1-c_1) - d_2)| \geq |\mu_{m_2}(d_2)| - |\mu_{m_2}(1-c_1)|$$

$$\geq |\mu_{m_2}(d_2)| - \text{Sup}_{n \geq 1} |\mu_n(1-c_1)| > 3.$$

If $t_{(1-c_1)-d_2} = \infty$, take $c_2 = d_2$. Now, $|\mu_{m_2}(c_2)| = |\mu_{m_2}(d_2)| > 3$. In any case, we observe that $m_2 > m_1$, $c_1 \wedge c_2 = 0$, $|\mu_{m_1}(c_1)| > 2$, $|\mu_{m_2}(c_2)| > 3$ and $t_{1-(c_1 \vee c_2)} = \infty$. Continuing this procedure, we obtain a sequence c_n, $n \geq 1$ of pairwise disjoint elements in $\mathbb{B}$ and an increasing sequence m_k, $k \geq 1$ of positive integers such that $\lim_{k \to \infty} |\mu_{m_k}(c_k)| = \infty$.

For each $k \geq 1$, we define λ_k on $\mathcal{P}(\Omega)$, where $\Omega = \{1, 2, 3, \ldots\}$, by

$$\lambda_k(A) = \mu_{m_k}\left(\bigvee_{j \in A} c_j\right) / \mu_{m_k}(c_k)$$

for $A \subset \Omega$. Note that each λ_k is a bounded charge on $\mathcal{P}(\Omega)$. Further, $\lim_{k \to \infty} \lambda_k(A) = 0$ for every $A \subset \Omega$. This follows from $\lim_{k \to \infty} |\mu_{m_k}(c_k)| = \infty$ and $\text{Sup}_{k \geq 1} |\mu_{m_k}(\bigvee_{j \in A} c_j)| < \infty$ for any $A \subset \Omega$. By Phillips' lemma 8.3.3, we conclude that $\lim_{k \to \infty} \lambda_k(\{k\}) = 0$. But $\lambda_k(\{k\}) = 1$ for every $k \geq 1$. This contradiction proves the result. □

Now, we generalize Theorem 8.4.4 to Boolean algebras with Seever property.

8.4.5 Theorem. *Let $\mathbb{B}$ be a Boolean algebra with Seever property and μ_n, $n \geq 1$ a sequence of bounded charges on $\mathbb{B}$. Suppose for every b in $\mathbb{B}$, $\text{Sup}_{n \geq 1} |\mu_n(b)| < \infty$. Then $\text{Sup}_{n \geq 1} \|\mu_n\| < \infty$.*

Proof. Define λ on $\mathbb{B}$ by

$$\lambda(b) = \sum_{n \geq 1} \frac{1}{2^n} \frac{|\mu_n|(b)}{1 + |\mu_n|(1)}$$

for b in $\mathbb{B}$. Then λ is a bounded charge on $\mathbb{B}$ and $\mu_n \ll \lambda$ for every $n \geq 1$. Let $\mathcal{N} = \{a \in \mathbb{B}; \lambda(a) = 0\}$. Then, by Theorem 8.4.3, the quotient Boolean algebra $\mathbb{B}/\mathcal{N}$ is complete. For each $n \geq 1$, define $\tilde{\mu}_n$ on $\mathbb{B}/\mathcal{N}$ by $\tilde{\mu}_n([b]) = \mu_n(b)$, $b \in \mathbb{B}$. Since $\mu_n \ll \lambda$, $\tilde{\mu}_n$ is well defined on $\mathbb{B}/\mathcal{N}$ for all $n \geq 1$. Note that $\tilde{\mu}_n$, $n \geq 1$ is a sequence of bounded charges on $\mathbb{B}/\mathcal{N}$ satisfying the hypothesis of Theorem 8.4.4. The conclusion of Theorem 8.4.4, now, gives the result. □

The following is the main result of this section which gives Nikodym theorem for sets of bounded charges on Boolean algebras having Seever property.

8. NIKODYM AND VITALI-HAHN-SAKS THEOREMS

8.4.6 Corollary. *Let M be any collection of bounded charges on a Boolean algebra* $\mathbb{B}$ *having Seever property. Suppose for every b in* $\mathbb{B}$, $\sup\{|\mu(b)|; \mu \in M\} < \infty$. *Then* $\sup\{\|\mu\|; \mu \in M\} < \infty$.

Proof. Suppose $\sup\{\|\mu\|; \mu \in M\} = \infty$. Then we can find a sequence μ_n, $n \geq 1$ in M such that $\sup_{n \geq 1}\|\mu_n\| = \infty$. By the given hypothesis, it is obvious that for every b in $\mathbb{B}$, $\sup_{n \geq 1}|\mu_n(b)| < \infty$. This contradicts the validity of Theorem 8.4.5. □

Of course, the classical Nikodym Theorem 8.1.1 follows.

8.4.7 Corollary. *Let M be any collection of bounded charges on a σ-field* $\mathfrak{A}$ *of subsets of a set* Ω. *Suppose for every A in* $\mathfrak{A}$, $\sup\{|\mu(A)|; \mu \in M\} < \infty$. *Then* $\sup\{\|\mu\|; \mu \in M\} < \infty$. *In particular, Theorem 8.1.1 holds.* □

8.5 NORM BOUNDED SETS IN THE PRESENCE OF UNIFORM ABSOLUTE CONTINUITY

Nikodym theorem (Corollary 8.4.6) gives a set of necessary and sufficient conditions under which a given subset M of $\text{ba}(\Omega, \mathcal{F})$ is norm bounded when $\mathcal{F}$ is a field of subsets of a set Ω having Seever property. Now, we give a set of sufficient conditions based on the notion of uniform absolute continuity and strongly continuous charges, for a given set $M \subset \text{ba}(\Omega, \mathcal{F})$ to be norm bounded without any conditions on $\mathcal{F}$.

We need some preliminary results on absolute continuity, singularity and strongly continuous charges. We say that a bounded charge ν on a field $\mathcal{F}$ of subsets of a set Ω is *atomic* if ν is a countable sum of two-valued charges on $\mathcal{F}$.

8.5.1 Proposition. *Let $\mathcal{F}$ be a field of subsets of a set Ω.*
(i). *If μ, ν and τ are bounded charges on $\mathcal{F}$ such that $\mu \ll \nu + \tau$ and $\mu \perp \tau$, then $\mu \ll \nu$.*
(ii). *If μ, ν and τ are bounded charges on $\mathcal{F}$ such that $\mu \ll \nu + \tau$, then we can write $\mu = \mu_1 + \mu_2$ such that $\mu_1 \ll \nu$ and $\mu_2 \ll \tau$.*
(iii). *If μ, ν and τ are bounded charges on $\mathcal{F}$ such that $\mu \ll \nu + \tau$ and $\nu \perp \tau$, then we can write $\mu = \mu_1 + \mu_2$ such that $\mu_1 \ll \nu$ and $\mu_2 \ll \tau$, and such a decomposition is unique.*
(iv). *If ν is a two-valued bounded charge on $\mathcal{F}$ and μ is any charge on $\mathcal{F}$ such that $\mu \ll \nu$, then μ is two-valued.*
(v). *If μ is a strongly continuous charge on $\mathcal{F}$ and ν is a bounded two-valued charge on $\mathcal{F}$, then $\mu \perp \nu$.*

(vi). If μ is a strongly continuous charge on $\mathscr{F}$ and ν is a bounded atomic charge on $\mathscr{F}$, then $\mu \perp \nu$.
(vii). If μ and ν are two bounded charges on $\mathscr{F}$ such that ν is atomic and $\mu \ll \nu$, then μ is atomic.
(viii). If μ and ν are two distinct two-valued charges on $\mathscr{F}$, then $\mu \perp \nu$. In fact, there exists A in $\mathscr{F}$ such that $|\mu|(A) = |\mu|(\Omega)$ and $|\nu|(A^c) = |\nu|(\Omega)$. More generally, if ν_i, $i \geq 1$ is a sequence of distinct two-valued charges on $\mathscr{F}$, then $\mu \perp \lambda$, where $\mu = \sum_{i \in D} \alpha_i \nu_i$ and $\lambda = \sum_{j \in E} \beta_j \nu_j$, with D and E being disjoint subsets of $\{1, 2, 3, \ldots\}$, $\sum_{i \in D} |\alpha_i| < \infty$ and $\sum_{j \in E} |\beta_j| < \infty$.

Proof. Without loss of generality, we assume that all the charges involved are positive.
(i). Let $\varepsilon > 0$. There exists $\delta > 0$ such that $\mu(B) < \varepsilon/2$ whenever $B \in \mathscr{F}$ and $(\nu + \tau)(B) < \delta$. Assume, without loss of generality, that $\delta < \varepsilon$. Since $\mu \perp \tau$, there is a set D in $\mathscr{F}$ such that $\mu(D) < \delta/2$ and $\tau(D^c) < \delta/2$. Now, we show that $\mu \ll \nu$. Let C be any set in $\mathscr{F}$ such that $\nu(C) < \delta/2$. Note that $\mu(C \cap D) < \delta/2 < \varepsilon/2$ and $(\nu + \tau)(C \cap D^c) = \nu(C \cap D^c) + \tau(C \cap D^c) < \delta/2 + \delta/2$. So, $\mu(C \cap D^c) < \varepsilon/2$. Thus we see that $\mu(C) < \varepsilon$.
(ii). By Lebesgue Decomposition Theorem 6.2.4, we write $\mu = \mu_1 + \mu_2$ such that $\mu_1 \ll \nu$ and $\mu_2 \perp \nu$. Clearly, μ_1 and μ_2 are positive. Now, $\mu_2 \leq \mu_1 + \mu_2 \ll \nu + \tau$. From (i), $\mu_2 \ll \tau$.
(iii). By (ii), we can write $\mu = \mu_1 + \mu_2$ such that $\mu_1 \ll \nu$ and $\mu_2 \ll \tau$. Since $\nu \perp \tau$, $\mu_2 \perp \nu$. Now, the uniqueness of the decomposition follows from the uniqueness of the Lebesgue decomposition.
(iv). This is obvious.
(v). Since μ is strongly continuous, it is bounded. Let $\varepsilon > 0$. There is a partition $\{A_1, A_2, \ldots, A_n\}$ of Ω in $\mathscr{F}$ such that $\mu(A_i) < \varepsilon$ for $i = 1, 2, \ldots, n$. Since ν is bounded and two-valued, $\nu(A_i) = 0$ for all but one i in $\{1, 2, \ldots, n\}$, i.e. there is i_0 in $\{1, 2, \ldots, n\}$ such that

$$\nu\left(\bigcup_{\substack{i=1 \\ i \neq i_0}}^{n} A_i\right) = 0 < \varepsilon \quad \text{and} \quad \mu\left(\left(\bigcup_{\substack{i=1 \\ i \neq i_0}}^{n} A_i\right)^c\right) = \mu(A_{i_0}) < \varepsilon.$$

Hence $\mu \perp \nu$.
(vi). Since ν is atomic, positive and bounded, we can write $\nu = \sum_{i \geq 1} \alpha_i \nu_i$ such that $\alpha_i > 0$ for every $i \geq 1$, $\sum_{i \geq 1} \alpha_i < \infty$, ν_i is 0–1 valued for every $i \geq 1$ and ν_i, $i \geq 1$ is a distinct sequence of charges. Let $\varepsilon > 0$. Choose $N \geq 1$ such that $\sum_{i \geq N+1} \alpha_i < \varepsilon/2$. By (v), $\mu \perp \sum_{i=1}^{N} \alpha_i \nu_i$. So, we can find D in $\mathscr{F}$ such that $\mu(D) < \varepsilon/2$ and $\sum_{i=1}^{N} \alpha_i \nu_i(D^c) < \varepsilon/2$. Now, $\nu(D^c) = \sum_{i=1}^{N} \alpha_i \nu_i(D^c) + \sum_{i \geq N+1} \alpha_i \nu_i(D^c) < \varepsilon/2 + \varepsilon/2 = \varepsilon$. This completes the proof.
(vii). By Sobczyk–Hammer Decomposition Theorem 5.2.7, we can write $\mu = \mu_1 + \mu_2$ such that μ_1 is strongly continuous and μ_2 is atomic. Since

8. NIKODYM AND VITALI-HAHN-SAKS THEOREMS

$\mu_1 + \mu_2 \ll \nu$ by hypothesis, $\mu_1 \ll \nu$. But by (v), $\mu_1 \perp \nu$. Hence $\mu_1 = 0$. Therefore, $\mu = \mu_2$ is atomic.

(viii). This is not too difficult to prove and so we omit the proof. □

We need another result on uniform absolute continuity.

8.5.2 Proposition. *Let $\mathcal{F}$ be a field of subsets of a set Ω and $M \subset \mathrm{ba}(\Omega, \mathcal{F})$ uniformly absolutely continuous with respect to some $\nu \in \mathrm{ba}(\Omega, \mathcal{F})$. Let $\nu = \nu_1 + \nu_2$, where ν_1 is a strongly continuous charge and ν_2 an atomic charge on $\mathcal{F}$.*

(i). *If every μ in M is a strongly continuous charge on $\mathcal{F}$, then M is uniformly absolutely continuous with respect to ν_1.*

(ii). *More generally, write every μ in M as $\mu_1 + \mu_2$, where μ_1 is a strongly continuous charge and μ_2 is an atomic charge on $\mathcal{F}$. Let $M_1 = \{\mu_1; \mu \in M\}$ and $M_2 = \{\mu_2; \mu \in M\}$. Then (a) M_1 is uniformly absolutely continuous with respect to ν_1 and (b) M_2 is uniformly absolutely continuous with respect to ν_2.*

Proof. We prove only (a) of (ii). Without loss of generality, we assume that all charges involved are positive.

Let $\varepsilon > 0$. Since M is uniformly absolutely continuous with respect to ν, there exists $\delta > 0$ such that $\mu(A) < \varepsilon/2$ for every μ in M whenever $A \in \mathcal{F}$ and $\nu(A) < \delta$. We take $\delta < \varepsilon$. We show that $\mu_1(A) < \varepsilon$ for every μ_1 in M_1 whenever $A \in \mathcal{F}$ and $\nu_1(A) < \delta/2$. Let $A \in \mathcal{F}$ and $\nu_1(A) < \delta/2$. Let $\mu_1 \in M_1$. Note that, by Proposition 8.5.1(vi), $\mu_1 \perp \nu_2$. So, there exists D in $\mathcal{F}$ such that $\mu_1(D) < \delta/2$ and $\nu_2(D^c) < \delta/2$. Note that $\mu_1(A \cap D) < \delta/2 < \varepsilon/2$. Also, $\nu(A \cap D^c) = \nu_1(A \cap D^c) + \nu_2(A \cap D^c) < \delta/2 + \delta/2 = \delta$. Consequently, $\mu(A \cap D^c) < \varepsilon/2$. So, $\mu_1(A \cap D^c) < \varepsilon/2$. Therefore, $\mu_1(A) = \mu_1(A \cap D) + \mu_1(A \cap D^c) < \varepsilon/2 + \varepsilon/2 = \varepsilon$. This completes the proof. □

Now, we prove the main theorem of this section.

8.5.3 Theorem. *Let $\mathcal{F}$ be a field of subsets of a set Ω and $M \subset \mathrm{ba}(\Omega, \mathcal{F})$ be uniformly absolutely continuous with respect to a $\nu \in \mathrm{ba}(\Omega, \mathcal{F})$.*

(i). *If ν is a strongly continuous charge on $\mathcal{F}$, then M is norm bounded, i.e. $\mathrm{Sup}\{\|\mu\|; \mu \in M\} < \infty$.*

(ii). *If every μ in M is a strongly continuous charge on $\mathcal{F}$, then M is norm bounded.*

Proof. (i). Without loss of generality, assume that all charges involved are positive. Since M is uniformly absolutely continuous with respect to ν, for $\varepsilon = 1$, there exists $\delta > 0$ such that $\mu(B) < 1$ for every μ in M whenever $B \in \mathcal{F}$ and $\nu(B) < \delta$. Since ν is strongly continuous, there exists a partition $\{B_1, B_2, \ldots, B_k\}$ of Ω in $\mathcal{F}$ such that $\nu(B_i) < \delta$ for $i = 1, 2, \ldots, k$. So, for any μ in M, $\mu(\Omega) = \sum_{i=1}^{k} \mu(B_i) < k$. Hence M is norm bounded.

(ii). By Sobczyk–Hammer Decomposition Theorem 5.2.7, write $\nu = \nu_1 + \nu_2$, where ν_1 is a strongly continuous charge on $\mathcal{F}$ and ν_2 is an atomic charge on $\mathscr{F}$. By Proposition 8.5.2(ii), M is uniformly absolutely continuous with respect to ν_1. (i) completes the proof now. □

8.6 A DECOMPOSITION THEOREM

Let $\mathscr{F}$ be a field of subsets of a set Ω and $M \subset ba(\Omega, \mathscr{F})$ be uniformly absolutely continuous with respect to a ν in $ba(\Omega, \mathscr{F})$. In this section, we make an attempt to isolate that part of M which is responsible for the failure of M to be norm bounded. More precisely, we obtain a decomposition of M into a norm bounded part and a finite dimensional part. We start with a definition.

8.6.1 Definition. Let $\lambda = \sum_{i=1}^{N} \alpha_i \nu_i$, where $\nu_1, \nu_2, \ldots, \nu_N$ are distinct 0–1 valued charges on $\mathscr{F}$ and $\alpha_1, \alpha_2, \ldots, \alpha_N$ are any non-zero real numbers. A_λ is defined to be the collection of all bounded charges on $\mathscr{F}$ absolutely continuous with respect to λ. Any subset of A_λ is called a *finite dimensional set*.

8.6.2 Remark. One can show that any μ in A_λ is of the form $\sum_{i=1}^{N} \beta_i \nu_i$ for some real numbers $\beta_1, \beta_2, \ldots, \beta_N$. One can use Proposition 8.5.1 to establish the veracity of this statement. This is the reason for the use of the term "finite dimensional set".

The following proposition is instrumental in proving the main result of this section.

8.6.3 Proposition. *Let $\mathscr{F}$ be a field of subsets of a set Ω and M a subset of $ba(\Omega, \mathscr{F})$ uniformly absolutely continuous with respect to an atomic charge ν in $ba(\Omega, \mathscr{F})$. Then there exist two subsets M_1 and M_2 of $ba(\Omega, \mathscr{F})$ satisfying the following properties.*
  (i). *M_1 and M_2 are both uniformly absolutely continuous with respect to ν.*
  (ii). *$M \subset M_1 + M_2 = \{\mu + \tau; \mu \in M_1, \tau \in M_2\}$.*
  (iii). *$M_1 \subset A_\lambda$ for some finite linear sum λ of 0–1 valued charges on $\mathscr{F}$, i.e. M_1 is a finite dimensional set.*
  (iv). *M_2 is norm bounded.*

Proof. The proof is carried out in the following steps.
1°. Without loss of generality, we can assume ν to be positive. Since ν is atomic, we can write $\nu = \sum_{i \geq 1} \alpha_i \nu_i$, where $\alpha_i > 0$ for every $i \geq 1$, $\sum_{i \geq 1} \alpha_i < \infty$ and $\nu_i, i \geq 1$ is a sequence of distinct 0–1 valued charges on $\mathscr{F}$. ($\sum_{i \geq 1} \alpha_i \nu_i$ could be a finite sum.)

8. NIKODYM AND VITALI-HAHN-SAKS THEOREMS

2°. Since M is uniformly absolutely continuous with respect to ν, for $\varepsilon = 1$, there exists $\delta > 0$ such that $|\mu|(B) < 1$ for every μ in M whenever $B \in \mathscr{F}$ and $\nu(B) < \delta$.

3°. There exists $N \geq 1$ such that $\sum_{i \geq N+1} \alpha_i < \delta$. Let $\lambda = \sum_{i=1}^{N} \alpha_i \nu_i$ and $\tau = \sum_{i \geq N+1} \alpha_i \nu_i$. Note that $\tau(\Omega) < \delta$ and $\nu = \lambda + \tau$.

4°. Let $\{A_1, A_2, \ldots, A_N\}$ be a partition of Ω in $\mathscr{F}$ such that $\nu_i(A_i) = 1$ for $i = 1, 2, \ldots, N$ and $\nu_j(A_i) = 0$ for every $i \neq j$, $i, j = 1, 2, \ldots, N$.

5°. Since the charges ν_i, $i \geq 1$ are distinct, $\lambda \perp \tau$. See Proposition 8.5.1(viii).

6°. Let $\mu \in M$. We can write $\mu = \mu_1 + \mu_2$, where μ_1 and μ_2 are charges on $\mathscr{F}$, $\mu_1 \ll \lambda$ and $\mu_2 \ll \tau$. See Proposition 8.5.1(ii). Since $\lambda \perp \tau$, this decomposition is unique. See Proposition 8.5.1(iii). Let $M_1 = \{\mu_1; \mu \in M\}$ and $M_2 = \{\mu_2; \mu \in M\}$.

7°. It is clear that $M \subset M_1 + M_2$. This proves (ii). Further, $M_1 \subset A_\lambda$. This proves (iii).

8°. Clearly, both M_1 and M_2 are uniformly absolutely continuous with respect to ν. This proves (i).

9°. Finally, we show that M_2 is norm bounded. Let $\mu \in M$ be arbitrary. Since $\mu_2 \ll \tau$ and $\nu_i \perp \tau$ for $i = 1, 2, \ldots, N$, it follows that $\nu_i \perp \mu_2$ for $i = 1, 2, \ldots, N$. So, we can find V_i in $\mathscr{F}$ such that $|\mu_2|(V_i) < 1$ and $\nu_i(V_i^c) < 1$ for $i = 1, 2, \ldots, N$. This means that $\nu_i(V_i) = 1$ and $\nu_i(F_i) = 1$, where $F_i = V_i \cap A_i$, $i = 1, 2, \ldots, N$. Further, $\nu_i(A_i - F_i) = 0$ for $i = 1, 2, \ldots, N$. Now,

$$|\mu_2|(\Omega) = \sum_{i=1}^{N} |\mu_2|(A_i)$$

$$= \sum_{i=1}^{N} |\mu_2|(F_i) + \sum_{i=1}^{N} |\mu_2|(A_i - F_i)$$

$$< N + \sum_{i=1}^{N} |\mu_2|(A_i - F_i).$$

Note that for each $i = 1, 2, \ldots, N$,

$$\nu(A_i - F_i) = \sum_{j=1}^{N} \alpha_j \nu_j(A_i - F_i) + \sum_{j \geq N+1} \alpha_j \nu_j(A_i - F_i)$$

$$< 0 + \delta = \delta.$$

By 2°, $|\mu|(A_i - F_i) < 1$ for each $i = 1, 2, \ldots, N$. Since $|\mu| = |\mu_1| + |\mu_2|$, $|\mu_2|(A_i - F_i) < 1$ for each $i = 1, 2, \ldots, N$. Thus $|\mu|(\Omega) < N + N = 2N$. Hence M_2 is norm bounded. □

Now, we prove the main decomposition theorem.

8.6.4 Theorem. *Let $\mathscr{F}$ be a field of subsets of a set Ω and $M \subset \text{ba}(\Omega, \mathscr{F})$ be uniformly absolutely continuous with respect to a ν in $\text{ba}(\Omega, \mathscr{F})$. Then* *there exist two sets* M_1 *and* M_2 *contained in* $ba(\Omega, \mathscr{F})$ *having the following properties.*

(i). M_1 *and* M_2 *are both uniformly absolutely continuous with respect to* ν.

(ii). $M \subset M_1 + M_2$.

(iii). $M_1 \subset A_\lambda$ *for some finite linear sum* λ *of 0–1 valued charges on* $\mathscr{F}$, *i.e.* M_1 *is a finite dimensional set.*

(iv). M_2 *is norm bounded.*

Proof. Assume, without loss of generality, that ν is positive. Write $\nu = \nu_1 + \nu_2$, where ν_1 is a strongly continuous charge on $\mathscr{F}$ and ν_2 is an atomic charge on $\mathscr{F}$. Write each μ in M also as $\mu_1 + \mu_2$, where μ_1 is a strongly continuous charge on $\mathscr{F}$ and μ_2 is an atomic charge on $\mathscr{F}$. Let $H_1 = \{\mu_1; \mu \in M\}$ and $H_2 = \{\mu_2; \mu \in M\}$. H_1 and H_2 are both absolutely continuous with respect to ν and $M \subset H_1 + H_2$. By Proposition 8.5.2(ii), H_1 is uniformly absolutely continuous with respect to ν_1 and H_2 is uniformly absolutely continuous with respect to ν_2. By applying Proposition 8.6.3 to H_2 and ν_2, we obtain two sets H_3 and H_4 contained in $ba(\Omega, \mathscr{F})$ such that $H_2 \subset H_3 + H_4$, $H_3 \subset A_\lambda$ for some finite linear sum λ of 0–1 valued charges on $\mathscr{F}$ and H_4 is norm bounded. By Theorem 8.5.3(i), H_1 is norm bounded.

Take $M_1 = H_3$ and $M_2 = H_1 + H_4$. M_1 and M_2 are the desired sets. $\square$

8.7 WEAK CONVERGENCE

Let $\mathscr{F}$ be a field of subsets of a set Ω. The principal object of study in this section is the space $ba(\Omega, \mathscr{F})$ in its weak topology. It seems difficult, in general, to characterize the dual space $ba^*(\Omega, \mathscr{F})$, i.e. the space of all continuous linear functionals on the Banach space $ba(\Omega, \mathscr{F})$. In this section, we characterize weak convergence in $ba(\Omega, \mathscr{F})$. Recall that the weak topology on $ba(\Omega, \mathscr{F})$ is the smallest topology on $ba(\Omega, \mathscr{F})$ with respect to which all functions in $ba^*(\Omega, \mathscr{F})$ are continuous. The classical results on weak convergence in the space of bounded measures on a σ-field follow as simple corollaries of the results of this section.

First, we give two results which are useful in characterizing weak convergence.

8.7.1 Theorem. *Let* ψ, μ_n, $n \geq 1$ *be a sequence of bounded charges on a field* $\mathscr{F}$ *of subsets of a set* Ω *satisfying the following conditions.*

(i). $\{\mu_n, n \geq 1\}$ *is uniformly absolutely continuous with respect to* ψ.

(ii). $\sup_{n \geq 1} |\mu_n(F)| < \infty$ *for every* F *in* $\mathscr{F}$.

Then $\sup_{n \geq 1} \|\mu_n\| < \infty$.

8. NIKODYM AND VITALI-HAHN-SAKS THEOREMS

Proof. Suppose $\text{Sup}_{n\geq 1} \|\mu_n\| = \infty$. We exhibit a sequence $n_1 < n_2 < \cdots$ of positive integers and a sequence A_n, $n \geq 1$ of pairwise disjoint sets in $\mathcal{F}$ such that $|\mu_{n_k}(A_k)| \geq 1$ for every $k \geq 1$. Let $r = \text{Sup}_{n\geq 1} |\mu_n(\Omega)|$. Then $r < \infty$, and there exist $n_1 \geq 1$ and B_1 in $\mathcal{F}$ such that $|\mu_{n_1}(B_1)| \geq r+1 \geq 1$. So, $|\mu_{n_1}(B_1^c)| \geq \|\mu_{n_1}(\Omega)| - |\mu_{n_1}(B_1)\| \geq |\mu_{n_1}(B_1)| - |\mu_{n_1}(\Omega)| \geq 1$. Then either $\text{Sup}_{n\geq 1} |\mu_n|(B_1) = \infty$ or $\text{Sup}_{n\geq 1} |\mu_n|(B_1^c) = \infty$. If $\text{Sup}_{n\geq 1} |\mu_n|(B_1) = \infty$, take $A_1 = B_1^c$. Using the same argument given above for B_1, we can find $n_2 > n_1$ and B_2 in $\mathcal{F}$ such that $B_2 \subset B_1$, $|\mu_{n_2}(B_2)| \geq 1$ and $|\mu_{n_2}(B_1 - B_2)| \geq 1$ and so on. If $\text{Sup}_{n\geq 1} |\mu_n|(B_1^c) = \infty$, take $A_1 = B_1$ and then work with B_1^c as above. Thus we obtain two sequences $n_1 < n_2 < \cdots$ and A_n, $n \geq 1$ of pairwise disjoint sets in $\mathcal{F}$ such that $|\mu_{n_k}(A_k)| \geq 1$ for every $k \geq 1$. Since ψ is bounded, $\lim_{n\to\infty} \psi(A_n) = 0$. Hence μ_n, $n \geq 1$ cannot be uniformly absolutely continuous with respect to ψ. This contradiction proves the result. □

The above result can be labelled as Nikodym theorem for charges on fields. There are no conditions on the field but we impose an extra condition on the sequence μ_n, $n \geq 1$ to conclude that the set $\{\mu_n, n \geq 1\}$ is norm bounded.

If μ_n, $n \geq 1$ satisfies condition (i) only, it is not true that $\{\mu_n, n \geq 1\}$ is norm bounded. As an example, let $\Omega = \{1, 2, 3, \ldots\}$, $\mathcal{F} = \mathcal{P}(\Omega)$ and μ any 0-1 valued charge on $\mathcal{F}$ such that $\mu(A) = 0$ for every finite subset A of Ω. Then the sequence $\{n\mu, n \geq 1\}$ is uniformly absolutely continuous with respect to μ but $\text{Sup}_{n\geq 1} \|n\mu\| = \infty$.

The argument presented in the proof of the above theorem is similar to the one used in the proof of Lemma 2.1.5.

8.7.2 Theorem. *Let $(\Omega, \mathcal{F}, \mu)$ be a probability charge space and μ_n, $n \geq 1$ a sequence in $V_1(\Omega, \mathcal{F}, \mu)$ such that $\{\mu_n, n \geq 1\}$ is uniformly absolutely continuous with respect to μ and $\lim_{n\to\infty} \mu_n(F) = 0$ for every F in $\mathcal{F}$. Then μ_n, $n \geq 1$ converges to 0 weakly in $V_1(\Omega, \mathcal{F}, \mu)$.*

Proof. Let $\lambda \in V_\infty(\Omega, \mathcal{F}, \mu)$. We show that $\lim_{n\to\infty} T_\lambda(\mu_n) = 0$. (See the proof of Theorem 7.3.1 for the definition of T_λ.) Let $\varepsilon > 0$. There exists $\delta > 0$ such that $|\mu_n|(F) < \varepsilon$ for every $n \geq 1$ whenever $F \in \mathcal{F}$ and $\mu(F) < \delta$. Since $\lambda \in V_\infty(\Omega, \mathcal{F}, \mu)$, $\lambda \in V_1(\Omega, \mathcal{F}, \mu)$. By Theorem 7.2.12, there exists $P_0 = \{F_1, F_2, \ldots, F_m\}$ in $\mathcal{P}$ such that

$$\|\lambda - \lambda_{P_0}\|_1 = \text{Sup}\left\{\sum_{i=1}^{k} |\lambda(E_i) - \lambda_{P_0}(E_i)|; P = \{E_1, E_2, \ldots, E_k\} \in \mathcal{P}\right\} < \delta\varepsilon.$$

Let $r = \text{Sup}_{n\geq 1} \|\mu_n\|$. By Theorem 8.7.1, $r < \infty$. Let $P = \{E_1, E_2, \ldots, E_k\}$ be an arbitrary element in $\mathcal{P}$. Let $I(P) = \{1 \leq i \leq k; |\lambda(E_i) - \lambda_{P_0}(E_i)| < \varepsilon\mu(E_i)\}$ and $J(P) = \{1 \leq i \leq k; |\lambda(E_i) - \lambda_{P_0}(E_i)| \geq \varepsilon\mu(E_i)\}$. Note that

$$\sum_{j\in J(P)} \varepsilon\mu(E_j) \leq \sum_{j\in J(P)} |\lambda(E_j) - \lambda_{P_0}(E_j)|$$

$$\leq \|\lambda - \lambda_{P_0}\|_1 < \delta\varepsilon.$$

Therefore, $\mu(\bigcup_{j \in J(P)} E_j) = \sum_{j \in J(P)} \mu(E_j) < \delta$. Consequently, $|\mu_n|(\bigcup_{j \in J(P)} E_j) < \varepsilon$ for every $n \geq 1$. Also,

$$\left| \sum_{i=1}^{k} \frac{\mu_n(E_i)(\lambda(E_i) - \lambda_{P_0}(E_i))}{\mu(E_i)} \right|$$

$$= \left| \sum_{i \in I(P)} \mu_n(E_i) \left[\frac{\lambda(E_i) - \lambda_{P_0}(E_i)}{\mu(E_i)} \right] + \sum_{j \in J(P)} \mu_n(E_j) \left[\frac{\lambda(E_j) - \lambda_{P_0}(E_j)}{\mu(E_j)} \right] \right|$$

$$\leq \varepsilon \left| \mu_n \left(\bigcup_{i \in I(P)} E_i \right) \right| + \left| \mu_n \left(\bigcup_{j \in J(P)} E_j \right) \right| \|\lambda - \lambda_{P_0}\|_\infty$$

$$\leq \varepsilon r + \varepsilon \|\lambda - \lambda_{P_0}\|_\infty \leq \varepsilon [r + 2\|\lambda\|_\infty].$$

Hence

$$|T_{\lambda - \lambda_{P_0}}(\mu_n)| = \left| \lim_{P \in \mathscr{P}} \sum_{i=1}^{k} \frac{\mu_n(E_i)(\lambda(E_i) - \lambda_{P_0}(E_i))}{\mu(E_i)} \right|$$

$$\leq \varepsilon [r + 2\|\lambda\|_\infty].$$

Observe that

$$|T_\lambda(\mu_n)| \leq |T_\lambda(\mu_n - (\mu_n)_{P_0})| + |T_\lambda((\mu_n)_{P_0})|,$$

$$T_\lambda((\mu_n)_{P_0}) = \sum_{i=1}^{m} \frac{\lambda(F_i) \mu_n(F_i)}{\mu(F_i)} \longrightarrow 0$$

as $n \to \infty$, and

$$T_\lambda(\mu_n - (\mu_n)_{P_0}) = T_\lambda(\mu_n) - T_\lambda((\mu_n)_{P_0})$$
$$= T_\lambda(\mu_n) - T_{\lambda_{P_0}}(\mu_n)$$
$$= T_{\lambda - \lambda_{P_0}}(\mu_n).$$

Therefore,

$$\limsup_{n \to \infty} |T_\lambda(\mu_n)| \leq \limsup_{n \to \infty} |T_{\lambda - \lambda_{P_0}}(\mu_n)| + 0$$

$$\leq \varepsilon [r + 2\|\lambda\|_\infty].$$

Since $\varepsilon > 0$ is arbitrary, it follows that $\lim_{n \to \infty} T_\lambda(\mu_n) = 0$. This completes the proof. □

The following theorem gives a comprehensive list of equivalent conditions for weak convergence of a sequence in $\text{ba}(\Omega, \mathscr{F})$.

8.7.3 Theorem. *Let $\mathscr{F}$ be a field of subsets of a set Ω and μ_n, $n \geq 1$ a sequence in $\text{ba}(\Omega, \mathscr{F})$. Let ψ on $\mathscr{F}$ be defined by*

$$\psi(A) = \sum_{n \geq 1} \frac{1}{2^n} \frac{|\mu_n|(A)}{1 + |\mu_n|(\Omega)}$$

8. NIKODYM AND VITALI-HAHN-SAKS THEOREMS

for A in $\mathscr{F}$. Then the following statements are equivalent.

(i). μ_n, $n \geq 1$ converges to 0 weakly in $\mathrm{ba}(\Omega, \mathscr{F})$.

(ii). $\lim_{n \to \infty} \sum_{k \geq 1} \mu_n(A_k) = 0$ for every sequence A_k, $k \geq 1$ of pairwise disjoint sets in $\mathscr{F}$.

(iii). $\lim_{n \to \infty} \sum_{k \geq 1} |\mu_n(A_k)| = 0$ for every sequence A_k, $k \geq 1$ of pairwise disjoint sets in $\mathscr{F}$.

(iv). $\lim_{n \to \infty} \mu_n(A) = 0$ for every A in $\mathscr{F}$ and $\lim_{n \to \infty} \mu_n(A_n) = 0$ for every sequence A_k, $k \geq 1$ of pairwise disjoint sets in $\mathscr{F}$.

(v). $\lim_{n \to \infty} \mu_n(A) = 0$ for every A in $\mathscr{F}$ and $\{\mu_n, n \geq 1\}$ is uniformly absolutely continuous with respect to ψ.

(vi). $\lim_{n \to \infty} \mu_n(A) = 0$ for every A in $\mathscr{F}$ and $\{\mu_n, n \geq 1\}$ is uniformly absolutely continuous with respect to ν whenever $\nu \in \mathrm{ba}(\Omega, \mathscr{F})$ and $\mu_n \ll \nu$ for every $n \geq 1$.

(vii). $\lim_{n \to \infty} \mu_n(A) = 0$ for every A in $\mathscr{F}$ and there exists a ν in $\mathrm{ba}(\Omega, \mathscr{F})$ such that $\{\mu_n, n \geq 1\}$ is uniformly absolutely continuous with respect to ν.

(viii). μ_n, $n \geq 1$ converges to 0 weakly in $V_1(\Omega, \mathscr{F}, \psi)$.

(ix). μ_n, $n \geq 1$ converges to 0 weakly in $V_1(\Omega, \mathscr{F}, \nu)$ whenever $\nu \in \mathrm{ba}(\Omega, \mathscr{F})$ and $\mu_n \ll \nu$ for every $n \geq 1$.

(x). There exists a ν in $\mathrm{ba}(\Omega, \mathscr{F})$ such that $\mu_n \ll \nu$ for every $n \geq 1$ and μ_n, $n \geq 1$ converges to 0 weakly in $V_1(\Omega, \mathscr{F}, \nu)$.

Proof. (i) $\Rightarrow$ (ii). Let A_k, $k \geq 1$ be a sequence of pairwise disjoint sets in $\mathscr{F}$. Define T on $\mathrm{ba}(\Omega, \mathscr{F})$ by

$$T(\mu) = \sum_{k \geq 1} \mu(A_k)$$

for μ in $\mathrm{ba}(\Omega, \mathscr{F})$. Since $|T(\mu)| \leq \sum_{k \geq 1} |\mu|(A_k) \leq |\mu|(\Omega) = \|\mu\|$, T is a continuous linear functional on $\mathrm{ba}(\Omega, \mathscr{F})$. Since μ_n, $n \geq 1$ converges to 0 weakly in $\mathrm{ba}(\Omega, \mathscr{F})$,

$$\lim_{n \to \infty} T(\mu_n) = T(0) = 0 = \lim_{n \to \infty} \sum_{k \geq 1} \mu_n(A_k).$$

(ii) $\Rightarrow$ (iii). Let A_k, $k \geq 1$ be a sequence of pairwise disjoint sets in $\mathscr{F}$. Define, for each $n \geq 1$, λ_n on $\mathscr{P}(\mathbb{N})$ by

$$\lambda_n(D) = \sum_{k \in D} \mu_n(A_k)$$

for $D \subset \mathbb{N}$, where $\mathbb{N} = \{1, 2, 3, \ldots\}$. Note that each λ_n is a bounded charge on $\mathscr{P}(\mathbb{N})$. By (ii), $\lim_{n \to \infty} \lambda_n(D) = 0$ for every $D \subset \mathbb{N}$. By Phillips' lemma 8.3.3(i) $\Rightarrow$ (iii),

$$\lim_{n \to \infty} \sum_{k \geq 1} |\lambda_n(\{k\})| = 0.$$

But $\sum_{k \geq 1} |\lambda_n(\{k\})| = \sum_{k \geq 1} |\mu_n(A_k)|$. This proves (iii).

(iii)⇒(iv). Let $A \in \mathscr{F}$. Take $A_1 = A$, $A_2 = \emptyset$, $A_3 = \emptyset, \ldots$. So, by (iii), $\lim_{n \to \infty} \sum_{k \geq 1} |\mu_n(A_k)| = 0 = \lim_{n \to \infty} |\mu_n(A)|$. Now, let A_n, $n \geq 1$ be any sequence of pairwise disjoint sets in $\mathscr{F}$. By (iii),

$$0 \leq \limsup_{n \to \infty} |\mu_n(A_n)| \leq \lim_{n \to \infty} \sum_{k \geq 1} |\mu_n(A_k)| = 0.$$

This completes the proof of (iv).

(iv)⇒(v). Assume that (iv) holds and μ_n, $n \geq 1$ is not uniformly absolutely continuous with respect to ψ. This means that there exists $\varepsilon > 0$ such that for every $\delta > 0$, we can find A in $\mathscr{F}$ and $n \geq 1$ such that $\psi(A) < \delta$, but $|\mu_n(A)| \geq 2\varepsilon$.

For $\delta = 1$, we find A_1 in $\mathscr{F}$ and $n_1 \geq 1$ such that $|\mu_{n_1}(A_1)| \geq 2\varepsilon$ and $\psi(A_1) < 1$. Since $\mu_1, \mu_2, \ldots, \mu_{n_1}$ are absolutely continuous with respect to ψ, there exists $\delta_1 > 0$ such that $|\mu_j(A)| < \varepsilon/2^2$ for $j = 1, 2, \ldots, n_1$ whenever $A \in \mathscr{F}$ and $\psi(A) < \delta_1$. For $\delta = \delta_1$, we find A_2 in $\mathscr{F}$ and $n_2 \geq 1$ such that $|\mu_{n_2}(A_2)| \geq 2\varepsilon$ and $\psi(A_2) < \delta_1$. Obviously, $n_2 > n_1$. Also, $|\mu_{n_1}(B)| < \varepsilon/2^2$ whenever $B \in \mathscr{F}$ and $B \subset A_2$, since $\psi(B) \leq \psi(A_2) < \delta_1$. Since $\mu_1, \mu_2, \ldots, \mu_{n_2}$ are absolutely continuous with respect to ψ, there exists $\delta_2 > 0$ such that $|\mu_j(A)| < \varepsilon/2^3$ for $j = 1, 2, \ldots, n_2$ whenever $A \in \mathscr{F}$ and $\psi(A) < \delta_2$. For $\delta = \delta_2$, we find A_3 in $\mathscr{F}$ and $n_3 \geq 1$ such that $|\mu_{n_3}(A_3)| \geq 2\varepsilon$ and $\psi(A_3) < \delta_2$. Obviously, $n_3 > n_2$. Also, $|\mu_{n_1}(B)| < \varepsilon/2^3$ and $|\mu_{n_2}(B)| < \varepsilon/2^3$ whenever $B \in \mathscr{F}$ and $B \subset A_3$, since $\psi(B) \leq \psi(A_3) < \delta_2$.

Proceeding this way, we obtain a sequence A_k, $k \geq 1$ of sets in $\mathscr{F}$ and a sequence $n_1 < n_2 < \cdots$ of positive integers having the following properties.

(a). $|\mu_{n_k}(A_k)| \geq 2\varepsilon$, $k \geq 1$.

(b). $|\mu_{n_j}(B)| < \varepsilon/2^k$ for $j = 1, 2, \ldots, k-1$, $k \geq 2$ whenever $B \in \mathscr{F}$ and $B \subset A_k$.

We relabel $\mu_{n_1}, \mu_{n_2}, \ldots$ as $\mu_1, \mu_2, \ldots$. This is admissible since the property (iv) is hereditary, i.e. any subsequence of μ_n, $n \geq 1$ has property (iv).

Now, we construct a sequence H_k, $k \geq 1$ of pairwise disjoint sets in $\mathscr{F}$ and find a sequence $0 = p_0 < p_1 < p_2 < \cdots$ of positive integers such that

$$|\mu_{p_j+1}(H_{j+1})| > \varepsilon$$

for every $j \geq 0$. This then gives a contradiction as property (iv) fails to hold for the sequence $\mu_1, \mu_{p_1+1}, \mu_{p_2+1}, \ldots$, and the property (iv) is hereditary.

First, we give H_1 and p_1.

Let $F_1 = A_1$. If there is an integer $i > 1$ such that $|\mu_i(F_1 \cap A_i)| > \varepsilon/2$, let i_1 be the smallest positive such integer and $F_2 = F_1 - A_{i_1}$. If there is an integer $i > i_1$ such that $|\mu_i(F_2 \cap A_i)| > \varepsilon/2$, let i_2 be the smallest positive such integer and $F_3 = F_2 - A_{i_2}$. If this process does not stop at any finite stage, we get a sequence $F_1 \cap A_{i_1}, F_2 \cap A_{i_2}, \ldots$ of pairwise disjoint sets in $\mathscr{F}$ such that $|\mu_{i_k}(F_k \cap A_{i_k})| > \varepsilon/2$ for every $k \geq 1$. This contradicts the validity

8. NIKODYM AND VITALI-HAHN-SAKS THEOREMS

of property (iv) for the sequence μ_{i_k}, $k \geq 1$. So, let us assume that the above procedure stops at some finite stage. This means that there is a positive integer p_1 such that $\mu_i(F_{p_1} \cap A_i) \leq \varepsilon/2$ for every $i > p_1$. Take $H_1 = F_{p_1}$. We show that $|\mu_1(H_1)| \geq 2\varepsilon - \varepsilon/2$.

$$|\mu_1(H_1)| = |\mu_1(F_{p_1})| = |\mu_1(F_{p_1-1} - A_{i_{p_1-1}})|$$
$$= |\mu_1(F_{p_1-2} - A_{i_{p_1-2}} - A_{i_{p_1-1}})|$$
$$= \cdots$$
$$= |\mu_1(A_1 - A_{i_1} - A_{i_2} - \cdots - A_{i_{p_1-1}})|$$
$$\geq |\mu_1(A_1)| - |\mu_1(A_1 \cap A_{i_1})| - |\mu_1(A_1 \cap A_{i_2})|$$
$$\quad - \cdots - |\mu_1(A_1 \cap A_{i_{p_1-1}})|$$
$$\geq 2\varepsilon - \varepsilon/2^{i_1} - \varepsilon/2^{i_2} - \cdots - \varepsilon/2^{i_{p_1-1}}$$
$$\geq 2\varepsilon - \varepsilon/2 > \varepsilon.$$

Look at the sequence $A_i^{(1)} = A_{p_1+i} - F_{p_1}$, $i \geq 1$ and the sequence $\mu_i^{(1)} = \mu_{p_1+i}$, $i \geq 1$ of charges. We claim that these sequences have properties similar to those of (a) and (b) with 2ε replaced by $2\varepsilon - \varepsilon/2$. Note that

$$|\mu_i^{(1)}(A_i^{(1)})| = |\mu_{p_1+i}(A_{p_1+i} - A_{p_1+i} \cap H_1)|$$
$$\geq |\mu_{p_1+i}(A_{p_1+i})| - |\mu_{p_1+i}(A_{p_1+i} \cap H_1)|$$
$$\geq 2\varepsilon - \varepsilon/2 \quad \text{for every } i \geq 1.$$

To prove (b), let $B \in \mathscr{F}$ and $B \subset A_i^{(1)}$, $i \geq 2$. Then for $j = 1, 2, \ldots, i-1$,

$$|\mu_j^{(1)}(B)| \leq \varepsilon/2^{p_1+i} \leq 3\varepsilon/2^{i+1} = \frac{3}{2}\varepsilon \frac{1}{2^i}$$
$$= (2\varepsilon - \varepsilon/2)\frac{1}{2^i}.$$

Thus the sequences $A_i^{(1)}$, $i \geq 1$ and $\mu_i^{(1)}$, $i \geq 1$ have the following properties.

(a)'. $|\mu_i^{(1)}(A_i^{(1)})| \geq 2\varepsilon - \varepsilon/2$ for every $i \geq 1$.
(b)'. $|\mu_j^{(1)}(B)| \leq (2\varepsilon - \varepsilon/2)(1/2^i)$ for $j = 1, 2, \ldots, i-1$, whenever $B \in \mathscr{F}$ and $B \subset A_i^{(1)}$ for $i \geq 2$.

By using the argument given after (a) and (b) for the sequences figuring in (a)' and (b)' and replacing $\varepsilon/2$ by $\varepsilon/2^2$, we obtain an integer $p_2 > p_1$, a set H_2 in $\mathscr{F}$ and sequences $A_i^{(2)}$, $i \geq 1$ and $\mu_i^{(2)}$, $i \geq 1$ having properties similar to (a)' and (b)'. Further, $|\mu_1^{(1)}(H_2)| = |\mu_{p_1+1}(H_2)| \geq (2\varepsilon - \varepsilon/2) - \varepsilon/2^2$. Observe that $H_2 \subset A_1^{(1)} = A_{p_1+1} - H_1$ and so, $H_1 \cap H_2 = \varnothing$.

Continuing this procedure, we get the desired sequences H_k, $k \geq 1$ and p_k, $k \geq 0$. Hence (iv) $\Rightarrow$ (v).

(v)$\Rightarrow$(vi). It suffices to show that $\psi \ll \nu$. But this is evident from Theorem 6.1.13.
(vi)$\Rightarrow$(v). This is obvious.
(v)$\Rightarrow$(vii). This is also obvious.
(v)$\Rightarrow$(viii). This follows from Theorem 8.7.2.
(viii)$\Rightarrow$(ix). By Theorem 6.1.13, $V_1(\Omega, \mathcal{F}, \psi)$ is a subspace of $V_1(\Omega, \mathcal{F}, \nu)$. Since μ_n, $n \geq 1$ converges to 0 weakly in $V_1(\Omega, \mathcal{F}, \psi)$, it follows that μ_n, $n \geq 1$ converges to 0 weakly in $V_1(\Omega, \mathcal{F}, \nu)$.
(ix)$\Rightarrow$(viii). This is obvious.
(viii)$\Rightarrow$(x). This is also obvious.
(vii)$\Rightarrow$(x). This follows from Theorem 8.7.2.
(x)$\Rightarrow$(i). This is analogous to the proof given for the implication (viii)$\Rightarrow$(ix) because $V_1(\Omega, \mathcal{F}, \nu)$ is a subspace of $ba(\Omega, \mathcal{F})$.

This completes the proof that all the ten statements are equivalent. $\square$

Now, we formulate a theorem on weak convergence of a sequence of bounded charges analogous to the above theorem without specifying the limit.

8.7.4 Theorem. *Let $\mathcal{F}$ be a field of subsets of a set Ω and μ_n, $n \geq 1$ a sequence in $ba(\Omega, \mathcal{F})$. Let ψ be the charge defined on $\mathcal{F}$ by*

$$\psi(A) = \sum_{n \geq 1} \frac{1}{2^n} \frac{|\mu_n|(A)}{1+|\mu_n|(\Omega)}$$

for A in $\mathcal{F}$. Then the following statements are equivalent.

(i). $\mu_n \geq 1$ *is weakly convergent in* $ba(\Omega, \mathcal{F})$.

(ii). $\mu_n(A)$, $n \geq 1$ *converges to a real number for every A in $\mathcal{F}$ and $\{\mu_n, n \geq 1\}$ is uniformly absolutely continuous with respect to ψ.*

(iii). $\mu_n(A)$, $n \geq 1$ *converges to a real number $\mu(A)$ for every A in $\mathcal{F}$, $\mu \in ba(\Omega, \mathcal{F})$ and $\{\mu_n - \mu, n \geq 1\}$ is uniformly absolutely continuous with respect to ψ.*

Proof. (i)$\Rightarrow$(ii). Let μ_n, $n \geq 1$ converge to μ weakly in $ba(\Omega, \mathcal{F})$. Then $\mu_n - \mu$, $n \geq 1$ converges to 0 weakly in $ba(\Omega, \mathcal{F})$. By Theorem 8.7.3(i)$\Rightarrow$(vi), $\lim_{n \to \infty} (\mu_n - \mu)(A) = 0$ and $\{\mu_n - \mu, n \geq 1\}$ is uniformly absolutely continuous with respect to $\nu \in ba(\Omega, \mathcal{F})$ whenever $\mu_n - \mu \ll \nu$ for every $n \geq 1$. Note that $\{\mu_n, n \geq 1\} \subset V_1(\Omega, \mathcal{F}, \psi)$ and $V_1(\Omega, \mathcal{F}, \psi)$ is a norm closed subspace of $ba(\Omega, \mathcal{F})$. Consequently, $V_1(\Omega, \mathcal{F}, \psi)$ is a weakly closed subspace of $ba(\Omega, \mathcal{F})$. See Theorem 1.5.17. So, $\mu \in V_1(\Omega, \mathcal{F}, \psi)$. Thus we observe that $\mu_n - \mu$, $n \geq 1$ converges to 0 weakly in $V_1(\Omega, \mathcal{F}, \psi)$. By Theorem 8.7.3, $\{\mu_n - \mu, n \geq 1\}$ is uniformly absolutely continuous with respect to ψ. Since $\mu \ll \psi$, $\{\mu_n, n \geq 1\}$ is uniformly absolutely continuous with respect to ψ.

(ii)$\Rightarrow$(iii). Let $\mu(A) = \lim_{n \to \infty} \mu_n(A)$, $A \in \mathcal{F}$. Since $\{\mu_n, n \geq 1\}$ is uniformly absolutely continuous with respect to ψ, $\mu \ll \psi$. So, by Theorem 6.1.10(iv), μ is bounded. Consequently, $\{\mu_n - \mu, n \geq 1\}$ is uniformly absolutely continuous with respect to ψ.

(iii)$\Rightarrow$(i). By Theorem 8.7.3(v)$\Rightarrow$(i), $\mu_n - \mu$, $n \geq 1$ converges to 0 weakly in ba(Ω, $\mathcal{F}$). Consequently, μ_n, $n \geq 1$ is weakly convergent in ba(Ω, $\mathcal{F}$). □

Now, we show that if the field $\mathcal{F}$ has Seever property, characterization of weak convergence of sequences in ba(Ω, $\mathcal{F}$) becomes very simple.

8.7.5 Theorem. *Let $\mathcal{F}$ be a field of subsets of a set Ω having Seever property and μ_n, $n \geq 1$ a sequence in ba(Ω, $\mathcal{F}$). Then μ_n, $n \geq 1$ is weakly convergent in ba(Ω, $\mathcal{F}$) if and only if $\mu_n(A)$, $n \geq 1$ converges to a real number for every A in $\mathcal{F}$.*

Proof. If μ_n, $n \geq 1$ converges weakly in ba(Ω, $\mathcal{F}$), by Theorem 8.7.3, $\mu_n(A)$, $n \geq 1$ converges to a real number for every A in $\mathcal{F}$. This proves the "only if" part of the theorem.

Conversely, let $\mu(A) = \lim_{n \to \infty} \mu_n(A)$, $A \in \mathcal{F}$. It is obvious that μ is a charge on $\mathcal{F}$. Further, by Theorem 8.4.5, μ is bounded. For each $n \geq 1$, let $\nu_n = \mu_n - \mu$. For this sequence ν_n, $n \geq 1$, we check the validity of (iii) of Theorem 8.7.3. Let A_k, $k \geq 1$ be a sequence of pairwise disjoint sets in $\mathcal{F}$. For each $n \geq 1$, define τ_n on $\mathcal{P}(\mathbb{N})$ by

$$\tau_n(D) = \nu_n\left(\bigcup_{k \in D} A_k\right)$$

for $D \subset \mathbb{N}$, where $\mathbb{N} = \{1, 2, 3, \ldots\}$. Each τ_n is a bounded charge on $\mathcal{P}(\mathbb{N})$ and it follows that $\lim_{n \to \infty} \tau_n(D) = 0$ for every $D \subset \mathbb{N}$. By Phillips' lemma 8.3.3(i)$\Rightarrow$(iii),

$$\lim_{n \to \infty} \sum_{k \geq 1} |\tau_n(\{k\})| = 0.$$

But $\sum_{k \geq 1} |\tau_n(\{k\})| = \sum_{k \geq 1} |\nu_n(A_k)|$. Hence ν_n, $n \geq 1$ converges to 0 weakly in ba(Ω, $\mathcal{F}$). □

Now, we obtain as a corollary the classical result on weak convergence of sequences of bounded measures on σ-fields.

8.7.6 Corollary. *Let $\mathfrak{A}$ be a σ-field of subsets of a set Ω and μ_n, $n \geq 1$ a sequence in ba(Ω, $\mathfrak{A}$). Then μ_n, $n \geq 1$ converges weakly in ba(Ω, $\mathfrak{A}$) if and only if $\mu_n(A)$, $n \geq 1$ converges to a real number for every A in $\mathfrak{A}$. In particular, if $\{\mu_n, n \geq 1\} \subset$ ca(Ω, $\mathfrak{A}$), then μ_n, $n \geq 1$ is weakly convergent in ca(Ω, $\mathfrak{A}$) if and only if $\mu_n(A)$, $n \geq 1$ converges to a real number for every A in $\mathfrak{A}$.*

Proof. Note that every σ-field has Seever property. □

The results proved in this section give a new characterization of uniform absolute continuity.

8.7.7 Theorem. *Let $\mathcal{F}$ be a field of subsets of a set Ω and μ_n, $n \geq 1$ a sequence in $\mathrm{ba}(\Omega, \mathcal{F})$. Let ψ be the charge on $\mathcal{F}$ defined by*

$$\psi(A) = \sum_{n \geq 1} \frac{1}{2^n} \frac{|\mu_n|(A)}{1 + |\mu_n|(\Omega)}$$

for A in $\mathcal{F}$. Then $\{\mu_n, n \geq 1\}$ is uniformly absolutely continuous with respect to ψ if and only if for every sequence A_n, $n \geq 1$ of pairwise disjoint sets in $\mathcal{F}$, $\lim_{n \to \infty} \mu_n(A_n) = 0$.

Proof. "If" part of this theorem is essentially the proof of the implication (iv) $\Rightarrow$ (v) of Theorem 8.7.3. We prove the "only if" part.

Suppose for some sequence A_n, $n \geq 1$ of pairwise disjoint sets in $\mathcal{F}$, $\lim_{n \to \infty} \mu_n(A_n) \neq 0$. Let $\varepsilon > 0$ be such that $\limsup_{n \to \infty} |\mu_n(A_n)| > \varepsilon$. Since μ_n, $n \geq 1$ is uniformly absolutely continuous with respect to ψ, for the ε above, there exists $\delta > 0$ such that $|\mu_n(A)| < \varepsilon$ for every $n \geq 1$ whenever $A \in \mathcal{F}$ and $\psi(A) < \delta$. Since ψ is a bounded charge, $\lim_{n \to \infty} \psi(A_n) = 0$. So, there exists $N \geq 1$ such that $\psi(A_n) < \delta$ for every $n \geq N$. Consequently, $|\mu_n(A_n)| < \varepsilon$ for every $n \geq N$. This implies that $\limsup_{n \to \infty} |\mu_n(A_n)| \leq \varepsilon$. This contradiction proves the result. $\square$

8.8 VITALI–HAHN–SAKS THEOREM

In this section, we obtain Vitali–Hahn–Saks theorem in $\mathrm{ba}(\Omega, \mathcal{F})$ for fields $\mathcal{F}$ having Seever property. For this purpose, we introduce the notion of uniform additivity.

8.8.1 Definition. Let $\mathcal{F}$ be a field of subsets of a set Ω and $M \subset \mathrm{ba}(\Omega, \mathcal{F})$. M is said to be *uniformly additive* if for any sequence A_n, $n \geq 1$ of pairwise disjoint sets in $\mathcal{F}$,

$$\lim_{m \to \infty} \sup_{\mu \in M} \sum_{n \geq m} |\mu(A_n)| = 0.$$

The following theorem gives equivalent versions of uniform additivity.

8.8.2 Theorem. *Let $\mathcal{F}$ be a field of subsets of a set Ω and $M \subset \mathrm{ba}(\Omega, \mathcal{F})$. Then the following statements are equivalent.*
  (i). *M is uniformly additive.*
  (ii). *For any sequence A_n, $n \geq 1$ of pairwise disjoint sets in $\mathcal{F}$,*

$$\lim_{m \to \infty} \sup_{\mu \in M} \left| \sum_{n \geq m} \mu(A_n) \right| = 0.$$

8. NIKODYM AND VITALI-HAHN-SAKS THEOREMS

(iii). *For any sequence* A_n, $n \geq 1$ *of pairwise disjoint sets in* $\mathscr{F}$,

$$\lim_{m \to \infty} \sup_{\mu \in M} \sup_{\substack{D \subset \mathbb{N} \\ n \in D}} \left| \sum_{\substack{n \geq m \\ n \in D}} \mu(A_n) \right| = 0.$$

Proof. It can be seen easily that each of the properties (i), (ii) and (iii) is hereditary, i.e. if M has any of the properties (i), (ii) and (iii), then any subset of M has the same property.

(i) $\Rightarrow$ (ii). This is obvious.

(ii) $\Rightarrow$ (i). The proof is carried out in the following steps. Suppose (i) is not true. We show that (ii) is not true.

1°. Since (i) is assumed to be false, there exist $\varepsilon > 0$, a sequence E_n, $n \geq 1$ of pairwise disjoint sets in $\mathscr{F}$, a sequence μ_n, $n \geq 1$ of charges in M and a sequence $j_1 < j_2 < \cdots$ of positive integers such that

$$\sum_{j \geq j_n} |\mu_n(E_j)| > \varepsilon, \qquad n = 1, 2, \ldots.$$

Since for any bounded charge μ on $\mathscr{F}$, $\lim_{m \to \infty} \sum_{n \geq m} |\mu(E_n)| = 0$, we can take μ_n, $n \geq 1$ to be distinct.

2°. Look at μ_1. Since $\sum_{j \geq j_1} |\mu_1(E_j)| > \varepsilon$, we can find $m_1 > j_1$ such that $\sum_{j=j_1}^{m_1} |\mu_1(E_j)| > \varepsilon$. We can also find $p_1 > m_1$ such that $\sum_{j \geq p_1} |\mu_1(E_j)| < \varepsilon/4$. Now, for μ_{p_1}, we can find $m_2 > j_{p_1}$ such that $\sum_{j=j_{p_1}}^{m_2} |\mu_{p_1}(E_j)| > \varepsilon$ and $p_2 > m_2$ such that $\sum_{j \geq p_2} |\mu_{p_1}(E_j)| < \varepsilon/4$. Continuing this procedure we obtain two sequences $m_1, m_2, \ldots$ and $p_1, p_2, \ldots$ of positive integers satisfying the following conditions.

(a). $j_1 < m_1 < p_1 \leq j_{p_1} < m_2 < p_2 \leq j_{p_2} < m_3 < p_3 \leq \cdots$.

(b). $\sum_{j=j_{p_i}}^{m_{i+1}} |\mu_{p_i}(E_j)| > \varepsilon$, $i = 1, 2, 3, \ldots$.

(c). $\sum_{j \geq p_{i+1}} |\mu_{p_i}(E_j)| < \varepsilon/4$, $i = 1, 2, 3, \ldots$.

(We ignore μ_1.)

3°. Look at the sets E_j, $j = j_{p_1}, j_{p_1}+1, j_{p_1}+2, \ldots, m_2$ and μ_{p_1}. By segregating those E_j for which $\mu_{p_1}(E_j) < 0$ and those for which $\mu_{p_1}(E_j) \geq 0$, we can find $F_1, F_2, \ldots, F_{k_1}$, a subcollection of these E_j's, such that either $\sum_{i=1}^{k_1} \mu_{p_1}(F_i) < -\varepsilon/2$ with each $\mu_{p_1}(F_i) < 0$ or $\sum_{i=1}^{k_1} \mu_{p_1}(F_i) > \varepsilon/2$ with each $\mu_{p_1}(F_i) \geq 0$. We can achieve this because of (b) for $i = 1$.

Now, look at the sets E_j, $j = j_{p_2}, j_{p_2}+1, j_{p_2}+2, \ldots, m_3$ and μ_{p_2}. By the same technique as above, we can find $F_{k_1+1}, F_{k_1+2}, \ldots, F_{k_2}$, a subcollection of these E_j's, such that either $\sum_{i=k_1+1}^{k_2} \mu_{p_2}(F_i) < -\varepsilon/2$ with each $\mu_{p_2}(F_i) < 0$ or $\sum_{i=k_1+1}^{k_2} \mu_{p_2}(F_i) > \varepsilon/2$ with each $\mu_{p_2}(F_i) \geq 0$.

Continuing this procedure, we get a sequence $F_1, F_2, \ldots, F_{k_1}, F_{k_1+1}, F_{k_1+2}, \ldots, F_{k_2}, F_{k_2+1}, F_{k_2+2}, \ldots, F_{k_3} \ldots$ of pairwise disjoint sets in $\mathscr{F}$ such that for each $n \geq 0$,

$$\left| \sum_{i=k_n+1}^{k_{n+1}} \mu_{p_{n+1}}(F_i) \right| > \varepsilon/2,$$

with the understanding that $k_0 = 0$.

4°. We show that property (ii) fails to hold for the sequence $\mu_{p_1}, \mu_{p_2}, \ldots$ and the sequence $F_1, F_2, \ldots$ of pairwise disjoint sets from $\mathscr{F}$. This would prove the implication (ii) $\Rightarrow$ (i).

For every $n \geq 0$, note that

$$\left| \sum_{i \geq k_n+1} \mu_{p_{n+1}}(F_i) \right| = \left| \sum_{i=k_n+1}^{k_{n+1}} \mu_{p_{n+1}}(F_i) + \sum_{i > k_{n+1}} \mu_{p_{n+1}}(F_i) \right|$$

$$\geq \left| \sum_{i=k_n+1}^{k_{n+1}} \mu_{p_{n+1}}(F_i) \right| - \sum_{i > k_{n+1}} |\mu_{p_{n+1}}(F_i)|$$

$$> \varepsilon/2 - \varepsilon/4 = \varepsilon/4.$$

This follows from the elementary inequality $|a+b| \geq |a| - |b|$. Since $k_n \to \infty$ as $n \to \infty$, property (ii) fails to hold for the sequence $\mu_{p_1}, \mu_{p_2}, \ldots$ with respect to the sequence $F_1, F_2, \ldots$ of pairwise disjoint sets in $\mathscr{F}$.

(i) $\Rightarrow$ (iii). This is obvious.

(iii) $\Rightarrow$ (i). Let A_n, $n \geq 1$ be a sequence of pairwise disjoint sets in $\mathscr{F}$. Let $\varepsilon > 0$. Since (iii) is assumed to be true, we can find $m_0 \geq 1$ such that for every $m \geq m_0$,

$$\operatorname*{Sup}_{D \subset \mathbb{N}} \operatorname*{Sup}_{\mu \in M} \left| \sum_{\substack{n \geq m \\ n \in D}} \mu(A_n) \right| < \varepsilon/2.$$

Now, if $m \geq m_0$ and $\mu \in M$, we have

$$\sum_{n \geq m} |\mu(A_n)| = \left| \sum_{\substack{n \geq m \\ n \in D}} \mu(A_n) \right| + \left| \sum_{\substack{n \geq m \\ n \in E}} \mu(A_n) \right|$$

$$< \varepsilon/2 + \varepsilon/2 = \varepsilon,$$

where $D = \{n \geq 1; \mu(A_n) \geq 0\}$ and $E = \{n \geq 1; \mu(A_n) < 0\}$. This proves (i). $\square$

8.8.3 Remark. From the above theorem, it is clear that uniform countable additivity of a set $M \subset \text{ca}(\Omega, \mathfrak{A})$ is equivalent to uniform additivity of M, if $\mathfrak{A}$ is a σ-field of subsets of Ω.

Now, we prove Vitali–Hahn–Saks theorem for charges on Boolean algebras having Seever property.

8.8.4 Theorem. *Let $\mathbb{B}$ be a Boolean algebra having Seever property and μ_n, $n \geq 1$ a sequence of bounded charges on $\mathbb{B}$. Suppose that $\mu_n(b)$, $n \geq 1$ converges to a real number $\mu(b)$ for every b in $\mathbb{B}$. Then the following statements are true.*

(i). *$\{\mu_n, n \geq 1\}$ is uniformly absolutely continuous with respect to ν whenever $\mu_n \ll \nu$ for all $n \geq 1$ for any bounded charge ν on $\mathbb{B}$.*

(ii). *$\{\mu_n, n \geq 1\}$ is uniformly additive.*

(iii). *μ is a bounded charge.*

8. NIKODYM AND VITALI-HAHN-SAKS THEOREMS

Proof. (iii). It is clear that μ is a charge on $\mathbb{B}$. Since $\mu_n(b)$, $n \geq 1$ converges to a real number, $\sup_{n \geq 1} |\mu_n(b)| < \infty$. So, by Theorem 8.4.5, $\sup_{n \geq 1} \|\mu_n\| < \infty$. Hence μ is bounded.

(i). Since $\mathbb{B}$ has Seever property, by Theorem 8.7.5, μ_n, $n \geq 1$ is weakly convergent to some bounded charge τ on $\mathbb{B}$. Obviously, $\tau = \mu$. Define ψ on $\mathbb{B}$ by

$$\psi(b) = \sum_{n \geq 1} \frac{1}{2^n} \frac{|\mu_n|(b)}{1 + |\mu_n|(1)}$$

for b in $\mathbb{B}$. By Theorem 8.7.4(ii), $\{\mu_n, n \geq 1\}$ is uniformly absolutely continuous with respect to ψ. By Theorem 6.1.13, $\psi \ll \nu$. Hence $\{\mu_n, n \geq 1\}$ is uniformly absolutely continuous with respect to ν.

(ii). Since μ_n, $n \geq 1$ converges to μ weakly, $\mu_n - \mu$, $n \geq 1$ converges to 0 weakly. By Theorem 8.7.3(i) $\Rightarrow$ (iii),

$$\lim_{n \to \infty} \sum_{k \geq 1} |\mu_n(e_k) - \mu(e_k)| = 0$$

for any sequence e_k, $k \geq 1$ of pairwise disjoint elements in $\mathbb{B}$. Let $\varepsilon > 0$. There exists $m_1 \geq 1$ such that

$$\sum_{k \geq 1} |\mu_n(e_k) - \mu(e_k)| < \varepsilon/2$$

whenever $n \geq m_1$. There exists $m_2 \geq 1$ such that

$$\sum_{k \geq m} |\mu(e_k)| < \varepsilon/2$$

whenever $m \geq m_2$. Let $m_0 = \max\{m_1, m_2\}$. There exists $N \geq 1$ such that

$$\sum_{k \geq m} |\mu_i(e_k) - \mu(e_k)| < \varepsilon/2$$

whenever $m \geq N$ and $i = 1, 2, \ldots, m_0$. Now, let $m \geq N$. Then

$$\sum_{k \geq m} |\mu_n(e_k)| \leq \sum_{k \geq m} |\mu_n(e_k) - \mu(e_k)| + \sum_{k \geq m} |\mu(e_k)|$$

$$< \varepsilon/2 + \varepsilon/2 = \varepsilon$$

for every $n \geq 1$. This shows that $\{\mu_n, n \geq 1\}$ is uniformly additive. $\square$

The classical Vitali-Hahn-Saks theorem follows as a corollary.

8.8.5 Corollary. *Let $\mathfrak{A}$ be a σ-field of subsets of a set Ω and μ_n, $n \geq 1$ a sequence in $\text{ba}(\Omega, \mathfrak{A})$. Suppose $\mu_n(A)$, $n \geq 1$ converges to a real number $\mu(A)$ for every A in $\mathfrak{A}$. Then the following statements are true.*

(i). $\{\mu_n, n \geq 1\}$ *is uniformly absolutely continuous with respect to ν whenever $\mu_n \ll \nu$ for every $n \geq 1$ for any ν in $\text{ba}(\Omega, \mathfrak{A})$.*

(ii). $\{\mu_n, n \geq 1\}$ *is uniformly additive.*

(iii). μ *is a bounded charge on* $\mathfrak{A}$.

In particular, if $\{\mu_n, n \geq 1\} \subset \mathrm{ca}(\Omega, \mathfrak{A})$, *then, under the above condition, the following statements are true.*

(i). $\{\mu_n, n \geq 1\}$ *is uniformly absolutely continuous with respect to* ν *whenever* $\nu \in \mathrm{ca}(\Omega, \mathfrak{A})$ *and* $\mu_n \ll \nu$ *for every* $n \geq 1$.

(ii). $\{\mu_n, n \geq 1\}$ *is uniformly countably additive.*

(iii). μ *is a bounded measure on* $\mathfrak{A}$.

Proof. Every σ-field has Seever property. □

CHAPTER 9

The Dual of ba(Ω, $\mathscr{F}$) and The Refinement Integral

One of the important problems in the study of the Banach Space ba(Ω, $\mathscr{F}$), the space of all bounded charges on a field $\mathscr{F}$ of subsets of a set Ω, is to describe its dual ba*(Ω, $\mathscr{F}$). This problem seems to be difficult and no satisfactory representation of the elements of ba*(Ω, $\mathscr{F}$) is known. The purpose of this chapter is to present one of the attempts made to describe ba*(Ω, $\mathscr{F}$) using what is known as refinement integral. In Section 9.1, we present some basic ideas on refinement integral. In Section 9.2, for superatomic fields $\mathscr{F}$, we show that refinement integral representation exists for all elements of ba*(Ω, $\mathscr{F}$). We also show that the later property characterizes superatomic fields. We also obtain refinement integral representation for elements of $V_1^*(\Omega, \mathscr{F}, \mu)$ in this section.

9.1 REFINEMENT INTEGRAL

9.1.1 Definition. Let $\mathscr{F}$ be a field of subsets of a set Ω and μ a bounded charge on $\mathscr{F}$. Let f be any real valued function defined on $\mathscr{F}$ and $A \in \mathscr{F}$. Then the *refinement integral of f with respect to μ over A is said to exist* if there is a real number α with the property:
for any $\varepsilon > 0$, there is a partition P_0 in $\mathscr{P}_A$ such that for any partition $P = \{E_1, E_2, \ldots, E_m\}$ of A in $\mathscr{F}$ finer than P_0,

$$\left|\sum_{i=1}^m f(E_i)\mu(E_i) - \alpha\right| < \varepsilon.$$

We write $\alpha = \int_A f\mu$. (Recall that $\mathscr{P}_A$ stands for all finite partitions of A in $\mathscr{F}$.)

By the statement "$\int_A f\mu$ exists", we mean that the refinement integral of f with respect to μ over A exists.

If $f \equiv 1$, then $\int_A f\mu$ exists for any A in $\mathscr{F}$ and any bounded charge μ on $\mathscr{F}$ and $\int_A f\mu = \mu(A)$.

We give some basic properties of refinement integrals.

9.1.2 Proposition. *Let $\mathscr{F}$ be a field of subsets of a set Ω and f a real valued function on $\mathscr{F}$. Let μ and λ be two bounded charges on $\mathscr{F}$ and $A \in \mathscr{F}$.*
(i). *If $\int_A f\mu$ exists and β is a real number, then $\int_A f\beta\mu$ exists and $\int_A f\beta\mu = \beta \int_A f\mu$.*
(ii). *If $\int_A f\mu$ exists and $\int_A f\lambda$ exists, then $\int_A f\mu + \lambda$ exists and $\int_A f\mu + \lambda = \int_A f\mu + \int_A f\lambda$.*
(iii). *If f is a bounded function on $\mathscr{F}$ and $\int_A f\mu$ exists, then*

$$\left| \int_A f\mu \right| \leq \left(\sup_{B \in \mathscr{F}} |f(B)| \right) |\mu|(A).$$

(iv). *Let f be a bounded function on $\mathscr{F}$. Let μ_n, $n \geq 1$ be a sequence of bounded charges on $\mathscr{F}$ and a_n, $n \geq 1$ a sequence of real numbers such that $\sum_{n \geq 1} |a_n| |\mu_n|(\Omega) < \infty$. Let $\mu = \sum_{n \geq 1} a_n \mu_n$. If $\int_A f\mu_n$ exists for every $n \geq 1$, then $\int_A f\mu$ exists and*

$$\int_A f\mu = \sum_{n \geq 1} a_n \int_A f\mu_n.$$

Proof. (i) and (ii) are obvious from the definition of the refinement integral.
(iii). Let $\int_A f\mu = \alpha$. Let $\varepsilon > 0$. Then there exists a partition $P = \{E_1, E_2, \ldots, E_n\}$ of A in $\mathscr{F}$ such that $|\sum_{i=1}^n f(E_i)\mu(E_i) - \alpha| < \varepsilon$. This implies that

$$|\alpha| \leq \left| \sum_{i=1}^n f(E_i)\mu(E_i) \right| + \varepsilon$$

$$\leq \left(\sup_{B \in \mathscr{F}} |f(B)| \right) \sum_{i=1}^n |\mu(E_i)| + \varepsilon$$

$$\leq \left(\sup_{B \in \mathscr{F}} |f(B)| \right) |\mu|(A) + \varepsilon.$$

Since $\varepsilon > 0$ is arbitrary, we obtain

$$|\alpha| = \left| \int_A f\mu \right| \leq \left(\sup_{B \in \mathscr{F}} |f(B)| \right) |\mu|(A).$$

(iv). Let $\lambda_n = \sum_{i=1}^n a_i \mu_i$, $n \geq 1$. By the given condition, λ_n, $n \geq 1$ converges to μ in the norm of $ba(\Omega, \mathscr{F})$. Note that, by (iii),

$$\sum_{n \geq 1} \left| a_n \int_A f\mu_n \right| < \infty.$$

Let $M = \sup_{B \in \mathscr{F}} |f(B)|$ and $\varepsilon > 0$. We can find $N \geq 1$ such that $\|\lambda_N - \mu\| < \varepsilon/3M$ and

$$\left| \sum_{i=1}^N a_i \int_A f\mu_i - \sum_{n \geq 1} a_n \int_A f\mu_n \right| = \left| \sum_{i \geq N+1} a_i \int_A f\mu_i \right|$$

$$< \varepsilon/3.$$

9. THE DUAL OF ba(Ω, $\mathscr{F}$) AND REFINEMENT INTEGRAL

Further, we can find a single partition P_0 in $\mathscr{P}_A$ such that for any patition $P = \{F_1, F_2, \ldots, F_m\}$ of A in $\mathscr{F}$ finer than P_0,

$$\left|\sum_{j=1}^{m} f(F_j)a_i\mu_i(F_j) - a_i \int_A f\mu_i\right| < \varepsilon/3N$$

for $i = 1, 2, \ldots, N$. Now, if $P = \{F_1, F_2, \ldots, F_m\}$ in $\mathscr{P}_A$ is finer than P_0, then

$$\left|\sum_{j=1}^{m} f(F_j)\mu(F_j) - \sum_{n \geq 1} a_n \int_A f\mu_n\right| \leq \left|\sum_{j=1}^{m} f(F_j)\mu(F_j) - \sum_{j=1}^{m} f(F_j)\lambda_N(F_j)\right|$$

$$+ \left|\sum_{j=1}^{m} f(F_j)\lambda_N(F_j) - \sum_{i=1}^{N} a_i \int_A f\mu_i\right|$$

$$+ \left|\sum_{i=1}^{N} a_i \int_A f\mu_i - \sum_{n \geq 1} a_n \int_A f\mu_n\right|$$

$$< M \sum_{j=1}^{m} |\mu(F_j) - \lambda_N(F_j)|$$

$$+ \sum_{i=1}^{N} \left|\sum_{j=1}^{m} f(F_j)a_i\mu_i(F_j) - a_i \int_A f\mu_i\right| + \varepsilon/3$$

$$< \varepsilon/3 + N(\varepsilon/3N) + \varepsilon/3 = \varepsilon.$$

Hence $\int_A f\mu$ exists and

$$\int_A f\mu = \sum_{n \geq 1} a_n \int_A f\mu_n. \qquad \square$$

We now show that every bounded real valued function on $\mathscr{F}$ for which the refinement integral with respect to μ over Ω exists for every μ in ba(Ω, $\mathscr{F}$) defines a continuous linear functional on ba(Ω, $\mathscr{F}$).

9.1.3 Theorem. *Let $\mathscr{F}$ be a field of subsets of a set Ω and f a bounded real valued function on $\mathscr{F}$. Suppose $\int_\Omega f\mu$ exists for every μ in ba(Ω, $\mathscr{F}$). Let T be defined on ba(Ω, $\mathscr{F}$) by*

$$T(\mu) = \int_\Omega f\mu$$

for μ in ba(Ω, $\mathscr{F}$). Then T is a continuous linear functional on ba(Ω, $\mathscr{F}$).

Proof. This follows from Proposition 9.1.2(i), (ii) and (iii). $\square$

The main problem considered in the next section is to characterize those fields $\mathscr{F}$ for which every T in ba*(Ω, $\mathscr{F}$) admits a representation of the above type.

9.2 THE DUAL OF ba(Ω, $\mathscr{F}$)

Recall the definition of a superatomic field from Definition 5.3.4 for the following theorem.

9.2.1 Theorem. *Let $\mathscr{F}$ be a field of subsets of a set Ω. Then the following statements are equivalent.*

(i). *For any continuous linear functional T on ba(Ω, $\mathscr{F}$), i.e. $T \in$ ba*(Ω, $\mathscr{F}$), there exists a real valued bounded function f on $\mathscr{F}$ such that $\int_\Omega f\mu$ exists for all μ in ba(Ω, $\mathscr{F}$) and*

$$T(\mu) = \int_\Omega f\mu$$

for all μ in ba(Ω, $\mathscr{F}$).

(ii). *$\mathscr{F}$ is superatomic.*

Proof. (i)$\Rightarrow$(ii). The proof is carried out in the following steps.

1°. Suppose $\mathscr{F}$ is not superatomic. By Theorem 5.3.6, there exists a strongly continuous probability charge λ on $\mathscr{F}$. We define T on ba(Ω, $\mathscr{F}$) by $T(\mu) = \mu_1(\Omega)$ for μ in ba(Ω, $\mathscr{F}$), where $\mu = \mu_1 + \mu_2$ with $\mu_1 \ll \lambda$ and $\mu_2 \perp \lambda$. See Lebesgue Decomposition Theorem, i.e. Theorem 6.2.4. Since this decomposition is unique, T is a well defined linear functional. Continuity of T also follows from Theorem 6.2.4 if we observe that

$$|T(\mu)| = |\mu_1(\Omega)| \le |\mu_1|(\Omega) \le |\mu_1|(\Omega) + |\mu_2|(\Omega)$$
$$\le |\mu|(\Omega) = \|\mu\|.$$

2°. Since we are assuming that (i) holds, there exists a bounded function f on $\mathscr{F}$ such that $T(\mu) = \int_\Omega f\mu$ for all μ in ba(Ω, $\mathscr{F}$).

3°. For F in $\mathscr{F}$, let λ_F on $\mathscr{F}$ be defined by $\lambda_F(E) = \lambda(E \cap F)$ for E in $\mathscr{F}$. Then for any F in $\mathscr{F}$,

$$\lambda(F) = \lambda_F(\Omega) = T(\lambda_F) = \int_\Omega f\lambda_F = \int_F f\lambda.$$

(The Lebesgue decomposition of λ_F with respect to λ is $\lambda_F + 0$.)

4°. Define $\mathscr{B} = \{F \in \mathscr{F};$ every finite partition of F in $\mathscr{F}$ has a set B in $\mathscr{F}$ such that $\lambda(B) > 0$ and $f(B) \ge \frac{1}{2}\}$.

5°. We claim that for every A in $\mathscr{F}$ with $\lambda(A) > 0$, there exists F in $\mathscr{F}$ such that $F \subset A$ and $F \in \mathscr{B}$. Suppose the claim is false. Then there exists A in $\mathscr{F}$ with $\lambda(A) > 0$ such that $F \notin \mathscr{B}$ whenever $F \subset A$ and $F \in \mathscr{F}$, i.e. for every partition $\{B_1, B_2, \ldots, B_m\}$ of F in $\mathscr{F}$, either $\lambda(B_i) = 0$ or $f(B_i) < \frac{1}{2}$ for all i. By 3°, since $\lambda(A) = \int_A f\lambda$, we can find a partition $P = \{F_1, F_2, \ldots, F_m\}$ of A in $\mathscr{F}$ such that

$$\left|\sum_{i=1}^m f(F_i)\lambda(F_i) - \lambda(A)\right| < \lambda(A)/2.$$

9. THE DUAL OF $ba(\Omega, \mathcal{F})$ AND REFINEMENT INTEGRAL

Note that $f(F_i)\lambda(F_i) \leq \frac{1}{2}\lambda(F_i)$ for all i, so that $\sum_{i=1}^{m} f(F_i)\lambda(F_i) \leq \lambda(A)/2$. Therefore,

$$\lambda(A)/2 \leq \left| \sum_{i=1}^{m} f(F_i)\lambda(F_i) - \lambda(A) \right| < \lambda(A)/2.$$

This contradiction establishes the claim.

6°. Since λ is a strongly continuous probability charge on $\mathcal{F}$, we can find B_0 and B_1 in $\mathcal{F}$ such that $B_0 \cap B_1 = \varnothing$, $\lambda(B_0) > 0$ and $\lambda(B_1) > 0$. By 5°, we can find $A_0 \subset B_0$ and $A_1 \subset B_1$ such that $A_0, A_1 \in \mathcal{B}$. By the same argument, we obtain $A_{00}, A_{01} \subset A_0$; $A_{10}, A_{11} \subset A_1$; $A_{00} \cap A_{01} = \varnothing$; $A_{10} \cap A_{11} = \varnothing$ and $A_{00}, A_{01}, A_{10}, A_{11} \in \mathcal{B}$. Continuing this way, we obtain $\{A_{\varepsilon_1,\varepsilon_2,\ldots,\varepsilon_n}; \varepsilon_i = 0$ or 1 for all i, $n \geq 1\}$, a subcollection of $\mathcal{B}$ with the following properties.

(i). $A_{\varepsilon_1,\varepsilon_2,\ldots,\varepsilon_n} \supset A_{\varepsilon_1,\varepsilon_2,\ldots,\varepsilon_n,\varepsilon_{n+1}}$ for all $n \geq 1$.

(ii). $A_{\varepsilon_1,\varepsilon_2,\ldots,\varepsilon_n} \cap A_{\delta_1,\delta_2,\ldots,\delta_n} = \varnothing$ if $(\varepsilon_1, \varepsilon_2, \ldots, \varepsilon_n) \neq (\delta_1, \delta_2, \ldots, \delta_n)$, $n \geq 1$.

7°. Let $\varepsilon = (\varepsilon_1, \varepsilon_2, \ldots)$ be a sequence of 0's and 1's. Let $\mathcal{E}_\varepsilon = \{A_{\varepsilon_1}, A_{\varepsilon_1,\varepsilon_2}, A_{\varepsilon_1,\varepsilon_2,\varepsilon_3}, \ldots\}$. Let $\mathcal{F}_\varepsilon$ be a filter in $\mathcal{F}$ containing $\mathcal{E}_\varepsilon$ and maximal with respect to the following property:

(∗) "For every A in $\mathcal{F}_\varepsilon$, there is a B in $\mathcal{B} \cap \mathcal{F}_\varepsilon$ such that $B \subset A$."

Existence of $\mathcal{F}_\varepsilon$ can be established using Zorn's lemma as follows. We look at the collection $\mathcal{G}$ of all filters in $\mathcal{F}$ containing $\mathcal{E}_\varepsilon$ and having the property (∗). This collection is non-empty since the filter in $\mathcal{F}$ generated by $\mathcal{E}_\varepsilon$ (which exists by remark 1.1.23(4)) has the property (∗). In the usual partial order of inclusion, one can show that every chain in $\mathcal{G}$ has an upper bound. Hence by Zorn's lemma, there is a filter $\mathcal{F}_\varepsilon$ in $\mathcal{F}$ containing $\mathcal{E}_\varepsilon$ and maximal with respect to the property (∗).

8°. Now, we claim that $\mathcal{F}_\varepsilon$ is a maximal filter in $\mathcal{F}$. Suppose the claim is false. There exists C in $\mathcal{F}$ such that neither C nor $C^c \in \mathcal{F}_\varepsilon$. Let $\mathcal{F}_\varepsilon^1$ be the filter in $\mathcal{F}$ generated by $\mathcal{F}_\varepsilon$ and $\{C\}$ and $\mathcal{F}_\varepsilon^2$ the filter in $\mathcal{F}$ generated by $\mathcal{F}_\varepsilon$ and $\{C^c\}$. In fact,

$$\mathcal{F}_\varepsilon^1 = \{D \in \mathcal{F}; D \supset C \cap B \text{ for some B in } \mathcal{F}_\varepsilon\}$$

and

$$\mathcal{F}_\varepsilon^2 = \{D \in \mathcal{F}; D \supset C^c \cap B \text{ for some B in } \mathcal{F}_\varepsilon\}.$$

By the definition of $\mathcal{F}_\varepsilon$, $\mathcal{F}_\varepsilon^1$ and $\mathcal{F}_\varepsilon^2$ do not have the property (∗). So, there are sets B_1, B_2 in $\mathcal{F}_\varepsilon$ such that $C \cap B_1$ and $C^c \cap B_2$ do not contain any member of $\mathcal{B}$. We can assume that $B_1 = B_2 = B$, say, by considering $B_1 \cap B_2$. By 5°, $\lambda(C \cap B) = 0 = \lambda(C^c \cap B)$. This implies that $\lambda(B) = 0$. But, since $B \in \mathcal{F}_\varepsilon$, $\lambda(B) > 0$. This contradiction shows that $\mathcal{F}_\varepsilon$ is a maximal filter in $\mathcal{F}$.

9°. We claim that if $\varepsilon = (\varepsilon_1, \varepsilon_2, \ldots)$ and $\delta = (\delta_1, \delta_2, \ldots)$ are two sequences of 0's and 1's such that $\varepsilon \neq \delta$, then $\mathcal{F}_\varepsilon$ and $\mathcal{F}_\delta$ are distinct. Suppose $\mathcal{F}_\varepsilon = \mathcal{F}_\delta$. Since $\varepsilon \neq \delta$, there exists $n \geq 1$ such that $(\varepsilon_1, \varepsilon_2, \ldots, \varepsilon_n) \neq (\delta_1, \delta_2, \ldots, \delta_n)$.

So, $A_{\varepsilon_1,\varepsilon_2,...,\varepsilon_n} \cap A_{\delta_1,\delta_2,...,\delta_n} = \varnothing \in \mathscr{F}_\varepsilon$, a contradiction. Thus the claim is valid.

10°. We claim that for every $n \geq 1$, $\{\varepsilon \in \{0, 1\}^{\aleph_0}; \lambda(A) \geq 1/n$ for every A in $\mathscr{F}_\varepsilon\}$ has at most n elements. (Recall that $\{0, 1\}^{\aleph_0}$ is the space of all sequences of 0's and 1's.) Suppose the above set has more than n elements. Pick up any $n + 1$ distinct elements $\varepsilon^{(1)}, \varepsilon^{(2)}, \ldots, \varepsilon^{(n+1)}$ from this set. Since $\mathscr{F}_{\varepsilon^{(1)}}, \mathscr{F}_{\varepsilon^{(2)}}, \ldots, \mathscr{F}_{\varepsilon^{(n+1)}}$ are distinct, we can find B_i in $\mathscr{F}_{\varepsilon^{(i)}}, i = 1, 2, \ldots, n + 1$ such that $B_1, B_2, \ldots, B_{n+1}$ are pairwise disjoint. Since $\lambda(B_i) \geq 1/n$ for every $i = 1, 2, \ldots, n + 1$, $\lambda(\bigcup_{i=1}^{n+1} B_i) \geq (n + 1)/n$. But $\lambda(\Omega) = 1$. This contradiction establishes the claim.

11°. Since

$$\left\{\varepsilon \in \{0, 1\}^{\aleph_0}; \inf_{A \in \mathscr{F}_\varepsilon} \lambda(A) > 0\right\} = \bigcup_{n \geq 1} \{\varepsilon \in \{0, 1\}^{\aleph_0};$$

$$\lambda(A) \geq 1/n \text{ for every } A \in \mathscr{F}_\varepsilon\},$$

the set on the left is countable. Since $\{0, 1\}^{\aleph_0}$ is uncountable, there exists $\eta \in \{0, 1\}^{\aleph_0}$ such that $\text{Inf}_{A \in \mathscr{F}_\eta} \lambda(A) = 0$.

12°. Let ν on $\mathscr{F}$ be defined by

$$\nu(A) = 0, \quad \text{if } A \notin \mathscr{F}_\eta, A \in \mathscr{F},$$
$$ = 1, \quad \text{if } A \in \mathscr{F}_\eta.$$

ν is a 0–1 valued charge on $\mathscr{F}$.

13°. Observe that $\nu \perp \lambda$. See Proposition 8.5.1(v). Hence the Lebesgue decomposition of ν with respect to λ is $0 + \nu$. Therefore, $T(\nu) = 0$.

14°. On the other hand, we show that $\int_\Omega f\nu \geq \frac{1}{2}$. It suffices to show that given $P = \{E_1, E_2, \ldots, E_m\}$ in $\mathscr{P}$, there exists a finer partition $P' = \{F_1, F_2, \ldots, F_n\}$ in $\mathscr{P}$ such that $\sum_{i=1}^n f(F_i)\nu(F_i) \geq \frac{1}{2}$. For $P = \{E_1, E_2, \ldots, E_m\}$ in $\mathscr{P}$, there exists exactly one i, say $i = 1$, such that $\nu(E_i) = 1$, i.e. $E_1 \in \mathscr{F}_\eta$. We can find $B \in \mathscr{F}_\eta \cap \mathscr{B}$ such that $B \subset E_1$. Take $P' = \{B, E_1 - B, E_2, E_3, \ldots, E_m\}$.

Thus $0 = T(\nu) \neq \int_\Omega f\nu \geq \frac{1}{2}$. This contradiction shows that $\mathscr{F}$ is superatomic.

Now, we prove (ii) $\Rightarrow$ (i). This is carried out in the following steps.

1°. We, first, collect some basic facts about derived sets in topology. Let X be any topological space and $A \subset X$. Set $A^0 = A$, A^1 (the derived set of A^0) $= \{x \in A^0; x$ is an accumulation point of $A^0\}$, if α is a limit ordinal, set $A^\alpha = \bigcap_{\beta < \alpha} A^\beta$ and for any ordinal α, set $A^{\alpha+1} = (A^\alpha)^1$. Then A^α, $\alpha \geq 0$ is a decreasing net of sets and each A^α is a closed subset of A.

For what follows, we assume that X is a scattered compact Hausdorff totally disconnected space. Then, there exists an ordinal α_0 such that X^{α_0} is a non-empty finite set and $X^{\alpha_0+1} = \varnothing$. This can be proved as follows. Let β be the least ordinal such that $X^\beta = X^{\beta+1}$. Then $X^\beta = \varnothing$. For, if $X^\beta \neq \varnothing$, then $X^{\beta+1}$ is a proper subset of X^β. (Since X is scattered, the closed set X^β is not perfect and hence has isolated points in it.) Now, we claim that $\beta = \alpha_0 + 1$ for some ordinal α_0. If β is a limit ordinal, then $X^\beta = \bigcap_{\gamma < \beta} X^\gamma = \emptyset$ and this implies that $X^\gamma = \emptyset$ for some $\gamma < \beta$, contradicting the definition of β.

Now, we describe some properties of derived sets of clopen subsets V of X. For any clopen subset V of X and any ordinal α, $V^\alpha = V \cap X^\alpha$. Hence, by the same reasoning given above, for any given non-empty clopen subset V of X, there is an ordinal α such that V^α is a non-empty finite set and $V^{\alpha+1} = \emptyset$. As a last observation, we note that for any x in X, there is a clopen set C and an ordinal β such that $C^\beta = \{x\}$. To prove this, look at the net X^α, $\alpha \le \alpha_0 + 1$. Let β be the ordinal such that $x \in X^\beta - X^{\beta+1}$. This means that x is an isolated point of X^β. Consequently, there exists a clopen subset C of X such that $C \cap X^\beta = \{x\}$. Hence $C^\beta = C \cap X^\beta = \{x\}$.

2°. Let $\mathscr{F}$ be superatomic. Let T be a continuous linear functional on $ba(\Omega, \mathscr{F})$. We exhibit a bounded function f on $\mathscr{F}$ such that $T(\mu) = \int_\Omega f\mu$ for every μ in $ba(\Omega, \mathscr{F})$.

Let X be the Stone space of $\mathscr{F}$. We treat X as the collection of all 0–1 valued charges on $\mathscr{F}$. See Example 2.1.3(3). We assume, without loss of generality, that $\mathscr{F}$ is the field of all clopen subsets of X. Since $\mathscr{F}$ is superatomic, X is a scattered compact Hausdorff totally disconnected space. See Remarks 5.3.5.

3°. For any non-empty clopen set V in $\mathscr{F}$, we define $f(V)$ as follows. Let β be the ordinal such that V^β is a non-empty finite set. We set

$$f(V) = \text{Sup}\{T(\mu); \mu \in V^\beta\}.$$

If $V = \emptyset$, let $f(V) = 0$. f is obviously a bounded function on $\mathscr{F}$.

4°. First, we show that $\int_X f\mu = T(\mu)$ for any 0–1 valued charge μ on $\mathscr{F}$. By 1°, there exists a clopen set $C \subset X$ and an ordinal β_0 such that $C^{\beta_0} = \{\mu\}$. Consequently, $f(C) = T(\mu)$. C also has the property: if $E \in \mathscr{F}$ and $\mu \in E \subset C$, then $E^{\beta_0} = E \cap C^{\beta_0} = \{\mu\}$. Now, for the partition $\{C, C^c\}$, we have $f(C)\mu(C) + f(C^c)\mu(C^c) = T(\mu) + 0 = T(\mu)$. Further, if $P = \{E_1, E_2, \ldots, E_m\}$ is a refinement of $\{C, C^c\}$, then $\sum_{i=1}^m f(E_i)\mu(E_i) = T(\mu)$ from the property of C mentioned above. Hence $\int_X f\mu = T(\mu)$.

5°. Let μ be any bounded charge on $\mathscr{F}$. Since $\mathscr{F}$ is superatomic, $\mu = \sum_{n \ge 1} a_n \mu_n$ for some sequence μ_n, $n \ge 1$ of 0–1 valued charges on $\mathscr{F}$ and for some sequence a_n, $n \ge 1$ of real numbers satisfying $\sum_{n \ge 1} |a_n| < \infty$. See the comment following Theorem 5.3.6. The conditions of Proposition 9.1.2(iv) are met and so $\int_X f\mu$ exists and

$$\int_X f\mu = \sum_{n \ge 1} a_n \int_X f\mu_n = \sum_{n \ge 1} a_n T(\mu_n)$$

$$= \lim_{n \to \infty} T\left(\sum_{i=1}^n a_i \mu_i\right) = T(\mu)$$

as $\sum_{i=1}^{n} a_i \mu_i$, $n \geq 1$ converges to μ in the norm of $ba(\Omega, \mathcal{F})$ and T is a continuous linear functional on $ba(\Omega, \mathcal{F})$.

This completes the proof. □

Now, we obtain a refinement integral representation for continuous linear functionals on $V_1(\Omega, \mathcal{F}, \mu)$ for a probability charge space $(\Omega, \mathcal{F}, \mu)$. For relevant information on $V_1(\Omega, \mathcal{F}, \mu)$, see Chapter 7.

9.2.2 Definition. Let $\mathcal{F}$ be a field of subsets of a set Ω and μ a positive bounded charge on $\mathcal{F}$. A real valued function f on $\mathcal{F}$ is said to be *convex with respect to μ* if

$$f(A \cup B) = \frac{\mu(A)}{\mu(A \cup B)} f(A) + \frac{\mu(B)}{\mu(A \cup B)} f(B)$$

for every A, B in $\mathcal{F}$ with $A \cap B = \emptyset$. (Recall the convention that $0/0 = 0$.)

9.2.3 Theorem. *Let $(\Omega, \mathcal{F}, \mu)$ be a probability charge space and T a continuous linear functional on $V_1(\Omega, \mathcal{F}, \mu)$. Then there exists a bounded real valued convex function f on $\mathcal{F}$ such that $T(\nu) = \int_\Omega f \nu$ for every ν in $V_1(\Omega, \mathcal{F}, \mu)$.*

Proof. By Theorem 7.3.1, there exists a bounded charge λ on $\mathcal{F}$ in $V_\infty(\Omega, \mathcal{F}, \mu)$ such that $T(\nu) = T_\lambda(\nu)$ for every ν in $V_1(\Omega, \mathcal{F}, \mu)$. From Remark 7.3.2,

$$T_\lambda(\nu) = \lim_{P \in \mathcal{P}} \sum_{i=1}^{m} \frac{\nu(E_i) \lambda(E_i)}{\mu(E_i)},$$

where $P = \{E_1, E_2, \ldots, E_m\}$ is a generic element of $\mathcal{P}$. For F in $\mathcal{F}$, define $f(F) = \lambda(F)/\mu(F)$. Since $\lambda \in V_\infty(\Omega, \mathcal{F}, \mu)$,

$$\|\lambda\|_\infty = \sup_{F \in \mathcal{F}} \left| \frac{\lambda(F)}{\mu(F)} \right| = \sup_{F \in \mathcal{F}} |f(F)| < \infty.$$

Now, we check that f is a convex function with respect to μ. For A, B in $\mathcal{F}$ with $A \cap B = \emptyset$,

$$f(A \cup B) = \frac{\lambda(A \cup B)}{\mu(A \cup B)}$$

$$= \frac{\lambda(A)}{\mu(A)} \frac{\mu(A)}{\mu(A \cup B)} + \frac{\lambda(B)}{\mu(B)} \frac{\mu(B)}{\mu(A \cup B)}$$

$$= \frac{\mu(A)}{\mu(A \cup B)} f(A) + \frac{\mu(B)}{\mu(A \cup B)} f(B).$$

This shows that f is a convex function with respect to μ. Now, for any ν in $V_1(\Omega, \mathscr{F}, \mu)$,

$$T(\nu) = T_\lambda(\nu) = \lim_{P \in \mathscr{P}} \sum_{i=1}^{m} f(E_i)\nu(E_i) = \int_\Omega f\nu,$$

where $P = \{E_1, E_2, \ldots, E_m\}$ is a generic element in $\mathscr{P}$. This completes the proof. $\square$

9.2.4 Remark. We can obtain a refinement integral representation for elements of $V_P^*(\Omega, \mathscr{F}, \mu)$ for any $1 < p < \infty$. The method is the same as the one described above.

CHAPTER 10

Pure Charges

A charge on a field may not be pure in the sense that there may be a part of it which is countably additive. If we isolate successfully a maximal countably additive part of a given charge, what is left may be termed as purely finitely additive. This chapter is devoted to the pursuit of these ideas. After presenting the preliminaries in Section 10.1, we prove a decomposition theorem due to Yosida and Hewitt in Section 10.2. In Section 10.3, we look at pure charges on σ-fields. In Section 10.4, we discuss some examples which illuminate some aspects of pure charges. Finally, in Section 10.5, pure charges on Boolean algebras are studied.

10.1 DEFINITIONS AND PROPERTIES

In this section, we introduce pure charges and give some of their properties.

10.1.1 Definition. A positive charge μ defined on a field $\mathscr{F}$ of subsets of a set Ω is called a *pure charge* if there is no non-zero positive measure λ on $\mathscr{F}$ satisfying $\lambda \leq \mu$. More generally, a charge μ on $\mathscr{F}$ is said to be a pure charge if $|\mu|$ is a pure charge.

Before characterizing bounded pure charges, we recall some results from Section 1.5 and Chapter 2. ba($\Omega, \mathscr{F}$) stands for the vector lattice of all bounded charges on $\mathscr{F}$. ca($\Omega, \mathscr{F}$) stands for the collection of all bounded measures on $\mathscr{F}$ and is a vector sublattice of ba($\Omega, \mathscr{F}$). We have already seen that ba($\Omega, \mathscr{F}$) is a boundedly complete vector lattice and that ca($\Omega, \mathscr{F}$) is a normal vector sublattice of ba($\Omega, \mathscr{F}$). See Theorems 2.2.1 and 2.4.2. ca($\Omega, \mathscr{F}$)$^\perp$ stands for the set $\{\lambda \in \text{ba}(\Omega, \mathscr{F}); \lambda \perp \tau \text{ for every } \tau \text{ in ca}(\Omega, \mathscr{F})\}$. Note that $\lambda \perp \tau$ if and only if $|\lambda| \wedge |\tau| = 0$.

The following is a characterization of bounded pure charges.

10.1.2 Theorem. *Let* $\mu \in \text{ba}(\Omega, \mathscr{F})$. *Then* μ *is a pure charge if and only if* $\mu \in \text{ca}(\Omega, \mathscr{F})^\perp$.

Proof. Suppose μ is a pure charge. Let $\tau \in \mathrm{ca}(\Omega, \mathscr{F})$. Note that $|\mu| \wedge |\tau| \leq |\mu|$ and $|\mu| \wedge |\tau| \leq |\tau|$. Since $|\tau|$ is a bounded positive measure on $\mathscr{F}$, it follows that $|\mu| \wedge |\tau|$ is a measure. Since μ is a pure charge, $|\mu| \wedge |\tau| = 0$. Consequently, $\mu \perp \tau$. Hence $\mu \in \mathrm{ca}(\Omega, \mathscr{F})^{\perp}$. Conversely, let $\mu \in \mathrm{ca}(\Omega, \mathscr{F})^{\perp}$ and $\lambda \in \mathrm{ca}(\Omega, \mathscr{F})$ be such that $|\lambda| \leq |\mu|$. Since $|\mu| \perp |\lambda|$, i.e. $|\mu| \wedge |\lambda| = 0$, it follows that $\lambda = 0$. Hence μ is a pure charge. □

The following is a simple characterization of a pure charge in terms of its positive and negative variations.

10.1.3 Corollary. *Let $\mu \in \mathrm{ba}(\Omega, \mathscr{F})$. Then μ is a pure charge if and only if μ^+ and μ^- are pure charges.*

Proof. Note that $\mathrm{ca}(\Omega, \mathscr{F})^{\perp}$ is a sublattice of $\mathrm{ba}(\Omega, \mathscr{F})$. See Theorem 1.5.8. So, if $\mu \in \mathrm{ca}(\Omega, \mathscr{F})^{\perp}$, μ^+ and $\mu^- \in \mathrm{ca}(\Omega, \mathscr{F})^{\perp}$. Conversely, if μ^+ and $\mu^- \in \mathrm{ca}(\Omega, \mathscr{F})^{\perp}$, then $\mu = \mu^+ - \mu^- \in \mathrm{ca}(\Omega, \mathscr{F})^{\perp}$. □

10.1.4 Corollary. *Let $\mu_1, \mu_2 \in \mathrm{ba}(\Omega, \mathscr{F})$ and α any real number. If μ_1 and μ_2 are pure charges, so are $\mu_1 + \mu_2$, $\mu_1 \vee \mu_2$, $\mu_1 \wedge \mu_2$ and $\alpha\mu_1$.*

Proof. $\mathrm{ca}(\Omega, \mathscr{F})^{\perp}$ is a vector sublattice of $\mathrm{ba}(\Omega, \mathscr{F})$. See Theorem 1.5.8. □

The following corollary shows that purity of charges is preserved under passage to limits.

10.1.5 Corollary. *Let $\mu, \mu_n, n \geq 1$ be a sequence in $\mathrm{ba}(\Omega, \mathscr{F})$. Suppose for each $n \geq 1$, μ_n is a pure charge and μ_n, $n \geq 1$ converges to μ under the total variation norm on $\mathrm{ba}(\Omega, \mathscr{F})$, i.e. $\lim_{n \to \infty} |\mu_n - \mu|(\Omega) = 0$. Then μ is a pure charge.*

Proof. Since $\mathrm{ca}(\Omega, \mathscr{F})^{\perp}$ is a normal vector sublattice of $\mathrm{ba}(\Omega, \mathscr{F})$, it is closed. See Theorem 1.5.19. Since each μ_n is a pure charge, $\mu_n \in \mathrm{ca}(\Omega, \mathscr{F})^{\perp}$. Consequently, the limit $\mu \in \mathrm{ca}(\Omega, \mathscr{F})^{\perp}$. Hence μ is a pure charge. □

10.2 A DECOMPOSITION THEOREM

The results developed so far yield the following decomposition theorem.

10.2.1 Theorem. *Any μ in $\mathrm{ba}(\Omega, \mathscr{F})$ can be written in the form $\mu = \mu_p + \mu_c$, where μ_p is a pure charge in $\mathrm{ba}(\Omega, \mathscr{F})$ and $\mu_c \in \mathrm{ca}(\Omega, \mathscr{F})$. Further, such a decomposition is unique.*

Proof. This follows from the Riesz Decomposition Theorem, i.e. Theorem 1.5.10 in which we take $L = \mathrm{ba}(\Omega, \mathscr{F})$ and $S = \mathrm{ca}(\Omega, \mathscr{F})$. From Theorem 2.4.2, it is clear that S is a normal vector sublattice of L. □

We give another description of the countably additive part μ_c of μ in the following theorem.

10.2.2 Theorem. *Let μ be a positive bounded charge on a field $\mathscr{F}$ of subsets of a set Ω. Then for any A in $\mathscr{F}$,*

$$\mu_c(A) = \operatorname{Inf}\left\{\lim_{n\to\infty} \mu(A_n); A_n, n \geq 1 \uparrow A, A_n \in \mathscr{F}, n \geq 1\right\}$$

$$= \operatorname{Inf}\left\{\sum_{n\geq 1} \mu(A_n); A_n, n \geq 1 \text{ is a sequence of}\right.$$

$$\left. \text{pairwise disjoint sets in } \mathscr{F} \text{ with } \bigcup_{n\geq 1} A_n = A\right\}.$$

Proof. The equality of the two expressions on the right above is clear. For A in $\mathscr{F}$, let

$$\tau(A) = \operatorname{Inf}\left\{\sum_{n\geq 1} \mu(A_n); A_n, n \geq 1 \text{ is a sequence of}\right.$$

$$\left. \text{pairwise disjoint sets in } \mathscr{F} \text{ with } \bigcup_{n\geq 1} A_n = A\right\}.$$

Suppose ν is any positive bounded measure on $\mathscr{F}$ such that $\nu \leq \mu$. Obviously, $\nu \leq \tau$. Now, we show that τ is a charge. Let $A, B \in \mathscr{F}$ be such that $A \cap B = \emptyset$. Let $\varepsilon > 0$. We can find two sequences $A_n, n \geq 1$ and $B_n, n \geq 1$ of pairwise disjoint sets in $\mathscr{F}$ such that $\bigcup_{n\geq 1} A_n = A$, $\bigcup_{n\geq 1} B_n = B$ and

$$\sum_{n\geq 1} \mu(A_n) \leq \tau(A) + \varepsilon/2,$$

$$\sum_{n\geq 1} \mu(B_n) \leq \tau(B) + \varepsilon/2.$$

Consequently,

$$\tau(A \cup B) \leq \sum_{n\geq 1} \mu(A_n) + \sum_{n\geq 1} \mu(B_n)$$

$$\leq \tau(A) + \tau(B) + \varepsilon.$$

Since $\varepsilon > 0$ is arbitrary, it follows that $\tau(A \cup B) \leq \tau(A) + \tau(B)$. To prove the reverse inequality, we proceed as follows. Let $\varepsilon > 0$. We can find a sequence $C_n, n \geq 1$ of pairwise disjoint sets in $\mathscr{F}$ such that $\bigcup_{n\geq 1} C_n = A \cup B$ and

$$\sum_{n\geq 1} \mu(C_n) \leq \tau(A \cup B) + \varepsilon.$$

But
$$\tau(A)+\tau(B) \le \sum_{n\ge 1} \mu(A\cap C_n) + \sum_{n\ge 1} \mu(B\cap C_n)$$
$$= \sum_{n\ge 1} \mu(C_n) \le \tau(A\cup B) + \varepsilon.$$

Since $\varepsilon > 0$ is arbitrary, it follows that $\tau(A)+\tau(B)\le \tau(A\cup B)$. Consequently, τ is a charge on $\mathcal{F}$. Next, we show that τ is a measure. Let A_n, $n\le 1$ be a sequence of pairwise disjoint sets in $\mathcal{F}$ such that $\bigcup_{n\ge 1} A_n = A \in \mathcal{F}$. Since τ is a positive charge on $\mathcal{F}$, $\sum_{n\ge 1} \tau(A_n) \le \tau(A)$. See Proposition 2.1.2(viii). For any $\varepsilon > 0$ and $n\ge 1$, we can find a sequence A_{ni}, $i\ge 1$ of pairwise disjoint sets in $\mathcal{F}$ such that $\bigcup_{i\ge 1} A_{ni} = A_n$ and
$$\sum_{i\ge 1} \mu(A_{ni}) \le \tau(A_n) + \varepsilon/2^n.$$
Consequently,
$$\tau(A) \le \sum_{n\ge 1}\sum_{i\ge 1} \mu(A_{ni}) \le \sum_{n\ge 1} \tau(A_n) + \varepsilon.$$
Since $\varepsilon > 0$ is arbitrary, we obtain $\tau(A)\le \sum_{n\ge 1}\tau(A_n)$. Hence
$$\tau\left(\bigcup_{n\ge 1} A_n\right) = \sum_{n\ge 1} \tau(A_n).$$
This implies that τ is a measure on $\mathcal{F}$.

Since $\tau \le \mu$, we can write $\mu = \tau + (\mu - \tau)$. We claim that $\mu - \tau$ is a pure charge. If ν is any positive bounded measure on $\mathcal{F}$ such that $\nu \le \mu - \tau$, then $\nu + \tau \le \mu$. Since $\nu + \tau$ is a measure, by what we have remarked above, $\nu + \tau \le \tau$. Hence $\nu \le 0$. Since ν is positive, $\nu = 0$. This shows that $\mu - \tau$ is a pure charge. Since, in the Riesz Decomposition Theorem, the decomposition is uniquely achieved, it follows that $\mu_c = \tau$. This proves the result. □

10.3 PURE CHARGES ON σ-FIELDS

In this section, we obtain further characterizations of pure charges defined on σ-fields.

The following theorem implies that any positive bounded pure charge and any positive bounded measure are singular.

10.3.1 Theorem. *Let μ be a positive bounded charge on a field $\mathcal{F}$ of subsets of a set Ω. Then μ is a pure charge if and only if for every positive bounded measure λ on $\mathcal{F}$, A in $\mathcal{F}$ and $\alpha > 0$, there exists B in $\mathcal{F}$ such that $B \subset A$,*
$$\lambda(B) < \alpha \quad \text{and} \quad \mu(A-B) < \alpha.$$

Proof. By Theorem 10.1.2, μ is a pure charge if and only if $\mu \in \text{ca}(\Omega, \mathscr{F})^{\perp}$. Equivalently, μ is a pure charge if and only if $\mu \wedge \lambda = 0$ for every positive bounded measure λ on $\mathscr{F}$. □

The above theorem can be strengthened if $\mathscr{F}$ is a σ-field.

10.3.2 Theorem. *Let $\mathscr{F}$ be a σ-field of subsets of a set Ω and μ a positive bounded charge on $\mathscr{F}$. Then μ is a pure charge if and only if for every bounded measure λ on $\mathscr{F}$ and $\varepsilon > 0$, there exists B in $\mathscr{F}$ such that*

$$\mu(B) = 0 \quad \text{and} \quad \lambda(B^c) < \varepsilon.$$

Proof. Suppose μ is a pure charge. Let λ be a positive bounded measure on $\mathscr{F}$. Let $\varepsilon > 0$. By Theorem 10.3.1, there exists a set B_n in $\mathscr{F}$ such that

$$\mu(B_n) < \varepsilon/2^n \quad \text{and} \quad \lambda(B_n^c) < \varepsilon/2^n$$

for every $n \geq 1$. Let $B = \bigcap_{n \geq 1} B_n$. Note that $B \in \mathscr{F}$. For every $n \geq 1$, $\mu(B) \leq \mu(B_n) < \varepsilon/2^n$. Consequently, $\mu(B) = 0$. Also, $\lambda(B^c) = \lambda(\bigcup_{n \geq 1} B_n^c) \leq \sum_{n \geq 1} \lambda(B_n^c) \leq \sum_{n \geq 1} \varepsilon/2^n = \varepsilon$, as λ is countably subadditive. The converse is a consequence of Theorem 10.3.1. □

Here is another characterization of pure charges on σ-fields useful in constructing examples.

10.3.3 Theorem. *Let λ be a positive bounded measure on a σ-field $\mathscr{F}$ of subsets of a set Ω. Let μ be a bounded charge on $\mathscr{F}$ such that μw $\ll \lambda$, i.e. $\mu(A) = 0$ whenever $A \in \mathscr{F}$ and $\lambda(A) = 0$. Then μ is a pure charge if and only if there exists a decreasing sequence A_n, $n \geq 1$ of sets in $\mathscr{F}$ such that $\lambda(A_n)$, $n \geq 1$ converges to zero and $|\mu|(A_n^c) = 0$ for every $n \geq 1$.*

Proof. The necessity part of this theorem follows from Theorem 10.3.2. To prove the sufficiency part, let ψ be any positive bounded measure on $\mathscr{F}$ satisfying $\psi \leq |\mu|$. From the given hypothesis on μ and λ, it follows that $\psi \ll \lambda$. Let A_n, $n \geq 1$ be the given decreasing sequence of sets in $\mathscr{F}$ such that $\lim_{n \to \infty} \lambda(A_n) = 0$ and $|\mu|(A_n^c) = 0$ for every $n \geq 1$. Since $\psi \ll \lambda$, $\lim_{n \to \infty} \psi(A_n) = 0 = \psi(\bigcap_{n \geq 1} A_n)$. Since $|\mu|(A_n^c) = 0$ for every $n \geq 1$, $\psi(A_n^c) = 0$ for every $n \geq 1$. Since ψ is a measure, $\psi(\bigcup_{n \geq 1} A_n^c) = 0$. Consequently, $\psi(\Omega) = \psi(\bigcap_{n \geq 1} A_n) + \psi(\bigcup_{n \geq 1} A_n^c) = 0$. Hence $\psi = 0$. Hence μ is a pure charge. This completes the proof. □

10.4 EXAMPLES

The aim of this section is to construct some interesting examples of pure charges.

10. PURE CHARGES

10.4.1 Example. Let $\mathscr{B}$ be the Borel σ-field on the real line, R, and let μ be a bounded charge on $\mathscr{B}$ such that $\mu(B) = 0$ for any bounded set B in $\mathscr{B}$. Then μ is a pure charge. (Indeed, if μ vanishes for all bounded Borel subsets of R, so does $|\mu|$.)

Proof. Let ψ be any positive measure on $\mathscr{B}$ such that $\psi \le |\mu|$. Then

$$\psi(R) = \sum_{n=-\infty}^{n=+\infty} \psi\{[n, n+1)\} \le \sum_{n=-\infty}^{n=+\infty} |\mu|\{[n, n+1)\} = 0.$$

Hence $\psi = 0$. □

Using the result that any ideal in $\mathscr{B}$ is contained in a maximal ideal in $\mathscr{B}$, one can construct non-trivial charges μ of the above type. One simply has to look at the ideal $\mathscr{I}$ of all bounded Borel subsets of R and then find a maximal ideal in $\mathscr{B}$ containing $\mathscr{I}$. See Section 1.2. It seems impossible to construct a pure charge on a σ-field without taking recourse to some axiom related to the Axiom of Choice.

10.4.2 Example. Let $\mathscr{F}$ be the field of all clopen subsets of the Cantor set $\{0, 1\}^{\aleph_0}$. There is no non-zero pure charge on $\mathscr{F}$. More generally, let $\mathscr{F}$ be the field of all clopen subsets of a compact Hausdorff space Ω. There is no non-zero pure charge on $\mathscr{F}$. Much more generally, there is no non-zero pure charge on a field $\mathscr{F}$ of subsets of a set Ω if and only if $\mathscr{F}$ is a compact class, i.e. for any decreasing sequence A_n, $n \ge 1$ of sets in $\mathscr{F}$, $\bigcap_{n \ge 1} A_n = \varnothing$ implies that $A_n = \varnothing$ for some $n \ge 1$.

Proof. We will prove the most general statement. If $\mathscr{F}$ is a compact class, every charge on $\mathscr{F}$ is a measure. Hence there is no non-zero pure charge on $\mathscr{F}$. If $\mathscr{F}$ is not a compact class, we can find an infinite sequence A_n, $n \ge 1$ of pairwise disjoint non-empty sets in $\mathscr{F}$ whose union is Ω. Let $\mathscr{F}_0 = \{A \subset \Omega; A \text{ is a finite union of sets from } \{A_n, n \ge 1\} \text{ or } A^c \text{ is a finite union of sets from } \{A_n, n \ge 1\}\}$. $\mathscr{F}_0$ is a subfield of $\mathscr{F}$. On $\mathscr{F}_0$, let μ be defined by

$\mu(A) = 0$, if A is a finite union of sets from $\{A_n, n \ge 1\}$,

$\quad\quad\quad = 1$, otherwise.

μ is a 0–1 valued charge on $\mathscr{F}_0$ and is certainly not a measure on $\mathscr{F}_0$. By Corollary 3.3.5, there is a 0–1 valued charge $\bar\mu$ on $\mathscr{F}$ which is an extension of μ from $\mathscr{F}_0$. It is obvious that $\bar\mu$ is a non-zero pure charge on $\mathscr{F}$. □

10.4.3 Example. Let $\mathscr{F}$ be a field of subsets of a set Ω. Any 0–1 valued charge μ on $\mathscr{F}$ is either a pure charge or a measure on $\mathscr{F}$. Indeed, if ψ is a non-zero measure on $\mathscr{F}$ satisfying $0 \le \psi \le \mu$, then μ is a measure. □

10.4.4 Example. Let $\Omega = [0, 1)$ and $\mathscr{F}$ the field of all sets each of which is a finite disjoint union of sets of the form $[a, b)$ with $0 \le a \le b \le 1$. For t in $(0, 1]$, define μ_t on $\mathscr{F}$ by

$\mu_t(A) = 1$, if there exists a $\delta > 0$ such that $[t - \delta, t) \subset A$,

$= 0$, otherwise.

μ_t is a 0–1 valued charge on $\mathscr{F}$. Since μ_t is not a measure, μ_t is a pure charge.

Conversely, if μ is any 0–1 valued pure charge on $\mathscr{F}$, then $\mu = \mu_t$ for some t in $(0, 1]$. This can be proved as follows. Since μ is a 0–1 valued pure charge, we can find a decreasing sequence A_n, $n \geq 1$ of sets in $\mathscr{F}$ such that $\bigcap_{n \geq 1} A_n = \emptyset$ and $\mu(A_n) = 1$ for every $n \geq 1$. For each $n \geq 1$, we can find an interval $[a_n, b_n) \subset A_n$ such that $\mu\{[a_n, b_n)\} = 1$ and $[a_1, b_1) \supset [a_2, b_2) \supset \cdots$. Obviously, $\bigcap_{n \geq 1} [a_n, b_n) = \emptyset$. Let $t = \lim_{n \to \infty} a_n = \lim_{n \to \infty} b_n$. Since $\bigcap_{n \geq 1} [a_n, b_n) = \emptyset$, $b_n = t$ for some $n \geq 1$. Consequently, $\mu = \mu_t$.

Now, we obtain explicitly the decomposition of a given positive bounded charge on $\mathscr{F}$ as a sum of a pure charge and a measure. Let μ be any positive bounded charge on $\mathscr{F}$. Define a real-valued function m on $[0, 1)$ by $m(t) = \mu\{[0, t)\}$ for $0 \leq t < 1$. Then m is a monotonically non-decreasing function on $[0, 1)$ and $\lim_{t \uparrow 1} m(t) < \infty$. m has at most countably many discontinuity points. Write $\{t \in [0, 1); m(t-0) < m(t)\} = \{t_1, t_2, \ldots\}$. Let $a_n = m(t_n) - m(t_n - 0)$, $n \geq 1$. We write

$$\mu = \sum_{n \geq 1} a_n \mu_{t_n} + \left(\mu - \sum_{n \geq 1} a_n \mu_{t_n}\right).$$

Since each μ_{t_n} is a 0–1 valued pure charge, $\sum_{n \geq 1} a_n \mu_{t_n}$ is a pure charge. The m function (as defined above for μ) of $\mu - \sum_{n \geq 1} a_n \mu_{t_n}$ is continuous, and consequently, $\mu - \sum_{n \geq 1} a_n \mu_{t_n}$ is a measure on $\mathscr{F}$. The above is the desired decomposition of μ into a pure charge and a measure. $\square$

10.5 PURE CHARGES ON BOOLEAN ALGEBRAS

The notion of a measure on a field $\mathscr{F}$ of subsets of a set Ω is not the same when $\mathscr{F}$ is viewed as a field of sets and when $\mathscr{F}$ is viewed as a Boolean algebra. If $\mathbb{B}$ is a Boolean algebra and b_n, $n \geq 1$ is a sequence of elements in $\mathbb{B}$, $\bigvee_{n \geq 1} b_n$ denotes the supremum of $\{b_n, n \geq 1\}$, if it exists. If $\mathscr{F}$ is a field of sets on a set Ω, B_n, $n \geq 1$ is a sequence of sets in $\mathscr{F}$ satisfying the condition that $\bigcup_{n \geq 1} B_n \in \mathscr{F}$, then $\bigvee_{n \geq 1} B_n = \bigcup_{n \geq 1} B_n$. If $\bigcup_{n \geq 1} B_n$ fails to belong to $\mathscr{F}$, it is possible that $\bigvee_{n \geq 1} B_n$ exists in $\mathscr{F}$ when $\mathscr{F}$ is viewed as a Boolean algebra. This is the feature that distinguishes the notion of a measure on a field and the notion of a measure on a Boolean algebra. We give an example to amplify this point. Before that, let us formalize the notion of a measure on a Boolean algebra.

10. PURE CHARGES

10.5.1 Definition. Let $\mathbb{B}$ be a Boolean algebra or $\mathscr{F}$ a field of subsets of a set Ω viewed as a Boolean algebra. An extended real-valued function μ on $\mathscr{F}$ (or on $\mathbb{B}$) is said to be a measure if for every sequence B_n, $n \geq 1$ of pairwise disjoint sets in $\mathscr{F}$ with $\bigvee_{n\geq 1} B_n$ in $\mathscr{F}$, $\mu(\bigvee_{n\geq 1} B_n) = \sum_{n\geq 1} \mu(B_n)$ holds.

The following example demonstrates the difference in the outlook that persists when dealing with measures on fields as well as on Boolean algebras.

10.5.2 Example. Let $\Omega = \{1, 2, 3, \ldots, \infty\}$, and $\mathscr{F}$, the collection of all finite subsets of $\{1, 2, 3, \ldots\}$ and their complements. $\mathscr{F}$ is a field on Ω. $\mathscr{F}$ is also a compact class. (See Example 10.4.2.) Consequently, every charge on $\mathscr{F}$ is a measure. Now, let us view $\mathscr{F}$ as a Boolean algebra. Consider the following set function μ on $\mathscr{F}$.

$$\mu(A) = 0, \quad \text{if A is a finite subset of } \{1, 2, 3, \ldots\},$$
$$= 1, \quad \text{otherwise.}$$

μ is a charge on $\mathscr{F}$ but μ is not a measure on $\mathscr{F}$ when $\mathscr{F}$ is viewed as a Boolean algebra. For, let $B_n = \{n\}$, $n \geq 1$. B_n, $n \geq 1$ is a sequence of pairwise disjoint elements in $\mathscr{F}$ with $\bigvee_{n\geq 1} B_n = \Omega$. But $\mu(\Omega) = (\bigvee_{n\geq 1} B_n) = 1$ and $\sum_{n\geq 1} \mu(B_n) = 0$.

The decomposition theorem, developed above for a bounded charge defined on a field $\mathscr{F}$ of subsets of a set Ω as a sum of a pure charge and a measure, takes the following form when $\mathscr{F}$ is viewed as a Boolean algebra. Let μ be a bounded charge on $\mathscr{F}$. Can we write $\mu = \mu_1 + \mu_2$, where μ_1 is a pure charge on $\mathscr{F}$ and μ_2 is a measure on $\mathscr{F}$? We will show subsequently that such a decomposition theorem is possible. Before that, we characterize measures and pure charges on $\mathscr{F}$ when $\mathscr{F}$ is viewed as a Boolean algebra.

10.5.3 Theorem. *Let $\mathscr{F}$ be a field of subsets of a set Ω. Let Y be the Stone space of $\mathscr{F}$, $\mathscr{C}$ the field of all clopen subsets of Y, T an isomorphism from $\mathscr{F}$ to $\mathscr{C}$ and $\mathscr{B}_0$ the Baire σ-field on Y. Let μ be a positive bounded charge on $\mathscr{F}$ and $\tilde{\mu}$ the positive bounded measure on $\mathscr{C}$ given by $\tilde{\mu}(C) = \mu(T^{-1}C)$, $C \in \mathscr{C}$. Let $\hat{\mu}$ be the extension of $\tilde{\mu}$ from $\mathscr{C}$ to $\mathscr{B}_0$ as a measure. (See Theorem 3.5.2.) Then the following statements are true.*

(i). *μ is a measure on $\mathscr{F}$ when $\mathscr{F}$ is viewed as a Boolean algebra if and only if $\hat{\mu}(C) = 0$ for every nowhere dense closed G_δ subset C of Y. Equivalently, μ is a measure on $\mathscr{F}$ when $\mathscr{F}$ is viewed as a Boolean algebra if and only if $\hat{\mu}(D) = 0$ for every first category Baire F_σ set $D \subset Y$.*

(ii). *μ is a pure charge on $\mathscr{F}$ when $\mathscr{F}$ is viewed as a Boolean algebra if and only if there is a first category Baire F_σ set $D_0 \subset Y$ such that $\hat{\mu}(D_0^c) = 0$.*

Proof. In the argument given below, we view $\mathscr{F}$ as a Boolean algebra.
(i). Let μ be a measure on $\mathscr{F}$. Let C be any nowhere dense closed G_δ

subset of Y. We can write $C = \bigcap_{n \geq 1} C_n$, where each C_n is a clopen subset of Y and $C_1 \supset C_2 \supset C_3 \supset \cdots$. This is possible because in any compact Hausdorff totally disconnected space Y, for any closed set C contained in an open set U, one can find a clopen set V such that $C \subset V \subset U$. Since C is a nowhere dense set, $\bigwedge_{n \geq 1} T^{-1} C_n = \varnothing$ in $\mathscr{F}$. Since μ is a measure on $\mathscr{F}$, $\lim_{n \to \infty} \mu(T^{-1} C_n) = 0$. Consequently, $\lim_{n \to \infty} \hat{\mu}(C_n) = 0$. Hence $\hat{\mu}(C) = 0$. The converse can be proved by retracing these steps.

(ii). Suppose μ is a pure charge on $\mathscr{F}$. Let $r = \operatorname{Sup} \hat{\mu}(D)$, where the supremum is taken over all first category Baire F_σ subsets D of Y. It is obvious that $r < \infty$. Let D_n, $n \geq 1$ be an increasing sequence of first category Baire F_σ subsets of Y such that $\lim_{n \to \infty} \hat{\mu}(D_n) = r$. Let $D_0 = \bigcup_{n \geq 1} D_n$. Then $\hat{\mu}(D_0) = r$ and D_0 is a first category Baire F_σ subset of Y. Further, D_0 has the following property. If D is any first category Baire F_σ subset of Y, then $\hat{\mu}(D - D_0) = 0$. Let $\hat{\tau}$ on $\mathscr{B}_0$ be defined as

$$\hat{\tau}(A) = \hat{\mu}(A) - \hat{\mu}(A \cap D_0)$$

for A in $\mathscr{B}_0$. $\hat{\tau}$ is a measure on $\mathscr{B}_0$. Let τ be the charge on $\mathscr{F}$ whose corresponding measure on $\mathscr{B}_0$ as explained in the statement of this theorem is the $\hat{\tau}$ given above. If C is a first category Baire F_σ set, then $\hat{\tau}(C) = 0$. By (i), τ is a measure on $\mathscr{F}$. Since $\tau \leq \mu$ and μ is a pure charge on $\mathscr{F}$, $\tau = 0$. This shows that for any A in $\mathscr{B}_0$,

$$\hat{\mu}(A) = \hat{\mu}(A \cap D_0).$$

Consequently, $\hat{\mu}(D_0^c) = 0$.

For the converse, we proceed as follows. Let μ be a charge on $\mathscr{F}$ such that $\hat{\mu}(D_0^c) = 0$ for some first category Baire F_σ subset D_0 of Y. Let τ be a measure on $\mathscr{F}$ such that $0 \leq \tau \leq \mu$. By (i), $\hat{\tau}(C) = 0$ for any first category Baire F_σ subset of Y. In particular, $\hat{\tau}(D_0) = 0$. Consequently, $\hat{\tau}(Y) = \hat{\tau}(D_0) + \hat{\tau}(D_0^c) = 0$. Hence $\tau = 0$. This shows that μ is a pure charge. □

Now, we give the desired decomposition theorem for charges on Boolean algebras. Let $\mathscr{F}$ be a field of subsets of a set Ω. Let $\widetilde{\mathrm{ca}}(\Omega, \mathscr{F})$ denote the collection of all bounded measures on $\mathscr{F}$ when $\mathscr{F}$ is viewed as Boolean algebra. Let $P(\Omega, \mathscr{F})$ denote the collection of all bounded pure charges on $\mathscr{F}$ with $\mathscr{F}$ being viewed as a Boolean algebra.

10.5.4 Theorem. *Let $\mathscr{F}$ be a field of subsets of a set Ω and μ a bounded charge on $\mathscr{F}$. Then there exists μ_1 in $P(\Omega, \mathscr{F})$ and μ_2 in $\widetilde{\mathrm{ca}}(\Omega, \mathscr{F})$ such that*

$$\mu = \mu_1 + \mu_2.$$

Further, this decomposition is unique.

Proof. A proof of this theorem can be modelled on the one for Theorem 10.2.1 by establishing that $\widetilde{\mathrm{ca}}(\Omega, \mathscr{F})$ is a normal vector sublattice of $\mathrm{ba}(\Omega, \mathscr{F})$. □

CHAPTER 11

Ranges of Charges

The range of a measure on a σ-field of subsets of a set Ω is a very well understood phenomenon. For example, the range of a real measure μ on a σ-field of sets is a closed subset of the real line and further, if μ is nonatomic, then its range is an interval. In this chapter, we study ranges of charges. In Section 11.1, we make some general comments on the ranges of bounded charges on fields. In Section 11.2, we show that the range of a bounded charge on a σ-field is either a finite set or contains a perfect set. In Section 11.3, we examine the cardinalities of the ranges of charges. In Section 11.4, the validity of Liapounoff's theorem for strongly continuous charges on fields is examined. In Section 11.5, we construct a positive bounded charge on a field such that its range is neither Lebesgue measurable nor has the property of Baire.

11.1 RANGES OF BOUNDED CHARGES ON FIELDS

To begin with, we introduce some basic definitions and establish notation.

11.1.1 Definition. Let μ be a charge defined on a field $\mathscr{F}$ of subsets of a set Ω. The *range of* μ is denoted by $R(\mu)$ and is defined to be the set

$$\{\mu(F); F \in \mathscr{F}\} = R(\mu).$$

11.1.2 Definition. A charge μ defined on a field $\mathscr{F}$ of subsets of a set Ω is said to be *finitely many valued* if its range $R(\mu)$ is a finite set. μ is said to be *infinitely many valued* if $R(\mu)$ is an infinite set.

The range of a finitely many valued charge is easy to describe. For this, we need a lemma.

11.1.3 Lemma. *Let μ be a finitely many valued real charge defined on a field $\mathscr{F}$ of subsets of a set Ω. Then we can find pairwise disjoint sets $A_1, A_2, \ldots, A_n$ in $\mathscr{F}$, non-zero real numbers $a_1, a_2, \ldots, a_n$, 0–1 valued charges*

$\mu_1, \mu_2, \ldots, \mu_n$ on $\mathscr{F}$ such that $\mu_i(A_i) = 1$ for $i = 1, 2, \ldots, n$ and

$$\mu = \sum_{i=1}^{n} a_i \mu_i.$$

Proof. Let $R(\mu)$ consist of k points. Let

$$\Theta = \{\{B_1, B_2, \ldots, B_m\}; B_i\text{'s are pairwise disjoint,}$$

B_i's are in $\mathscr{F}$, $\mu(B_i) \neq 0$ for every i, and $m \geq 1\}$.

Note that if $\{B_1, B_2, \ldots, B_m\} \in \Theta$, then $m \leq k$. Let $\{A_1, A_2, \ldots, A_n\}$ be a maximal family in Θ, i.e. if $\{B_1, B_2, \ldots, B_m\} \in \Theta$, then $m \leq n$. Each A_i has the following property. If $B \in \mathscr{F}$ and $B \subset A_i$, then $\mu(B) = 0$ or $\mu(B) = \mu(A_i)$. If this is not true, then there is a B in $\mathscr{F}$ such that $B \subset A_i$ and $\mu(B) \neq 0$, $\mu(A_i - B) \neq 0$. Then $\{A_1, A_2, \ldots, A_{i-1}, B, A_i - B, A_{i+1}, \ldots, A_n\} \in \Theta$ contradicting the maximality of $\{A_1, A_2, \ldots, A_n\}$. Let $a_i = \mu(A_i)$ and μ_i on $\mathscr{F}$ be defined by $\mu_i(B) = (1/a_i)\mu(A_i \cap B)$ for B in $\mathscr{F}$ and $i = 1, 2, \ldots, n$. It is clear that each μ_i is a 0–1 valued charge on $\mathscr{F}$. Further,

$$\mu = \sum_{i=1}^{n} a_i \mu_i.$$

For this, use the fact that if $B \in \mathscr{F}$ and $B \subset (A_1 \cup A_2 \cup \cdots \cup A_n)^c$, then $\mu(B) = 0$. $\square$

From the above lemma, the following proposition is obvious.

11.1.4 Proposition. *Let μ be a finitely many valued charge defined on a field $\mathscr{F}$ of subsets of a set Ω having the representation*

$$\mu = \sum_{i=1}^{n} a_i \mu_i,$$

where μ_i's are 0–1 valued charges sitting on disjoint sets in $\mathscr{F}$ and $a_1, a_2, \ldots, a_n$ are non-zero real numbers. Then

$$R(\mu) = \{0\} \cup \left\{ \sum_{i=1}^{k} a_{j_i} ; \{j_1, j_2, \ldots j_k\} \right.$$

$$\left. \subset \{1, 2, \ldots, n\} \text{ and } 1 \leq k \leq n \right\}.$$

Now, we look at infinitely many valued bounded charges defined on fields. We need a preliminary result.

11.1.5 Proposition. *If μ is any bounded charge defined on a field $\mathscr{F}$ of subsets of a set Ω and if there is a real number $c > 0$ such that $|\mu(A)| > c$ whenever $A \in \mathscr{F}$ and $\mu(A) \neq 0$, then μ is finitely many valued.*

Proof. We attempt to write μ in the form $\sum_{i=1}^{n} a_i \mu_i$ in the spirit of Lemma 11.1.3. Let $\text{Sup}_{F \in \mathscr{F}} |\mu(F)| = k$. Since μ is bounded, $k < \infty$. Find the largest positive integer N such that $Nc \leq k$. Let $\{A_\alpha; \alpha \in \Gamma\}$ be any family of pairwise disjoint sets in $\mathscr{F}$ such that $\mu(A_\alpha) \neq 0$ for every α in Γ. We show that Γ is a finite set and in fact, the cardinality of Γ is $\leq 2N$. Suppose the cardinality of Γ is $>2N$. Let $\Gamma_1 = \{\alpha \in \Gamma; \mu(A_\alpha) > 0\}$ and $\Gamma_2 = \{\alpha \in \Gamma; \mu(A_\alpha) < 0\}$. Then, either the cardinality of Γ_1 is $>N$ or the cardinality of Γ_2 is $>N$. Assume, without loss of generality, that the cardinality of Γ_1 is $>N$. Select $N+1$ distinct sets $A_{\alpha_1}, A_{\alpha_2}, \ldots, A_{\alpha_{N+1}}$ with $\alpha_1, \alpha_2, \ldots, \alpha_{N+1}$ in Γ_1. Let $A = \bigcup_{i=1}^{N+1} A_{\alpha_i}$. Then $\mu(A) > (N+1)c > k$. But $\mu(A) \leq k$. This contradiction establishes the claim. As in the proof of Lemma 11.1.3, let

$$\Theta = \{\{B_1, B_2, \ldots, B_m\}; B_i\text{'s are pairwise disjoint,}$$

B_i's are in $\mathscr{F}$, $\mu(B_i) \neq 0$ for every i, and $m \geq 1\}$.

Let $\{A_1, A_2, \ldots, A_n\}$ be any maximal family in Θ with n being the largest possible integer. Following the argument in the proof of Lemma 11.1.3, we can write

$$\mu = \sum_{i=1}^{n} a_i \mu_i,$$

where $a_1, a_2, \ldots, a_n$ are non-zero real numbers and $\mu_1, \mu_2, \ldots, \mu_n$ are 0-1 valued charges on $\mathscr{F}$ sitting on disjoint sets in $\mathscr{F}$. Hence μ is finitely many valued. □

As a consequence of the above proposition, we prove the following result which tells us about the nature of $R(\mu)$ when μ is infinitely many valued.

11.1.6 Theorem. *Let μ be an infinitely many valued bounded charge on a field $\mathscr{F}$ of subsets of a set Ω. Then 0 is an accumulation point of $R(\mu)$. More strongly, $R(\mu)$ is a dense-in-itself set, i.e. $R(\mu)$ has no isolated points.*

Proof. From Proposition 11.1.5, it follows that 0 is an accumulation point of $R(\mu)$, i.e. in every open interval $(-c, c)$ with $c > 0$, there exists A in $\mathscr{F}$ such that $\mu(A) \neq 0$ and $\mu(A) \in (-c, c)$. To prove the second part, let $A \in \mathscr{F}$ be such that $\mu(A) \neq 0$. We show that $\mu(A)$ is an accumulation point of $R(\mu)$. Now, we note that either μ is infinitely many valued on A or μ is infinitely many valued on A^c.
Case (i). μ is infinitely many valued on A. By the first part of this theorem, since μ is also bounded on A, we can find a sequence B_n, $n \geq 1$ in $\mathscr{F}$ such that each $B_n \subset A$, $\mu(B_n) \neq 0$ and $\lim_{n \to \infty} \mu(B_n) = 0$. Then $\mu(A - B_n) = \mu(A) - \mu(B_n)$, $n \geq 1$ converges to $\mu(A)$. Note that $\mu(A - B_n) \neq \mu(A)$ for all but a finite number of n's. Hence $\mu(A)$ is an accumulation point of $R(\mu)$.
Case (ii). μ is infinitely many valued on A^c. By an argument similar to

the above, we can find a sequence B_n, $n \geq 1$ of sets in $\mathscr{F}$ such that $B_n \subset A^c$, $\mu(B_n) \neq 0$ for every $n \geq 1$ and $\lim_{n \to \infty} \mu(B_n) = 0$. Note that $\lim_{n \to \infty} \mu(A \cup B_n) = \mu(A) + \lim_{n \to \infty} \mu(B_n) = \mu(A)$. Also, $\mu(A \cup B_n) \neq \mu(A)$ for all but a finite number of n's. This shows that $\mu(A)$ is an accumulation point of $R(\mu)$. □

The results established so far can be summed up in the following theorem.

11.1.7 Theorem. *Let μ be a bounded charge on a field $\mathscr{F}$ of subsets of a set Ω. Then either every point of its range $R(\mu)$ is an isolated point in which case $R(\mu)$ is a finite set or every point of $R(\mu)$ is an accumulation point of $R(\mu)$.*

Now, we take up the case of real charges (not necessarily bounded) defined on fields of sets. Theorems 11.1.6 and 11.1.7 are no longer true for such charges. We present a couple of examples to amplify this point.

11.1.8 Example. Let $\Omega = \{1, 2, 3, \ldots\}$, $\mathscr{F}$ = Finite–cofinite field on Ω and μ on $\mathscr{F}$ be defined by

$$\mu(A) = \#A, \quad \text{if A is finite,}$$
$$= -\#A^c, \quad \text{if A is cofinite,}$$

where $\#A$ denotes the number of points in A.

μ is a real charge on $\mathscr{F}$. μ is also infinitely many valued. But every point in the range $R(\mu) = \{\ldots, -3, -2, -1, 0, 1, 2, 3, \ldots\}$ is an isolated point of $R(\mu)$.

11.1.9 Example. Let $\Omega = \{0, 1, 2, 3, \ldots\}$, $\mathscr{F}$ the finite–cofinite field on Ω and μ on $\mathscr{F}$ be defined by

$$\mu(A) = \sum_{n \in A} \left(1 + \frac{1}{n}\right), \quad \text{if A is finite,}$$
$$= -\mu(A^c), \quad \text{if A is cofinite.}$$

(If $n = 0$, $1 + 1/n$ is interpreted as equal to 1.) μ is a real charge on $\mathscr{F}$ taking infinitely many values. 0 is not an accumulation point of $R(\mu)$. Note that 1 is an accumulation point of $R(\mu)$.

11.2 RANGES OF CHARGES ON σ-FIELDS

Ranges of bounded charges on σ-fields exhibit some additional properties over what we have established in the previous section. The main result of this section is that the range of any bounded charge on a σ-field is either a finite set or contains a perfect set.

We need some preliminary results.

11.2.1 Definition. Let μ be a charge defined on a σ-field $\mathfrak{A}$ of subsets of a set Ω. Let A_n, $n \geq 1$ be a sequence of pairwise disjoint sets in $\mathfrak{A}$. We say that μ is σ-*additive across* A_n, $n \geq 1$ if for any B in $\mathfrak{A}$, $B \subset \bigcup_{n \geq 1} A_n$,

$$\mu(B) = \sum_{n \geq 1} \mu(A_n \cap B)$$

holds.

The following proposition characterizes this notion.

11.2.2 Proposition. *Let μ be a bounded charge on a σ-field $\mathfrak{A}$ of subsets of a set Ω. Let A_n, $n \geq 1$ be a sequence of pairwise disjoint sets in $\mathfrak{A}$. Then the following statements are equivalent.*
(i). μ *is σ-additive across* A_n, $n \geq 1$.
(ii). μ^+ *is σ-additive across* A_n, $n \geq 1$ *and* μ^- *is σ-additive across* A_n, $n \geq 1$.
(iii). $|\mu|$ *is σ-additive across* A_n, $n \geq 1$.
(iv). $|\mu|(\bigcup_{n \geq 1} A_n) = \sum_{n \geq 1} |\mu|(A_n)$.
(v). $|\mu|(B_n)$, $n \geq 1$ *converges to* 0, *where* $B_n = \bigcup_{m \geq n} A_m$, $n \geq 1$.

Proof. (i)$\Rightarrow$(ii). Let $A \in \mathfrak{A}$ and $A \subset \bigcup_{n \geq 1} A_n$. We show that $\mu^+(A) = \sum_{n \geq 1} \mu^+(A \cap A_n)$. Let $B \in \mathfrak{A}$ and $B \subset A$. By (i), $\mu(B) = \sum_{n \geq 1} \mu(B \cap A_n) \leq \sum_{n \geq 1} \mu^+(A \cap A_n)$. Since this inequality is true for every B in $\mathfrak{A}$ with $B \subset A$, it follows that $\mu^+(A) \leq \sum_{n \geq 1} \mu^+(A \cap A_n)$. But $\sum_{n \geq 1} \mu^+(A \cap A_n) \leq \mu^+(A)$ is always true for positive charges. See Proposition 2.1.2 (viii). Hence the desired equality follows. Since μ and μ^+ are σ-additive across A_n, $n \geq 1$, it follows that μ^- is σ-additive across A_n, $n \geq 1$.
(ii)$\Rightarrow$(iii). Since $|\mu| = \mu^+ + \mu^-$, (iii) follows from (ii).
(iii)$\Rightarrow$(iv). This is obvious.
(iv)$\Rightarrow$(v). We show that $\sum_{m \geq n} |\mu|(A_m) = |\mu|(B_n) = |\mu|(\bigcup_{m \geq n} A_m)$ for any $n \geq 1$. It is always true that $\sum_{m \geq n} |\mu|(A_m) \leq |\mu|(B_n) = |\mu|(\bigcup_{m \geq n} A_m)$ for any $n \geq 1$. Note that

$$\sum_{m \geq 1} |\mu|(A_m) = |\mu|\left(\bigcup_{m \geq 1} A_m\right) = |\mu|\left(\bigcup_{m=1}^{n-1} A_m \cup \bigcup_{m \geq n} A_m\right)$$

$$= \sum_{m=1}^{n-1} |\mu|(A_m) + |\mu|(B_n) \geq \sum_{m=1}^{n-1} |\mu|(A_m)$$

$$+ \sum_{m \geq n} |\mu|(A_m) = \sum_{m \geq 1} |\mu|(A_m).$$

Hence, equality should prevail throughout, and consequently, $|\mu|(B_n) = \sum_{m \geq n} |\mu|(A_m)$. By (iv), it follows that

$$\lim_{n \to \infty} |\mu|(B_n) = \lim_{n \to \infty} \sum_{m \geq n} |\mu|(A_m) = 0.$$

(v) $\Rightarrow$ (i). Let $B \in \mathfrak{A}$ and $B \subset \bigcup_{n \geq 1} A_n$. Then for any $n \geq 1$, $|\sum_{m \geq n} \mu(B \cap A_m)| \leq |\mu|(\bigcup_{m \geq n} A_m)$. By (v), $\lim_{n \to \infty} \mu(\bigcup_{m \geq n} B \wedge A_m) = 0$. Hence

$$\mu(B) = \sum_{m \geq 1} \mu(B \cap A_m).$$

Thus μ is σ-additive across A_n, $n \geq 1$. □

The following proposition gives another property of the above notion.

11.2.3 Proposition. *Let μ be any charge on a σ-field $\mathfrak{A}$ of subsets of a set Ω. Let A_n, $n \geq 1$ be a sequence of pairwise disjoint sets in $\mathfrak{A}$ such that μ is σ-additive across A_n, $n \geq 1$. Let B_n, $n \geq 1$ be any sequence of sets in $\mathfrak{A}$ such that $B_n \subset A_n$ for every $n \geq 1$. Then μ is σ-additive across B_n, $n \geq 1$.*

Proof. This follows directly from Definition 11.2.1. □

In the following proposition, for any given bounded charge μ which is not finitely many valued, we exhibit a sequence of pairwise disjoint sets across which μ is σ-additive.

11.2.4 Proposition. *Let μ be a positive bounded charge on a σ-field $\mathfrak{A}$ of subsets of a set Ω. Suppose the range of μ is not finite. Then there exists a sequence W_n, $n \geq 1$ of pairwise disjoint sets in $\mathfrak{A}$ such that $\mu(W_n) > 0$ for every $n \geq 1$ and μ is σ-additive across W_n, $n \geq 1$.*

Proof. First, we exhibit a sequence Z_n, $n \geq 1$ of pairwise disjoint sets in $\mathfrak{A}$ such that $\mu(Z_n) > 0$ for every $n \geq 1$. Since $R(\mu)$ is not a finite set, we can find A_1 in $\mathfrak{A}$ such that $0 < \mu(A_1) < \mu(\Omega)$ and the range of μ on A_1 is an infinite set. By similar argument, we can find A_2 in $\mathfrak{A}$, $A_2 \subset A_1$ such that $0 < \mu(A_2) < \mu(A_1)$ and the range of μ on A_2 is an infinite set. Continuing this procedure, we obtain a sequence $A_1 \supset A_2 \supset A_3 \supset \cdots$ in $\mathfrak{A}$ with the property that $\mu(A_n) > \mu(A_{n+1})$ for every $n \geq 1$. Take $Z_n = A_n - A_{n+1}$, $n \geq 1$. Now, we construct the desired sequence W_n, $n \geq 1$. Let $Y_1 = \bigcup_{n \geq 1} Z_n$. Write $\{1, 2, 3, \ldots\}$ as a disjoint union of two sets N_1 and N_2 such that both are infinite. Then either $\mu(\bigcup_{n \in N_1} Z_n) \leq \frac{1}{2}\mu(Y_1)$ or $\mu(\bigcup_{n \in N_2} Z_n) \leq \frac{1}{2}\mu(Y_1)$. Let $Y_2 = \bigcup_{n \in N_1} Z_n$ if $\mu(\bigcup_{n \in N_1} Z_n) \leq \frac{1}{2}\mu(Y_1)$, and $Y_2 = \bigcup_{n \in N_2} Z_n$ otherwise. Adopting the above procedure for Y_2, we can obtain a set Y_3 in $\mathfrak{A}$ such that $Y_3 \subset Y_2$, Y_3 is an infinite union of sets from $\{Z_n, n \geq 1\}$ and $\mu(Y_3) \leq \frac{1}{2}\mu(Y_2)$. Continuing this procedure, we obtain a sequence $Y_1 \supset Y_2 \supset Y_3 \supset \cdots$ such that $0 < \mu(Y_n) \leq \frac{1}{2}\mu(Y_{n-1})$ for every $n \geq 2$ and each Y_n is an infinite union of sets from $\{Z_n,\ n \geq 1\}$. It follows that $\lim_{n \to \infty} \mu(Y_n) = 0$. Consequently, $\bigcap_{n \geq 1} Y_n = \emptyset$. Define $W_n = Y_n - Y_{n+1}$, $n \geq 1$. W_n, $n \geq 1$ is a sequence of pairwise disjoint sets in $\mathfrak{A}$, $\mu(W_n) > 0$ for every $n \geq 1$ and μ is σ-additive across W_n, $n \geq 1$ by Proposition 11.2.2. □

The following theorem is the main result of this section.

11.2.5 Theorem. *Let μ be a bounded charge on a σ-field $\mathfrak{A}$ of subsets of a set Ω. Then the range $R(\mu)$ of μ is either a finite set or contains a perfect set. Consequently, $R(\mu)$ is either finite or has the power of the continuum.*

Proof. Suppose the range of μ is not finite. We prove the theorem, first, when μ is positive. By Proposition 11.2.4, we can find a sequence W_n, $n \geq 1$ of pairwise disjoint sets in $\mathfrak{A}$ such that $\mu(W_n) > 0$ for every $n \geq 1$ and μ is σ-additive across W_n, $n \geq 1$. Then $\{\sum_{n \in C} \mu(W_n);\ C \subset \{1, 2, 3, \ldots\}\} \subset R(\mu)$. But $\{\sum_{n \in C} \mu(W_n);\ C \subset \{1, 2, 3, \ldots\}\}$ is a perfect set.

Now, let μ be any bounded charge on $\mathfrak{A}$. Since $R(\mu)$ is infinite, $R(|\mu|)$ is infinite. For, if $R(|\mu|)$ is finite, since $|\mu| = \mu^+ + \mu^-$, then μ^+ takes finitely many values and μ^- takes finitely many values. Consequently, $\mu = \mu^+ - \mu^-$ takes finitely many values which is a contradiction. Since $R(|\mu|)$ is infinite, by Proposition 11.2.4, we can find a sequence W_n, $n \geq 1$ of pairwise disjoint sets in $\mathfrak{A}$ such that $|\mu|(W_n) > 0$ for every $n \geq 1$ and $|\mu|$ is σ-additive across W_n, $n \geq 1$. Since $|\mu|(W_n) > 0$, we can find B_n in $\mathfrak{A}$ such that $\mu(B_n) \neq 0$ and $B_n \subset W_n$ for every $n > 1$. By Propositions 11.2.2 and 11.2.3, μ is σ-additive across B_n, $n \geq 1$. Either we can find an infinite subset $N_1 \subset \{1, 2, 3, \ldots\}$ such that $\mu(B_n) > 0$ for every $n \in N_1$ or we can find an infinite subset $N_2 \subset \{1, 2, 3, \ldots\}$ such that $\mu(B_n) < 0$ for every $n \in N_2$. Without loss of generality, assume that the former holds. By Proposition 11.2.3, μ is σ-additive across B_n, $n \in N_1$. Note that the perfect set $\{\sum_{n \in C} \mu(B_n);\ C \subset N_1\} \subset R(\mu)$. This proves the theorem. □

Now, we strengthen the above theorem under the same conditions.

11.2.6 Theorem. *Let μ be any bounded charge on a σ-field $\mathfrak{A}$ of subsets of a set Ω. Then either every point in the range $R(\mu)$ of μ is an isolated point of $R(\mu)$ in which case μ is finitely many valued or every neighbourhood of any point a in $R(\mu)$ contains a perfect set $C \subset R(\mu)$.*

Proof. Assume that μ is not finitely many valued. For the point $0 \in R(\mu)$ and the neighbourhood $(-\varepsilon, \varepsilon)$ with $\varepsilon > 0$, we exhibit a perfect set $C \subset R(\mu)$ such that $C \subset (-\varepsilon, \varepsilon)$. Since $R(|\mu|)$ is infinite, as in the proof of Theorem 11.2.5, we can find a sequence C_n, $n \geq 1$ of pairwise disjoint sets in $\mathfrak{A}$ such that μ is σ-additive across C_n, $n \geq 1$, $|\mu|(\bigcup_{m \geq n} C_m)$, $n \geq 1$ converges to zero and $\mu(C_n) > 0$ for all $n \geq 1$ or $\mu(C_n) < 0$ for all $n \geq 1$. Ignoring the first few sets if necessary, we can assume that $|\mu|(\bigcup_{m \geq 1} C_m) < \varepsilon$. It follows that for any subset $D \subset \{1, 2, 3, \ldots\}$, $|\mu(\bigcup_{m \in D} C_m)| < \varepsilon$. Thus we find that the perfect set $\{\sum_{m \in D} \mu(C_m); D \subset \{1, 2, 3, \ldots\}\} \subset (-\varepsilon, \varepsilon)$.

Now, let a be a non-zero point in $R(\mu)$. Let $A \in \mathfrak{A}$ be such that $\mu(A) = a$. Suppose μ when restricted to A is infinitely many valued. Then, by what we have proved above, for a given $\varepsilon > 0$, there exists $\mathscr{C} \subset A \cap \mathfrak{A}$ such that $\{\mu(C); C \in \mathscr{C}\}$ is a perfect set contained in $(-\varepsilon, \varepsilon)$. Note that $\{\mu(A - C) = \mu(A) - \mu(C); C \in \mathscr{C}\}$ is a perfect set contained in $(\mu(A) - \varepsilon, \mu(A) + \varepsilon) = (a - \varepsilon, a + \varepsilon)$. If, on the other hand, μ happens to be finitely many valued on A, then μ when restricted to A^c is infinitely many valued. By what we have established above, we can find $\mathscr{D} \subset A^c \cap \mathfrak{A}$ such that $\{\mu(D); D \in \mathscr{D}\}$ is a perfect set contained in $(-\varepsilon, \varepsilon)$. Now, note that $\{\mu(A \cup D) = \mu(A) + \mu(D); D \in \mathscr{D}\}$ is a perfect set and is contained in $(\mu(A) - \varepsilon, \mu(A) + \varepsilon) = (a - \varepsilon, a + \varepsilon)$ as well as in $R(\mu)$.

This completes the proof of the theorem. □

11.3 CARDINALITIES OF RANGES OF CHARGES

In this section, we make some remarks on the cardinalities of ranges of charges. In Section 11.2, we saw that if μ is a bounded charge on a σ-field of sets, then its range $R(\mu)$ is either a finite set or has the cardinality of the continuum. If μ is a bounded charge on a field $\mathscr{F}$ of sets, then its range $R(\mu)$ can be of any cardinal number. It is an amusing and instructive exercise for the reader to construct a charge on a suitable field of sets so that its range is of cardinality n, a given integer >1. The following theorem goes beyond finite cardinals.

11.3.1 Theorem. *Let $\aleph$ be any infinite cardinal less than or equal to the cardinality of the continuum. Then there is a set Ω, a field $\mathscr{F}$ of subsets of Ω and a real charge μ on $\mathscr{F}$ such that the cardinality of the range $R(\mu)$ of μ is $\aleph$.*

Proof. Let X be any subset of R having the following properties.
 (i). Cardinality of $X = \aleph$.
 (ii). If $x, y \in X$ and α, β are rational numbers, then $\alpha x + \beta y \in X$.
Such a set X can be constructed as follows. Let B be any subset of the real line R with cardinality $\aleph$. Let

$$X = \{\alpha_1 x_1 + \alpha_2 x_2 + \cdots + \alpha_n x_n; x_1, x_2, \ldots, x_n \in B,$$

$$\alpha_1, \alpha_2, \ldots, \alpha_n \text{ rational numbers and } n \geq 1\}.$$

Then the set X has the above properties (i) and (ii).

Take $\Omega = \mathbb{R}$. Let $\mathscr{F}$ be the collection of all sets A of the form $\bigcup_{i=1}^{n} [a_i, b_i)$, where $[a_1, b_1), [a_2, b_2), \ldots, [a_n, b_n)$ are pairwise disjoint intervals, $a_1 \leq b_1$, $a_2 \leq b_2, \ldots, a_n \leq b_n$, $a_1, a_2, \ldots, a_n \in X$, $b_1, b_2, \ldots, b_n \in X$ and $n \geq 1$, and their complements. $\mathscr{F}$ is clearly a field on Ω. Define μ on $\mathscr{F}$ by

$$\mu(A) = \sum_{i=1}^{n} (b_i - a_i), \quad \text{if A is of the form } \bigcup_{i=1}^{n} [a_i, b_i),$$

$$= -\mu(A^c), \quad \text{if } A^c \text{ is of the above form.}$$

μ is a real charge on $\mathscr{F}$ and $R(\mu) = X$. This shows that $R(\mu)$ has cardinality $\aleph$. □

11.3.2 Remarks. (i). One can construct a bounded charge μ with cardinality of $R(\mu) = \aleph$ in Theorem 11.3.1. Further, one could have μ to be positive as well.
(ii). If μ is allowed to take infinite values, the construction of a positive charge μ with cardinality of its range being a prescribed infinite cardinal number could be made much simpler.

11.4 CHARGES WITH CLOSED RANGE

If μ is a bounded charge on a field $\mathscr{F}$ of subsets of a set Ω, then its range $R(\mu)$ need not be a closed subset of the real line $\mathbb{R}$. See Theorem 11.3.1. In the following, we give an example of a bounded charge μ on a σ-field $\mathfrak{A}$ of subsets of a set Ω such that its range $R(\mu)$ is not a closed set.

11.4.1 Example. Let $\Omega = \{1, 2, 3, \ldots\}$ and $\mathfrak{A} = \mathscr{P}(\Omega)$, the class of all subsets of Ω. Let μ_0 be any probability charge on $\mathfrak{A}$ such that $\mu_0(A) = 0$ for any finite subset A of Ω. For each $n \geq 1$, let μ_n on $\mathfrak{A}$ be the measure defined by

$$\mu_n(A) = 0, \quad \text{if } n \notin A,$$

$$= 1, \quad \text{if } n \in A \text{ and } A \subset \Omega.$$

Let $\mu = \sum_{n \geq 0} (1/2^{n+1}) \mu_n$. Note that $\frac{1}{2} \notin R(\mu)$ but $\frac{1}{2}$ is an accumulation point of $R(\mu)$. Hence $R(\mu)$ is not a closed set.

In view of the above example, it is of interest to derive a set of sufficient conditions under which $R(\mu)$ is a closed set. Sobczyk and Hammer Decomposition theorem (See Theorem 5.2.7.) provides a basis for further exploration in this direction. We need a definition, to begin with.

11.4.2 Definition. Let $\mathscr{F}$ be a field of subsets of a set Ω. A sequence μ_n, $n \geq 1$ of 0-1 valued charges is said to be *discrete* if for any given positive integer n, there exists a set A in $\mathscr{F}$ such that $\mu_n(A) = 1$ and $\mu_m(A) = 0$ for every $m \neq n$.

Let us state a lemma about discrete sequences of charges on fields.

11.4.3 Lemma. *If μ_n, $n \geq 1$ is a discrete sequence of 0-1 valued charges on a field $\mathscr{F}$ of subsets of a set Ω, then there exists a sequence A_n, $n \geq 1$ of pairwise disjoint sets in $\mathscr{F}$ such that $\mu_n(A_n) = 1$ and $\mu_m(A_n) = 0$ for all m and n such that $m \neq n$.*

Proof. If B_n, $n \geq 1$ is a sequence of sets from $\mathscr{F}$ such that $\mu_n(B_n) = 1$ and $\mu_m(B_n) = 0$ for $m \neq n$, then the sequence A_n, $n \geq 1$ defined by

$$A_1 = B_1, \quad \text{and} \quad A_n = B_n - \bigcup_{m=1}^{n-1} B_m, \quad n \geq 2$$

serves the purpose of the lemma. $\square$

The notion of discreteness introduced above is weaker than the notion of infinite disjointness. See Remark 5.2.3(i). The following is an example amplifying this point.

Let $\Omega = \{1, 2, 3, \ldots, \infty\}$ and $\mathscr{F}$ the collection of all finite subsets of $\{1, 2, 3, \ldots\}$ and their complements. On the field $\mathscr{F}$, for each $n \geq 1$, define μ_n by

$$\mu_n(A) = 1, \quad \text{if } n \in A,$$
$$= 0, \quad \text{if } n \notin A.$$

This sequence μ_n, $n \geq 1$ of distinct 0-1 valued charges is discrete but not infinitely disjoint. This is because for any countable partition $\{F_1, F_2, \ldots\}$ of Ω in $\mathscr{F}$, all but a finite number of sets among $\{F_1, F_2, \ldots\}$ are empty.

The following is an instance when the range is a compact set.

11.4.4 Theorem. *Let α_n, $n \geq 1$ be a sequence of real numbers such that $\sum_{n \geq 1} |\alpha_n| < \infty$. Let $\mathfrak{A}$ be a σ-field of subsets of a set Ω and μ_n, $n \geq 1$ a discrete sequence of 0-1 valued charges on $\mathfrak{A}$. Let*

$$\mu = \sum_{n \geq 1} \alpha_n \mu_n.$$

Then the range $R(\mu)$ of μ is a compact subset of R and hence is closed.

Proof. By Lemma 11.4.3, we can find a sequence A_n, $n \geq 1$ of pairwise disjoint sets in $\mathfrak{A}$ such that $\mu_n(A_n) = 1$ for every $n \geq 1$ and $\mu_m(A_n) = 0$ for all $m \neq n$. Define a real valued map h on the Cantor set $\{0, 1\}^{\aleph_0}$ by

$$h(x_1, x_2, \ldots) = \sum_{n \geq 1} x_n \alpha_n$$

for $(x_1, x_2, \ldots)$ in $\{0, 1\}^{\aleph_0}$. ($\{0, 1\}^{\aleph_0}$ is equipped with the usual product topology.) Since $\sum_{n \geq 1} |\alpha_n| < \infty$, h is a continuous function on $\{0, 1\}^{\aleph_0}$. Let D be the range of the function h. Since $\{0, 1\}^{\aleph_0}$ is a compact space, D is a compact subset of R. Now, we claim that $D = R(\mu)$. Let $a \in D$. Then there exists $(x_1, x_2, \ldots)$ in $\{0, 1\}^{\aleph_0}$ such that $h(x_1, x_2, \ldots) = a$. Let $A = \bigcup_{n \in C} A_n$, where $C = \{n \geq 1; x_n = 1\}$. Since $\mathfrak{A}$ is a σ-field, $A \in \mathfrak{A}$. Further,

$$\mu(A) = \sum_{n \geq 1} \alpha_n \mu_n(A) = \sum_{n \in C} \alpha_n = \sum_{n \geq 1} \alpha_n x_n$$

$$= h(x_1, x_2, \ldots) = a.$$

Consequently, $a \in R(\mu)$. Conversely, let $b \in R(\mu)$. Then there is a set A in $\mathfrak{A}$ such that $\mu(A) = b = \sum_{n \geq 1} \alpha_n \mu_n(A) = \sum_{n \geq 1} \alpha_n \mu_n(A \cap A_n)$. Define the sequence x_n, $n \geq 1$ by

$$x_n = 1, \quad \text{if } \mu_n(A \cap A_n) = 1,$$
$$= 0, \quad \text{if } \mu_n(A \cap A_n) = 0, n \geq 1.$$

Consequently,

$$h(x_1, x_2, \ldots) = \sum_{n \geq 1} \alpha_n x_n$$

$$= \sum_{n \geq 1} \alpha_n \mu_n(A \cap A_n)$$

$$= \mu(A) = b \in D.$$

This proves that $D = R(\mu)$. □

The following result is an analogue of one dimensional Liapounoff's theorem for nonatomic measures on σ-fields. See Theorem 5.1.6.

11.4.5 Theorem. *Let μ be a positive bounded strongly continuous charge defined on a σ-field $\mathfrak{A}$ of subsets of a set Ω. Then the range $R(\mu)$ of μ is a closed interval.*

Proof. Let $a \in (0, \mu(\Omega))$. We show that there exists a set A in $\mathfrak{A}$ such that $\mu(A) = a$. Since μ is strongly continuous, for $\varepsilon = \frac{1}{2}$, we can find a finite partition of Ω in $\mathscr{F}$ such that each set in the partition has μ-value $< \frac{1}{2}$. Let A_1 be the largest possible union of sets from this partition satisfying the condition that $\mu(A_1) \leq a$. $A_1 \neq \Omega$ because $0 < a < \mu(\Omega)$. Let B_1 be any set in the partition which is not in this union. Then $0 < \mu(B_1) < \frac{1}{2}$ and $\mu(A_1 \cup B_1) > a$. If $\mu(A_1) = a$, the desired claim is established. Suppose $\mu(A_1) < a$. Since μ restricted to B_1 is strongly continuous, we can find a finite partition of B_1 in $\mathfrak{A}$ such that each set in the partition has μ-value $< 1/2^2$. Let A_2 be the largest possible union of sets from this partition such that $\mu(A_2) \leq a - \mu(A_1)$. Since $A_2 \neq B_1$, we can find a set B_2 from this partition which is not in this union A_2. Then $0 < \mu(B_2) < 1/2^2$ and $\mu(A_2 \cup B_2) > a - \mu(A_1)$, or equivalently, $\mu(A_1) + \mu(A_2) + \mu(B_2) > a$. If $\mu(A_2) = a - \mu(A_1)$, then $A_1 \cup A_2$ is the desired set. Otherwise, we proceed to obtain A_3 and B_3 as above using B_2. If this procedure terminates at a finite stage, we obtain a set A in $\mathfrak{A}$ with μ-value equal to a. Otherwise, we get a sequence $A_1, B_1; A_2, B_2; \ldots$ of sets in $\mathfrak{A}$ having the following properties for every $n \geq 1$.

(i). $A_n \cap B_n = \varnothing$.
(ii). $A_{n+1} \cup B_{n+1} \subset B_n$.
(iii). $\mu(A_1) + \mu(A_2) + \cdots + \mu(A_n) \leq a$.
(iv). $\mu(A_1) + \mu(A_2) + \cdots + \mu(A_n) + \mu(B_n) > a$.
(v). $0 < \mu(B_n) < 1/2^n$.

It is now clear that $\sum_{n \geq 1} \mu(A_n) = a$. We now show that $\mu(\bigcup_{n \geq 1} A_n) = \sum_{n \geq 1} \mu(A_n)$. Note that

$$\bigcup_{i \geq 1} A_i \subset \left(\bigcup_{i=1}^{n+1} A_i\right) \cup B_{n+1}$$

for every $n \geq 1$. Consequently,

$$\mu\left(\bigcup_{i \geq 1} A_i\right) \leq \sum_{i=1}^{n+1} \mu(A_i) + \mu(B_{n+1}) \quad \text{for every } n \geq 1.$$

So,

$$\mu\left(\bigcup_{i \geq 1} A_i\right) \leq \sum_{n \geq 1} \mu(A_n).$$

But the inequality

$$\sum_{n \geq 1} \mu(A_n) \leq \mu\left(\bigcup_{n \geq 1} A_n\right)$$

is always true for positive charges. Hence

$$\mu\left(\bigcup_{n \geq 1} A_n\right) = \sum_{n \geq 1} \mu(A_n) = a.$$

This proves the result. □

Now, it is natural to attempt to extend Theorem 11.4.5 to cover the case when we have a general strongly continuous bounded charge on a σ-field. We give an example to show that Theorem 11.4.5 fails in general. Before that, we prove a proposition which will be useful in the construction of the desired example.

11.4.6 Proposition. *Let μ be a bounded charge on a field $\mathcal{F}$ of subsets of a set Ω and $R(\mu)$ its range. Let $\alpha = \mathrm{Sup}_{B \in \mathcal{F}} \mu(B)$ and $\beta = \mathrm{Inf}_{A \in \mathcal{F}} \mu(A)$. Then the following statements are equivalent.*

(i). μ admits a Hahn set, i.e. a ε-Hahn set for $\varepsilon = 0$. (See Definition 2.6.1.
(ii). $\alpha \in R(\mu)$.
(iii). $\beta \in R(\mu)$.
Consequently, $R(\mu)$ is not a closed set if μ does not admit a Hahn set.

Proof. We show that (i) and (ii) are equivalent.
(i)$\Rightarrow$(ii). If A is a Hahn set for μ, then $\mu(A) = \alpha$.
(ii)$\Rightarrow$(i). If $A \in \mathscr{F}$ and $\mu(A) = \alpha$, then A is a Hahn set for μ. $\square$

Now, we construct the desired example. In view of Proposition 11.4.6, it suffices to construct a strongly continuous bounded charge on any given infinite σ-field not admitting a Hahn set.

11.4.7 Example. Let $\mathfrak{A}$ be a given infinite σ-field of subsets of a set Ω. Let ν be any strongly continuous probability charge on $\mathfrak{A}$. See Corollary 5.3.3. Using Theorem 11.4.5, we obtain a tree $\{A_{\delta_1,\delta_2,\ldots,\delta_n}; \delta_1, \delta_2, \ldots, \delta_n$ a finite sequence of 0's and 1's, $n \geq 1\}$ having the following properties.

(i). $A_{\delta_1,\delta_2,\ldots,\delta_n,0} \cap A_{\delta_1,\delta_2,\ldots,\delta_n,1} = \varnothing$ for any $n \geq 1$ and any finite sequence $\delta_1, \delta_2, \ldots, \delta_n$ of 0's and 1's.
(ii). $A_{\delta_1,\delta_2,\ldots,\delta_n,0} \cup A_{\delta_1,\delta_2,\ldots,\delta_n,1} = A_{\delta_1,\delta_2,\ldots,\delta_n}$ for any $n \geq 1$ and any finite sequence $\delta_1, \delta_2, \ldots, \delta_n$ of 0's and 1's.
(iii). $A_0 \cap A_1 = \varnothing$.
(iv). $A_0 \cup A_1 = \Omega$.
(v). $\nu(A_{\delta_1,\delta_2,\ldots,\delta_n}) = \alpha_1(\delta_1)\alpha_2(\delta_2) \cdots \alpha_n(\delta_n)$ for any $n \geq 1$ and any sequence $\delta_1, \delta_2, \ldots, \delta_n$ of 0's and 1's, where $\alpha_n(0) = 1/(n+1)$ and $\alpha_n(1) = n/(n+1)$, $n \geq 1$.

Let $\mathscr{N} = \{A \in \mathfrak{A}; \nu(A) = 0\}$. Look at the quotient Boolean algebra $\mathfrak{A}/\mathscr{N}$. For A in $\mathfrak{A}$, $[A]$ denote the equivalence class in $\mathfrak{A}/\mathscr{N}$ containing A. Note that $\{[A_{\delta_1,\delta_2,\ldots,\delta_n}]; \delta_1, \delta_2, \ldots, \delta_n$ a sequence of 0's and 1's, $n \geq 1\}$ is a tree in $\mathfrak{A}/\mathscr{N}$. As in the proof of (ii)$\Rightarrow$(iii) of Theorem 5.3.2, one can construct a strongly continuous probability charge $\tilde{\tau}$ on $\mathfrak{A}/\mathscr{N}$ such that

$$\tilde{\tau}([A_{\delta_1,\delta_2,\ldots,\delta_n}]) = \alpha_1(1-\delta_1)\alpha_2(1-\delta_2) \cdots \alpha_n(1-\delta_n),$$

for any finite sequence $\delta_1, \delta_2, \ldots, \delta_n$ of 0's and 1's, and for any $n \geq 1$. Now, define τ on $\mathfrak{A}$ by $\tau(A) = \tilde{\tau}([A])$ for A in $\mathfrak{A}$. τ is a strongly continuous probability charge on $\mathfrak{A}$ because for any finite sequence $\delta_1, \delta_2, \ldots, \delta_n$ of 0's and 1's,

$$\tau(A_{\delta_1,\delta_2,\ldots,\delta_n}) \leq 1/n,$$

for every $n \geq 1$. $\nu \wedge \tau = 0$ because for any $n \geq 1$,

$$\bigcup A_{\delta_1,\delta_2,\ldots,\delta_n,0} \quad \text{and} \quad \bigcup A_{\delta_1,\delta_2,\ldots,\delta_n,1},$$

where both the unions are taken over all $\delta_1, \delta_2, \ldots, \delta_n$ in $\{0, 1\}$, are disjoint with union equal to Ω, ν-value of the first set and τ-value of the second set are each equal to $1/(n+2)$. τ also has the property that if $A \in \mathfrak{A}$ and $\nu(A) = 0$, then $\tau(A) = 0$.

Now, define μ on $\mathfrak{A}$ by $\mu = \nu - \tau$. Since $\nu \wedge \tau = 0$, $\mu^+ = \nu$ and $\mu^- = \tau$. See Theorem 2.2.2(4). It is clear that ν and τ are distinct and μ is a strongly continuous bounded charge on $\mathfrak{A}$. We note that μ does not admit a Hahn set. If μ admits a Hahn set A in $\mathfrak{A}$, then $\mu^+(A^c) = \nu(A^c) = 0$. So, $\tau(A^c) = 0$. Also, $\mu^-(A) = \tau(A) = 0$. This implies that $\tau(\Omega) = 0$ which is a contradiction.

Theorem 11.4.5 says that the range of a positive bounded strongly continuous charge defined on a σ-field is convex and closed. The convexity part can be generalized to any finitely many positive bounded strongly continuous charges defined on a σ-field whereas the closedness cannot be generalized.

11.4.8 Example. Let ν and τ be positive bounded strongly continuous charges on an infinite σ-field $\mathfrak{A}$ as in Example 11.4.7. $R(\nu, \tau) = \{(\nu(A), \tau(A)); A \in \mathfrak{A}\}$ is not a closed set because $(1, 0) \notin R(\nu, \tau)$ but belongs to the closure of $R(\nu, \tau)$.

11.4.9 Theorem. *Let $\mu_1, \mu_2, \ldots, \mu_n$ be positive bounded strongly continuous charges defined on a σ-field $\mathfrak{A}$ of subsets of a set Ω. Then the range $R(\mu_1, \mu_2, \ldots, \mu_n) = \{(\mu_1(A), \mu_2(A), \ldots, \mu_n(A)); A \in \mathfrak{A}\}$ of $\mu_1, \mu_2, \ldots, \mu_n$ is a convex subset of the n-dimensional Euclidean space $\mathbf{R}^n$.*

Proof. For $n = 1$, the result was already established in Theorem 11.4.5. Let us assume the result to be true for $n = k$ and prove the result for $n = k+1$. Let $\mu_1, \mu_2, \ldots, \mu_{k+1}$ be positive bounded strongly continuous charges on $\mathfrak{A}$. The proof is divided into several steps as follows.

1°. Let $\tau_i = \mu_i + \mu_{i+1} + \cdots + \mu_{k+1}$ for $i = 1, 2, \ldots, k+1$. Then it is easily seen that $R(\mu_1, \mu_2, \ldots, \mu_{k+1})$ is convex if and only if $R(\tau_1, \tau_2, \ldots, \tau_{k+1})$ is convex. So, we prove that $R(\tau_1, \tau_2, \ldots, \tau_{k+1})$ is convex.

2°. If we show that for every A in $\mathfrak{A}$, there exists a set B in $\mathfrak{A}$ such that $B \subset A$ and $\tau_i(B) = \frac{1}{2}\tau_i(A)$ for $i = 1, 2, \ldots, k+1$, then it would follow that $R(\tau_1, \tau_2, \ldots, \tau_{k+1})$ is convex. Let us see why.

For any given set A in $\mathfrak{A}$, by repeated application of the above assertion, for every dyadic rational r between 0 and 1, we can find a set A_r with the following properties.

(i). $A_0 = \varnothing$.
(ii). $A_1 = A$.
(iii). $A_r \subset A_s$ whenever the dyadic rationals r and s satisfy $0 \leq r < s \leq 1$.
(iv). $\tau_i(A_r) = r\tau_i(A)$ for $i = 1, 2, \ldots, k+1$ and for any dyadic rational $0 \leq r \leq 1$.

(v). For any real number a between 0 and 1, if we define $A_a = \bigcup A_r$, where the union is taken over all dyadic rational numbers $r \leq a$, then $A_a \in \mathfrak{A}$ and $\tau_i(A_a) = a\tau_i(A)$ for $i = 1, 2, \ldots, k+1$.

Let C, D $\in \mathfrak{A}$ and $0 \leq a \leq 1$. Then

$$a\tau_i(C) + (1-a)\tau_i(D) = \tau_i((C-D)_a \cup (C \cap D) \cup (D-C)_{1-a})$$

for every $i = 1, 2, \ldots, k+1$. Using the properties (i), (ii), (iii), (iv) and (v) given above, one can establish the above equality. Consequently, $R(\tau_1, \tau_2, \ldots, \tau_{k+1})$ is convex.

Thus it suffices to exhibit a set B in $\mathfrak{A}$, for a given A in $\mathfrak{A}$, such that $B \subset A$ and $\tau_i(B) = \tfrac{1}{2}\tau_i(A)$ for every $i = 1, 2, \ldots, k+1$.

3°. Let $C \in \mathfrak{A}$ be a set obtained by the induction hypothesis satisfying $C \subset A$ and $\tau_i(C) = \tfrac{1}{2}\tau_i(A)$ for $i = 1, 2, \ldots, k$. If $\tau_{k+1}(C) = \tfrac{1}{2}\tau_{k+1}(A)$, then we have finished. If not, let $\tau_{k+1}(C) < \tfrac{1}{2}\tau_{k+1}(A) < \tau_{k+1}(A-C)$. Now, look at the sets C_a, $0 \leq a \leq 1$ and $(A-C)_a$, $0 \leq a \leq 1$ obtained as in Step 2° with respect to $\tau_1, \tau_2, \ldots, \tau_k$. The existence of these sets is assured by the induction hypothesis. Since $\tau_{k+1} \leq \tau_k$, we have that $\tau_{k+1}(C_a - C_b) \leq \tau_k(C_a - C_b) = (a-b)\tau_k(C)$ if $a \geq b$. Hence $\tau_{k+1}(C_a)$ is a continuous function of a. Similarly, $\tau_{k+1}((A-C)_a)$ is also a continuous function of a. Hence for $0 \leq a \leq 1$, $\tau_{k+1}(C_a \cup (A-C)_{1-a})$ is a continuous function of a taking the value $\tau_{k+1}(C)$ at $a = 1$ and the value $\tau_{k+1}(A-C)$ at $a = 0$. Hence there exists a real number a_0 in $[0, 1]$ such that $\tau_{k+1}(C_{a_0} \cup (A-C)_{1-a_0}) = \tfrac{1}{2}\tau_{k+1}(A)$. However, for $1 \leq i \leq k$,

$$\tau_i(C_{a_0} \cup (A-C)_{1-a_0}) = a_0 \tau_i(C) + (1-a_0)\tau_i(A-C)$$
$$= a_0 \tfrac{1}{2}\tau_i(A) + (1-a_0)\tfrac{1}{2}\tau_i(A)$$
$$= \tfrac{1}{2}\tau_i(A).$$

Thus $C_{a_0} \cup (A-C)_{1-a_0} = B$ serves the purpose. $\square$

The analogue of Theorem 11.4.9 for bounded strongly continuous charges which are not necessarily positive is also true and is an easy consequence of Theorem 11.4.9. Recall that a bounded charge μ is strongly continuous if $|\mu|$ is strongly continuous.

11.4.10 Theorem. *Let $\mu_1, \mu_2, \ldots, \mu_n$ be a finite number of bounded strongly continuous charges defined on a σ-field $\mathfrak{A}$ of subsets of a set Ω. Then the range $R(\mu_1, \mu_2, \ldots, \mu_n) = \{(\mu_1(A), \mu_2(A), \ldots, \mu_n(A)); A \in \mathfrak{A}\}$ of $\mu_1, \mu_2, \ldots, \mu_n$ is a convex subset of $\mathbf{R}^n$.*

Proof. Using the Jordan Decomposition Theorem, write $\mu_i = \mu_i^+ - \mu_i^-$ for $i = 1, 2, \ldots, n$. Then all the μ_i^+'s and μ_i^-'s are strongly continuous. So, $R(\mu_1^+, \mu_1^-, \mu_2^+, \mu_2^-, \ldots, \mu_n^+, \mu_n^-)$ is a convex set by Theorem 11.4.9. From this, it follows easily that $R(\mu_1, \mu_2, \ldots, \mu_n)$ is convex. $\square$

11.5 CHARGES WHOSE RANGES ARE NEITHER LEBESGUE MEASURABLE NOR HAVE THE PROPERTY OF BAIRE

In this section, we examine whether the range of any probability charge on a σ-field is a Borel subset of the real line R. Given any infinite σ-field $\mathfrak{A}$ on a set Ω, we will exhibit a family of probability charges on $\mathfrak{A}$ the range of each of which is neither Lebesgue measurable nor has the property of Baire. For this purpose, we need some notions and results from Topology and Measure Theory.

Let $\mathfrak{A}$ be a σ-field of subsets of a set Ω and τ a positive measure on $\mathfrak{A}$. The charge space $(\Omega, \mathfrak{A}, \tau)$ is said to be *complete* if $B \in \mathfrak{A}$ whenever $B \subset A$ for some A in $\mathfrak{A}$ with $\tau(A) = 0$. If $(\Omega, \mathfrak{A}, \tau)$ is not complete, we can enlarge $\mathfrak{A}$ so as to make the resultant triplet complete. More precisely, let

$$\tilde{\mathfrak{A}} = \{B \Delta A; B \subset C \text{ for some C in } \mathfrak{A} \text{ with } \tau(C) = 0 \text{ and } A \in \mathfrak{A}\}.$$

One can show that $\tilde{\mathfrak{A}}$ is a σ-field on Ω and contains $\mathfrak{A}$. Define $\tilde{\tau}$ on $\tilde{\mathfrak{A}}$ by

$$\tilde{\tau}(B \Delta A) = \tau(A),$$

where $B \subset C$ for some C in $\mathfrak{A}$ with $\tau(C) = 0$ and $A \in \mathfrak{A}$. $\tilde{\tau}$ is unambiguously defined on $\tilde{\mathfrak{A}}$, agrees with τ on $\mathfrak{A}$ and is indeed a measure on $\tilde{\mathfrak{A}}$. The charge space $(\Omega, \tilde{\mathfrak{A}}, \tilde{\tau})$ is complete. Sets in $\tilde{\mathfrak{A}}$ are usually called *τ-measurable sets*.

The following is a particularly interesting case of the above. Let $\mathscr{B}$ be the Borel σ-field on the real line R and λ Lebesgue measure on $\mathscr{B}$. Then $(R, \mathscr{B}, \lambda)$ is not complete. Let $\tilde{\mathscr{B}}$ be the completion of $\mathscr{B}$ with respect to λ as described above. The sets in $\tilde{\mathscr{B}}$ are called Lebesgue measurable sets. Obviously, $\tilde{\mathscr{B}}$ contains the Borel σ-field $\mathscr{B}$ properly.

Let X be a topological space. Recall that a subset A of X is said to have *the property of Baire* if there exists an open subset U of X such that $A \Delta U$ is a set of first category. Let $\mathscr{B}^*$ be the collection of all subsets of X each of which has the property of Baire. Then $\mathscr{B}^*$ is a σ-field on X and contains the Borel σ-field of X, i.e. the smallest σ-field on X containing all open subsets of X.

Let X be a complete separable metric space and $\mathscr{B}$ its Borel σ-field. Let $X^{\aleph_0}$ be the product space $X \times X \times \cdots = \{(x_1, x_2, \ldots); x_n \in X \text{ for all } n \geq 1\}$. Let $\mathscr{B}^\infty$ be the product σ-field on $X^{\aleph_0}$, i.e. the smallest σ-field on $X^{\aleph_0}$ containing all finite dimensional cylinder sets $\{A_1 \times A_2 \times \cdots \times A_n \times X \times X \times \cdots; A_1, A_2, \ldots, A_n \in \mathscr{B}, n \geq 1\}$. Let μ_n, $n \geq 1$ be a sequence of probability measures on $\mathscr{B}$. Then there exists a unique probability measure μ on $\mathscr{B}^\infty$ with the following property.

$$\mu(A_1 \times A_2 \times \cdots \times A_n \times X \times X \times \cdots) = \mu_1(A_1)\mu_2(A_2) \cdots \mu_n(A_n)$$

for any $A_1, A_2, \ldots, A_n$ in $\mathscr{B}$ and $n \geq 1$. μ is called the *product probability measure* of μ_n, $n \geq 1$ and is denoted by $\prod_{n \geq 1} \mu_n$.

In the above framework of product spaces, we introduce the following notion. A set $D \subset X^{\aleph_0}$ is said to be a *tail set* if $(y_1, y_2, \ldots, y_k, x_{k+1}, x_{k+2}, \ldots) \in D$ whenever $(x_1, x_2, \ldots, x_k, x_{k+1}, x_{k+2}, \ldots) \in D$ for some $x_1, x_2, \ldots, x_k$ in X, for any $y_1, y_2, \ldots, y_k$ in X and for any $k \geq 1$. Now, we state, in this connection, a useful result.

Kolmogorov's 0-1 Law. Let D be a μ-measurable tail subset of $X^{\aleph_0}$. Then $\mu(D) = 0$ or 1.

Now, we look at the space $X^{\aleph_0}$ in its product topology, where X is a complete separable metric space. The following result is a topological analogue of the above law.

Oxtoby's Category Analogue of Kolmogorov's 0-1 Law. Let D be a tail subset of $X^{\aleph_0}$ with the property of Baire. Then either D is of first category or D^c is of first category.

Now, we are in a position to construct the desired family of probability charges.

11.5.1 Definition. Let $\mathscr{F}$ be a field of subsets of a set Ω and μ_n, $n \geq 0$ a sequence of 0–1 valued charges on $\mathscr{F}$. μ_0 is said to be an *accumulation point* of $\{\mu_n, n \geq 1\}$ if for every $A \in \mathscr{F}$ with $\mu_0(A) = 1$, there exists an infinite subset D of $\{1, 2, 3, \ldots\}$ such that $\mu_n(A) = 1$ for every $n \in D$.

The above notion has a simple interpretation in terms of the Stone space Y of $\mathscr{F}$. Each μ_n can be identified as a point in Y and the above definition is equivalent to saying that μ_0 is an accumulation point of the subset $\{\mu_n, n \geq 1\}$ of Y in the usual topological sense.

11.5.2 Theorem. *Let $\mathfrak{A}$ be a σ-field of subsets of a set Ω and μ_n, $n \geq 0$ be a sequence of 0–1 valued charges on $\mathfrak{A}$ such that μ_n, $n \geq 1$ is a discrete sequence and μ_0 is an accumulation point of $\{\mu_n, n \geq 1\}$. Define μ on $\mathfrak{A}$ by*

$$\mu = \sum_{n \geq 0} \frac{1}{2^{n+1}} \mu_n.$$

Then the range $R(\mu)$ of μ is neither Lebesgue measurable nor has the property of Baire.

Proof. Let $Z = \{0, 1\}$ and ν the measure on the discrete σ-field on Z given by $\nu(\{0\}) = \nu(\{1\}) = \frac{1}{2}$. Let $C = Z^{\aleph_0}$ equipped with the product σ-field and τ the product probability measure $\nu \times \nu \times \nu \cdots$ on this σ-field. Sets in this product σ-field are called Borel subsets of C. (C is the Cantor set.) Define a real valued function h on C by

$$h(x_1, x_2, \ldots) = \sum_{n \geq 1} \frac{x_n}{2^n}$$

for $(x_1, x_2, \ldots)$ in C. This function has many interesting properties. It is one-to-one except for a countable set of points; it is a homeomorphism except for a countable set of points; $h(A)$ is a Borel subset of $[0, 1]$ if and only if A is Borel subset of C; h preserves the measures τ and the Lebesgue measure λ on $[0, 1]$, i.e. $\tau(h^{-1}(B)) = \lambda(B)$ for every Borel subset B of $[0, 1]$; $h(A)$ is a Lebesgue measurable subset of $[0, 1]$ if and only if A is a τ-measurable subset of C; $h(A)$ has the property of Baire in $[0, 1]$ if and only if A has the property of Baire in C. See Kuratowski (1966).

In view of these properties of h, if we show that $F = \{(\mu_0(A), \mu_1(A), \mu_2(A), \ldots); A \in \mathfrak{A}\}$ is neither τ-measurable nor has the property of Baire in C, then it will follow that $R(\mu)$ is neither Lebesgue measurable nor has the property of Baire in $[0, 1]$. This is because $h(F) = R(\mu)$.

Let $D = \{(\mu_1(A), \mu_2(A), \ldots); A \in \mathfrak{A}, \mu_0(A) = 1\}$. So, if we show that D is neither τ-measurable nor has the property of Baire in C, the desired conclusion about F follows.

First, we show that D is not τ-measurable. We prove that D is a tail set. Since μ_n, $n \geq 1$ is a discrete sequence, by Lemma 11.4.3, we can find a sequence A_n, $n \geq 1$ of pairwise disjoint sets in $\mathfrak{A}$ such that $\mu_n(A_n) = 1$ for every $n \geq 1$ and $\mu_m(A_n) = 0$ for all $m \neq n$. Since μ_0 is an accumulation point of $\{\mu_n, n \geq 1\}$, μ_0 is distinct from all μ_n, $n \geq 1$. One can assume, without loss of generality, that $\mu_0(A_n) = 0$ for every $n \geq 1$. This follows from the fact that if ξ and η are two distinct 0-1 valued charges on $\mathfrak{A}$, then there is a set A in $\mathfrak{A}$ such that $\xi(A) = 0 = \eta(A^c)$. Let $(x_1, x_2, \ldots) = (\mu_1(A), \mu_2(A), \ldots) \in D$ for some A in $\mathfrak{A}$ with $\mu_0(A) = 1$. Let k be any positive integer and $y_1, y_2, \ldots, y_k$ be any finite sequence of 0's and 1's. Let $E_1 = \{1 \leq i \leq k; y_i = 0\}$ and $E_2 = \{1 \leq i \leq k; y_i = 1\}$. Let $B = (A - \bigcup_{i \in E_1} A_i) \cup (\bigcup_{i \in E_2} A_i)$. It is obvious that $\mu_0(B) = 1$, $\mu_1(B) = y_1$, $\mu_2(B) = y_2, \ldots, \mu_k(B) = y_k$, $\mu_{k+1}(B) = \mu_{k+1}(A)$, $\mu_{k+2}(B) = \mu_{k+2}(A), \ldots$. Consequently, $(y_1, y_2, \ldots, y_k, x_{k+1}, x_{k+2}, \ldots) \in D$. Hence D is a tail set.

Suppose D is τ-measurable. By the Kolmogorov's 0-1 law, $\tau(D) = 0$ or 1. Let us look at the map ψ from C to C defined by $\psi(x_1, x_2, \ldots) = (1 - x_1, 1 - x_2, \ldots)$ for $(x_1, x_2, \ldots)$ in C. We claim that $\psi(D) \cap D = \emptyset$ and $\psi(D) \cup D = C$. Suppose $\psi(D) \cap D \neq \emptyset$. Let $(x_1, x_2, \ldots) \in \psi(D) \cap D$. Then we can find two sets A and B in $\mathfrak{A}$ such that $\mu_0(A) = 1 = \mu_0(B)$ and

$$(x_1, x_2, \ldots) = (\mu_1(A), \mu_2(A), \ldots),$$
$$(1 - x_1, 1 - x_2, \ldots) = (\mu_1(B), \mu_2(B), \ldots).$$

11. RANGES OF CHARGES   267

Note that $\mu_0(A \cap B) = 1$ and $(\mu_1(A \cap B), \mu_2(A \cap B), \ldots) = (0, 0, \ldots)$. This is a contradiction to the fact that μ_0 is an accumulation point of $\{\mu_n, n \geq 1\}$. Therefore, $\psi(D) \cap D = \emptyset$. To show that $\psi(D) \cup D = C$, let $(x_1, x_2, \ldots) \in C$. Let $E = \{n \geq 1; x_n = 1\}$ and $A = \bigcup_{n \in E} A_n$. Then $(\mu_1(A), \mu_2(A), \ldots) = (x_1, x_2, \ldots)$. If $\mu_0(A) = 1$, then $(x_1, x_2, \ldots) \in D$. If $\mu_0(A) = 0$, then $(x_1, x_2, \ldots) \in \psi(D)$. This shows that $\psi(D) \cup D = C$. Note that ψ preserves the measure τ, i.e. $\tau(\psi(G)) = \tau(G)$ for every τ-measurable set G contained in C. Now, if $\tau(D) = 1$, then $\tau(\psi(D)) = 1$ and consequently, $\tau(C) = 2$ which is a contradiction. If $\tau(D) = 0$, then $\tau(\psi(D)) = 0$ which is again a contradiction since it works out that $\tau(C) = 0$. Thus D is not τ-measurable.

To prove that D does not have the property of Baire, one can repeat the above argument and use Oxtoby's category analogue of Kolmogorov's 0–1 law and Baire Category theorem. □

11.5.3 Remark. If $\mathfrak{A}$ is an infinite σ-field on a set Ω, one can always find a sequence μ_n, $n \geq 0$ of 0–1 valued charges on $\mathfrak{A}$ satisfying the conditions imposed in Theorem 11.5.2.

CHAPTER 12

On Lifting

The purpose of this chapter is to present some ideas on lifting in the setting of charges on fields. In the following, we elucidate the concept of lifting.

Let $\mathscr{F}$ be a field of subsets of a set Ω and $\mathscr{I}$ an ideal in $\mathscr{F}$. For the following definition, recall the definition of the quotient Boolean algebra $\mathscr{F}/\mathscr{I}$ and the natural homomorphism h from $\mathscr{F}$ to $\mathscr{F}/\mathscr{I}$ which takes each set A in $\mathscr{F}$ to its equivalence class [A] in $\mathscr{F}$. (See Section 1.4.)

12.1 Definition. The natural homomorphism h from $\mathscr{F}$ to $\mathscr{F}/\mathscr{I}$ is said to admit a *lifting* if there exists a subfield $\mathscr{F}_0$ of $\mathscr{F}$ such that the map h restricted to $\mathscr{F}_0$ is an isomorphism from $\mathscr{F}_0$ to $\mathscr{F}/\mathscr{I}$.

A lifting is tantamount to selecting one set from each equivalence class of $\mathscr{F}$ so that the resulting collection of sets becomes a field on Ω and is isomorphic to $\mathscr{F}/\mathscr{I}$. This notion of lifting can be reformulated as follows. Suppose the natural homomorphism h from $\mathscr{F}$ to $\mathscr{F}/\mathscr{I}$ admits a lifting. Let ψ be the induced isomorphism from $\mathscr{F}/\mathscr{I}$ to $\mathscr{F}_0$.

$$\mathscr{F} \xrightarrow{h} \mathscr{F}/\mathscr{I} \xrightarrow{\psi} \mathscr{F}_0.$$

Let ρ be the composition of h and ψ, i.e., $\rho = \psi \circ h$. Then this map ρ has the following properties.

(i). ρ is a homomophism from $\mathscr{F}$ to $\mathscr{F}$.
(ii). ρ is onto $\mathscr{F}_0$.
(iii). If $A, B \in \mathscr{F}$ and $A \sim B$, then $\rho(A) = \rho(B)$.
(iv). $A \sim \rho(A)$ for every A in $\mathscr{F}$.

For ease of expression, we call ρ the lifting of $\mathscr{F}$ with respect to the ideal $\mathscr{I}$.

If $\Omega = \{1, 2, 3, \ldots\}$, $\mathscr{F} = \mathscr{P}(\Omega)$ and $\mathscr{I}$ the ideal of all finite subsets of Ω, it is not difficult to see that $\mathscr{F}$ does not admit a lifting with respect to $\mathscr{I}$.

A well-known result of von Neumann and Maharam establishes the existence of a lifting in the case when $(\Omega, \mathscr{F}, \mu)$ is a complete measure space, i.e. $\mathscr{F}$ is a σ-field of subsets of Ω, μ is a positive bounded measure on $\mathscr{F}$ with the property that $B \in \mathscr{F}$ whenever $B \subset A$, $A \in \mathscr{F}$ and $\mu(A) = 0$, and $\mathscr{I} = \{A \in \mathscr{F}; \mu(A) = 0\}$ the ideal of all sets in $\mathscr{F}$ with μ-measure zero. See Maharam (1958).

It is natural to enquire about the existence of a lifting of $\mathscr{F}$ with respect to $\mathscr{I}$ in the above if we assume μ to be a charge only. But lifting fails to exist in the generality mentioned above and the following theorem due to Maharam-Erdös amplifies this point. (Recall the notion of a density charge from Example 2.1.3(10).)

12.2 Theorem. Let $\Omega = \{1, 2, 3, \ldots\}$, $\mathscr{F} = \mathscr{P}(\Omega)$ and μ any density charge on $\mathscr{F}$. Let $\mathscr{I} = \{A \in \mathscr{F}; \mu(A) = 0\}$. Then there is no lifting of $\mathscr{F}$ with respect to the ideal $\mathscr{I}$.

Proof. The proof is carried out in the following steps.
1°. Suppose there exists a lifting ρ of $\mathscr{F}$ with respect to $\mathscr{I}$.
2°. Let $1 < p_1 < p_2 < \cdots$ be a sequence of positive integers such that p_i and p_j are mutually prime for every $i \neq j$ and $\sum_{i \geq 1} 1/p_i < \infty$. It follows that this sequence has the following properties. (i) $\lim_{n \to \infty} n/p_n = 0$. (ii) $\prod_{i \geq 1} (1 - 1/p_i)$ converges (to a non-zero real number).
3°. For $1 \leq j \leq p_i$ and $i = 1, 2, 3, \ldots$, let

$$A_{ij} = \{j, j + p_i, j + 2p_i, \ldots\}.$$

Note that, for each $i \geq 1$, $A_{i1}, A_{i2}, \ldots, A_{ip_i}$ are pairwise disjoint sets with union Ω. So, $\rho(A_{i1}), \rho(A_{i2}), \ldots, \rho(A_{ip_i})$ are pairwise disjoint sets with union Ω. Let R_i be the set among $\{\rho(A_{i1}), \rho(A_{i2}), \ldots, \rho(A_{ip_i})\}$ containing i and denote the corresponding set A_{ij} by S_i for each $i = 1, 2, 3, \ldots$.
4°. Obviously, $\bigcup_{i \geq 1} \rho(S_i) = \bigcup_{i \geq 1} R_i = \Omega$. Since $\rho(\bigcup_{i \geq 1} S_i)$ contains S_j for every $j \geq 1$, it follows that $\rho(\bigcup_{i \geq 1} S_i) = \Omega$.
5°. Let T_i be the set obtained from S_i by removing the first element of S_i, $i \geq 1$. Let $T^* = \bigcup_{i \geq 1} T_i$. Since $\rho(T^*) \supset \rho(T_j) = \rho(S_j)$ for every $j \geq 1$, it follows that $\rho(T^*) = \Omega$. This implies, by the definition of lifting, that $T^* \sim \bigcup_{i \geq 1} S_i$, i.e. $\mu(T^*) = 1$. We show that

$$\mu(T^*) = 1 - \prod_{i \geq 1} \left(1 - \frac{1}{p_i}\right) \quad (<1)$$

which will then lead to a contradiction.
6°. We use the notation $\#A$ for the number of elements in A. Let $\varepsilon > 0$. Let r be a positive integer such that

$$\left|\prod_{i \geq 1} \left(1 - \frac{1}{p_i}\right) - \prod_{i=1}^{r} \left(1 - \frac{1}{p_i}\right)\right| < \varepsilon/3,$$

$$\sum_{i > r} \frac{1}{p_i} < \varepsilon/6 \quad \text{and} \quad \frac{i}{p_i} < \varepsilon/6 \quad \text{for every } i \geq r.$$

Let k_0 be a positive integer such that

$$k_0 > p_1 p_2 \cdots p_r \quad \text{and} \quad \frac{2^r}{k_0} < \varepsilon/6.$$

We show that for any integer $k > k_0$,

$$\left| \frac{\#T^* \cap \{1, 2, \ldots, k\}}{k} - \left[1 - \prod_{i \geq 1} \left(1 - \frac{1}{p_i}\right)\right] \right| < \varepsilon.$$

(Since μ is a density charge on $\mathscr{F}$, we will then have

$$\mu(T^*) = \lim_{k \to \infty} \frac{\#T^* \cap \{1, 2, \ldots, k\}}{k}$$

$$= 1 - \prod_{i \geq 1} \left(1 - \frac{1}{p_i}\right).)$$

7°. Let $k > k_0$ be fixed. Let s be the smallest integer i such that $p_i > k$, i.e. $p_{s-1} \leq k < p_s$. Obviously, $s > r$. Note that every element in T_j is greater than p_j. If $j \geq s$, then $p_j \geq p_s$. Consequently, $T_j \cap \{1, 2, \ldots, k\} = \varnothing$ if $j \geq s$. It suffices to estimate $\#(\bigcup_{i=1}^{s-1} T_i) \cap \{1, 2, \ldots, k\}$.

8°. First, we estimate $\#(\bigcup_{i=1}^{r} T_i) \cap \{1, 2, \ldots, k\}$. Here, we introduce the following notation; for any two numbers a and b, $a \simeq b$ means $|a - b| \leq 1$. By the inclusion–exclusion principle,

$$\#\left(\bigcup_{i=1}^{r} T_i\right) \cap \{1, 2, \ldots, k\} = \sum_{i=1}^{r} \#T_i \cap \{1, 2, \ldots, k\}$$

$$- \sum_{\substack{i=1 \\ i<j}}^{r} \sum_{j=1}^{r} \#T_i \cap T_j \cap \{1, 2, \ldots, k\}$$

$$+ \sum_{\substack{i=1 \\ i \leq j < u}}^{r} \sum_{j=1}^{r} \sum_{u=1}^{r} \#T_i \cap T_j \cap T_u \cap \{1, 2, \ldots, k\} - \cdots$$

$$+ (-1)^{r-1} \#T_1 \cap T_2 \cap \cdots \cap T_r \cap \{1, 2, \ldots, k\}.$$

Since T_i is of the form $\{r_i + p_i, r_i + 2p_i, \ldots\}$ for some $1 \leq r_i < p_i$, it follows that

$$T_i \cap \{1, 2, \ldots, k\} \simeq \frac{k}{p_i}$$

for every $i \geq 1$. We now show that, for $i \neq j$,

$$T_i \cap T_j \subset \{r_{ij} + p_i p_j, r_{ij} + 2p_i p_j, r_{ij} + 3p_i p_j, \ldots\}$$

for some $1 \leq r_{ij} < p_i p_j$. For this, it suffices to show that if $r_i + up_i = r_j + u'p_j$ and $r_i + mp_i = r_j + m'p_j$, then the difference $(r_i + up_i) - (r_i + mp_i)$ is divisible by $p_i p_j$. But this follows from the fact that p_i and p_j are mutually prime. Consequently,

$$\#T_i \cap T_j \cap \{1, 2, \ldots, k\} \simeq \frac{k}{p_i p_j}.$$

12. ON LIFTING

Pursuing this argument, we obtain

$$\left|\#\left(\bigcup_{i=1}^{r}T_i\right)\cap\{1,2,\ldots,k\}-\sum_{i=1}^{r}\frac{k}{p_i}+\sum_{\substack{i=1\\i<j}}^{r}\sum_{j=1}^{r}\frac{k}{p_ip_j}\right.$$

$$\left.-\sum_{\substack{i=1\\i<j<u}}^{r}\sum_{j=1}^{r}\sum_{u=1}^{r}\frac{k}{p_ip_jp_u}+\cdots+(-1)^r\frac{k}{p_1p_2\cdots p_r}\right|$$

$$=\left|\#\left(\bigcup_{i=1}^{r}T_i\right)\cap\{1,2,\ldots,k\}-k\left(1-\prod_{i=1}^{r}\left(1-\frac{1}{p_i}\right)\right)\right|$$

$$\leq r+\binom{r}{2}+\cdots+\binom{r}{r}=2^r-1<2^r.$$

9°. Using the same argument as above, we note that

$$\#T_j\cap\{1,2,\ldots,k\}\leq 1+\frac{k}{p_j}$$

for every $r+1\leq j\leq s-1$.

10°. Thus, finally, we have

$$\left|\frac{\#T^*\cap\{1,2,\ldots,k\}}{k}-\left(1-\prod_{i\geq 1}\left(1-\frac{1}{p_i}\right)\right)\right|$$

$$=\left|\frac{\#\left(\bigcup_{i=1}^{s-1}T_i\right)\cap\{1,2,\ldots,k\}}{k}-\left(1-\prod_{i\geq 1}\left(1-\frac{1}{p_i}\right)\right)\right|$$

$$=\left|\left[\frac{\#\left(\bigcup_{i=1}^{r}T_i\right)\cap\{1,2,\ldots,k\}}{k}-\left(1-\prod_{i=1}^{r}\left(1-\frac{1}{p_i}\right)\right)\right]\right.$$

$$+\left[\left(1-\prod_{i=1}^{r}\left(1-\frac{1}{p_i}\right)\right)-\left(1-\prod_{i\geq 1}\left(1-\frac{1}{p_i}\right)\right)\right]$$

$$\left.+\frac{\#\left(\bigcup_{i=r+1}^{s-1}T_i\right)\cap\{1,2,\ldots,k\}}{k}\right|$$

$$<\frac{2^r}{k}+\varepsilon/3+\left(\frac{s-1}{k}+\sum_{i>r}\frac{1}{p_i}\right),$$

by 8°, 6° and 9° respectively.

$$<\varepsilon/6+\varepsilon/3+\frac{s-1}{p_{s-1}}+\varepsilon/6$$

$$<\varepsilon/6+\varepsilon/3+\varepsilon/6+\varepsilon/6<\varepsilon.$$

This completes the proof. □

APPENDIX 1

Notes and Comments

CHAPTER 1

Sections 1.2, 1.3 and 1.4 offer fairly standard, though cursory, treatment of the topics covered. For set theoretical notions dealt in Section 1.2, one could refer to Kamke (1950) or Kelley (1955) or Dunford and Schwartz (1954) for more details. Kelley (1955) is a standard reference for topological concepts covered in Section 1.3. Halmos (1963) and Sikorski (1969) are excellent sources for ideas on Boolean Algebras. Some parts of Section 1.1 may be found in Halmos (1950). Theorem 1.1.9(1) is due to Pettis (1951). Corollary 1.1.12 is inspired by Section 4 of Sikorski (1969). A substantial part of Section 1.5 on Functional Analytic concepts is inspired by the fundamental paper of Bochner and Phillips (1941).

CHAPTER 2

Bochner and Phillips (1941) identify the space ba(Ω, $\mathscr{F}$) of all bounded charges on the field $\mathscr{F}$ of subsets of the set Ω with the space ca(Ω', $\mathscr{F}'$) of all bounded measures on a suitable σ-field $\mathscr{F}'$ of subsets of a suitable set Ω'. From this it follows that ba(Ω, $\mathscr{F}$) is a boundedly complete vector lattice. Theorem 2.2.1 arrives at the same conclusion more directly using a bank of ideas from vector lattices.

Construction of invariant charges using Banach limits is a standard practice in Ergodic Theory. See Example 2.1.3(8). We have used Banach limits to show the existence of Density charges. See Example 2.1.3(10).

The comprehensive Jordan Decomposition Theorem proved in Section 2.5 is new.

Existence of ε-Hahn Decomposition for charges for every $\varepsilon > 0$ was first established by Darst (1962a).

Topsøe (1979) gave some conditions under which a charge becomes a measure. This result is a generalization of Theorem 2.3.4. First, we need a definition.

Definition. A collection $\mathscr{C}$ of subsets of a set Ω is called a *monocompact class* if it has the following property: if C_n, $n \geq 1$ is a decreasing sequence of sets in $\mathscr{C}$ with $\bigcap_{n \geq 1} C_n = \varnothing$, then there exists $m \geq 1$ such that $C_m = \varnothing$.

Theorem A.1. *Let $\mathscr{F}$ be a field of subsets of a set Ω and $\mathscr{C}$ a monocompact class of subsets of Ω. Let μ be a positive bounded charge on $\mathscr{F}$ having the following approximation property: for any F in $\mathscr{F}$ and $\varepsilon > 0$ there exist C in $\mathscr{C}$ and G in $\mathscr{F}$ such that $G \subset C \subset F$ and $\mu(F - G) < \varepsilon$. Then μ is a measure on $\mathscr{F}$.*

Christensen (1971) gave some conditions under which a charge becomes a measure. We present some of his results.

Let $\mathscr{F}$ be a σ-field of subsets of a set Ω. For ν in $\text{ca}(\Omega, \mathscr{F})$, let h_ν be the map on $\mathscr{F}$ defined by

$$h_\nu(F) = \nu(F), \qquad F \in \mathscr{F}.$$

Let $\mathscr{C}$ be the smallest σ-field on $\mathscr{F}$ with respect to which each of the maps h_ν is measurable for ν in $\text{ca}(\Omega, \mathscr{F})$.

Theorem A.2. *Let μ be a real charge on $\mathscr{F}$. If μ is measurable with respect to the σ-field $\mathscr{C}$ on $\mathscr{F}$, i.e.*

$$h_\mu^{-1}(B) = \{F \in \mathscr{F}; \mu(F) \in B\} \in \mathscr{C}$$

for every Borel subset B of the real line R, then μ is a measure on $\mathscr{F}$.

Another result in the context of Polish spaces, i.e. complete separable metric spaces, can be described as follows. Let Ω be a Polish space and $\mathscr{F}$ its Borel σ-field, i.e. the smallest σ-field on Ω containing all open subsets of Ω. Let μ be a probability charge on $\mathscr{F}$. Let Ω^* be the collection of all closed subsets of Ω. The gist of the following result is that if μ restricted to Ω^* is a *decent* function on Ω^*, then μ is a measure on $\mathscr{F}$. We now elaborate this statement. We can introduce a suitable metric d^* on Ω^* so that (Ω^*, d^*) becomes a separable metric space. Let d be a metric on Ω compatible with its topology. Since Ω is separable, one can always choose d to be a precompact metric on Ω. Define a metric d^* on Ω^* by

$$d^*(A, B) = \sup_{\substack{a \in A \\ b \in B}} \{\max\{d(a, B), d(A, b)\}\}, \qquad A, B \in \Omega^*.$$

Then (Ω^*, d^*) is a separable metric space. Let $\mathscr{F}^*$ be the Borel σ-field on Ω^*.

Theorem A.3. *If the map μ restricted to Ω^* is measurable with respect to $\mathscr{F}^*$, then μ is a measure on $\mathscr{F}$.*

Rao (1971) gave a sufficient condition under which a charge becomes a measure. Let μ be a positive real charge defined on a field $\mathscr{F}$ of subsets of a set Ω. A subfield $\mathscr{F}_0$ of $\mathscr{F}$ is said to be μ-pure if the following conditions are met.

(i). $\mu(A_N) = 0$ for some $N \geq 1$ whenever A_n, $n \geq 1$ is a sequence in $\mathscr{F}_0$ satisfying $A_1 \supset A_2 \supset A_3 \supset \cdots$ and $\bigcap_{n \geq 1} A_n = \varnothing$.

(ii) $\mu(A) = \text{Inf}\{\sum_{n \geq 1} \mu(A_n); \{A_n, n \geq 1\} \subset \mathscr{F}_0 \text{ and } \bigcup_{n \geq 1} A_n \supset A\}$ for every A in $\mathscr{F}$. (This condition means that the Caratheodory measure induced by μ on $\mathscr{F}_0$ coincides with μ on $\mathscr{F}$.)

Theorem A.4. *If there exists a μ-pure subfield $\mathscr{F}_0$ of $\mathscr{F}$, then μ is a measure on $\mathscr{F}$.*

Rao (1971) stated that the converse of the above theorem is true. This is not correct, however, as the following discussion demonstrates.

Theorem A.5. *Let $\mathscr{F}$ be a σ-field of subsets of a set Ω and μ a nonatomic probability measure on $\mathscr{F}$. Let $\mathscr{F}_0$ be a μ-pure subfield of $\mathscr{F}$ and $\mathscr{F}_1$ the smallest σ-field on Ω containing $\mathscr{F}_0$. Then μ is nonatomic on $\mathscr{F}_1$. (For the definition of a nonatomic measure, see Chapter 5.)*

Proof. Obviously, $\mathscr{F}_1 \subset \mathscr{F}$. We show that $\mathscr{F}$ and $\mathscr{F}_1$ are μ-equivalent, i.e. given A in $\mathscr{F}$ there exists B in $\mathscr{F}_1$ such that $\mu(A \triangle B) = 0$. To begin with, given $\varepsilon > 0$ we show that there exists a set B_ε in $\mathscr{F}_1$ such that $\mu(A \triangle B) < \varepsilon$. Since $\mathscr{F}_0$ is a μ-pure subfield of $\mathscr{F}$, there exists a sequence E_n, $n \geq 1$ in $\mathscr{F}_0$ such that $\bigcup_{n \geq 1} E_n \supset A$ and $\sum_{n \geq 1} \mu(E_n) < \mu(A) + \varepsilon$. Clearly, $\bigcup_{n \geq 1} E_n \in \mathscr{F}_1$. Take $B_\varepsilon = \bigcup_{n \geq 1} E_n$. Thus for each $n \geq 1$, we can find B_n in $\mathscr{F}_1$ such that $\mu(A \triangle B_n) < 1/2^n$. Take $B = \lim \sup_{n \to \infty} B_n$. Since $A \triangle (\lim \sup_{n \to \infty} B_n) \subset \lim \sup_{n \to \infty} (A \triangle B_n)$ and $\mu(\lim \sup_{n \to \infty} (A \triangle B_n)) = 0$ (Borel–Cantelli Lemma), it follows that $\mu(A \triangle B) = 0$. Finally, since $\mathscr{F}$ and $\mathscr{F}_1$ are μ-equivalent, μ is nonatomic on $\mathscr{F}_1$. $\square$

Theorem A.6. *Let $\mathscr{F}$ be a σ-field of subsets of a set Ω and μ a nonatomic probability measure on $\mathscr{F}$. Let $\mathscr{F}_0$ be a μ-pure subfield of $\mathscr{F}$. Then μ is strongly continuous on $\mathscr{F}_0$.*

Proof. This is a consequence of Theorem A5 above and Proposition 5.3.7.

Theorem A.7. *Let $\mathscr{F}$ be a σ-field of subsets of a set Ω and μ a nonatomic probability measure on $\mathscr{F}$. Let $\mathscr{F}_0$ be a μ-pure subfield of $\mathscr{F}$. Then there exists a set A in $\mathscr{F}$ of cardinality greater than or equal to the cardinality of the continuum c such that $\mu(A) = 0$.*

Proof. By Theorem A6, μ is a strongly continuous probability charge on $\mathscr{F}_0$. So, there exist two sets B_0 and B_1 in $\mathscr{F}_0$ such that $B_0 \cap B_1 = \varnothing$, $0 < \mu(B_0) < 1/1(2)$ and $0 < \mu(B_1) < 1/1(2)$. By a similar reasoning, we can find sets $B_{00}, B_{01}, B_{10}, B_{11}$ in $\mathscr{F}_0$ such that $B_{00} \cap B_{01} = \varnothing$, $B_{00} \cup B_{01} \subset B_0$, $0 <$

$\mu(B_{00}) < 1/2(2^2)$, $0 < \mu(B_{01}) < 1/2(2^2)$, $B_{10} \cap B_{11} = \emptyset$, $B_{10} \cup B_{11} \subset B_1$, $0 < \mu(B_{10}) < 1/2(2^2)$ and $0 < \mu(B_{11}) < 1/2(2^2)$. Continuing this process, we obtain a family $\{B_{i_1,i_2,...,i_n}; i_1, i_2, ..., i_n = 0 \text{ or } 1, n \geq 1\}$ of sets in $\mathscr{F}_0$ with the following properties.

(i) $B_{i_1,i_2,...,i_n} \cap B_{i_1,i_2,...,i_{n-1},1-i_n} = \emptyset$,
(ii) $B_{i_1,i_2,...,i_{n+1}} \subset B_{i_1,i_2,...,i_n}$ and
(iii) $0 < \mu(B_{i_1,i_2,...,i_n}) < 1/n(2^n)$

for all $i_1, i_2, ..., i_n, i_{n+1}$ in $\{0, 1\}$ and $n \geq 1$.

For each $n \geq 1$, let $A_n = \bigcup B_{i_1,i_2,...,i_n}$, where the union is taken over all $i_1, i_2, ..., i_n$ in $\{0, 1\}$. It is obvious that $\mu(A_n) < 1/n$ for every $n \geq 1$. Let $A = \bigcap_{n \geq 1} A_n$. It now follows that $\mu(A) = 0$. Since $\mathscr{F}_0$ is a μ-pure subfield of $\mathscr{F}$, it follows that $\bigcap_{n \geq 1} B_{i_1,i_2,...,i_n} \neq \emptyset$ for any sequence $i_1, i_2, ...$ of 0's and 1's. (Actually, here, we use the fact that $\mathscr{F}_0$ satisfies (i) on p. 274 and that $B_{i_1,i_2,...,i_n}$'s satisfy (iii) above.) Consequently, the cardinality of $A \geq c$.

Sierpiński showed with the aid of continuum hypothesis the existence of a set Ω, a σ-field $\mathscr{F}$ on Ω and a nonatomic probability measure μ on $\mathscr{F}$ such that $\mu(A) = 0$ if and only if A is at most countable. In this case, in view of Theorem A7, there is no μ-pure subfield $\mathscr{F}_0$ of $\mathscr{F}$. For a discussion on the above type of measure, see Marczewski (1953, 7(iv), p. 123). In fact, continuum hypothesis is not needed at all for an example. Any non-compact measure on a countably generated σ-field would serve the purpose. See Bhaskara Rao, Bhaskara Rao and Rao (1982) and Frolik and Pachl (1973).

Theorems A.1, A.2, A.3 and A.4 stated above are analogous to Proposition 2.3.2 and Theorem 2.3.4 in spirit.

CHAPTER 3

The results of Sections 3.1 and 3.2 are due to Tarski (1938) and Horn and Tarski (1948). Our treatment is slightly different from the one given in Horn and Tarski (1948). The results of Section 3.3 are due to Los and Marczewski (1949). Section 3.4 contains some ideas of Los and Marczewski (1949) in its development. Some of the results of Section 3.5 are due to Pettis (1951). For the results of Section 3.6, see Guy and Maharam (1972).

Extending some of the results of this chapter to group-valued charges, Bhaskara Rao and Aversa (1982) have proved the following theorem.

Theorem A.8. *Let G be an algebraically compact abelian group. If μ is any G-valued charge defined on a subfield $\mathscr{C}$ of a field $\mathscr{F}$ of subsets of a set Ω, then there is a G-valued charge $\bar{\mu}$ on $\mathscr{F}$ which extends μ.*

Carlson and Prikry (1982) have proved that the above result is true for all groups using a result on Specker groups. For Specker groups see Fuchs (1970).

CHAPTER 4

The notion of a.e. $[\mu]$ introduced in Definition 4.2.4 is slightly different from the standard one adopted in Measure Theory. For example, we say $f \leq g$ a.e. $[\mu]$ if there exists a null function h such that $f \leq g + h$. The definition prevalent in Measure Theory is that $f \leq g$ a.e. $[\mu]$ if and only if $|\mu|^*(\{\omega \in \Omega; f(\omega) > g(\omega)\}) = 0$. Of course, if $|\mu|^*(\{\omega \in \Omega; f(\omega) > g(\omega)\}) = 0$, then $f \leq g$ a.e. $[\mu]$ according to Definition 4.2.4 though the converse is not true. If a D-integrable function f has the property that

$$D \int_E f \, d\mu \geq 0 \quad \text{for every E in } \mathscr{F},$$

then $f \geq 0$ a.e. $[\mu]$ in the sense of Definition 4.2.4 but $|\mu|^*(\{\omega \in \Omega; f(\omega) < 0\})$ may not be equal to zero.

In the definition of D-integral, we worked with simple functions only, whereas Dunford and Schwartz used μ-simple functions. See Definition III.2.17 of Dunford and Schwartz (1954, p. 112). There is virtually no difference between the two approaches. Chapter 4 sets out D-integral and its ramifications in greater detail than that covered by Dunford and Schwartz.

In connection with Section 4.6 on L_p-spaces, it is worth mentioning the following facts. If $\mathscr{F}$ is a σ-field of subsets of a set Ω and μ a bounded measure on $\mathscr{F}$, then $L_p(\Omega, \mathscr{F}, \mu)$ equipped with the pseudo-norm $\|\cdot\|_p$ is complete for every $1 \leq p \leq \infty$. This result is not true if either $\mathscr{F}$ is a field only or μ a charge only on $\mathscr{F}$. However, if μ is a 0–1 valued charge on $\mathscr{F}$, then $L_p(\Omega, \mathscr{F}, \mu)$ is complete. In general, necessary and sufficient conditions for the completeness of $L_p(\Omega, \mathscr{F}, \mu)$ are not known. Green (1970/71) made an attempt in this direction but his results are not complete. See Notes and Comments in Chapter 7.

The last section is based on the papers by Yosida and Hewitt (1952), Leader (1955) and Darst (1961).

Some ideas of Gould (1965) were used in the development of Section 4.5.

Finally, we end this section with an interesting example.

Example. Let $\Omega = \{1, 2, 3, \ldots, \infty\}$, $\mathscr{F} = \{A \subset \Omega; A \text{ or } A^c \text{ is a finite subset of } \{1, 2, 3, \ldots\}\}$ and μ on $\mathscr{F}$ be defined by

$$\mu(A) = 0 \quad \text{if } A \in \mathscr{F}, A \text{ is finite},$$
$$= 1 \quad \text{otherwise}.$$

Let $\mathfrak{A}$ be the smallest σ-field on Ω containing $\mathscr{F}$. Indeed, $\mathfrak{A} = \mathscr{P}(\Omega)$. Let $\tilde{\mu}$ be the extension of μ from $\mathscr{F}$ to $\mathfrak{A}$ as a measure. In fact, for $A \subset \Omega$,

$$\tilde{\mu}(A) = 1 \quad \text{if } \infty \in A,$$
$$= 0 \quad \text{if } \infty \notin A.$$

It can be observed that $L_1(\Omega, \mathscr{F}, \mu)$ and $L_1(\Omega, \mathfrak{A}, \tilde{\mu})$ are both complete but $L_1(\Omega, \mathscr{F}, \mu) \neq L_1(\Omega, \mathfrak{A}, \tilde{\mu})$. For, obviously, $I_{\{\infty\}} \in L_1(\Omega, \mathfrak{A}, \tilde{\mu})$ but $I_{\{\infty\}} \notin L_1(\Omega, \mathscr{F}, \mu)$. (Observe that $I_{\{\infty\}}$ is not T_2-measurable in the framework of $(\Omega, \mathscr{F}, \mu)$.)

CHAPTER 5

The main decomposition theorem presented in Section 5.2 is due to Sobczyk and Hammer (1944a). Other results in this chapter are due to the authors (1973) and (1978). It is possible to derive Theorem 5.2.7 using Riesz Decomposition Theorem on vector lattices. See Section 1.5.

CHAPTER 6

The treatment of absolute continuity, singularity and Lebesgue Decomposition Theorem given here follows closely that of Bochner and Phillips (1941). Darst (1962a), (1962b) and (1963) has worked extensively on extensions of Lebesgue Decomposition Theorem in the unbounded case. The following are some of his results.

Theorem A.9. *Let μ and ν be two charges on a field $\mathscr{F}$ of subsets of a set Ω such that one of μ and ν is bounded. Then ν admits a Lebesgue Decomposition with respect to μ if and only if there exists a sequence E_n, $n \geq 1$ of decreasing sets in $\mathscr{F}$ such that $\lim_{n \to \infty} |\mu|(E_n) = 0$ and $\lim_{n \to \infty} |\nu|(E_n^c)$ is finite.*

A special case of this theorem is the following result.

Theorem A.10. *Let μ and ν be two charges on a field $\mathscr{F}$ of subsets of a set Ω such that ν is bounded. Then ν admits a Lebesgue Decomposition with respect to μ.*

The Radon–Nikodym Theorem presented here was originally formulated and proved by Bochner (1939). Later, it was proved again by various people at various times: Bochner and Phillips (1941), Dunford and Schwartz (1964), Fefferman (1967), Darst and Green (1968), Dubins (1969), Darst (1970b) and Pachl (1972). The proof presented here is due to Pachl (1972). This proof, even though it looks lengthy, is elementary and in our opinion, the simplest. Another proof of Radon–Nikodym theorem appears in Chapter 7 as a simple consequence of a result on V_1 spaces. This proof is due to Bochner and Phillips (1941). Maynard (1979) gave necessary and sufficient conditions for the existence of exact Radon–Nikodym derivatives.

Dunford and Schwartz (1964) use Stone Representation Theorem for Boolean algebras in their proof of Radon–Nikodym theorem.

CHAPTER 7

V_p-spaces were introduced by Bochner (1939). Leader (1953) studied these spaces in great detail. Our presentation follows closely that of Leader (1953) with simplifications. Remark 7.5.2 follows from Fefferman (1968) and Bhaskara Rao and Halevy (1977). In the later paper, the results on V_p-spaces were obtained using Stone Representation Theorem on Boolean Algebras.

Green (1970/71) gave necessary and sufficient conditions for the completeness, in particular, of $\mathcal{L}_1(\Omega, \mathcal{F}, \mu)$ for a given charge space $(\Omega, \mathcal{F}, \mu)$. Let $(\Omega, \mathcal{F}, \mu)$ be a probability charge space. For A, B in $\mathcal{F}$, say $A \sim B$ if $\mu(A \Delta B) = 0$. $\sim$ is an equivalence relation. Let $\mathcal{F}^*$ be the collection of all equivalence classes of $\mathcal{F}$. $\mathcal{F}^*$ is a Boolean Algebra. Let Ω' be the Stone space of $\mathcal{F}^*$, $\mathcal{F}'$ the class of all clopen subsets of Ω' and $\mathcal{B}'$ the Borel σ-field on Ω'. There is a natural probability measure μ' on $\mathcal{B}'$ which corresponds to the probability charge μ on $\mathcal{F}$ via the correspondences

$$\mathcal{F} \to \mathcal{F}^* \leftrightarrow \mathcal{F}' \leftrightarrow \mathcal{B}'.$$

Green's necessary and sufficient conditions for the completeness of $\mathcal{L}_1(\Omega, \mathcal{F}, \mu)$ are that Ω' be extremely disconnected (i.e. the closure of every open subset of Ω' is open) and every open subset of Ω' is equivalent to its closure under μ'. These conditions are not correct. An example can be constructed from the following theorem of Bhaskara Rao and Aversa (1982).

Theorem A.11. *Let μ be a positive bounded measure on a field $\mathcal{F}$ of subsets of a set Ω. Let $\mathfrak{A}$ be the smallest σ-field on Ω containing $\mathcal{F}$ and $\bar{\mu}$ the positive bounded measure on $\mathfrak{A}$ which extends μ. Then*

$$\mathcal{L}_1(\Omega, \mathcal{F}, \mu) = \mathcal{L}_1(\Omega, \mathfrak{A}, \bar{\mu})$$

if for any $\varepsilon > 0$ and for any A in $\mathfrak{A}$ there exist B and C in $\mathcal{F}$ such that $B \subset A \subset C$ and $\mu(C - B) < \varepsilon$.

This theorem gives the following two examples.

Example 1. Let Ω be any set and $\mathcal{F}$ be the finite-cofinite field on Ω. Then for any positive bounded measure μ on $\mathcal{F}$, $\mathcal{L}_1(\Omega, \mathcal{F}, \mu)$ is complete.

Example 2. Let $\mathscr{F}$ be the field generated by all the open subsets of the real line R. Then for any positive bounded measure μ on $\mathscr{F}$, $\mathscr{L}_1(R, \mathscr{F}, \mu)$ is complete.

CHAPTER 8

With the exceptions of Sections 8.5 and 8.6, the treatment presented here follows that of Phillips (1940a), Darst (1966), Seever (1968), Porcelli (1960), Brooks and Jewett (1970) and Leader (1953). In Sections 8.5 and 8.6, we have included the results of Bell (1979). The proof of Phillips' Lemma presented here is essentially the original proof of Phillips (1940a). The results on Nikodym theorem and Vitali–Hahn–Saks theorem use essentially the ideas of Seever (1968). We now make some observations on Theorem 8.7.3 which gives equivalent conditions for weak convergence. Condition (ii) is due to Hildebrandt (1934). Condition (iii) is due to Porcelli (1960) The proof of (iv) $\Rightarrow$ (v) is due to Darst (1966). Condition (v) is due to Leader (1953). The results on weak convergence in $ba(\Omega, \mathscr{F})$ are scatteed in the literature and we made an attempt to bring some semblance of order by unifying them in our Theorem 8.7.3. Some equivalent conditions in Theorem 8.7.3 and quite a few results in this chapter are new.

CHAPTER 9

The basic ideas in the development of refinement integrals are due to Kolmogoroff (1930).

The main result (Theorem 9.2.1) in Section 9.2 is due to Keisler (1979). Our proof of this result is slightly different from the one presented by Keisler (1979) in that, we avoid using a deep result in refinement integrals due to Kolmogoroff.

CHAPTER 10

The decomposition theorem presented in Section 10.2 is due to Yosida and Hewitt (1952) (and Kakutani, as Yosida and Hewitt mention). Their original proof is a working-out of the proof of Riesz Decomposition

Theorem for the vector lattice ba(Ω, $\mathscr{F}$) and the normal sublattice ca(Ω, $\mathscr{F}$). Some alternative proofs, extensions and generalizations of this decomposition theorem can also be found in the literature. See Ranga Rao (1958), Plachky (1971), Traynor (1972) and Huff (1973). Pure charges on Boolean algebras are characterized by Lloyd (1963). See Theorem 10.5.3.

CHAPTER 11

A substantial part of these results is from a paper of K. P. S. Bhaskara Rao (1981). Theorem 11.2.5 is due to Sobczyk and Hammer (1944b, Theorem 3.3, p. 849) when μ is nonnegative. They gave an example of a bounded charge on a σ-field of subsets of a set Ω whose range is countably infinite, thus negating the validity of their Theorem 3.3 for general bounded charges on σ-fields. See Sobczyk and Hammer (1944b, Theorem 3.4, p. 850). This example is incorrect. Their Theorem 3.3 is true for any bounded charge. See Theorem 11.2.5. See also K. P. S. Bhaskara Rao (1981).

Here is an amusing example of a real charge whose range is the set of all rational numbers. Let $\Omega = \{1, 2, 3, \ldots\}$, $\mathscr{F}$ the finite-cofinite field on Ω and μ on $\mathscr{F}$ is defined by

$$\mu(A) = \sum_{n \in A} \frac{1}{n}, \qquad \text{if A is finite,}$$

$$= -\sum_{n \in A^c} \frac{1}{n}, \qquad \text{if A is cofinite, } A \subset \Omega.$$

The range $R(\mu)$ of μ is the set of all rational numbers. This follows from the Egyptian Fraction Theorem in Number Theory, that, every positive rational number can be written as $\sum_{i=1}^{k} 1/n_i$, where $n_1, n_2, \ldots, n_k$ are distinct positive integers.

We now give an example of a bounded charge μ taking positive and negative values such that 0 is not a two-sided accumulation point of $R(\mu)$. Let $\Omega = \{1, 2, 3, \ldots\}$, $\mathscr{F}$ the finite–cofinite field on Ω and μ on $\mathscr{F}$ is defined by

$$\mu(A) = -\sum_{n \in A} \frac{1}{2^n}, \qquad \text{if A is finite,}$$

$$= 1 - \mu(A^c), \qquad \text{if A is cofinite.}$$

μ is a bounded charge on $\mathscr{F}$ and 0 is not a two-sided accumulation point of $R(\mu)$.

Theorem 11.4.5 is due to Sobczyk and Hammer (1944a, Theorem 5.1, p. 843). It was also rediscovered by Maharam (1976, Theorem 2, p. 49).

According to Theorem 11.4.4, if μ_n, $n \geq 1$ is a discrete sequence of 0–1 valued charges on a σ-field $\mathfrak{A}$ of subsets of a set Ω, then the range $R(\mu)$ of $\mu = \sum_{n \geq 1} (1/2^n) \mu_n$ is compact. Erdös (see Maharam (1976) posed the problem of whether the range $R(\mu)$ of $\mu = \sum_{n \geq 1} (1/2^n) \mu_n$ is a non-Borel set when $\{\mu_1, \mu_2, \ldots\}$ is not a discrete set. The answer to this question is affirmative in some special cases as pointed out in Theorem 11.5.2. If the answer to the Erdös' problem is in the affirmative whenever $\{\mu_n : n \geq 1\}$ is a dense-in-itself subset of the Stone space Y of $\mathfrak{A}$, then the answer for the general case is also in the affirmative.

In connection with the above problem, we establish the following assertions of Maharam (1976).

Let μ_n, $n \geq 1$ be a sequence of 0–1 valued charges on a σ-field $\mathfrak{A}$. Assume that $\{\mu_n : n \geq 1\}$ is a dense-in-itself subset of the Stone space Y of $\mathfrak{A}$.

(a) If $\{(\mu_1(A), \mu_2(A), \ldots): A \in \mathfrak{A}\}$ is a τ-measurable subset of $\{0, 1\}^{\aleph_0}$, then it has τ-measure zero. (For the definition of τ see the proof of Theorem 11.5.2.)

(b) If $\{(\mu_1(A), \mu_2(A), \ldots): A \in \mathfrak{A}\}$ has the property of Baire, then it is of first category.

(a) and (b) can be interpreted as follows. Let $\mu = \sum_{n \geq 1} (1/2^n) \mu_n$. If $R(\mu)$ is Lebesgue measurable, then it has Lebesgue measure zero. If $R(\mu)$ has the property of Baire, then it is of first category.

(a) can be established as follows. The set $E = \{(\mu_1(A), \mu_2(A), \ldots): A \in \mathfrak{A}\}$ is a subgroup of the abelian group $C = \{0, 1\}^{\aleph_0}$ under coordinate addition modulo 2, and is disjoint with the set $\{(x_1, x_2, \ldots) \in C : x_i = 0$ for all but a finite number of i's$\}$. Now, if E is τ-measurable and $\tau(E) > 0$, then $F = E + E = \{(x_1 + y_1, x_2 + y_2, \ldots): (x_1, x_2, \ldots) \in E$ and $(y_1, y_2, \ldots) \in E\}$ should contain a point of $\{(x_1, x_2, \ldots) \in C : x_i = 0$ for all but a finite number of i's$\}$. See Oxtoby (1971) and Bhaskara Rao and Bhaskara Rao (1974). But $F = E$. This contradiction proves (a).

In Theorem 11.5.2, the proof that D is not τ-measurable is essentially due to Sierpinski (1938).

CHAPTER 12

The main result of this chapter is from Maharam (1976). See also Weissacker (1982) and Talagrand (1981) for further related results.

APPENDIX 2

Selected Annotated Bibliography

BOOKS

To begin with, we give a list of books we have consulted at one time or the other in our study of finitely additive measures.

1. BIRKHOFF, G. "Lattice Theory" American Mathematical Society Colloquium Publications, New York, 1948.
2. DUBINS, L. E. and SAVAGE, L. J. "How to Gamble If You Must (Inequalities for Stochastic Processes)". McGraw-Hill, London, 1965.
3. DUNFORD, N. and SCHWARTZ, J. T. "Linear Operators, Part I: General Theory". Wiley–Interscience, London, 1954.
4. FUCHS, L. "Infinite Abelian Groups," Vol. 1. Academic Press, London and New York, 1970.
5. HALMOS, P. R. "Measure Theory". Van Nostrand, London, 1950.
6. HALMOS, P. R. "Lectures on Boolean Algebras". Van Nostrand, London, 1963.
7. KAMKE, E. "Theory of Sets". Dover Publications, New York, 1950.
8. KELLEY, J. L. "General Topology". Van Nostrand, London, 1955.
9. KURATOWSKI, K. "Topology", Vol. 1. Academic Press, London and New York, 1966.
10. OXTOBY, J. C. "Measure and Category". Springer-Verlag, New York, 1971.
11. PFANZAGL, J. and PIERLO, W. "Compact Systems of Sets", Lecture Notes in Mathematics No. 16. Springer-Verlag, New York, 1966.
12. SCHAEFER, H. H. "Banach Lattices and Positive Operators". Springer-Verlag, New York, 1974.
13. SIKORSKI, R. "Boolean Algebras", Third Edition. Springer-Verlag, New York, 1969.

PAPERS

We now give a list of research papers which we have come across in our quest to achieve a good understanding of the world of finitely additive measures. This list is by no means exhaustive on this subject. We provide a brief description of some of the salient features of some of the papers which we think are relevant to the main theme of this book. Most of the papers contain a lot more information than the cursory annotation we provide here.

ALBANO, L. (1974). Teoremi di decompozione per funzioni finitamente additive in un reticolo relativamente complementato, *Ricerche Mat.* **23**, 63–86.
Lebesgue, Jordan, Yosida–Hewitt Decomposition theorems are discussed for charges taking values in a complete vector lattice.

ALEKSANDROV, I. I. (1973). The decomposition of a finitely additive set function (in Russian), *Comment. Math. Univ. Carolinae* **14**, 87–93.
Using results on vector lattices, Lebesgue Decomposition theorem for charges is proved. See Section 6.2.

ALIĆ, M. and KRONFELD, B. (1969). A remark on finitely additive measures, *Glasnik Mat., Ser. III* **4**(24), 197–200.
The problem of embedding a charge space into a measure space is considered. See also Fefferman (1968).

ANDO, T. (1961). Convergent sequences of finitely additive measures, *Pacific J. Math.* **11**, 395–404.
Vitali–Hahn–Saks theorem for sequences of charges defined on σ-fields is proved. See Chapter 8.

ARMSTRONG, T. and PRIKRY, K. (1978). Residual measures, *Illinois J. Math.* **22**, 64–78.

ARMSTRONG, T. and PRIKRY, K. (1981). Liapounoff's theorem for non-atomic finitely-additive, finite-dimensional vector-valued measures, *Trans. Amer. Math. Soc.* **266**, 499–514.
We came across this paper at the proof-reading stage of this book. Ranges of charges defined on fields of sets is the main theme of this paper. See Chapter 11. There is some overlap of results between this paper and that of Bhaskara Rao (1981).

ARMSTRONG, T. and PRIKRY, K. (1982). On the semimetric of a Boolean algebra induced by a finitely additive probability measure, *Pacific J. Math.* **99**, 249–263.
Let μ be a probability charge on a field $\mathscr{F}$ of subsets of a set Ω and $\mathscr{N}_\mu$ the ideal of all μ-null sets. On the quotient Boolean algebra $\mathscr{F}/\mathscr{N}_\mu$, there is a natural metric d_μ defined by $d_\mu([A],[B]) = \mu(A \Delta B)$ for $[A], [B]$ in $\mathscr{F}/\mathscr{N}_\mu$. The completion of the metric space $(\mathscr{F}/\mathscr{N}_\mu, d_\mu)$ is studied in detail in this paper. See also Bhaskara Rao and Bhaskara Rao (1977).

AUSTIN, D. G. (1955). An isomorphism for finitely additive measures, *Proc. Amer. Math. Soc.* **6**, 205–208.
An isomorphism theorem for charge spaces analogous to the classical Halmos and von Neumann (1942) theorem for measure spaces is proved. See also Buck and Buck (1947) for a similar result.

BANACH, S. (1948). On measures in independent fields, *Studia Math.* **10**, 159–177.
Let $(\Omega, \mathscr{F}_\alpha, \mu_\alpha)$, $\alpha \in \Gamma$ be a collection of probability charge spaces in which each μ_α is a probability measure. Let $\mathscr{F}$ be the field on Ω generated by $\{\mathscr{F}_\alpha, \alpha \in \Gamma\}$. A common extension of all these probability measures to $\mathscr{F}$ as a probability measure with a special property is sought. See also Marczewski (1951).

BARONE, E. (1978). Sulle misure semplicimenten additive non continue, *Atti Sem. Mat. Fis. Univ. Modena* **27**, 39–44.
An example of a nonatomic charge which is not strongly nonatomic is given.

BARONE, E. and BHASKARA RAO, K. P. S. (1981). Misure di probabilita finitamente additive e continue invarianti per transformazioni, *Boll. Un. Mat. Ital.* **18**, 175–184.
Existence of a nonatomic probability charge invariant with respect to a transformation is discussed.

BARONE, E. and BHASKARA RAO, K. P. S. (1981). Poincaré recurrence theorem for finitely additive measures, *Rendiconti di Matematica* **1**, 521–526.
The classical Poincaré recurrence theorem in Ergodic theory is discussed in the context of a charge space.

BARONE, E., GIANNONE, A. and SCOZZAFAVA, R. (1980). On some aspects of the theory and applications of finitely additive probability measures, *Pubbl. Istit. Mat. Appl. Fac. Univ. Stud. Roma Quaderno* **16**, 43–53.
Sobczyk–Hammer Decomposition theorem for charges on σ-fields is proved. See Section 5.2.

BAUER, H. (1955). Darstellung additiver Funktionen auf Booleschen Algebren als Mengenfunktionen, *Archiv der Math.* **6**, 215–222.
Let $\mathbb{B}^*$ be a Boolean algebra and $\mathbb{B}$ a subalgebra of $\mathbb{B}^*$. The notion of a positive bounded charge on $\mathbb{B}$ being a measure relative to $\mathbb{B}^*$ is introduced and some of the results of Yosida and Hewitt (1952) and Hewitt (1953) are generalized.

BELL, W. C. (1977). A decomposition of additive set functions, *Pacific J. Math.* **72**, 305–311.
Every positive bounded charge μ on a field $\mathscr{F}$ of subsets of a set Ω can be written as a sum of positive bounded charges μ_1 and μ_2 on $\mathscr{F}$ with the following properties. (i) μ_1 and μ_2 are mutually singular. (ii) The linear functional induced by the Lebesgue Decomposition of charges with respect to μ_1 has a refinement integral representation. See Chapter 9.

BELL, W. C. (1979). Unbounded uniformly absolutely continuous sets of measures, *Proc. Amer. Math. Soc.* **77**, 58–62.
A uniformly absolutely continuous set of charges can be decomposed into bounded and finite dimensional parts. See Section 8.6.

BELL, W. C. (1979). Hellinger integrals and set function derivatives, *Houston J. Math.* **5**, 465–481.

Using the concept of a refinement integral (see Chapter 9), the author introduces the notion of derivative of a bounded charge on a field $\mathscr{F}$ of sets with respect to a real valued function on $\mathscr{F}$ and studies some of its properties.

BELL, W. C. (1981). Approximate Hahn decompositions, uniform absolute continuity and uniform integrability, *J. Math. Anal. Appl.*, **80**, 393–405.

A sequence μ_n, $n \geq 1$ of bounded charges on a field $\mathscr{F}$ of subsets of a set Ω is said to be disjoint if $|\mu_n| \wedge |\mu_m| = 0$ for all $n \neq m$. A subset G of $ba(\Omega, \mathscr{F})$ is uniformly absolutely continuous if and only if each disjoint sequence in $(\hat{G})^+$ is norm convergent to zero, where $(\hat{G})^+$ is the set of positive elements in $\hat{G} = \{\eta \in ba(\Omega, \mathscr{F}); |\eta| < |\mu|$ for some μ in G$\}$. See Theorem 8.7.7 for a related result.

BELL, W. C. and KEISLER, M. (1979). A characterization of the representable Lebesgue Decomposition Projections, *Pacific J. Math.* **84**, 185–186.

Representability of the linear functional induced by the Lebesgue Decomposition of charges with respect to a fixed charge is studied.

BHASKARA RAO, K. P. S. (1981). Remarks on ranges of charges, to appear in *Illinois J. Math.*

See Chapter 11 and Armstrong and Prikry (1981). See also Notes and Comments on Chapter 11.

BHASKARA RAO, K. P. S. and AVERSA, V. (1982). On Tarski's extension theorem for group valued charges, a pre-print.

See Notes and Comments on Chapter 3. See also Carlson and Prikry (1982).

BHASKARA RAO, K. P. S. and AVERSA, V. (1982). A remark on E. Green's paper "Completeness of L_p-spaces over finitely additive set functions", to appear in *Coll. Math.*

See Notes and Comments on Chapter 7.

BHASKARA RAO, K. P. S. and BHASKARA RAO, M. (1973). Charges on Boolean algebras and almost discrete spaces, *Mathematika* **20**, 214–223.

A systematic study of nonatomic, strongly continuous and strongly nonatomic charges is made. Superatomic Boolean algebras are characterized. See Chapter 5.

BHASKARA RAO, K. P. S. and BHASKARA RAO, M. (1974). A category analogue of the Hewitt–Savage zero-one law, *Proc. Amer. Math. Soc.* **44**, 497–499.

See Notes and Comments on Chapter 11.

BHASKARA RAO, K. P. S. and BHASKARA RAO, M. (1977). Topological properties of charge algebras, *Rev. Roum. Math. Pures et Appl.* **22**, 363–375.

Let μ be a positive bounded charge on a field $\mathscr{F}$ of subsets of a set Ω. μ induces a natural semi-metric or pseudo-metric d_μ on $\mathscr{F}$ by $d_\mu(A, B) = \mu(A \Delta B)$ for A, B in $\mathscr{F}$. This paper studies some topological properties of the semi-metric space $(\mathscr{F}, d_\mu)$. See also Armstrong and Prikry (1982).

BHASKARA RAO K. P. S. and BHASKARA RAO, M. (1978). Existence of nonatomic charges, *J. Austral. Math. Soc.* **25** (Series A), 1–6.

A set of necessary and sufficient conditions for the existence of a nonatomic charge on a given Boolean algebra is provided. See Chapter 5.

BHASKARA RAO, K. P. S. and BHASKARA RAO, M. (1981). On the separating number of a finite family of charges, *Math. Nachr.* **101**, 215–217.

Given any finite number of distinct charges on a field $\mathscr{F}$ of subsets of a set Ω, a partition of Ω in $\mathscr{F}$ with minimal number of sets is sought which separates the charges.

BHASKARA RAO, K. P. S., BHASKARA RAO, M. and RAO, B. V. (1982). A note on μ-pure sub-fields, a pre-print.

Let μ be a probability measure on a countably generated σ-field of subsets of a set Ω. The following are equivalent. (i) There exists a μ-pure sub-field of $\mathscr{F}$. (ii) μ is perfect. (iii) μ is compact. For the notions of compactness and perfectness of measures, see Ryll–Nardzewski (1953). This result was anticipated by Frolik and Pachl (1973). See also Notes and Comments on Chapter 2.

BHASKARA RAO, M. and HALEVY, A. (1977). On Leader's V_p-spaces of finitely additive measures, *J. Reine Angew. Math.* **293/294**, 204–216.

V_p-spaces (Leader (1953)) are shown to be isometrically isomorphic to L_p-spaces of a measure space using the Stone Representation Theorem for Boolean algebras. See Notes and Comments on Chapter 7.

BOCHNER, S. (1939). Additive set functions on groups, *Ann. Math.* **40**, 769–799.

V_p-spaces ($1 \le p \le \infty$) in the setting of charges are introduced. Radon–Nikodym theorem for charges is also proved. See Chapters 7 and 6. See also Notes and Comments on Chapters 6 and 7.

BOCHNER, S. (1940). Finitely additive integral, *Ann. Math.* **41**, 495–504.

Representation of positive linear functionals on vector lattices is provided.

BOCHNER, S. (1946). Finitely additive set functions and stochastic processes, *Proc. Nat. Acad. Sci., U.S.A.* **32**, 259–261.

This paper introduces a notion called stochastic phenomenon. Let P be a probability measure on a σ-field $\mathfrak{A}$ of subsets of a set S and $\mathscr{F}$ a field of subsets of a set Ω. A real valued function f defined on $\mathscr{F} \times S$ is called a stochastic phenomenon if $f(\bigcup_{i=1}^n E_i, \cdot) = \sum_{i=1}^n f(E_i, \cdot)$ a.e. $[P]$ for every finite number of pairwise disjoint sets $E_1, E_2, \ldots, E_n$ in $\mathscr{F}$. A stochastic phenomenon can be regarded as a general type of stochastic process and it includes many known processes.

BOCHNER, S. and PHILLIPS, R. S. (1941). Additive set functions and vector lattices, *Ann. Math.* **42**, 316–324.

This is a fundamental paper on vector lattices. Riesz Decomposition Theorem in the general setting of vector lattices is proved. See Section 1.5. Lebesgue Decomposition Theorem in the setting of charges is observed. See Section 6.2.

BOGDAN, V. and OBERE, R. A. (1978). Topological rings of sets and the theory of vector measures, *Dissert. Math.* **154**.

Nikodym and Vitali–Hahn–Saks type of theorems for finitely additive vector measures on rings of sets are presented along the lines initiated by Drewnowski (1972a, b, c).

BROOKS, J. K. (1969). On the Vitali–Hahn–Saks and Nikodym theorems, *Proc. Nat. Acad. Sci., U.S.A.* **64**, 468–471.
Simplified proofs of Vitali–Hahn–Saks and Nikodym theorems for measures on σ-fields are presented.

BROOKS, J. K. (1972). Weak compactness in the space of vector measures, *Bull. Amer. Math. Soc.*, **78**, 284–287.
A set of necessary and sufficient conditions are given for a subset of ba$(\Omega, \mathscr{F})$ to be conditionally weakly compact in a general setting.

BROOKS, J. K. (1973). Equicontinuity, Absolute continuity and weak compactness in Measure Theory, a paper in "Vector and Operator valued Measures and Applications", (D. H. Tucker and H. B. Maynard, eds). pp. 51–61. Academic Press, London and New York.
Some extensions of the result in Brooks (1972) are dealt with.

BROOKS, J. K. (1974). Interchange of limit theorems for finitely additive measures, *Rev. Roumaine Math. Pures et Appl.* **19**, 731–744.
Let $\mathscr{F}$ be a field of subsets of a set Ω and $\mathscr{F}^*$ the smallest σ-field on Ω containing $\mathscr{F}$. Let $K \subset \text{ba}(\Omega, \mathscr{F}^*)$. Equivalence of uniform s-boundedness of K over $\mathscr{F}^*$ and uniform s-boundedness of K over $\mathscr{F}$ is examined. See also Brooks and Dinculeanu (1974).

BROOKS, J. K. and DINCULEANU, N. (1974). Strong additivity, absolute continuity and compactness in spaces of measures, *J. Math. Anal. Appl.* **45**, 156–175.
The notion of strong additivity of a charge studied in this paper is the same as s-boundedness we have used in this book. See Definition 2.1.4. Uniform s-boundedness of a collection of bounded charges is characterized in terms of uniform absolute continuity. See Theorem 8.7.7 for another characterization of uniform absolute continuity of a sequence of charges.

BROOKS, J. K. and JEWETT, R. S. (1970). On finitely additive vector measures, *Proc. Nat. Acad. Sci., U.S.A.* **67**, 1294–1298.
Vitali–Hahn–Saks and Nikodym theorems for charges on σ-fields are proved.

BUCK, R. C. (1946). The measure theoretic approach to density, *Amer. J. Math.* **68**, 560–580.
Density charges are constructed from simple set functions defined on a particular class of subsets of $\{1, 2, 3, \ldots\}$. See Section 2.1.

BUCK, E. F. and BUCK, R. C. (1947). A note on finitely additive measures, *Amer. J. Math.* **69**, 413–420.
Isomorphism of the charge spaces $(\Omega, \mathscr{F}, \mu)$ and $(\Omega', \mathscr{D}_0^*, m^*)$, where $\Omega' = \{1, 2, 3, \ldots\}$, $\mathscr{D}_0^*$ contains all arithmetic progressions and m^* is a density-like charge on $\mathscr{D}_0^*$, is investigated.

BUMBY, R. and ELLENTUCK, E. (1969). Finitely additive measures and the first digit problem, *Fund. Math.* **65**, 33–42.

A class S of probability charges on the power set of the set of all natural numbers is constructed such that for any μ in S, $\mu(P_n) = \log_{10}(n+1)$, where P_n is the set of all natural numbers whose first significant digit lies between 1 and n, $1 \le n \le 9$.

CANDELORO, D. and SACCHETTI, A. M. (1978). Su alcuni problemi relativi a misura scalari sub additive e applicazionial caso dell'additivita finita, *Atti. Sem. Mat. Fis. Univ. Modena* **27**, 284–296.

Connectedness of the range of a bounded charge is studied.

CARLSON, T. and PRIKRY, K. (1982). Ranges of Signed Measures, a pre-print.

Theorem A.8 is true for all abelian groups. See Notes and Comments on Chapter 3. This paper came to our notice at the proofreading stage of this book.

CHENEY, C. A. and de KORVIN, A. (1976/77). The representation of linear operators on spaces of finitely additive set functions, *Proc. Edinburgh Math. Soc.* **2**(20), 233–242.

An integral (Kolmogorov–Burkill type) representation of a continuous linear operator from $V_1(\Omega, \mathscr{F}, \mu)$ to a Banach space is provided. See also Edwards and Wayment (1974).

CHERSI, F. (1978). Finitely additive invariant measures, *Boll. Un. Mat. Ital. A*(5) **15**, 176–179.

Existence of invariant probability charges is shown. See Section 2.1.

CHRISTENSEN, J. P. R. (1971). Borel structures and a topological zero-one law, *Math. Scand.* **29**, 245–255.

A probability charge μ on the Borel σ-field of a complete separable metric space X is a measure if μ is measurable as a function on the space of all closed subsets of X equipped with a natural distance (metric) function. See Notes and Comments on Chapter 2.

COBZAS, S. (1976). Hahn Decompositions of finitely additive measures, *Arch. Math.* **27**, 620–621.

Let $\mathscr{F}$ be a field of subsets of a set Ω. Let ba$(\Omega, \mathscr{F})$ and $\mathscr{C}(\Omega, \mathscr{F})$ be as in Sections 2.2 and 4.7 respectively. ba$(\Omega, \mathscr{F})$ is equipped with the total variation norm and $\mathscr{C}(\Omega, \mathscr{F})$ is equipped with the supremum norm. ba$(\Omega, \mathscr{F})$ is the dual of $\mathscr{C}(\Omega, \mathscr{F})$. In this paper, it is proved that a μ in ba$(\Omega, \mathscr{F})$ admits an exact Hahn decomposition if and only if μ attains its norm on the unit ball of $\mathscr{C}(\Omega, \mathscr{F})$.

DARST, R. B. (1961). A note on abstract integration, *Trans. Amer. Math. Soc.* **99**, 292–297.

A real valued function on a set Ω is $\mathscr{F}$-continuous if and only if f is integrable with respect to every bounded charge on $\mathscr{F}$, where $\mathscr{F}$ is a field on Ω. See Section 4.7. See also Leader (1955).

DARST, R. B. (1962a). A decomposition of finitely additive set functions, *J. Reine Angew. Math.* **210**, 31–37.
Lebesgue Decomposition Theorem for bounded charges is proved. See Section 6.2.

DARST, R. B. (1962b). A decomposition for complete normed abelian groups with applications to spaces of additive set functions, *Trans. Amer. Math. Soc.* **103**, 549–558.
A Lebesgue type decomposition theorem is proved in a general setting. Validity of Lebesgue Decomposition Theorem for unbounded charges is examined. See Section 6.2. See also Notes and Comments on Chapter 6.

DARST, R. B. (1963). The Lebesgue Decomposition, *Duke Math. J.* **30**, 553–556.
An extension of a result in Darst (1962b) is established.

DARST, R. B. (1966). A direct proof of Porcelli's condition for weak convergence, *Proc. Amer. Math. Soc.* **17**, 1094–1096.
See Section 8.7 and Notes and Comments on Chapter 8.

DARST, R. B. (1967). On a theorem of Nikodym with applications to weak convergence and von Neumann algebras, *Pacific J. Math.* **23**, 473–477.
Nikodym theorem for sequences of charges on a σ-field is proved. See Section 8.4.

DARST, R. B. (1970a). The Vitali–Hahn–Saks and Nikodym theorems for additive set functions, *Bull. Amer. Math. Soc.* **76**, 1297–1298.
Vitali–Hahn–Saks and Nikodym theorems are proved for charges on σ-fields. See Sections 8.4 and 8.8.

DARST, R. B. (1970b). The Lebesgue Decomposition, Radon–Nikodym derivative, conditional expectation and martingale convergence for lattice of sets, *Pacific J. Math.* **35**, 581–600.
The Lebesgue Decomposition Theorem and the Radon–Nikodym theorem are considered in a general setting.

DARST, R. B. and GREEN, E. (1968). On a Radon–Nikodym theorem for finitely additive set functions, *Pacific J. Math.* **27**, 255–259.
Radon–Nikodym theorem for finitely additive bounded complex valued functions on a field of sets is proved. See Fefferman (1967). See also Notes and Comments on Chapter 6.

DIESTEL, J. and UHL, J. J. Jr. (1977). Vector measures, *American Mathematical Society Math. Surveys* **15**, Providence.
A sharper version of Phillips' lemma due to Rosenthal is presented.

DREWNOWSKI, L. (1972a). Topological rings of sets, continuous set functions, integration I, *Bull. Acad. Polon. Sci., Ser. Sci. Math. Astronom. Phys.* **20**, 269–276.
Rings equipped with a topology such that the operations Δ and $\cap$ become continuous are presented. Vitali–Hahn–Saks theorem for charges taking values in a topological group is proved.

DREWNOWSKI, L. (1972b). Topological rings of sets, continuous set functions, integration II, *Bull. Acad. Polon. Sci., Ser. Sci. Math. Astronom. Phys.* **20**, 277–286.
This is a continuation of the paper of Drewnowski (1972a) in which extensions of s-bounded group-valued charges on a ring of sets to the σ-ring generated by the ring are sought.

DREWNOWSKI, L. (1972c). Topological rings of sets, continuous set functions, integration III, *Bull. Acad. Polon. Sci., Ser. Sci. Math. Astronom. Phys.* **20**, 439–445.
Nikodym theorem for group-valued measures is proved.

DREWNOWSKI, L. (1972d). Equivalence of Brooks–Jewett, Vitali–Hahn–Saks and Nikodym Theorems, *Bull. Acad. Polon. Sci., Ser. Sci. Math. Astronom. Phys.* **20**, 725–731.
See the following paper of Drewnowski (1973).

DREWNOWSKI, L. (1973). Decomposition of set functions, *Studia Math.*, **48**, 23–48.
This paper and the above paper give analogues of Vitali–Hahn–Saks and Nikodym theorems for sequences of strongly bounded charges defined on σ-rings of sets taking values in a commutative Hausdorff topological group.

DREWNOWSKI, L. (1973a). Uniform boundedness principle for finitely additive vector measures, *Bull. Acad. Polon. Sci., Ser. Sci. Math. Astronom. Phys.* **21**, 115–118.
Nikodym theorem for s-bounded charges on a σ-ring of sets taking values in a normed group is proved.

DOLGUSEV, A. N. (1981). Remark on finitely additive measures, *Sibirsk. Mat. Z.* **22**, 105–120.

DUBINS, L. E. (1969). An elementary proof of Bochner's finitely additive Radon–Nikodym Theorem. *Amer. Math. Monthly* **76**, 520–523.
See Notes and Comments on Chapter 6.

EDWARDS, J. R. and WAYMENT, S. G. (1971). Representations for transformations continuous in the BV norm, *Trans. Amer. Math. Soc.* **154** 251–265.
An integral representation theorem for continuous linear functionals on $V_1(\Omega, \mathscr{F}, \mu)$, where $\Omega = [0, 1]$, can be deduced using v-integrals.

EDWARDS, J. R. and WAYMENT, S. G. (1974). Extensions of the v-integral, *Trans. Amer. Math. Soc.* **191**, 165–184.
An integral (with respect to a charge) representation of continuous linear operators on $V_1(\Omega, \mathscr{F}, \mu)$ into a Banach space can be deduced. See also Cheney and de Korvin (1976/77).

FAIRES, B. F. (1970). On Vitali–Hahn–Saks–Nikodym type theorems, *Ann. Insti. Fourier, Grenoble* **26**, No. 4, 99–114.
Vitali–Hahn–Saks and Nikodym type theorems are studied in the setting of

Boolean algebras with interpolation property (which are same as Boolean algebras with Seever property) for Banach-space-valued s-bounded charges. See Seever (1968) and Chapter 8.

FEFFERMAN, C. (1967). A Radon–Nikodym theorem for finitely additive set functions, *Pacific J. Math.* **23**, 35–45.

Radon–Nikodym theorem for bounded complex valued charges on a field of sets is proved. See also Darst and Green (1968).

FEFFERMAN, C. (1968). L_p-spaces over finitely additive measures, *Pacific J. Math.* **26**, 265–271.

The problem of embedding a charge space into a measure space is considered. See also Alić and Kronfeld (1969).

de FINETTI, B. (1955). La Struttura delle Distribuzioni in un insieme astratto qualsiasi, *Giorn. Ist. Ital. Attuari* **18**, 15–28.

A decomposition theorem similar to the one given by Sobczyk and Hammer (1944) is proved.

FROLIK, Z. and PACHL, J. (1973). Pure measures, *Comment. Math. Univ. Carolinae* **14**, 279–293.

Properties of charges μ which admit μ-pure subfields of $\mathscr{F}$ are studied. Pure measures discussed here are different from pure charges studied in Chapter 10. This paper pointed out an error in M. M. Rao's (1971) paper. See Bhaskara Rao, Bhaskara Rao and Rao (1982) and also Notes and Comments on Chapter 2.

GAIFMAN, H. (1964). Concerning measures on Boolean algebras, *Pacific J. Math.* **14**, 61–73.

Existence of a strictly positive charge on a field $\mathscr{F}$ of subsets of a set Ω is related to some conditions in Set Theory. Most importantly, he exhibited a Boolean algebra satisfying countable chain condition having no strictly positive charge on it. See also Kelley (1959).

GOULD, G. G. (1965). Integration over vector-valued measures, *Proc. London Math. Soc.* **15**, 193–225.

Integration of scalar-valued functions with respect to vector-valued charges is developed. See Section 4.5.

GRECO, G. H. (1981). The continuous measures defined on a Boolean algebra (Italian), *Ann. Univ. Ferrara Sez. VII (N.S.)* **26**, 213–218.

A characterization of superatomic Boolean algebras $\mathbb{B}$ in terms of exact Hahn Decomposition of bounded charges on $\mathbb{B}$ is provided.

GREEN, E. (1970/71). Completeness of L_p-spaces over finitely additive set functions, *Coll. Math.* **22**, 257–261.

See Notes and Comments on Chapter 7. See also Bhaskara Rao and Aversa (1982).

GUY, D. L. (1961). Common extensions of finitely additive probability measures, *Portugal. Math.* **20**, 1–5.

A necessary and sufficient condition is given for the existence of a common extension of two probability charges defined on two different fields on the same set to any field containing these two fields as a probability charge. See Section 3.6.

HALMOS, P. R. (1947). The set of values of a finite measure, *Bull. Amer. Math. Soc.* **53**, 138–141.

A simple proof of a result of Liapounoff on the range of a measure is given.

HALMOS, P. R. (1948). The range of a vector measure, *Bull. Amer. Math. Soc.* **54**, 416–421.

A simple proof of two results of Liapounoff on the range of a measure with values in a finite dimensional vector space is provided.

HALMOS, P. R. and von NEUMANN, J. (1942). Operator methods in classical mechanics II, *Ann. Math.* **43**, 332–350.

Isomorphism between two measure spaces is abstractly characterized.

HATTA, L. and WAYMENT, S. G. (1973). A Radon–Nikodym theorem for the v-integral, *J. Reine Angew. Math.* **259**, 137–146.

An analogue of the classical Radon–Nikodym theorem is considered in the setting of v-integrals for charges.

an der HEIDEN, U. (1978). On the representatation of linear functionals by finitely additive set functions, *Arch. Math.* **30**, 210–214.

Necessary and sufficient conditions for the existence of a charge μ for a given linear functional on a Stonean lattice of functions to be expressed as an integral with respect to μ are derived.

HEWITT, E. (1951). A problem concerning finitely additive measures, *Mat. Tidsskr. B* 81–94.

The structure of all bounded charges on the field $\mathscr{F}$ on $\Omega = [0, 1)$ generated by all intervals of the type $[a, b)$ with $0 \le a \le b \le 1$ is determined. See Section 10.4.

HEWITT, E. (1953). A note on measures on Boolean algebras, *Duke Math. J.* **20**, 253–256.

Distinction between measures on fields and measures on Boolean algebras is pointed out. See Section 10.5.

HILDEBRANDT, T. H. (1934). On bounded linear functional operations, *Trans. Amer. Math. Soc.* **36**, 868–875.

The dual of the Banach space of all $\mathscr{F}$-continuous functions is shown to be $ba(\Omega, \mathscr{F})$, where $\mathscr{F}$ is a field on Ω. See Section 4.7.

HILDEBRANDT, T. H. (1938). Linear operations of functions of bounded variation, *Bull. Amer. Math. Soc.* **44**, 75.

Integral representation of continuous linear functionals on a subspace of ba(Ω, $\mathcal{F}$) is given, where $\Omega = [0, 1]$.

HILDEBRANDT, T. H. (1940). On unconditional convergence in normed vector spaces, *Bull. Amer. Math. Soc.* **46**, 959–962.

Properties of unconditional convergence in normed linear spaces are used to define some simple measures on $\mathcal{P}(\Omega)$, where $\Omega = \{1, 2, 3, \dots\}$.

HILDEBRANDT, T. H. (1958). On a theorem in the space ℓ_1 of absolutely convergent sequences with applications to completely additive set functions, *Math. Research Center Report No. 62* Madison, Wisconsin.

HODGES, J. L. Jr. and HORN, A. (1948). On Maharam's conditions for measure, *Trans. Amer. Math. Soc.* **64**, 594–595.

One of the conditions in the set of necessary and sufficient conditions given by Maharam (1947) for a Boolean σ-algebra to admit a strictly positive bounded measure is shown to be redundant.

HORN, A. and TARSKI, A. (1948). Measures on Boolean algebras, *Trans. Amer. Math. Soc.* **64**, 467–497.

Extension of set functions defined on a collection $\mathscr{C}$ of subsets of a set Ω to a field $\mathcal{F}$ on Ω containing $\mathscr{C}$ as charges are sought. See Chapter 3. See also Notes and Comments on Chapter 3.

HUFF, R. E. (1973). The Yosida–Hewitt Decomposition as an Ergodic theorem, a paper in "Vector and Operator Valued Measures And Applications", (D. H. Tucker and H. B. Maynard, eds), pp. 133–139. Academic Press, London and New York.

The Yosida–Hewitt (1952) Decomposition of a charge as a sum of a pure charge and a measure is obtained using an ergodic theorem for commutative semigroup of idempotent linear operators on a Banach space. This approach covers both the scalar valued and vector valued charges.

JECH, T. and PRIKRY, K. (1979). On projections of finitely additive measures, *Proc. Amer. Math. Soc.* **74**, 161–165.

There exists a translation invariant charge μ on $\mathcal{P}(\Omega)$, where $\Omega = \{1, 2, 3, \dots\}$ and a function f from Ω to Ω such that $\mu = \mu f^{-1}$ and $\mu(A) \leq \frac{1}{2}$ if f is one-to-one on $A \subset \Omega$.

JØRSBOE, O. G. (1966). Set transformations and Invariant measures, A Survey, *Math. Inst. Aarhus Universitet Various Publications Series*, No. 3, Aarhus, Denmark.

Invariant charges are constructed using Banach limits. See Section 2.1.

KEISLER, M. (1979). Integral representation for elements of the dual of ba($\mathcal{S}$, Σ), *Pacific J. Math.* **83**, 177–183.

If $\mathcal{F}$ is a superatomic Boolean algebra, then every continuous linear functional on ba(Ω, $\mathcal{F}$) has a refinement integral representation. See Chapter 9. See also Notes and Comments on Chapter 9.

KELLEY, J. L. (1959). Measures on Boolean algebras, *Pacific J. Math.* **9**, 1165–1177.

Necessary and sufficient conditions for a Boolean algebra to admit a strictly positive charge are given.

KELLEY, J. L. and SRINIVASAN, T. P. (1970/71). Pre-measures on lattices of sets, *Math. Ann.* **190**, 233–241.

Necessary and sufficient conditions are given for a positive bounded charge defined on a lattice of sets closed under countable intersections admits an extension as a measure to the σ-field generated by the lattice.

KELLEY, J. L., NAYAK, M. K. and SRINIVASAN, T. P. (1973). Pre-measures on lattice of sets II. "Proceedings of a Symposium on Vector and Operator valued measures and Applications" held at University of Utah, August 7–12, 1972, (D. H. Tucker and H. B. Maynard, eds) Academic Press, London and New York.

Some improvements of the results of Kelley and Srinivasan (1970/71) are presented.

KHURANA, S. S. (1978). A note on Radon–Nikodym theorem for finitely additive measures, *Pacific J. Math.* **74**, 103–104.

Radon–Nikodym theorem for charges is proved using the corresponding result for measures. The argument is essentially that of Dunford and Schwartz (1954), p. 315.

KINGMAN, J. F. C. (1967). Additive set functions and the theory of probability, *Proc. Camb. Phil. Soc.* **63**, 767–775.

A certain notion dense subset of a set Ω in the context of a field of subsets of Ω is introduced and its ramifications are studied.

KISYŃSKI, J. (1968). Remark on strongly additive set functions, *Fund. Math.* **63**, 327–332.

Smiley's (1944) result on the extension of a strongly additive set function defined on a lattice of sets containing the null set to the ring generated by the lattice is reproved. See Section 3.5.

LADUBA, I. (1972). Sur quelques généralisations de théorèmes de Nikodym et de Vitali–Hahn–Saks, *Bull. Acad. Polon. Sci., Ser. Sci. Math. Astronom. Phys.* **20**, 447–456.

Some generalizations of Nikodym and Vitali–Hahn–Saks theorems are presented for charges on σ-fields taking values in a specified space of functions.

LEADER, S. (1953). The theory of L_p-spaces for finitely additive set functions, *Ann. Math.* **58**, 528–543.

A systematic study of V_p-spaces is presented. See Chapter 7. See also Notes and Comments on Chapter 7.

LEADER, S. (1955). On universally measurable functions, *Proc. Amer. Math. Soc.* **6**, 232–234.

A real valued function f on a set Ω is $\mathscr{F}$-continuous if and only if f is integrable with respect to every bounded charge on $\mathscr{F}$, where $\mathscr{F}$ is a field on Ω. See Section 4.7. See also Darst (1961).

LEMBCKE, J. (1970). Konservative Abbildungen und Fortsetzung regularer Masse, *Z. Wahrscheinlichkeitstheorie und Verw. Gebiete* **15**, 57–96.

A certain order relation on the set of all real measures on a ring of sets is introduced and the maximal elements in this order are identified.

LEMBCKE, J. (1972). Gemeinsame Urbilder endlich additiver Inhalte, *Math. Ann.* **198**, 239–258.

LIPECKI, Z. (1971). On strongly additive set functions, *Coll. Math.* **22**, 255–256.

Another proof of a result of Smiley (1944) is presented. See Section 3.5.

LIPECKI, Z. (1974). Extensions of additive set functions with values in a topological group, *Bull. Acad. Polon. Sci., Ser. Sci. Math. Astronom. Phys.* **22**, 19–27.

Extensions of group-valued charges are discussed.

LIPECKI, Z. (1982). On unique extensions of positive additive set functions, a pre-print.

LIPECKI, Z. (1982). Maximal-valued extensions of positive operators, a pre-print.

LIPECKI, Z. (1982). Conditional and simultaneous extensions of group-valued quasi-measures, a pre-print.

LIPECKI, Z., PLACHKY, D. and THOMSEN, W. (1979). Extensions of positive operators and extreme points I, *Coll. Math.* **42**, 279–284.

The result of Plachky (1976) concerning extreme points of a certain convex subsets of ba$(\Omega, \mathscr{F})$ is generalized. Extensions of results of Łos and Marczewski (1949) are derived in a Functional Analytic setting.

LLOYD, S. P. (1963). On finitely additive set functions, *Proc. Amer. Math. Soc.* **14**, 701–704.

Pure charges on Boolean algebras are characterized in terms of measures on the Stone space of the Boolean algebras. See Section 10.5.

LOMNICKI, Z. and ULAM, S. (1934). Sur la théorie de la mesure dans les espaces combinatoires et son application au calcul des probabilités I. Variables indépendantes, *Fund. Math.* **23**, 237–278.

ŁOS, J. and MARCZEWSKI, E. (1949). Extensions of measures, *Fund. Math.* **36**, 267–276.

The problem of extending a charge from a subfield of a field $\mathscr{F}$ of subsets of a set Ω to $\mathscr{F}$ as a charge is tackled. See Section 3.3.

LUXEMBURG, W. A. J. (1963/64). On finitely additive measures in Boolean algebras, *J. Reine Angew. Math.* **213**, 165–173.

A special class of Boolean algebras in which every charge is a measure when restricted to some suitable ideal is studied.

MAHARAM, D. (1947). An algebraic characterization of Measure algebras, *Ann. Math.* **48**, 154–167.

Necessary and sufficient conditions are given for a Boolean σ-algebra to admit a strictly positive bounded measure. See also Hodges and Horn (1948).

MAHARAM, D. (1958). On a theorem of von Neumann, *Proc. Amer. Math. Soc.* **9**, 987–994.
Lifting exists in complete measure spaces. See Chapter 12.

MAHARAM, D. (1972). Consistent extensions of linear functionals and of probability measures, "Proceedings of the Sixth Berkeley Symposium on Mathematical Statistics and Probability (University of California, Berkeley, 1970/71)", Vol. 2, Probability theory, p. 127–147, Univ. California Press, Berkeley.
Let $\mathscr{F}_\alpha$, $\alpha \in \Gamma$ be a collection of fields on a set Ω and $\mathscr{F}$ the smallest field on Ω containing this collection. For each α in Γ, let μ_α be a bounded charge on $\mathscr{F}_\alpha$. The existence of a charge on $\mathscr{F}$ agreeing with μ_α on $\mathscr{F}_\alpha$ for every $\alpha \in \Gamma$ is discussed. A simple case of this problem is studied in Section 3.6.

MAHARAM, D. (1976). Finitely additive measures on the integers, *Sankhya*, Series A **38**, 44–59.
Lifting fails to exist in the setting of charge spaces. See Chapter 12.

MAHARAM, D. (1977). "Category, Boolean algebras and measures, General Topology and its relation to modern analysis and algebra", pp. 124–135. Springer-Verlag, Berlin.

MARCZEWSKI, E. (1947). Sur les mesures à deux valeurs et les idéaux premiers dans les corps d'ensembles, *Ann. Soc. Polon. Math.* **19**, 232–233.

MARCZEWSKI, E. (1947). Two-valued measures and prime ideals in fields of sets, *Soc. Sci. Lett. Varsovie C. R. Cl. III. Sci. Math. Phys.* **40**, 11–17.
Let $\mathscr{F}$ be the smallest field on $[0, 1]$ containing all sub-intervals of $[0, 1]$. There is no non-trivial two-valued measure on $\mathscr{F}$.

MARCZEWSKI, E. (1947). Indépendance d'ensembles et prolongement de mesures (Résultats et Problèmes), *Coll. Math.* **1**, 122–132.

MARCZEWSKI, E. (1948). Ensembles d'indépendants et leurs applications a la théorie de la mesure, *Fund. Math.* **25**, 13–28.

MARCZEWSKI, E. (1951). Measures in almost independent fields, *Fund. Math.*, **38**, 217–229.
This paper and the two papers above deal with the following problem in all its facets. Let $\mathscr{F}_\alpha$, $\alpha \in \Gamma$ be a collection of fields on a set Ω and $\mathscr{F}$ a field on Ω containing all $\mathscr{F}_\alpha$'s. Let μ_α be a probability charge on $\mathscr{F}_\alpha$ for each α in Γ. Is there a common extension μ (with a special property) of all μ_α's to $\mathscr{F}$ as a probability charge? This problem is linked with the notion of almost-independence of the fields $\mathscr{F}_\alpha$'s. This problem was also studied by Banach (1948) in the setting of probability measures.

MARCZEWSKI, E. (1953). On Compact measures, *Fund. Math.* **40**, 113–124.
See Notes and Comments on Chapter 2.

MAYNARD, H. B. (1972). A Radon–Nikodym theorem for operator valued measures, *Trans. Amer. Math. Soc.* **173**, 449–463.

MAYNARD, H. B. (1979). A Radon–Nikodym theorem for finitely additive bounded measures, *Pacific J. Math.* **83**, 401–413.

Necessary and sufficient conditions for the existence of exact Radon–Nikodym derivative in the setting of charges are presented. See Section 6.3. See also Notes and Comments on Chapter 6.

MOLTÓ, A. (1981a). On the Vitali–Hahn–Saks theorem, *Proc. Roy. Soc. Edinburgh, Sec. A* **90**, 163–173.

Boolean rings with property (f) are introduced. These include Boolean algebras with Seever property strictly. See Seever (1968) and Definition 8.4.1. Let G be a commutative Hausdorff topological group. It is proved that if μ_n, $n \geq 1$ is a sequence of G-valued s-bounded charges defined on a Boolean ring with property (f), pointwise convergent and E_n, $n \geq 1$ is a sequence of pairwise disjoint elements in the ring, then $\lim_{p \to \infty} \mu_n(E_p) = 0$ uniformly in n. See also Faires (1976).

MOLTÓ, A. (1981b). On Uniform boundedness properties in exhaustive additive set function spaces, *Proc. Roy. Soc. Edinburgh, Sec. A* **90**, 175–184.

Uniform boundedness of a family of s-bounded G-valued charges defined on a Boolean ring having the property (f) is discussed. See Molto (1981a).

NAYAK, M. K. and SRINIVASAN, T. P. (1975). Scalar and Vector-valued premeasures, *Proc. Amer. Math. Soc.* **48**, 391–396.

Let $\mathscr{F}$ be a lattice of subsets of a set Ω and $\mathscr{F}^*$ the smallest σ-field on Ω containing $\mathscr{F}$. Conditions under which a charge on $\mathscr{F}$ taking values either in R or in a Banach space is extendable as a measure to $\mathscr{F}^*$ are presented.

NAYAK, M. K. and SRINIVASAN, T. P. (1976). Vector-valued inner-measures, "Lecture Notes in Mathematics", Vol. 541, pp. 107–116. Springer-Verlag, Berlin.

Extension of a vector valued charge defined on a lattice of sets to the σ-field generated by the lattice as a measure is discussed. See also Nayak and Srinivasan (1975).

NUNKE, R. J. and SAVAGE, L. (1952). On the set of values of a nonatomic, finitely additive, finite measure, *Proc. Amer. Math. Soc.* **3**, 217–218.

A nonatomic charge whose range is not convex is exhibited. See Section 11.4.

OLEJČEK, V. (1977). Darboux properties of finitely additive measures on a δ-ring, *Math. Slovaca* **27**, 195–201.

An example of a nonatomic charge defined on a δ-ring which is not strongly nonatomic is given. See Definition 5.1.5, Theorem 5.1.6 and Remarks 5.1.7.

OLEJČEK, V. (1981). Ultrafilters and Darboux property of finitely additive measure, *Math. Slovaca* **31**, 263–276.

The notion of an ultrafilter-atom is introduced in the setting of a charge space and some of its properties are studied.

PACHL, J. (1972). An elementary proof of a Radon–Nikodym theorem for finitely additive set functions, *Proc. Amer. Math. Soc.* **32**, 225–228.

See Notes and Comments on Chapter 6.

PACHL, J. (1972). On projective limits of probability spaces, *Comment. Math. Univ. Carolinae* **13**, 685–691.

Let μ be a non-atomic probability measure on a σ-field $\mathscr{F}$ of subsets of a set Ω. If there exists a μ-pure sub-field of $\mathscr{F}$, then there is a set A in $\mathscr{F}$ such that $\mu(A) = 0$ and the cardinality of A is at least that of the continuum. See Notes and Comments on Chapter 2.

PACHL, J. (1975). Every weakly compact probability is compact, *Bull. Acad. Polon. Sci., Ser. Sci. Math. Astronom. Phys.* **23**, 401–405.

Let μ be a probability measure on a σ-field $\mathscr{F}$ of subsets of a set Ω. If there is a μ-pure sub-field of $\mathscr{F}$, then μ is compact. See Ryll-Nardzewski (1953) for the notion of a compact measure.

PETTIS, B. J. (1951). On the extension of measures, *Ann. Math.* **54**, 186–197.

Various extensions of set functions are dealt with. See Section 3.5.

PHILLIPS, R. S. (1940). On linear transformations, *Trans. Amer. Math. Soc.* **48**, 516–541.

Lemma 3.3 of this paper is Phillips' lemma. See Section 8.3.

PHILLIPS, R. S. (1940a). A decomposition of additive set functions, *Bull. Amer. Math. Soc.* **46**, 274–277.

Let $\mathscr{F}$ be a σ-field of subsets of a set Ω and $\aleph$ an infinite cardinal number not greater than the cardinal of Ω. Every μ in ba($\Omega, \mathscr{F}$) can be written as a sum $\mu_1 + \mu_2$ uniquely with μ_1, μ_2 in ba($\Omega, \mathscr{F}$) and μ_2 vanishing on every set of cardinal $\leq \aleph$ in $\mathscr{F}$.

PIERCE, R. S. (1970). Existence and uniqueness theorems for extensions of zero-dimensional compact metric spaces, *Trans. Amer. Math. Soc.* **148**, 1–21.

Some comments on countable superatomic Boolean algebras are made. See Chapter 5.

PLACHKY, D. (1971). Decomposition of Additive Set Functions, "Transactions of the Sixth Prague Conference on Information theory, Statistical Decision Functions, Random Processes", pp. 715–719. Publishing House of the Czechoslovak Academy of Sciences, Prague.

A general decomposition theorem is proved from which the Yosida–Hewitt Decomposition and the Lebesgue Decomposition of bounded charges follow as corollaries.

PLACHKY, D. (1976). Extremal and monogenic additive set functions, *Proc. Amer. Math. Soc.* **54**, 193–196.

Let $\mathscr{F}$ be a field of subsets of a set Ω and $\mathscr{F}_0$ a sub-field of $\mathscr{F}$. Let ν be a probability charge on $\mathscr{F}_0$. The extreme points of the convex set of all probability charges μ on $\mathscr{F}$ which agree with ν on $\mathscr{F}_0$ are characterized.

PLACHKY, D. (1980). Darboux property of measures and contents, *Math. Slovaca* **30**, pp. 243–246.

Let $\mathscr{F}_0$ and $\mathscr{F}_1$ be two σ-fields on a set such that $\mathscr{F}_0 \subset \mathscr{F}_1$. Let μ_0 be a positive bounded charge on $\mathscr{F}_0$. Then μ_0 is strongly continuous if and only if every positive bounded charge μ defined on $\mathscr{F}_1$ whose restriction to $\mathscr{F}_0$ is μ_0 is strongly continuous.

PORCELLI, P. (1958a). On weak convergence in the space of functions of bounded variation, *Math. Research Center Reports* No. 39, Madison, Wisconsin.
See Porcelli (1960).

PORCELLI, P. (1958b). On weak convergence in the space of functions of bounded variation II, *Math. Research Center Reports*, No. 68, Madison, Wisconsin.
See Porcelli (1960).

PORCELLI, P. (1960). Two embedding theorems with applications to weak convergence and compactness in spaces of additive type functions, *J. Math. Mech.* **9**, 273-292.
Weak convergence in ba(Ω, $\mathscr{F}$) is characterized using Porcelli (1958a and 1958b). See Section 8.7. See also Notes and Comments on Chapter 8.

PORCELLI, P. (1966). Adjoint spaces of Abstract L_p-spaces, *Port. Math.* **25**, 105-122.
V_p-spaces are studied from another angle. See Chapter 7. See also Leader (1953).

PTÁK, V. (1969). Simultaneous extension of two functionals, *Czechoslovak Math. J.* **3**, 553-569.
The results of this paper are relevant to the problem studied by Maharam (1972).

PYM, J. S. and VASUDEVA, H. L. (1975). An algebra of finitely additive measures, *Studia Math.* **54**, 29-40.
Maximal ideals in the algebra ba(Ω, $\mathscr{F}$) are determined, where Ω is a discrete semigroup which is a totally ordered set with multiplication as max.

RAMACHANDRAN, D. (1972). A note on finitely additive set functions, *Proc. Amer. Math. Soc.* **31**, 314-315.
A counterexample is presented to a conjecture of Yosida and Hewitt (1952) concerning the correspondence between charges on a Boolean algebra and the measures on the Stone space of the Boolean algebra.

RANGA RAO, R. (1958). A note on finitely additive measures, *Sankhya* **19**, 27-28.
Another proof of the Yosida-Hewitt (1952) Decomposition of a charge as a sum of a pure charge and a measure is presented. See Chapter 10.

RAO, M. M. (1971). Projective limits of Probability spaces, *J. Multivariate Anal.* **1**, 28-57.
Some conditions are given for a charge to be a measure. See Notes and Comments on Chapter 2.

RICKART, C. E. (1943). Decomposition of Additive Set Functions, *Duke Math. J.* **10**, 653-665.
Generalizations of a result of Phillips (1940a) are presented.

RIEFFEL, M. A. (1968). The Radon–Nikodym theorem for the Bochner integral, *Trans. Amer. Math. Soc.* **131**, 466–487.
Hahn Decomposition Theorem for measures on σ-fields is presented using Banach space methods.

RYLL-NARDZEWSKI, C. (1953). On quasi-compact measures, *Fund. Math.* **40**, 125–130.
Perfect and compact measures are discussed.

SASTRY, A. S. and SASTRY, K. P. R. (1977). Measure extensions of set functions over lattices of sets, *J. Indian Math. Soc.* **41**, 317–330.
Extension of vector-valued set functions from a lattice of sets to the ring generated by the lattice is examined.

SCOZZAFAVA, R. (1978). On finitely additive probability measures, "Transactions of the Eighth Prague Conference on Information theory, Statistical Decision functions, Random Processes, (Prague 1978)", Vol. C, pp 175–180. Reidel, Dordrecht.
Let μ be a strongly continuous probability charge on the power set $\mathcal{P}(\Omega)$ of an infinite set Ω. Given $0 < \alpha < 1$, there exists a sequence F_n, $n \geq 1$ of pairwise disjoint subsets of Ω such that $\alpha = \mu(\bigcup_{n \geq 1} F_n) = \sum_{n \geq 1} \mu(F_n)$.

SCOZZAFAVA, R. (1979). Complete additivity, on suitable sequences of sets, of a simply additive and strongly nonatomic probability measure (Italian), *Boll. Un. Mat. Ital. B* **5**, 16, 639–648.
Sobczyk–Hammer Decomposition theorem for nonconcentrated charges μ, i.e. $\mu(\{\omega\}) = 0$ for every ω in Ω, on the power set $\mathcal{P}(\Omega)$ of Ω is proved.

SEEVER, G. L. (1968). Measures on F-spaces, *Trans. Amer. Math. Soc.* **133**, 267–280.
Nikodym and Vitali–Hahn–Saks theorems are presented for Boolean algebras having Seever property. See Sections 8.4 and 8.8. See also Notes and Comments on Chapter 8.

SIERPIŃSKI, W. (1938). Fonctions additives non complètement additives et fonctions non mesurables, *Fund. Math.* **30**, 96–99.
A non-Lebesgue measurable function on the unit interval $[0, 1]$ is constructed based on a charge on $\mathcal{P}(\Omega)$, where $\Omega = \{1, 2, 3, \ldots\}$. See Notes and Comments on Chapter 11.

SINCLAIR, G. E. (1974). A finitely additive generalization of the Fichtenholz–Lichtenstein theorem, *Trans. Amer. Math. Soc.* **193**, 359–374.
An analogue of Fubini's theorem is established in the setting of charges.

SMILEY, M. F. (1944). An extension of metric distributive lattices with an application in general analysis, *Trans. Amer. Math. Soc.* **56**, 435–447.
Every strongly additive set function defined on a lattice of sets containing the empty set can be extended in a unique manner as a charge on the smallest ring containing this lattice. See Section 3.5.

SOBCZYK, A. and HAMMER, P. C. (1944). A decomposition of additive set functions, *Duke Math. J.* **11**, 839–846.
Sobczyk–Hammer Decomposition Theorem is proved. See Section 5.2. See also Notes and Comments on Chapter 5.

SOBCZYK, A. and HAMMER, P. C. (1944). The ranges of additive set functions, *Duke Math. J.* **11**, 847–851.
Some results on the ranges of charges are obtained. See Chapter 11. See also Notes and Comments on Chapter 11.

SRINIVASAN, T. P. (1955). On extensions of measures, *J. Indian Math. Soc.*, (*N.S.*) **19**, 31–60.
Extension of measures is discussed using inner measures.

STRATIGOS, P. D. (1980). Extensions of additive set functions, *Serdica* **6**, 197–201.
Extension of regular bounded charges on fields of sets generated by σ-topological spaces is discussed.

SUCHESTON, L. (1967). Banach limits, *Amer. Math. Monthly* **74**, 308–311.
Existence of Banach limits is shown using an old-fashioned version of the Hahn–Banach theorem. See Section 2.1.

TALAGRAND, M. (1981). Non existence de relèvement pour certaines mesures finiement additives et retractés de βN, *Math. Ann.* **256**, 63–66.
Under continuum hypothesis, the author constructs a separable subset of βN$-$N which is not a retract of βN, where N is the set of all natural numbers with discrete topology and βN its Stone–Cech compactification. This example is used to show non-existence of a lifting in the setting of charges. See Maharam (1976) and Chapter 12.

TARSKI, A. (1930). Une contribution à la théorie de la mesure, *Fund. Math.* **15**, 42–50.

TARSKI, A. (1938). Algebraische Fassung des Massproblems, *Fund. Math.* **31**, 47–66.

TARSKI, A. (1939). Ideale in Völlstandigen Mengenkörpern I, *Fund. Math.* **32**, 45–63.
Weak and strong accessibility of cardinals are discussed and existence of measures on some quotient Boolean algebras is considered.

TARSKI, A. (1945). Ideale in Völlstandigen Mengenkörpern II, *Fund. Math.* **33**, 51–65.
There exists a $0-1$ valued charge on a Boolean algebra $\mathbb{B}$ vanishing on all atoms of $\mathbb{B}$ if and only if $\mathbb{B}$ contains a countable set of disjoint elements.

THOMSEN, W. (1978). On a Fubini-type theorem and its application in game theory, *Math. Operations forsch. Statist. Ser. Statist.* **9**, 419–423.
Sinclair's (1974) analogue of Fubini's theorem for measures in the setting of charges is generalized.

THOMSEN, W. (1979). The common domain of uniqueness of the products of finitely additive probability measures, "Transactions of the Eighth Prague Conference on Information Theory, Statistical Decision Functions, Random Processes, (Prague, 1978)", Vol. C, pp. 311–316, Reidel, Dordrecht.

Let B(X) be the Banach space of all bounded real valued functions defined on a set X, F a subset of B(X), B(X, F) the closed subspace of B(X) generated by F and B*(X) the dual of B(X). Let μ be a real valued function on F such that there exists a probability charge $\hat{\mu}$ on $\mathcal{P}(X)$ such that $\mu(f) = \int f \, d\hat{\mu}$ for all f in F. Let $U(\mu) = \{f \in B(X); \rho_1(f) = \rho_2(f)$ for all ρ_i in $B_+^*(X)$ with $\rho_i/F = \mu\}$ which is the domain uniqueness of μ. It is proved that $\bigcap_\mu U(\mu) = B(X, F)$. An extension to product spaces is also considered.

TOPSØE, F. (1978). On construction of measures, "Proceedings of the Conference on Topology and Measure I (Zinnowitz 1974)", Part 2, pp. 343–381, Ernst–Maritz–Arndt Univ., Griefswald.

A general result is proved in Section 8 of this paper from which the result of Smiley (1944) on the extension of strongly additive functions on a lattice of sets containing the null set to the ring generated by the lattice as a charge follows.

TOPSØE, F. (1979). Approximate pavings and construction of measures, *Coll. Math.*, **42**, 377–385.

A condition under which a positive bounded charge on a field of sets becomes a measure is given. See Notes and Comments on Chapter 2.

TRAYNOR, T. (1972). Decomposition of Group-valued Additive Set Functions, *Ann. Inst. Fourier, Grenoble* **22**, Part 3, 131–140.

Lebesgue-type decomposition theorem is obtained for group-valued charges.

TRAYNOR, T. (1972). A general Hewitt–Yosida Decomposition, *Can. J. Math.* **24**, 1164–1169.

Yosida–Hewitt (1952) Decomposition of a group-valued charge is presented using Caratheodory process.

TUCKER, D. H. and WAYMENT, S. G. (1970). Absolute continuity and the Radon–Nikodym theorem, *J. Reine Angew. Math.* **244**, 1–19.

A general discussion about Radon–Nikodym Theorem in various settings is presented.

TULIPANI, S. (1979). On continuous and invariant measures for a transformation (Italian), *Rend. Mat.* **12**, 249–256.

Let Ω be a set and T a map from Ω to Ω. Existence of a nonatomic, T-invariant probability charge on the power set $\mathcal{P}(\Omega)$ of Ω is discussed.

UHL, J. J. (1967). Orlicz spaces of finitely additive set functions, *Studia Math.* **29**, 19–58.

Spaces of set functions more general than the V_p-spaces (Leader (1953)) are studied.

VOROB'EV, N. N. (1962). Consistent families of measures and their extensions, *Theory Prob. Appl.* **7**, 147–162.

This paper treats the problem described in the annotation of Maharam's (1972) paper in the setting of probability measures on σ-fields. Some combinatorial methods are used to solve the problem.

WAJDA, L. (1972). Remarks on infinite products of finitely additive measures, *Coll. Math.* **25**, 269–271.

Product charge of a sequence of probability charge spaces is shown to exist.

WALKER, H. D. (1975). Uniformly additive families of measures, *Bull. Math. Soc. Sci. Math. R. S. Roumanie (N.S.)* **18**, 217–222.

Let $\mathscr{F}$ be a field of subsets of a set Ω and $K \subset \mathrm{ba}(\Omega, \mathscr{F})$. If K is uniformly s-bounded and pointwise bounded, then K is a bounded subset of $\mathrm{ba}(\Omega, \mathscr{F})$. For some more results in this direction, see Section 8.5. See also Brooks (1974).

WEBER, H. (1982). Unabhängige Topologien, Zerlegung von Ringtopologien, *Math. Z.* **180**, 379–393.

WEBER, H. (1982). Vergleich monotoner Ringtopologien und absolute Stetigkeit von Inhalten, *Comment. Math. Univ. St. Pauli* **31**, 49–60.

WEBER, H. (1982). Die atomare Struktur topologischer Boolescher Ringe und s-beschränkter Inhalte, a pre-print.

WEBER, H. (1982). Der Verband der s-beschränkter monotoner Ringtopologien und Zerlegung s-beschränkter Inhalte, a preprint.

WEBER, H. and VOLKMER, H. (1982). Der Wertebereich atomloser Inhalte, a pre-print.

WEISSÄCKER, H. U. (1982). The non-existence of liftings for arithmetical density, a pre-print.

The argument presented in Maharam's (1976) paper is clearly explained. See Chapter 12.

WILHELM, M. (1976). Existence of additive functionals on semi-groups and the von Neumann minimax theorem *Coll. Math.* **35**, 267–274.

A general result which may be considered as a common generalization of a result on charges due to Kelley (1959) and of the von Neumann minimax theorem in Game theory is presented.

WOODBURY, M. A. (1950). A decomposition theorem for finitely additive set functions, Abstract presented in *Bull. Amer. Math. Soc.* **56**, 171.

A forerunner of Yosida–Hewitt (1952) Decomposition Theorem was announced.

YASUMOTO, M. (1979). Finitely additive measures on N, *Proc. Japan Acad.* **55**, Ser. A, 81–84.

An improved version of a theorem of Jech and Prikry (1979) is established.

YOSIDA, K. (1941). Vector lattices and additive set functions, *Proc. Imp. Acad. Tokyo* **17**, 228–232.

ba$(\Omega, \mathscr{F})$ is studied from the point of view as a vector lattice.

YOSIDA, K. and HEWITT, E. (1952). Finitely additive measures, *Trans. Amer. Math. Soc.* **72**, 46–66.

Yosida–Hewitt Decomposition of a charge into a pure charge and a measure is presented. See Chapter 10.

APPENDIX 3

Some Set Theoretic Nomenclature

1. *Empty set* or *null set* is denoted by $\emptyset$.
2. The symbol Ω is used to denote an "abstract space" or "whole space" or "master set" which is a nonempty set of elements. The members of Ω are denoted generically by ω. The sets in a collection of sets we consider are usually subsets of Ω.
3. *Membership.* If ω is a member of a set E, we use the notation $\omega \in E$. If a set E is a member of a collection of sets $\mathscr{A}$, we use the symbol $E \in \mathscr{A}$.
4. *Inclusion.* For any two sets E and F, $E \subset F$ indicates that E is a subset of F, i.e. every member of E is a member of F.
5. *Union.* If $\{E_\alpha ; \alpha \in \Gamma\}$ is a nonempty collection of sets, we denote the union of these sets by $\bigcup_{\alpha \in \Gamma} E_\alpha$ and is defined to be the the set $\{\omega ; \omega \in E_\alpha$ for some α in $\Gamma\}$.
6. *Intersection.* If $\{E_\alpha ; \alpha \in \Gamma\}$ is a nonempty collection of sets, the intersection of these sets is denoted by $\bigcap_{\alpha \in \Gamma} E_\alpha$ and is defined to be the set $\{\omega ; \omega \in E_\alpha$ for every α in $\Gamma\}$.
7. *Difference.* If E and F are any two sets, the difference of E and F is denoted by $E - F$ and is defined to be the set $\{\omega ; \omega \in E$ and $\omega \notin F\}$.
8. *Complement.* If E is any subset of Ω, the complement of E is denoted by E^c and is defined to be the set $\Omega - E$.
9. *Symmetric difference.* If E and F are any two sets, the symmetric difference of E and F is denoted by $E \triangle F$ and is defined to be the set $(E - F) \cup (F - E)$.

Index of Symbols and Function Spaces

Function Spaces

$ba(\Omega, \mathscr{F})$	= The space of all bounded charges defined on the field $\mathscr{F}$ of subsets of Ω.	43
$ba(\Omega, \mathfrak{A}, \mathscr{I})$	= The space of all bounded charges defined on the σ-field $\mathfrak{A}$ of subsets of Ω vanishing on the σ-ideal $\mathscr{I}$ in $\mathfrak{A}$.	140
$B(\Omega, \mathscr{F}, \mu)$	= The space of all essentially bounded real valued functions defined on Ω.	90
$ca(\Omega, \mathscr{F})$	= The space of all bounded measures defined on the field $\mathscr{F}$ of subsets of Ω.	50
$\widetilde{ca}(\Omega, \mathscr{F})$	= The space of all bounded measures defined on the field $\mathscr{F}$ of subsets of Ω, when $\mathscr{F}$ is viewed as a Boolean algebra.	248
$\mathscr{C}(\Omega, \mathscr{F})$	= The space of all $\mathscr{F}$-continuous functions defined on Ω.	133
$C(\Omega, \mathscr{F}, \mu)$	= The space of all real valued functions defined on Ω. (This space is topologized in such a way that convergence in this space is precisely equivalent to hazy convergence, see p. 94.)	88
$C_\infty(\Omega, \mathscr{F}, \mu)$	= The space of all equivalence classes of $B(\Omega, \mathscr{F}, \mu)$ formed under the equivalence relation $f \sim g$ if $f = g$ a.e. $[\mu]$.	90
$L_p(\Omega, \mathscr{F}, \mu)$	= The space of all T_1-measurable functions defined on Ω such that $\|f\|^p$ is D-integrable, $1 \leq p < \infty$.	121, 178
$\mathscr{L}_p(\Omega, \mathscr{F}, \mu)$	= The space of all equivalence classes of $L_p(\Omega, \mathscr{F}, \mu)$ formed under the equivalence relation $f \sim g$ if $f = g$ a.e. $[\mu]$.	178
$L_\infty(\Omega, \mathscr{F}, \mu)$	= The space of all essentially bounded T_1-measurable functions defined on Ω.	122, 178
$L_\infty(\Omega, \mathfrak{A}, \mathscr{I})$	= The space of all essentially bounded measurable functions defined on Ω, where $\mathfrak{A}$ is a σ-field on Ω and $\mathscr{I}$ is a σ-ideal in $\mathfrak{A}$.	137
$\mathscr{L}_\infty(\Omega, \mathfrak{A}, \mathscr{I})$	= The space of all equivalence classes of $L_\infty(\Omega, \mathfrak{A}, \mathscr{I})$ formed under the equivalence relation $f \sim g$ if $f - g$ is a null function.	138
$P(\Omega, \mathscr{F})$	= The space of all bounded pure charges defined on the field $\mathscr{F}$ of subsets of Ω.	248
$S(\Omega, \mathscr{F}, \mu)$	= The space of all simple functions.	101
$SC(\Omega, \mathscr{F}, \mu)$	= The space of all simple charges defined on the field $\mathscr{F}$ of subsets of Ω.	188

INDEX OF SYMBOLS AND FUNCTION SPACES   307

$\mathrm{Sim}(\Omega, \mathscr{F}, \mu)$	= The space of all D-integrable simple functions defined on Ω.	132
$V_p(\Omega, \mathscr{F}, \mu)$	= The space of all bounded charges λ on $\mathscr{F}$ absolutely continuous with respect to μ and satisfying $\|\lambda\|_p < \infty$.	185
$(\Omega, \mathscr{F}, \mu)$	: Charge space, i.e. μ is a charge on the field $\mathscr{F}$ of subsets of Ω.	87

Operations on Boolean Algebras

$\mathbb{B}$	: Boolean Algebra	18
$\mathbb{B}/\mathscr{I}$	: Quotient Boolean algebra	20

Operations on Charges

A_λ	= The collection of all bounded charges on $\mathscr{F}$ absolutely continuous with respect to $\lambda = \alpha_1 \nu_1 + \alpha_2 \nu_2 + \cdots + \alpha_n \nu_n$, where $\alpha_1, \alpha_2, \ldots, \alpha_n$ are real numbers and $\nu_1, \nu_2, \ldots, \nu_n$ are $0-1$ valued charges on $\mathscr{F}$.	216
λ_B	: $\lambda_B(A) = \lambda(A \cap B)$, $A \in \mathscr{F}$, $B \in \mathscr{F}$ fixed.	180
μ_P	= $\max_{1 \le i \le n} \mu(F_i)$, $P = \{F_1, F_2, \ldots, F_n\}$ is a partition of Ω in $\mathscr{F}$.	145
μ^+	= Positive variation of μ.	45
μ^-	= Negative variation of μ.	45
$|\mu|$	= Total variation of μ.	45
μ^*	= Outer charge induced by μ.	86
$\tilde{\mu}$	= Semi-variation of μ.	206
μ_i	: See Definition 3.2.6.	66
μ_e	: See Definition 3.2.6.	66
$\mu \vee \nu$	: See Definition 2.5.2.	52
$\mu \wedge \nu$	: See Definition 2.5.2.	52
$\mu \ll \nu$	: μ is absolutely continuous with respect to ν.	159
$\mu_w \ll \nu$	: μ is weakly absolutely continuous with respect to ν.	159
$\mu_s \ll \nu$	: μ is strongly absolutely continuous with respect to ν.	159
$\mu \perp \nu$	: μ and ν are singular.	164
$\mu_s \perp \nu$	: μ and ν are strongly singular.	164

Operations on Functions

f^+	= Positive part of f.	11
f^-	= Negative part of f.	11
$|f|$	= Modulus of f.	11
$f \vee g$	= Maximum of f and g.	11
$f \wedge g$	= Minimum of f and g.	11
$f = g$ a.e. $[\mu]$	: See Definition 4.2.4.	88
$f \le g$ a.e. $[\mu]$	: See Definition 4.2.4.	88

I_A	= Indicator function of the set A.	12
$O(f, F)$	: See the proof of Theorem 4.7.3.	134

Operations on Sets

A^1	= A	6
A^0	= A^c, the complement of A.	6
$\bar{A}$	= Closure of A.	15
A^0	= Interior of A.	15
#A	= Number of points in the set.	41
$\mathcal{P}(\Omega)$	= The class of all subsets of Ω.	3
$\mathcal{P}$	= The collection of all finite partitions of Ω in $\mathcal{F}$.	15
$\mathcal{P}_F$	= The collection of all finite partitions of F in $\mathcal{F}$ for F in $\mathcal{F}$.	15
~	: Equivalence relation on a set.	14
$\leq$	: Partial order on a set.	13
$\geq$	: Relation directing a set.	13
c	: The cardinality of the continuum.	192

Operations in Vector Lattices

$x \vee y$	: Lattice supremum of x and y.	24
$x \wedge y$	: Lattice infimum of x and y.	24
x^+	: Positive part of x.	24
x^-	: Negative part of x.	24
$|x|$	: Modulus of x.	24
$x \perp y$	: x and y are orthogonal.	24
$S^\perp$	: The orthogonal complement of S.	29

Miscellaneous Symbols

l_∞	: The space of all bounded sequences of real numbers.	39
$\|\cdot\|$	: Norm on a linear space.	33
$\|\cdot\|_p$, $1 \leq p \leq \infty$	: Norms on $\mathcal{L}_p(\Omega, \mathcal{F}, \mu)$-spaces or on $V_p(\Omega, \mathcal{F}, \mu)$-spaces.	121, 122, 178, 180, 183
$D\int f \, d\mu$	: See Definition 4.4.11.	104
$S\int f \, d\mu$	: See Definition 4.5.5.	116
$\int f \mu$	: Refinement Integral of f with respect to μ.	231
$a \simeq b$	: a and b are numbers satisfying $|a - b| \leq 1$.	270
$\{0, 1\}^{\aleph_0}$	: The space of all sequences of 0's and 1's.	17

INDEX

A

Accumulation point, 15
  of a sequence of charges, 265
Additive-class, 2
Additivity
  uniform, 226
  uniform countable, 204
Antisymmetric relation, 13
Atom
  of a Boolean algebra, 22
  of a charge (μ-atom), 141
  of a field, 7
Axiom of choice, 14

B

Baire
  category theorem, 17, 267
  property of, 17
  σ-field, 17
Banach
  lattice, 34
  limit, 39
  space, 33
Base
  topological, 16
  filter, 134
Boolean algebra, 18
  atomic, 22
  complete, 19
  nonatomic, 22
  pairwise disjoint elements in a, 21
  quotient, 20
Boolean algebras
  atomic, 22
  homomorphism between, 19
  isomorphic, 19
  isomorphism between, 19
  Stone representation theorem for, 20
Boolean σ-algebra, 19
Borel–Cantelli lemma, 274
Borel σ-field on R, 12

C

Cantor set, 17, 265
Caratheodory
  extension theorem, 81
  measure, 274
Cardinal number, 14
Cartesian product space, 13
Cauchy–Schwartz inequality, 123
Cauchy sequence, 16
  weak, 33
Chain, 13
Chain condition
  countable, 21, 211
Charge, 35
  atomic, 213
  bounded, 35
  convex function with respect to a, 238
  density, 41
  finitely many valued, 249
  general invariant, 41
  infinitely many valued, 245
  modular, 36, 60
  negative variation of a, 45
  nonatomic, 141
  0-a valued, 35
  outer, 86
  positive, 35
  positive bounded, 35
  positive real partial, 64
  positive variation of a, 45
  probability, 35

Charge (cont.)
  pure, 240
  range of a, 249
  real, 35
  real partial, 64
  s-bounded, 41
  shift-invariant, 39
  simple, 188
  strongly continuous, 142
  strongly nonatomic, 142
  total variation of a, 45
  unbounded, 42
Charge space, 87
  probability, 179
  complete, 265
Chebychev's inequality, 127
Class
  additive-, 2
  compact, 49, 245
  equivalence, 14
  monocompact, 273
Clopen set, 16
Closed
  under complementation, 5, 6
  under countable intersections, 3
  under countable unions, 2
  under differences, 3
  under finite disjoint unions, 5
  under finite intersections, 3
  under finite unions, 3
  under proper differences, 3
  under symmetric differences, 3
Closure of a set, 15
Cofinite set, 3
Compact class, 49, 245
Compact topological space, 16
Condition
  countable chain, 21, 211
Continuous function, 16
  $\mathscr{F}$-, 133
Continuity
  absolute, 99, 159
  strong absolute, 159
  uniform absolute, 127, 204
  weak absolute, 159
Convergence
  hazy, 99
  in measure, 92
  of a net, 15
Convergence theorem
  dominated, 88
  Lebesgue dominated, 131
Convex function with respect to a charge, 238
Cover
  open, 16
  sub-, 16

D

D-integral, 96
Decomposition
  ε-Hahn, 56
  exact Hahn, 57
Decomposition theorem
  general Jordan, 52
  Hahn, 56
  Jordan, 52
  Lebesgue, 168
  Riesz, 29
  Sobczyk–Hammer, 146
  Yosida–Hewitt, 240, 241
Decomposition theorem for measures on σ-fields
  Hahn, 165
  Jordan, 56
Dense-in-itself set, 16, 251
Dense set, 17
Density charge, 41
Derived set, 236
Determining sequence, 104
Directed set, 13
Dominated convergence theorem, 88
  Lebesgue, 131
Dual space, 33
Dual of V_p-space, 193

E

ε-Hahn decomposition, 56
Egyptian fraction theorem, 280
Equivalence class, 14
Equivalence relation, 14
Essential boundedness, 89
Exact Hahn decomposition, 57
Exhaustion,
  principle of, 143
Extension theorem
  Caratheodory, 81
Extremely disconnected topological space, 278

F

$\mathcal{F}$-continuous function, 133
F_σ-set, 15
Field, 2
  atomic, 8
  discrete, 3
  finite-cofinite, 49
  μ-pure sub-, 274
  nonatomic, 8
  superatomic, 151
Filter
  base, 134
  in a Boolean algebra, 19
  in a field, 10
Finite-cofinite field, 49
Finite dimensional set, 216
Finite intersection property, 16
Finite partition, 8, 14
Finitely disjoint sequence of charges, 144
Finitely many valued charge, 249
First category set, 17
Function
  continuous, 16
  $\mathcal{F}$-continuous, 133
  indicator, 12
  μ-measurable, 91
  measurable, 12
  modular, 61
  null, 88
  simple, 90
  smooth, 91
  strongly additive, 61
  T_1-measurable, 101
  T_2-measurable, 101
Functional
  induced by a real valued set function, 59
  linear, 31
Functions equal almost everywhere, 88

G

G_δ-set, 15
General invariant charge, 41
General Jordan decomposition theorem, 53
Generator, 4

H

Hahn–Banach theorem, 32
Hahn decomposition
  ε-, 56
  exact, 57
Hahn decomposition theorem, 56
  for measures, 165
Hamel basis, 32
Hausdorff topological space, 16
Hazy convergence, 92
Hölder's inequality, 122
Homomorphism between Boolean algebras, 19

I

Ideal
  in a Boolean algebra, 19
  in a field, 10
  σ-, 137
Image of a set under a map, 16
Indicator function, 12
Inequality
  Cauchy–Schwartz, 123
  Chebychev's, 127
  Hölder's, 122
  Minkowski's, 124
Infinitely disjoint sequence of charges, 145, 258
Infinitely many valued charge, 249
Integral
  D-, 96
  lower, 116
  refinement, 231
  S-, 116
  upper, 116
Interior of a set, 15
Invariant charge
  general, 41
  shift-, 39
Isolated point, 15
Isomorphic Boolean algebras, 19
Isomorphism between Boolean algebras, 19

J

Jordan decomposition theorem, 52
  for measures on σ-fields, 56

K

Kolmogorov's Zero-One law, 265

L

L_p-space, 121, 178
Lattice
  Banach, 34
  boundedly complete vector, 29
  modulus of an element in a vector, 24
  negative part of an element in a vector, 24
  normal sub-, 28
  normed vector, 34
  of sets, 1
  orthogonal complement of a subset of a vector, 29
  orthogonal elements in a vector, 24
  positive part of an element in a vector, 24
  sub-, 28
  vector, 24
Lebesgue
  decomposition theorem, 168
  dominated convergence theorem, 131
  measurable set, 264
  measure, 49
Lifting, 268
Limit
  Banach, 39
  infimum, 11
  supremum, 11
Linear functional, 31
Linear order, 13
Linear space
  complete pseudo-normed, 33
  normed, 33
Linearly independent set, 32
Linearly ordered set, 13
Lower integral, 116
Lower sum, 115

M

μ-atom, 141
μ-measurable function, 91
μ-null set, 87
μ-pure sub-field, 274
Maximal filter
  in a Boolean algebra, 19
  in a field, 10
Maximal ideal
  in a Boolean algebra, 19
  in a field, 10

Measurable function, 12
  μ-, 91
  T_1-, 101
  T_2-, 101
Measurable set
  Lebesgue, 264
  τ-, 264
Measure, 47
  bounded, 47
  Caratheodory, 274
  convergence in, 92
  Lebesgue, 49
  nonatomic, 141
  positive, 47
  product probability, 265
  real, 47
Metric
  pre-compact, 273
Metric space, 16
  pseudo-, 16
Minkowski's inequality, 124
Modular charge, 36, 60
Modular function, 61
Modulus of an element in a vector lattice, 24
Monocompact class, 273

N

Negative part of an element in a vector lattice, 24
Negative variation of a charge, 45
Net, 15
  convergence of a, 15
  sub-, 15
  weakly convergent, 33
Nikodym theorem, 204
Nonatomic charge, 141
Nonatomic Boolean algebra, 22
Nonatomic field, 8
Nonatomic measure, 141
Norm, 33
  pseudo-, 33
Norm bounded set, 33
Normal sub-lattice, 28
Normed linear space, 33
Normed vector lattice, 34
Nowhere dense set, 17
Null function, 88

INDEX

Number
  cardinal, 14
  ordinal, 14

O

0-a valued charge, 35
Open cover, 15
Open set, 15
Order
  linear, 13
  partial, 13
Ordered set
  linearly, 13
  partially, 13
  well-, 13
Ordered vector space, 23
Ordinal number, 14
Orthogonal complement of a subset of a vector lattice, 29
Orthogonal elements in a vector lattice, 24
Outer charge, 87
Oxtoby's category analogue of Kolmogorov's zero-one law, 265

P

Pairwise disjoint elements in a Boolean algebra, 21
Partial order, 13
Partially ordered set, 13
Partition
  finite, 8, 14
  refinement of a, 15
Perfect set, 16
Phillips' lemma, 206
Polish space, 273
Positive bounded charge, 35
Positive charge, 35
Positive measure, 47
Positive part of an element in a vector lattice, 24
Positive real partial charge, 64
Positive variation of a charge, 45
Power set, 3
Pre-compact metric, 273
Principle of exhaustion, 143
Probability charge, 35
Probability charge space, 179
Product probability measure, 265

Property of Baire, 17, 264
Pseudo-metric space, 16
  complete, 16
  completion of a, 17
Pseudo-norm, 33
Pure charge, 240

Q

Quotient Boolean algebra, 20

R

Radon-Nikodym theorem, 174, 191
Range of a charge, 249
Real charge, 35
Real measure, 47
Real partial charge, 64
Refinement integral, 231
Refinement of a partition, 15
Reflexive relation, 13
Relation
  antisymmetric, 13
  equivalence, 14
  reflexive, 13
  symmetric, 13
  transitive, 13
Relative topology, 16
Riesz decomposition theorem, 29
Riesz representation theorem, 136
Ring, 2

S

s-bounded charge, 41
S-integral, 116
Scattered set, 16, 236
Seever property, 210
Semi-field, 2
Semi-ring, 1
Semi-variation, 206
Sequence
  Cauchy, 16
  determining, 104
  weak Cauchy, 33
Sequence of charges
  accumulation point of a, 265
  discrete, 258
  finitely disjoint, 144
  infinitely disjoint, 145, 258
Set
  Cantor, 17, 265
  clopen, 16

INDEX

Set (cont.)
  closed, 15
  closure of a, 15
  cofinite, 3
  dense, 17
  dense-in-itself, 16, 251
  derived, 236
  F_σ-, 15
  finite dimensional, 216
  first category, 17
  G_δ-, 15
  image of a, 16
  interior of a, 15
  Lebesgue measurable, 264
  linearly independent, 32
  linearly ordered, 13
  μ-null, 87
  norm bounded, 33
  nowhere dense, 17
  open, 15
  partially ordered, 13
  perfect, 16
  scattered, 16, 236
  τ-measurable, 264
  tail, 265
  weakly closed, 33
  well-ordered, 13
  with the property of Baire, 17, 264
σ-additivity across a sequence of sets, 253
σ-class, 2
σ-field, 2
  Baire, 17
  Borel, 12, 17
  discrete, 3
σ-ideal, 137
σ-ring, 2
Shift-invariant charge, 39
Simple charge, 188
Simple function, 90
Singularity, 164
  strong, 164
Smooth function, 91
Sobczyk–Hammer decomposition theorem, 146
Space
  Banach, 33
  Cartesian product, 13
  charge, 87
  compact topological, 16
  complete charge, 264
  complete pseudo-metric, 16
  completion of a pseudo-metric, 17
  dual, 33
  extremely disconnected topological, 278
  Hausdorff topological, 16
  L_p-, 121, 178
  metric, 16
  normed linear, 33
  ordered vector, 23
  Polish, 273
  probability charge, 179
  pseudo-metric, 16
  Stone, 21
  topological, 15
  totally disconnected topological, 16
  V_p-, 185
  vector, 31
  weakly complete, 33
Stone representation theorem for Boolean algebras, 20
Stone space, 21
Strong absolute continuity, 159
Strong singularity, 164
Strong topology, 33
Strongly additive function, 61
Strongly continuous charge, 142
Strongly nonatomic charge, 142
Subcover, 16
Sub-field,
  μ-pure, 247
Sublattice, 28
  normal, 28
Subnet, 15
Sum
  lower, 115
  upper, 115
Superatomic field, 151
Symmetric relation, 13

T

T_1-measurable function, 101
T_2-measurable function, 101
τ-measurable set, 264
Tail set, 265
Theorem
  Baire, 17, 267
  Caratheodory extension, 81

Theorem (cont.)
  dominated convergence, 88
  Egyptian fraction, 280
  general Jordan decomposition, 53
  Hahn–Banach, 32
  Hahn decomposition, 56, 165
  Jordan decomposition, 52
  Lebesgue decomposition, 168
  Lebesgue dominated convergence, 131
  Nikodym, 204
  Radon–Nikodym, 174, 191
  Riesz decomposition, 29
  Riesz representation, 136
  Sobczyk–Hammer decomposition, 146
  Stone representation, 20
  Vitali–Hahn–Saks, 204
  Yosida–Hewitt decomposition, 240, 241
Topological space, 15
  compact, 16
  extremely disconnected, 278
  Hausdorff, 16
  totally disconnected, 16
Topology
  relative, 16
  strong, 33
  weak, 33
  weak*, 158
Total variation of a charge, 45
Totally disconnected topological space, 16
Transfinite induction, 14
Transitive relation, 13
Tree, 150

U

Uniform absolute continuity, 127, 204
Uniform additivity, 226
Uniform countable additivity, 204
Upper integral, 116

Upper sum, 115
Urysohn's lemma, 17

V

V_p-space, 185
Vector lattice, 24
  boundedly complete, 29
  modulus of an element in a, 24
  negative part of an element in a, 24
  normed, 34
  orthogonal complement of a subset of a, 29
  orthogonal elements in a, 24
  positive part of an element in a, 24
Vector space
  ordered, 23
  over the real line R, 31
  over the field of rational numbers, 31
Vitali–Hahn–Saks theorem, 204

W

Weak absolute continuity, 159
Weak Cauchy sequence, 33
Weak topology, 33
Weak* topology, 158
Weakly closed set, 33
Weakly complete space, 34
Weakly convergent net, 33
Well-ordered set, 13
Well-ordering, 13

Y

Yosida–Hewitt decomposition theorem, 240, 241

Z

Zero-one law
  Kolmogorov's, 265
  Oxtoby's category analogue of Kolmogorov's, 265
Zorn's lemma, 14

Pure and Applied Mathematics

A Series of Monographs and Textbooks

Editors **Samuel Eilenberg and Hyman Bass**

Columbia University, New York

RECENT TITLES

CARL L. DEVITO. Functional Analysis
MICHIEL HAZEWINKEL. Formal Groups and Applications
SIGURDUR HELGASON. Differential Geometry, Lie Groups, and Symmetric Spaces
ROBERT B. BURCKEL. An Introduction to Classical Complex Analysis: Volume 1
JOSEPH J. ROTMAN. An Introduction to Homological Algebra
C. TRUESDELL AND R. G. MUNCASTER. Fundamentals of Maxwell's Kinetic Theory of a Simple Monatomic Gas: Treated as a Branch of Rational Mechanics
BARRY SIMON. Functional Integration and Quantum Physics
GRZEGORZ ROZENBERG AND ARTO SALOMAA. The Mathematical Theory of L Systems.
DAVID KINDERLEHRER and GUIDO STAMPACCHIA. An Introduction to Variational Inequalities and Their Applications.
H. SEIFERT AND W. THRELFALL. A Textbook of Topology; H. SEIFERT. Topology of 3-Dimensional Fibered Spaces
LOUIS HALLE ROWEN. Polynominal Identities in Ring Theory
DONALD W. KAHN. Introduction to Global Analysis
DRAGOS M. CVETKOVIC, MICHAEL DOOB, AND HORST SACHS. Spectra of Graphs
ROBERT M. YOUNG. An Introduction to Nonharmonic Fourier Series
MICHAEL C. IRWIN. Smooth Dynamical Systems
JOHN B. GARNETT. Bounded Analytic Functions
EDUARD PRUGOVEČKI. Quantum Mechanics in Hilbert Space, Second Edition
M. SCOTT OSBORNE AND GARTH WARNER. The Theory of Eisenstein Systems
K. A. ZHEVLAKOV, A. M. SLIN'KO, I. P. SHESTAKOV, AND A. I. SHIRSHOV. Translated by HARRY SMITH. Rings That Are Nearly Associative
JEAN DIEUDONNÉ. A Panorama of Pure Mathematics; Translated by I. Macdonald
JOSEPH G. ROSENSTEIN. Linear Orderings
AVRAHAM FEINTUCH AND RICHARD SAEKS. System Theory: A Hilbert Space Approach
ULF GRENANDER. Mathematical Experiments on the Computer
HOWARD OSBORN. Vector Bundles: Volume 1, Foundations and Stiefel-Whitney Classes
BHASKARA RAO AND BHASKARA RAO. Theory of Charges

IN PREPARATION

ROBERT B. BURCKEL. An Introduction to Classical Complex Analysis: Volume 2
RICHARD V. KADISON AND JOHN R. RINGROSE. Fundamentals of the Theory of Operator Algebras, Volume 1, Volume 2
BARRETT O'NEILL. Semi-Riemannian Geometry: With Applications to Relativity
EDWARD B. MANOUKIAN. Renormalization
E. J. MCSHANE. Unified Integration
A. P. MORSE. A Theory of Sets, Revised and Enlarged Edition